FIRE INVESTIGATOR

THIRD EDITION

Wayne Chapdelaine
Writer

Libby Snyder
Lead Editor

Reinaldo Sanchez
Angel Muzik
Lead Instructional Developers

Ben Brock
Graphic Designer

Validated by the International Fire Service Training Association
Published by Fire Protection Publications • Oklahoma State University

Cover photo courtesy of Hugh Graham

The International Fire Service Training Association (IFSTA) was established in 1934 as a *nonprofit educational association of fire fighting personnel dedicated to the advancement of fire fighting techniques and firefighter safety through training.* To carry out the IFSTA mission, Fire Protection Publications (FPP) was established at Oklahoma State University in the College of Engineering, Architecture and Technology. The primary purpose of FPP is to publish and distribute training materials as proposed, developed, and validated by IFSTA. FPP also develops state of the art teaching and learning materials to support the adoption and use of the manuals. The print products are supported by a full array of eBooks, online learning, and apps.

IFSTA holds two meetings each year, one in January and one in July. During these meetings, selected committees of subject matter experts, who are acknowledged leaders in their fields, review draft materials, and ensure that the content is accurate and meets the National Fire Protection Association professional qualification standards. These assemblies bring together individuals from several related and allied occupations:

- Fire department executives, training officers, and line personnel
- Educators from colleges and universities
- Representatives from governmental agencies
- Delegates of firefighter associations and industrial organizations

Validation committee members receive no fees for their work and their travel expenses are only partially funded by FPP. Members participate because of a personal commitment to the fire service and its future through training. Serving on a committee is prestigious in the fire and emergency response community. This unique process to validate the content of training textbooks creates a close relationship between IFSTA, the publications, and the fire service community.

The manuals and associated instructional materials are the primary source of training and education for most fire departments and training agencies in North America, the U.S. Department of Defense, fire service related higher education programs, and other emergency response organizations worldwide. Some of the manuals are available in other languages including French, Arabic, Japanese, and Chinese.

If you need additional information concerning the International Fire Service Training Association (IFSTA) or Fire Protection Publications, contact:

Customer Service, Fire Protection Publications, Oklahoma State University
930 N. Willis St., Stillwater, OK. 74078-8045
Phone: 800-654-4055   Fax: 405-744-8204   E-mail: customer.service@osufpp.org

For assistance with training materials, to recommend material for inclusion in an IFSTA manual, or to ask questions or comment on manual content, contact:

Editorial Department, Fire Protection Publications, Oklahoma State University
930 N. Willis St., Stillwater, OK. 74078-8045
Phone: 405-744-4111   Fax: 405-744-4112   E-mail: editors@osufpp.org

Chapters

Appendices 435

List of Key Terms 467

Index 471

Table of Contents

Table of Contents

Table of Contents

Acknowledgements

The third edition of **Fire Investigator** is designed to provide public and private sector Fire Investigators with basic information necessary to meet the job performance requirements (JPRs) of National Fire Protection Association (NFPA) 1033, *Standard for Professional Qualifications for Fire Investigator* (2022). To support that end, this manual also draws from NFPA 921, *Guide for Fire and Explosion Investigations*. Further, this manual addresses the outcomes for the National Fire Academy's (NFA's) Fire and Emergency Services Higher Education (FESHE) model curriculum:

- Associates (Non-Core): Fire Investigation I (C0283)
- Associates (Non-Core): Fire Investigation II (C0284)
- Bachelor's (Non-Core): Fire Investigation and Analysis (C0285)

Acknowledgement and special thanks are extended to the members of the IFSTA validation committee who contributed their time, wisdom, and knowledge to the development of this manual.

Contract Writer

The third edition of **Fire Investigator** could not have been completed without the skill and hard work of contracted technical writer Wayne Chapdelaine. Mr. Chapdelaine provided early drafts of all manual chapters to the validation committee for their review, revision, and approval. He also assisted the committee and project manager with changes to chapter content and validation discussions.

IFSTA Fire Investigator Third Edition, Validation Committee

Chair
Donny Howard
Oklahoma State Fire Marshal Special Agent
Oklahoma State Fire Marshal's Office
Owasso, OK

Vice Chair
Dennis Smith
President/Principal Fire Expert
Premier Fire Consulting Services
Orlando, FL

Secretary
Nicole Brewer
Fire/Arson Investigator
Portland Fire and Rescue
Canby, OR

Committee Members

Thomas Aurnhammer
Captain/Investigator
San Juan County Fire Department
Farmington, NM

Mark Dooley
Captain
Des Moines Fire Department
Des Moines, IA

Scott Barthelmass
Division Chief
Eureka Fire Protection District
St. Louis, MO

Hugh Graham
Owner/Investigator
Graham Investigations, LLC
Atoka, OK

JT Collier, Deputy Fire Marshal
Scottsdale Fire Department
Scottsdale, AZ

Dan Hrouda
Senior Forensic Consultant
Fire and Explosion Division
Envista Forensics
Denver, CO

Committee Members (continued)

Jennifer Johnson
Kansas Fire & Rescue Training Institute
Lawrence, KS

John Kitchens, Sr.
Fire Investigator,
Life Safety Consultant/Instructor
Liberty Fire Investigations/ProNet Group
Somis, CA

Dan Madrzykowski
Research Director
UL Fire Safety Research Institute
Columbia, MD

Mike Makela
Assistant Chief
North County Fire EMS
Stanwood, WA

Nelson Mascarenhas
Captain
Fairfax County Fire and Rescue Dept
Fairfax, VA

Mary Maurath
Firefighter/Paramedic, Fire Investigator
Huber Heights Fire Division,
Fire and Explosion Consultants
Miamisburg, OH

Jim Pendergast
Fire Chief
Penhold Fire Department
Red Deer, Alberta, Canada

Michael J. Schulz, M.S.
M.J. Schulz & Associates, Inc.
Roseville, CA

Vance Swisher
Deputy Fire Chief
Rochester Fire Department
Rochester, MN

Much appreciation is given to the following individuals and organizations for contributing photographs instrumental in the development of this manual:

AP/Wide World Photos
Captain James A. Sobota, Fairvax County (VA) Fire and Rescue
Dan Madrzykowski, NIST
Dave Coombs
David M. Smith, Associated Fire Consultants, Inc.
Dennis Smith
Donny Howard
Ed Prendergast, PE
Hugh W. Graham
Iowa State Fire Training Bureau
Jennifer Johnson
Jocelyn Augustino, FEMA News
Mary Maurath
McKinney (TX) Fire Department
Mike Makela
Nelson Mascarenhas
New South Wales Fire Brigades
Nicole Brewer

Noble County Courthouse, Perry, OK, and the Honorable Jennifer A. Brock
Rich Mahaney
Ron Jeffers, Union City, NJ
Ron Moore, McKinney (TX) FD
U.S. Air Force, photo by Senior Airman Sean Worrell
Wayne Chapdelaine
Yates & Associates

Special thanks go to Donny Howard, who provided a truly significant collection of investigation photos for use throughout this manual. The collection is above and beyond expectations, and will serve well. Last, but certainly not least, gratitude is extended to the following members of the Fire Protection Publications staff whose contributions made the final publication of this manual possible.

Fire Investigator, Third Edition, Project Team

Lead Editor
Libby Snyder

Contract Writers
Wayne Chapdelaine
Sr. Fire Investigator
Fire Dynamics Analysts
Hubley, Nova Scotia, Canada

Lead Instructional Developers
Reinaldo Sanchez, Curriculum Developer
Angel Muzik, Curriculum Developer

Director of Fire Protection Publications
Craig Hannan

Managing Editor
Colby Cagle

Editorial Manager
Clint Clausing

Curriculum Manager
Angel Muzik

Production Manager
Missy Hannan

eProducts Manager
Sarah Axtell

Editors
Beckie Bigler
Cindy Brakhage
Amy Marie Echo-Hawk
Kim Edwards
Jeff Fortney
Michael Fox

Illustrators and Layout Designers
Ben Brock, Senior Graphic Designer
David Hicks, Senior Graphic Designer

Photographers
Clint Parker, Senior Graphic Designer
Jeff Fortney, Lead Editor
Leslie Miller, Lead Editor
Libby Snyder, Lead Editor
Reinaldo Sanchez, Curriculum Developer

Editorial Staff
Tyler Auld, Student Worker

eProducts Staff
Janice French
Kelly Naas
Ryan Souders

Indexer
Meredith Murray

The IFSTA Executive Board at the time of validation of Fire Investigator was as follows:

IFSTA Executive Board

Executive Board Chair
Bradd Clark
Deputy Chief
Palm Coast Fire Department
Palm Coast, Florida

Vice Chair
Mary Cameli
Fire Chief
City of Mesa Fire Department
Mesa, Arizona

Executive Director
Mike Wieder
Associate Director
Fire Protection Publications
Oklahoma State University
Stillwater, Oklahoma

IFSTA Board Members

Stephen Ashbrock
Fire Chief (Retired)
Madeira & Indian Hill Fire Department
Cincinnati, Ohio

Steve Austin
Project Manager
Cumberland Valley Vol. FF Association
Newark, Delaware

Jim Brinkley
Assistant to the General President
International Association of Fire Fighters
Washington, DC

Dr. Larry Collins
Associate Dean
Safety, Security & Emergency Department
Eastern Kentucky University
Richmond, Kentucky

Dennis Compton
Fire Chief (Retired)
Mesa & Phoenix, Arizona

Elizabeth Hendel
Retired
Beacon OHSS
Phoenix, Arizona

John Hoglund
Director-Emeritus
Maryland Fire and Rescue Institute
New Carrollton, Maryland

Tonya Hoover
Superintendent
National Fire Academy
Frederick, Maryland

Tom Jenkins
Fire Chief
Rogers Fire Department
Rogers, Arkansas

Dr. Scott Kerwood
Fire Chief
Hutto Fire Rescue
Hutto, Texas

Brett Lacey
Fire Marshal
Colorado Springs Fire Department
Colorado Springs, Colorado

Jeffrey Morrissette
State Fire Administrator
CT Commission on Fire Prevention
Windsor Locks, Connecticut

Dan Ripley
Fire Captain
Lincoln Fire and Rescue
Lincoln, Nebraska

Dr. Demond Simmons
Battalion Chief
Oakland Fire Department
Oakland, California

Josh Stefancic
Fire Chief
Safety Harbor Fire Department
Safety Harbor, Florida

Dedication

This manual is dedicated to the men and women who hold devotion to duty above personal risk, who count on sincerity of service above personal comfort and convenience, who strive unceasingly to find better and safer ways of protecting people, homes and property from the ravages of fire, medical emergencies, and other disasters...

...The Firefighters of All Nations

Introduction

Fire investigation is the compilation and analysis of information related to fires and explosions. A fire investigation is usually conducted to determine and document the area of origin and cause of a fire and the factors that contributed to the ignition and subsequent growth. The information developed during the investigation is then used in many ways. In its basic form, it can be used to complete a fire incident report for the incident. The information, along with other data developed by a jurisdiction, can be used to identify trends, assist in the prevention of similar incidents, and develop public education programs. On a broader level, investigation data is used in the development of codes and standards aimed at reducing fire losses and preventing fire deaths.

Fire investigators also conduct investigations to determine whether a fire was intentionally set. In these cases, the origin and cause information developed by the fire investigator becomes an integral part of any criminal investigation and subsequent legal actions taken against the responsible parties. Fire investigations may also assist in civil litigations to provide a preponderance of evidence for liability cases or negligence cases.

To be successful, fire investigators must rely on scientific principles and available research to reach their conclusions. As a result of the increase in the information available on the ignition, growth, and development of fires, the information a fire investigator must possess has significantly increased in recent decades. Fire investigators must remain knowledgeable in many areas including the following:

- Fire science
- Evidence collection
- Explosion dynamics
- Scene documentation
- Postinvestigation analysis
- Hazardous materials recognition
- Building construction and systems
- Presentation of investigative findings

NFPA 1033, *Standard for Professional Qualifications for Fire Investigator*, defines the basic job performance requirements (JPRs) for fire investigators. This standard focuses on determining the origin of, cause of, and responsibility for fires. On some occasions, investigators may also assist in determining factors related to damage after a fire began (developing fire spread scenarios); identifying specific reasons for injuries or deaths (evaluating life safety issues); and assessing human fault or responsibility factors (determining facts related to potential negligence). These are often important functions; however, they are not the focus of this manual. If the department assigns such functions, the fire investigator should consult additional references and resources.

Purpose and Scope

The *purpose* of **Fire Investigator, Third Edition**, is two-fold. First, this manual provides fire and emergency services personnel and civilian investigators with basic information necessary to meet the job performance requirements (JPRs) of NFPA 1033, *Standard for Professional Qualifications for Fire Inves-*

tigator, 2022 Edition. Second, this manual meets the Fire and Emergency Services for Higher Education (FESHE) Course Outcomes for two Associates, Non-Core Courses (C0283 and C0284) and one Bachelors, Non-Core Course (C0285).

The *scope* of the manual addresses the basic duties assigned to a Fire Investigator in NFPA 1033 and the additional knowledge necessary to complete the FESHE courses. Also included in the scope is an emphasis on NFPA 921, *Guide for Fire and Explosion Investigations, 2021Edition.* NFPA 921 serves as the first reference source for all information contained in this manual and should be considered a companion text for training purposes.

Terminology

This manual is written with a global, international audience in mind. For this reason, we often use more universally-accepted terminology in place of regional- or agency-specific terminology (*jargon*). Additionally, in order to keep sentences uncluttered and easy to read, the word *state* is used to represent both state and provincial level governments (or their equivalent). This usage is applied to this manual for the purposes of brevity and is not intended to address or show preference for only one nation's method of identifying regional governments within its borders.

The list of key terms at the end of each chapter will assist the reader in understanding words that may not have their roots in the fire and emergency services. Definitions for terms related to fire and emergency services come from the IFSTA **Fire and Emergency Services Orientation and Terminology** manual.

NOTE: For the purpose of this manual, the term "agency" refers to a private sector investigator's person or entity in legal control.

Key Information

Various types of information in this book are given in shaded boxes marked by symbols or icons. See the following definitions:

Safety Alert Definition

Safety alert boxes are used to highlight information that is important for safety reasons. (In the text, the title of safety alerts will change to reflect the content.)

Information Box Definition

Information boxes give facts that are complete in themselves but belong with the text discussion. These boxes contain information that needs more emphasis or separation. (In the text, the title of information boxes will change to reflect the content.)

What This Means to You Box Definition

What This Means to You boxes take information presented in the text and synthesize it into an example of how the information is relevant to (or will be applied by) the intended audience, essentially answering the question, "What does this mean to you?"

A **key term** is designed to emphasize key concepts, technical terms, or ideas that fire investigators need to know. Key terms are emphasized in the text. Definitions are listed in alphabetical order at the end of the chapter. An example of a key term is:

Compartmentation — The way that the arrangement of compartments creates or does not create a series of barriers designed to keep flames, smoke, and heat from spreading from one room or floor to another.

Three key signal words are found in the book: **WARNING, CAUTION,** and **NOTE**. Definitions and examples of each are as follows:

- **WARNING** indicates information that could result in death or serious injury to fire investigators. See the following example:

> **WARNING:** A fire investigator must be thoroughly trained and knowledgeable about radiation emergencies before being involved in an investigation with this type of hazard. Under no circumstances should an investigator work at a scene with ionizing radiation without the proper PPE and training.

- **CAUTION** indicates important information or data that fire investigators need to be aware of in order to perform their duties safely. See the following example:

> **CAUTION:** Investigators should not use fire-resistance ratings to establish a timeline for the duration of a fire.

- **NOTE** indicates important operational information that helps explain why a particular recommendation is given or describes optional methods for certain procedures. See the following example:

NOTE: Public sector investigators should obtain legal advice regarding right-of-entry issues so that guidelines and procedures can be established to address legal right of entry from their local authority having jurisdiction or agency.

Chapter References refer the reader to sources referenced within the chapter. Not all chapters have Chapter References as they are only included when appropriate to do so.

Chapter References

"Hazard Prevention and Control." Recommended Practices for Safety and Health Programs. United States Department of Labor. Accessed online.

"Asbestos at Superfund Sites: Cleanup Examples." Superfund. United States Environmental Protection Agency, 2019. Accessed online.

"Fire Investigator Health and Safety Best Practices." International Association of Arson Investigators, Inc., Health and Safety Committee; June 15, 2018. Accessed online.

To find curriculum or study materials associated with this manual and its contents, please go to www.ifsta.org and use the search tool in the shop to find accompanying products. You can also search for IFSTA apps using your smartphone or tablet device.

Metric Conversions

Throughout this manual, U.S. units of measure are converted to metric units for the convenience of our international readers. We adhere to the following guidelines for metric conversions in this manual:

- Metric conversions are approximated unless the number is used in mathematical equations.
- Most measurements of fire service equipment are provided in millimeters; therefore millimeters convert to inches in this manual.
- Exact conversions are used when an exact number is necessary such as in construction measurements or hydraulic calculations.
- Set values such as hose diameter, ladder length, and nozzle size use their Canadian counterpart naming conventions and are not mathematically calculated. For example, 1½ inch hose is referred to as 38 mm hose.

The following two tables provide detailed information on IFSTA's conversions. The first table includes examples of our conversion factors for a number of measurements used in the fire service. The second shows examples of exact conversions beside the approximated measurements you will see in this manual.

U.S. to Canadian Measurement Conversion

Measurements	Customary (U.S.)	Metric (Canada)	Conversion Factor
Length/Distance	Inch (in)	Millimeter (mm)	1 in = 25 mm
	Foot (ft) [3 or less feet]	Millimeter (mm)	1 ft = 300 mm
	Foot (ft) [3 or more feet]	Meter (m)	1 ft = 0.3 m
	Mile (mi)	Kilometer (km)	1 mi = 1.6 km
Area	Square Foot (ft^2)	Square Meter (m^2)	1 ft^2 = 0.09 m^2
	Square Mile (mi^2)	Square Kilometer (km^2)	1 mi^2 = 2.6 km^2
Mass/Weight	Dry Ounce (oz)	gram	1 oz = 28 g
	Pound (lb)	Kilogram (kg)	1 lb = 0.5 kg
	Ton (T)	Ton (T)	1 T = 0.9 T
Volume	Cubic Foot (ft^3)	Cubic Meter (m^3)	1 ft^3 = 0.03 m^3
	Fluid Ounce (fl oz)	Milliliter (mL)	1 fl oz = 30 mL
	Quart (qt)	Liter (L)	1 qt = 1 L
	Gallon (gal)	Liter (L)	1 gal = 4 L
Flow	Gallons per Minute (gpm)	Liters per Minute (L/min)	1 gpm = 4 L/min
	Cubic Foot per Minute (ft^3/min)	Cubic Meter per Minute (m^3/min)	1 ft^3/min = 0.03 m^3/min
Flow per Area	Gallons per Minute per Square Foot (gpm/ft^2)	Liters per Square Meters Minute (L/(m^2.min))	1 gpm/ft^2 = 40 L/(m^2.min)
Pressure	Pounds per Square Inch (psi)	Kilopascal (kPa)	1 psi = 7 kPa
	Pounds per Square Foot (psf)	Kilopascal (kPa)	1 psf = .05 kPa
	Inches of Mercury (in Hg)	Kilopascal (kPa)	1 in Hg = 3.4 kPa
Speed/Velocity	Miles per Hour (mph)	Kilometers per Hour (km/h)	1 mph = 1.6 km/h
	Feet per Second (ft/sec)	Meter per Second (m/s)	1 ft/sec = 0.3 m/s
Heat	British Thermal Unit (Btu)	Kilojoule (kJ)	1 Btu = 1 kJ
Heat Flow	British Thermal Unit per Minute (BTU/min)	watt (W)	1 Btu/min = 18 W
Density	Pound per Cubic Foot (lb/ft^3)	Kilogram per Cubic Meter (kg/m^3)	1 lb/ft^3 = 16 kg/m^3
Force	Pound-Force (lbf)	Newton (N)	1 lbf = 0.5 N
Torque	Pound-Force Foot (lbf ft)	Newton Meter (N.m)	1 lbf ft = 1.4 N.m
Dynamic Viscosity	Pound per Foot-Second (lb/ft.s)	Pascal Second (Pa.s)	1 lb/ft.s = 1.5 Pa.s
Surface Tension	Pound per Foot (lb/ft)	Newton per Meter (N/m)	1 lb/ft = 15 N/m

Conversion and Approximation Examples

Measurement	U.S. Unit	Conversion Factor	Exact S.I. Unit	Rounded S.I. Unit
Length/Distance	10 in	1 in = 25 mm	250 mm	250 mm
	25 in	1 in = 25 mm	625 mm	625 mm
	2 ft	1 in = 25 mm	600 mm	600 mm
	17 ft	1 ft = 0.3 m	5.1 m	5 m
	3 mi	1 mi = 1.6 km	4.8 km	5 km
	10 mi	1 mi = 1.6 km	16 km	16 km
Area	36 ft²	1 ft² = 0.09 m²	3.24 m²	3 m²
	300 ft²	1 ft² = 0.09 m²	27 m²	30 m²
	5 mi²	1 mi² = 2.6 km²	13 km²	13 km²
	14 mi²	1 mi² = 2.6 km²	36.4 km²	35 km²
Mass/Weight	16 oz	1 oz = 28 g	448 g	450 g
	20 oz	1 oz = 28 g	560 g	560 g
	3.75 lb	1 lb = 0.5 kg	1.875 kg	2 kg
	2,000 lb	1 lb = 0.5 kg	1 000 kg	1 000 kg
	1 T	1 T = 0.9 T	900 kg	900 kg
	2.5 T	1 T = 0.9 T	2.25 T	2 T
Volume	55 ft³	1 ft³ = 0.03 m³	1.65 m³	1.5 m³
	2,000 ft³	1 ft³ = 0.03 m³	60 m³	60 m³
	8 fl oz	1 fl oz = 30 mL	240 mL	240 mL
	20 fl oz	1 fl oz = 30 mL	600 mL	600 mL
	10 qt	1 qt = 1 L	10 L	10 L
	22 gal	1 gal = 4 L	88 L	90 L
	500 gal	1 gal = 4 L	2 000 L	2 000 L
Flow	100 gpm	1 gpm = 4 L/min	400 L/min	400 L/min
	500 gpm	1 gpm = 4 L/min	2 000 L/min	2 000 L/min
	16 ft³/min	1 ft³/min = 0.03 m³/min	0.48 m³/min	0.5 m³/min
	200 ft³/min	1 ft³/min = 0.03 m³/min	6 m³/min	6 m³/min
Flow per Area	50 gpm/ft²	1 gpm/ft² = 40 L/(m².min)	2 000 L/(m².min)	2 000 L/(m².min)
	326 gpm/ft²	1 gpm/ft² = 40 L/(m².min)	13 040 L/(m².min)	13 000 L/(m².min)
Pressure	100 psi	1 psi = 7 kPa	700 kPa	700 kPa
	175 psi	1 psi = 7 kPa	1225 kPa	1200 kPa
	526 psf	1 psf = 0.05 kPa	26.3 kPa	25 kPa
	12,000 psf	1 psf = 0.05 kPa	600 kPa	600 kPa
	5 psi in Hg	1 psi = 3.4 kPa	17 kPa	17 kPa
	20 psi in Hg	1 psi = 3.4 kPa	68 kPa	70 kPa
Speed/Velocity	20 mph	1 mph = 1.6 km/h	32 km/h	30 km/h
	35 mph	1 mph = 1.6 km/h	56 km/h	55 km/h
	10 ft/sec	1 ft/sec = 0.3 m/s	3 m/s	3 m/s
	50 ft/sec	1 ft/sec = 0.3 m/s	15 m/s	15 m/s
Heat	1200 Btu	1 Btu = 1 kJ	1 200 kJ	1 200 kJ
Heat Flow	5 BTU/min	1 Btu/min = 18 W	90 W	90 W
	400 BTU/min	1 Btu/min = 18 W	7 200 W	7 200 W
Density	5 lb/ft³	1 lb/ft³ = 16 kg/m³	80 kg/m³	80 kg/m³
	48 lb/ft³	1 lb/ft³ = 16 kg/m³	768 kg/m³	770 kg/m³
Force	10 lbf	1 lbf = 0.5 N	5 N	5 N
	1,500 lbf	1 lbf = 0.5 N	750 N	750 N
Torque	100	1 lbf ft = 1.4 N.m	140 N.m	140 N.m
	500	1 lbf ft = 1.4 N.m	700 N.m	700 N.m
Dynamic Viscosity	20 lb/ft.s	1 lb/ft.s = 1.5 Pa.s	30 Pa.s	30 Pa.s
	35 lb/ft.s	1 lb/ft.s = 1.5 Pa.s	52.5 Pa.s	50 Pa.s
Surface Tension	6.5 lb/ft	1 lb/ft = 15 N/m	97.5 N/m	100 N/m
	10 lb/ft	1 lb/ft = 15 N/m	150 N/m	150 N/m

Photo courtesy of Nicole Brewer, Portland Fire Investigations.

Chapter Contents

NFPA 1033 JPRs addressed in this chapter

This chapter provides information that addresses the following job performance requirements (JPRs) of NFPA 1033, *Standard for Professional Qualifications for Fire Investigator* (2022):

4.1.1   4.1.2   4.1.4   4.1.5   4.1.6   4.1.7

This chapter provides information that addresses the following course outcomes for the FESHE courses:

Fire Investigation I (C0283)

2. Understand and demonstrate the process of conducting fire origin and cause investigation.

5. Define the common terms used in fire investigations.

Fire Investigation II (C0284)

1. Recognize the need for the use of the scientific method for investigations.

4. Analyze and determine the causes of fires and contributing factors.

Fire Investigation and Analysis (C0285)

1. Demonstrate a technical understanding of the characteristics and impacts of fire loss and the crime of arson, which is necessary to conduct competent fire investigation and analysis.

4. Analyze the legal foundation for conducting a systematic incendiary fire investigation and case preparation.

5. Design and integrate a variety of arson-related intervention and mitigation strategies.

1. Explain the general duties and responsibilities of a fire investigator. [NFPA 1033, 4.1.2, 4.1.6, 4.1.7]

2. Describe a fire investigator's role as a liaison with other agencies and organizations. [NFPA 1033, 4.1.4]

3. Describe fire investigator training and certification. [NFPA 1033, 4.1.1, 4.1.2, 4.1.5]

4. Describe legal procedures to which a fire investigator shall adhere. [NFPA 1033, 4.1.5]

Chapter 1
Organization, Responsibilities, and Authority

A starting point for discussing fire investigations is describing an investigator's general responsibilities. This chapter serves as that introduction with a brief overview of topics including:

- Introduction to fire investigator duties
- Investigator as liaison
- Fire investigator training and certification
- Legal and regulatory foundations

Introduction to Fire Investigator Duties

NFPA 1033 (2022): 4.1.2, 4.1.6, 4.1.7

The fire investigation determines the origin, focusing on *where*, and the cause, focusing on *how* and *why* the fire began. Unless that chain of events can be codified, the fire cause is *undetermined*. Where a fire investigator can determine a first fuel, ignition source, and chain of events, the following descriptors may apply:

- **Accidental** — Fire cause did not involve malicious or criminal intent.
- **Incendiary** — Fire was deliberately set. **Arson** is an incendiary cause.
- **Natural** — Fire did not include human intervention in the ignition process.

Accurately determining fire cause can provide long-term benefits. For example, the investigation of fires caused by defects in manufactured products provides a means to develop a knowledge base that could enable the elimination of that cause. Fire cause is further addressed after fire origin determination in this manual.

Investigation and analysis of any fire-related incident is a complex scientific endeavor. Therefore, the methodology of such an endeavor must include the comprehensive, objective, and accurate compilation and analysis of data. This manual is designed to assist the public and private sector fire investigator to determine the origin and cause of a fire.

Public sector fire investigators may be employed by a local, state/provincial, or federal agency such as a fire department or law enforcement agency. In many states and provinces, the head of the fire department (fire chief) is given the responsibility of investigating the cause of fires as part of the legislation that empowers the department **(Figure 1.1, p. 10)**. As a result, many departments are legally obligated to investigate the origin and cause of fires to which they respond. To meet these obligations, a department may assign fire investigations to firefighters or fire officers who have been cross-trained as investigators, or to members of the department specifically assigned as fire investigators.

Figure 1.1 Public sector fire investigators have authority to investigate per legislation.

Private sector fire investigators may be retained by insurance companies, lawyers, manufacturers or other **interested parties**. They operate primarily under civil procedure rules. There are cases that may require the private sector fire investigator to also become involved in the criminal investigation. Although private sector fire investigators are also responsible to investigate the fire to determine the origin and cause, they are primarily responsible to their client.

The initial fire investigation may be conducted by public sector fire investigators. In some instances, the private sector fire investigator may be the first fire investigator to examine the scene. The private sector fire investigator may also examine the scene after the public sector fire investigators. Both parties should recognize the need to work together when possible.

The fire investigator must have a wide range of knowledge and skills. Examples include:

- Knowledge in many subjects including building construction, human behavior, forensics and legal matters pertaining to fire investigation, fire science, fire dynamics, chemistry, physics, fire analysis, and fire investigation methodology.

- Ability to communicate clearly and write easily understood technical reports.

- Ability to perform strenuous physical labor and organize and analyze large volumes of information.

Fire investigators should also expect to have their methodology, findings, and professional qualifications vigorously challenged whenever they testify in legal proceedings. Thus, to perform their job, fire investigators must be appropriately educated and trained, remain up-to-date with current practices in their profession, and seek credentials that attest to their knowledge and abilities.

Requisite Knowledge for Fire Investigators

Section 4.1.7 of NFPA 1033, *Standard for Professional Qualifications for Fire Investigator* stipulates that fire investigators shall have and maintain at a minimum an up-to-date basic knowledge of the following topics "beyond the high school level:"

1. Fire science
   a. Fire chemistry
   b. Thermodynamics
   c. Fire dynamics
   d. Explosion dynamics
2. Fire investigation
   a. Fire analysis
   b. Fire investigation methodology
   c. Fire investigation technology
   d. Evidence documentation, collection, and preservation
   e. Failure analysis and analytical tools
3. Fire scene safety
   a. Hazard recognition, evaluation, and basic mitigation procedures
   b. Hazardous materials
   c. Safety regulations
4. Building systems
   a. Types of construction
   b. Fire protection systems
   c. Electricity and electrical systems
   d. Fuel gas systems

Most of these subjects are identified as **requisite knowledge** for the individual task listed in Chapter 4 of NFPA 1033. The investigator should have a basic understanding of these subjects to perform to the minimum job performance standards as a fire investigator.

In some areas, the investigation of fires may be the role of the county, state, or provincial fire marshal. In these cases, the local fire department may only play a support role in the investigation. Depending on the jurisdiction, the fire department may start the investigation and then turn it over to an agency with more specialized training. In certain circumstances, the legislation requires the involvement of other agencies, such as cases involving an arson or fatality.

Many jurisdictions have organized investigative units or teams that are composed of investigators from the fire department and law enforcement agencies. Multiagency or multijurisdictional investigative units bring together investigators with different areas of expertise, as well as providing sufficient resources for the investigation. They can also foster a better working relationship between departments and leverage wider access to resources **(Figure 1.2)**.

Figure 1.2 Multiagency investigative teams may have access to all the resources available at each included agency. *Courtesy of Wayne Chapdelaine, Metro-Rural Fire Forensics.*

No matter what organizational structure fire investigators work in, their primary function is to determine the origin and cause of fires that occur within their jurisdictions. The data collected from investigations is used to develop fire prevention priorities, public education programs, and product safety awareness and plan for fire protection needs in the community. The data is also combined with data from other jurisdictions to identify trends, and assist in the development and revision of fire prevention-related codes and standards. Another essential function of the fire investigator is the detection and investigation of incendiary fires. The thorough investigation of all fires within a jurisdiction is a significant deterrent to arson.

What This Means to You

Fire investigation is a forensic science (NIST). Forensic science is defined as "the application of a broad spectrum of sciences to answer questions of interest to the legal system, including both criminal and civil actions" (Houck and Siegel 2015).

The job of the forensic scientist is to provide scientific evidence, notably the analysis of scientific or engineering data, to the justice system in order to reduce uncertainty. Forensic scientists must interpret and present the significance of the evidence to the court (Taroni et al. 2010).

Investigator as Liaison

NFPA 1033 (2022): 4.1.4

A fire investigator regularly works with other agencies and organizations. These liaisons may be either very formal such as in a joint agency investigative unit or informal such as with a local construction contractor who can supply skilled labor to build shoring systems on an as-needed basis.

Some investigations, such as those involving large losses, or multiple fire deaths or injuries, commonly involve multiple agencies or interested parties conducting simultaneous, separate investigations. These investigations can involve agencies and interested parties consisting of both public sector and private sector investigators.

Public Agencies

Fire investigators most commonly form liaisons with agencies directly involved in their investigations. These agencies may include:

- State/provincial fire marshal's office

- Forensics laboratory that supports the jurisdiction

- Medical examiner/coroner that serves the jurisdiction

- Local, state/provincial, and federal law enforcement agencies **(Figure 1.3)**

In many jurisdictions, the fire marshal's office may need to be notified in the event of an incendiary fire, burn injury/death, or large property-loss fire. Should the incident under investigation involve a fatality, the jurisdiction's medical examiner/coroner

Figure 1.3 Law enforcement agencies can help fire investigators maintain scene security. *Courtesy of Donny Howard, Agent, Oklahoma State Fire Marshal's Office.*

will likely be involved in the investigation and may be required to respond to the scene before removal of the victim, or alterations are made to the fire scene. Incidents that may involve criminal activity result in the fire investigator working with law enforcement agencies. Incidents with special circumstances, such as explosions or fires in government facilities or government-regulated facilities or as a result of suspected terrorism, may also necessitate that the fire investigator work in collaboration with state/provincial or federal law enforcement and regulatory agencies. Fires involving hazardous materials may require public and private agencies.

During an investigation, the fire investigator often needs information from the municipal or state/provincial government. The fire investigator should have a working relationship with the town or city clerk's office, assessor's office, and treasurer's office. These governmental departments can assist in the identification and retrieval of records and files essential to an investigation. Often, municipal departments have counterparts at the state/provincial level. A working relationship with local officials can be helpful when additional information or resources are necessary at the higher governmental level.

Investigators may also work with the building department to obtain information regarding buildings involved in investigations or for expertise on construction and potential hazards on the **fireground**. For example, building or electrical inspectors may serve as expert resources to assist in the investigation by providing specialized knowledge of building or electrical systems. The local street or highway department may assist with information related to streets such as buried natural gas and fuel distribution lines and may be able to assist in securing large incident scenes with road barricades. The street or highway department may also be a source of heavy equipment to assist with debris removal during the examination of a scene.

Also, first-due fire departments may have prefire plans that show hazards in greater detail than may be possible on another municipal map. These resources can provide expertise on construction, design, installation, recent renovations, and potential hazards on the fireground.

Private Agencies

Fire investigators and the agencies they work for should also maintain liaisons with private businesses that can assist in the investigation of fires. One of the most important private-sector relationships is with the insurance industry. The resources and assistance available from the insurance industry are essential for many public agencies. They may be the source of the following information and services:

- Educational materials related to arson and fire prevention.

- Identification of alternate vehicle identification number (VIN) locations.

- Information regarding the insured property on which the fire investigator may be working.

- Special experts and laboratory resources that may not be normally available to the fire investigator.

- Provision of specialized equipment, such as cranes and other heavy equipment, needed to remove debris from large buildings.

- The insurance underwriter may be required to provide the investigation file. That file may contain statements and documents obtained from the person/entity holding the insurance policy. For more information on arson immunity laws, see Chapter 7.

- Data from central insurance claim databases that collect loss information as a result of fire, theft, and burglary claims. These databases contain information regarding the company that paid a claim, the name of the insured, the mortgagee or lien holder, and the type of claim.

Other private resources the fire investigator may want to maintain a liaison with include:

- **Local financial institutions** — Information regarding mortgages and business history

- **Local construction companies** — Sources of heavy equipment and expertise for debris removal and building stabilization

- **Utility companies** — Information regarding their distribution systems, customer databases and information regarding changes or increases in electricity, as well as disconnection and reconnection dates

In some cases, requests for information may require a warrant or subpoena. In Canada, the documentation may include a production order. For more information, see Chapter 7, Investigative Process.

Investigators may need support from persons who are experienced in the examination and analysis of:

- Financial records and data

- Building systems or components

- Electrical equipment or components

- Industrial or manufacturing processes

- Fire protection suppression and detection systems

Arson Task and Strike Forces

Fire investigators may occasionally need assistance or resources they do not normally have access to, such as in cases of "mass arson," or "serial arson." In these circumstances, jurisdictions may seek to establish a multi-jurisdictional arson **strike force**. The USFA defines an arson strike force as "... a special purpose, short-term mobilization of a team (or teams) of investigators together with allied resources that applies high intensity investigative efforts to a major arson incident or series of incidents." Arson strike forces help in the formal investigation and assist investigators by sharing information about arsonists, their methods of operation, and special investigative techniques that may be used to apprehend and prosecute them. For additional information see the USFA website.

Task forces can be legal or quasi-legal bodies or private advisory committees, but their purpose (for example, fighting arson in a community) is the same. These task forces normally have policy-setting and program-implementation responsibilities and are not investigative units. Thus, while arson task forces are policy-making bodies, an arson strike force is an operational investigative unit with the ability to bring additional resources and capabilities into a major investigation.

Arson task forces have the following functions:

- Set policy or guidelines
- Mobilize public and private resources **(Figure 1.4)**
- Identify and coordinate responsibilities
- Integrate the efforts of agencies, groups, and individuals who seek to prevent and control arson

Figure 1.4 Public or private resources may include portable scene lighting systems.

Fire Investigator Training and Certification

NFPA 1033 (2022): 4.1.1, 4.1.2, 4.1.5

The training and certification of fire service professionals has been centered on the NFPA professional qualifications standards first issued in the early 1970s. NFPA 1033 establishes the minimum job performance requirements (JPRs) for the fire investigator. As with many of the other professional qualifications standards produced by NFPA, this document is the basis for most fire investigator training programs in North America. The standard is periodically revised by NFPA to meet changing needs.

States, provinces, and other agencies such as the U.S. Department of Defense can adopt the professional qualifications standards such as NFPA 1033 and use them to develop training and certification programs.

Certification programs are designed to evaluate that an individual has the requisite knowledge and skills to perform a specific job. The professional qualifications standard for the specific job sets the criteria used for these evaluations — normally written tests and skill evaluations. As part of the professional qualifications system, organizations that provide certification to fire service professionals can seek to be accredited. Accreditation is the formal review by a third-party of the policies, procedures, and processes used by an organization providing certification.

Fire service accrediting organizations include the National Professional Qualifications Board (the ProBoard™) and the International Fire Service Accreditation Congress (IFSAC). While only a few states and provinces are accredited to certify fire investigators using NFPA 1033, accreditation systems are in place and growing; and it is expected that more entities will seek accreditation to certify for this position. Several states and provinces offer fire investigator certification that is based on training and criteria established by the state/province.

A non-fire service accrediting organization is the Forensic Specialties Accreditation Board (FSAB). Several fire investigation training providers are exploring this accreditation organization as they specialize in forensic specialties as an addition or enhancement to fire investigation training and certification.

Professional Development

Because of the rapid changes brought about by new technology and ever-changing codes and legal decisions, fire investigators must continually learn. To withstand the careful examination of their qualifications, which take place when they become involved with the legal system, fire investigators should also document their training and develop their professional credentials by seeking certifications in their field. Investigators may find it helpful to create electronic folders for each of the requisite knowledge topics outlined in NFPA 1033 and place an abstract or copy of their training certificate in each of the folders relating to that training. Investigators should maintain a personal **Curriculum Vitae (CV)** for reference as needed. This document can be used as a professional record of credentials, particularly ahead of legal testimony.

Resources in addition to 1033 and 921 include textbooks and references, **peer reviewed** journals, and technical handbooks, and can also offer in-depth information on specialized fields such as investigative techniques and fire dynamics. Investigators may benefit from networking with other investigators and seeking out higher education classes to improve their knowledge of resources and current information.

Formal training programs are offered by many jurisdictions, state and provincial fire service training academies, national law enforcement academies, the National Fire Academy (NFA), NFPA, colleges and universities, private companies and professional associations for fire investigators. The International Association of Arson Investigators (IAAI), the National Association of Fire Investigators (NAFI), the Canadian Association of Fire Investigators (CAFI), and other international associations of fire investigators also offer continuing education and certification programs based on NFPA 1033. Many of these training programs reference information provided in NFPA 921, *Guide for Fire and Explosion Investigations*. In addition, NFA has issued several packages developed around the NFPA 1033 requirements for the fire service. These materials should help to increase the delivery of fire investigator training programs at the local level.

Online training modules are available to fire investigators. One such resource is available at cfitrainer.net. Successful completion of these modules may count toward fire investigation certifications and continuing education.

Professional organizations serving fire investigators (NAFI, CAFI and IAAI) provide training and certification of investigators. These organizations provide training opportunities at various locations. These programs normally focus on one or more aspects of the fire investigator's position. Program requirements may be parallel to the requirements of NFPA 1033 and NFPA 921. Additional information regarding training and certification available from national organizations can be obtained from them directly **(Table 1.1)**.

Table 1.1	
Online Access to National Training Organizations	
Organization	**Web Presence**
National Association of Fire Investigators	www.nafi.org
International Association of Arson Investigators	www.firearson.com www.cfitrainer.net
Canadian Association of Fire Investigators	www.cafi.ca

Several organizations offer professional certification programs. Overall, competency is based on an accumulation of experience and training over time. Proficiency in completing the tasks identified in NFPA 1033 is measured during practical training and evaluation to identify a measure of performance of an investigator's skills.

Through its National Certification Board, NAFI administers the Certified Fire and Explosion Investigator (CFEI) Program. Certification by the National Certification Board is attained following an acceptable review of an applicant's documentation of professional education, training, and experience by the Board and the subsequent successful completion of a comprehensive proctored and written certification evaluation. Successful applicants must demonstrate to the Board that they have met certification requirements and possess professional training, education, and experience mandated by the CFEI program.

The IAAI administers the Certified Fire Investigator (CFI) Program. This program is aimed at identifying and recognizing an investigator's expertise. It is based on an evaluation of the individual's investigation training and related experience, and a comprehensive examination. The IAAI program is nationally accredited to certify fire investigators by the National Professional Qualifications Board.

The Canadian Association of Fire Investigators (CAFI) is a nonprofit national organization made up of both public and private sector investigators as well as law enforcement officials. CAFI offers the Canadian Certified Fire investigator (CCFI) designation to successful candidates. Different levels of the CCFI designation are available for obtainment.

Legal and Regulatory Foundations

NFPA 1033 (2022): 4.1.5

Fire investigators must have the legal authority to conduct an investigation, and they must conduct themselves and their investigations in accordance with legal procedures present in their jurisdictions. This adherence to legal procedures applies to public- and private-sector investigators and for criminal and civil cases. It is also true whether the investigation is conducted in the U.S. or in Canada. The sections that follow discuss important aspects of due process clauses, legal authority for investigations, and legal guidelines for private-sector investigators.

Difference between 1033 and 921

NFPA 1033 is a professional qualifications standard. NFPA 921, while a guide and not a standard, provides the procedures, methodologies, and information necessary to meet the job performance requirements in NFPA 1033. Investigators are cautioned that the material contained in NFPA 1033 may apply to scene investigations as well as for the job performance requirements identified within the standard.

Annex material at the end of each of these documents is increasingly referenced in court settings. NFPA 921 is recognized by many courts in the U.S. and Canada as a standard of care to be used during a fire investigation. According to NFPA 1033, Fire Investigators should employ all elements of the scientific method for all investigations, and be prepared to justify their approach to each investigation. See Chapter 7 for more information.

Authority

In the U.S. and Canada, public-sector investigations must be conducted within the bounds of constitutionally regulated authority. The constitution of each country defines the boundaries of lawmaking power, balancing it with the constitutionally protected rights of the individual.

The authority to conduct an investigation is provided by law, and the investigation must be conducted within the boundaries of that legal authority. One or more laws will apply to a given case and dictate the purpose of the investigation and the powers and jurisdiction of the investigator. If an investigator exceeds his or her authority, evidence from the investigation might be ruled inadmissible, charges might be dismissed, or in extreme cases an investigator might be subject to criminal penalties or civil liability.

Private-Sector Investigations

In the private sector, there are also laws governing legal procedures and the authority of investigators. Constitutional provisions such as due process clauses have little to do with private-sector investigations. The powers and legal authority of private-sector investigators are derived from civil laws governing rights and obligations between private persons. Rights or obligations arising from contracts, insurance policies, and tort laws, such as negligence issues, affect the authority of the private-sector investigator. State or provincial laws regulating the licensing of investigators also have an effect.

Failure to act within the bounds of legal authority can have negative consequences similar to those in criminal cases. Evidence or statements might be ruled inadmissible, pleadings may be stricken, or an investigator might be exposed to civil suit or criminal prosecution. In the insurance arena, for example, bad-faith cases may be founded on circumstances where an investigator exceeded his or her authority.

NOTE: While a detailed analysis of the legal foundations of private-sector fire investigations are beyond the scope of this manual, some of the central legal issues that affect fire investigators in civil cases are introduced in later chapters.

Review Questions

1. What are some differences between public fire investigators and private sector fire investigators?

2. What specific knowledge and skills should a fire investigator have in order to effectively perform his or her job?

3. What is the primary function of a fire investigator, regardless of whether he or she works in the public or private sector?

4. What are the agencies and organizations with which a fire investigator regularly works?

5. What are some of the organizations that offer training and certification programs for fire investigators?

6. From where are the powers and legal authority of public- and private-sector investigators derived?

7. What consequences may arise if a fire investigator acts outside of his or her legal authority?

Discussion Questions

1. In what areas do you feel most confident about your knowledge and skills relating to fire investigation? In what areas will you need to put in extra effort to gain that confidence?

2. What certification and training requirements does your jurisdiction have for fire investigators?

Chapter References

Houck, Max M, and Jay A Siegel. 2015. Fundamentals of Forensic Science, Third Edition. Elsevier Ltd, New York.

National Institute of Standards and Technology (NIST). Organization of Scientific Area Committees for Forensic Science (OSAC). Accessed online.

Taroni F, Bozzo S, Aitken C, Garbolino P, Biedermann A. 2010. Data Analysis in Forensic Science: A Bayesian Decision Perspective. John Wiley & Sons, Chichester (UK).

Chapter Key Terms

Arson — Crime of willfully, maliciously, and intentionally starting an incendiary fire or causing an explosion to destroy one's property or the property of another. Precise legal definitions vary among jurisdictions, wherein it is defined by statutes and judicial decisions.

Curriculum Vitae (CV) — Detailed listing of professional accomplishments, certifications, and credentials used to quantify an applicant's qualifications for a position or function.

Fireground — Area around a fire and occupied by fire fighting forces.

Interested Party — Any person, entity, or organization, including their representatives, with statutory obligations or whose legal rights or interest may be affected by the investigation of a specific incident. (Reproduced with permission from NFPA 921-2021, *Guide for Fire and Explosion Investigations*, Copyright 2020, National Fire Protection Association)

Peer Review — Process of quality control evaluation in which reviewers from a relevant field critique a publishable scientific, academic, or professional work for quality and accuracy.

Requisite Knowledge — Fundamental knowledge one must have in order to perform a specific task. (Reproduced with permission from NFPA 1033-2022, *Standard for Professional Qualifications for Fire Investigator*, Copyright 2021, National Fire Protection Association)

Strike Force — Special purpose, short-term mobilization of a team (or teams) of investigators together with allied resources that applies high intensity investigative efforts to a major incident or series of incidents. (FEMA/USFA: Establishing an Arson Strike Force, 1989.)

Task Force — Legal or advisory committees with policy-setting and program-implementation responsibilities and are not investigative units. Policy-making body.

Chapter Contents

NFPA 1033 JPRs addressed in this chapter

This chapter provides information that addresses the following job performance requirements (JPRs) of NFPA 1033, *Standard for Professional Qualifications for Fire Investigator* (2022):

4.1.3   4.1.6   4.1.7   4.2.1   4.2.2   4.2.8

This chapter provides information that addresses the following course outcomes for the FESHE courses:

Fire Investigation I (C0283)

    8. Recognize potential health and safety hazards.

1. Explain how a fire investigator should function within the incident management system. [NFPA 1033, 4.1.6]

2. Explain the necessity of a site safety assessment. [NFPA 1033, 4.1.3, 4.2.1, 4.2.2]

3. Describe the types of safety equipment used by fire investigators. [NFPA 1033, 4.1.3, 4.2.1]

4. Identify the various hazards that fire investigators may encounter at an incident. [NFPA 1033, 4.1.7, 4.2.1, 4.2.2]

5. Describe the types of utility systems that a fire investigator must understand in order to remain safe at an incident. [NFPA 1033, 4.1.7, 4.2.1, 4.2.8]

6. Skill Sheet 2-1: Don and doff personal protective equipment. [NFPA 1033, 4.1.3]

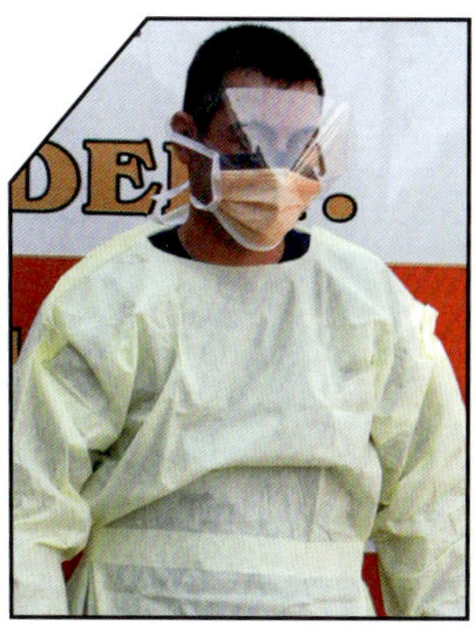

Chapter 2
Safety

Fire investigators conducting an investigation face many of the same hazards and dangers that firefighters encounter during fire suppression and overhaul operations. This chapter addresses safety considerations important to investigators including:

- Incident management systems
- Site safety assessments
- Safety equipment
- Scene safety hazards
- Utility systems

Incident Management Systems

NFPA 1033 (2022): 4.1.6

Management of fireground operations involves the use of an incident management system. An incident management system allows for the efficient management, safety, and accountability of all personnel or entities involved with an incident. Although most fire investigation activities occur during the de-escalation phase of a fire incident, fire investigators should operate as part of the incident management system whenever they arrive at an incident scene.

National Incident Management System – Incident Command System (NIMS-ICS)

In the United States, the Department of Homeland Security (DHS) has adopted the National Incident Management System-Incident Command System (NIMS-ICS) as a unifying organizational system for emergency responders and programs. The organizational structure and terms outlined in a number of IFSTA publications describing the Incident Management System (IMS) are also consistent with the guidelines stipulated by NIMS-ICS. For more information about NIMS-ICS, which is under the Federal Emergency Management Agency (FEMA), consult its website.

The on-scene investigation falls within the scope of the **incident management system (IMS)**. An investigator does *not* operate outside of or independently from the established incident management organization at the scene. From the moment the investigators arrive until the last unit is demobilized and returned to service, the incident scene is required to be within the control of an incident management system.

Upon arrival at any incident requiring fire investigative services, the fire investigator, whether public or private, must check in with the **incident commander (IC)** if it is still an active fire scene. Reporting to the incident command post brings notice to the IC that the request for investigative services has been satisfied and identifies the investigator as an accountable unit should he or she need to operate in a hazardous environment. Reporting to the IC is also an opportunity to gather information about a scene and activities conducted or completed. If fire investigation activities end before fire suppression activities, the investigator should notify the IC.

The IC is ultimately responsible for all functions on the fireground. Depending on fireground conditions, it is common for the IC to designate a **safety officer**. Before beginning a fire investigation, an investigator should consult the safety officer to obtain information concerning site safety and hazards **(Figure 2.1)**. In addition to consulting the safety officer, an investigator should verify that consultation by completing a **site safety assessment** as outlined later in this chapter. The IC should be consulted on these issues in the absence of a safety officer.

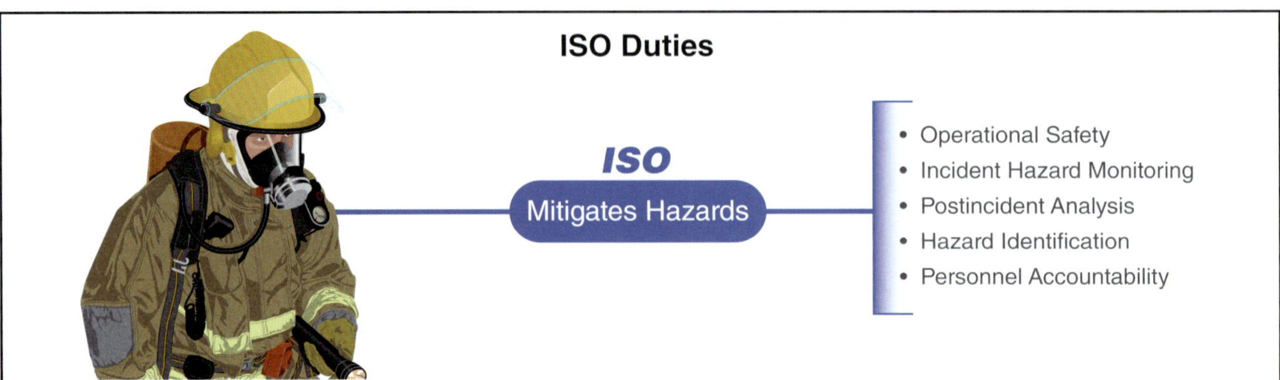

Figure 2.1 An incident safety officer (ISO), if still assigned, will be able to share information on the hazards and conditions present at the incident scene.

Some incidents may require the use of a **unified command** structure. The unified command structure ensures that multiple priorities from multiple agencies are considered during the resolution of an incident. Public safety agencies from multiple jurisdictions and disciplines would have input on activities used to mitigate an incident. Local, state/provincial, and federal law enforcement agencies are usually part of a unified command structure. All efforts should be made to ensure that fire investigation activities are coordinated with members of the unified command structure.

Site Safety Assessments

NFPA 1033 (2022): 4.1.3, 4.2.1, 4.2.2

The primary objective of a site safety assessment is to minimize the level of risk to personnel so that a safe, effective, and timely investigation can be conducted to determine the origin and cause of the incident. The site safety assessment begins upon arrival at the scene by noting environmental conditions such as weather, scene accessibility, and terrain **(Figure 2.2)**. Downed electrical wires, vehicle traffic, and poorly lighted areas all pose hazards for investigators making their way to the scene. Investigators should park in a manner that does not expose their vehicles to unnecessary risks during emergency activities or when emergency equipment departs the scene.

Site safety assessments shall be completed on all scenes. If the scene is still active, investigators should get instructions from the IC and safety officer. Each investigator on the scene has a responsibility to practice **situational awareness**, conduct individual site safety assessments, and communicate additional identified hazards to the IC. Regional and national safety standards should be followed and included in organizational policies and procedures. The investigator should be aware of information concerning structural issues, air quality, utility control, cut and puncture risks, and any other hazardous conditions.

> **CAUTION:** Investigators are responsible for their own personal safety and scene safety.

Figure 2.2 An investigator will be able to determine some hazards at an incident before leaving the vehicle.

Exterior and interior surveys of the scene will identify most safety hazards. Common hazards that surveys identify include but are not limited to the following:

- Inadequate illumination
- Energized utilities
- Blocked means of egress
- Holes in floors
- Potential collapse areas
- Presence of hazardous materials and atmospheres

CAUTION: Floor integrity issues and all collapse hazards should be clearly marked and identified before beginning fire investigation activities. In addition, all hazards and required safety precautions should be communicated to any person who attends the scene during the examination prior to allowing access.

If the site safety assessment results in an unsatisfactory determination, the on-scene investigation must stop until appropriate measures to address safety concerns are applied. Resources such as heavy equipment and construction crews may be necessary to remove unstable walls or perform shoring operations. If efforts to render the site safe fail and solutions are neither economically feasible nor practical, the scene may have to be secured and abandoned until such future time when resources become available. In certain circumstances, portions of the scene may be deemed to be unsafe while other areas may be safe to work in. The unsafe areas should be clearly marked and access restricted.

A site safety assessment is a continuous process and ends at the conclusion of the investigation. During the course of investigations, areas of a structure initially believed to be safe from collapse may later be in danger of a collapse due to changes in the conditions at the scene. Investigators must continually reassess safety as they conduct their duties. They should never assume that their initial safety assessment represents the conditions at the scene from that point forward.

Prior to any investigative activities being commenced, a site safety briefing shall be held with all attendees that are part of the fire investigation team. All efforts should be made to ensure that all attendees are familiar with the hazards and be given an opportunity to bring forward any safety concerns during the examination. This briefing shall include hazards, safety precautions, PPE, communications, and each party's duty to keep themselves and others safe.

Safety Equipment

NFPA 1033 (2022): 4.1.3, 4.2.1

Regardless of the severity of an incident, investigators must wear appropriate clothing and equipment to ensure their safety and well-being. Safety will be compromised if personal protective equipment (PPE) is not available or if it is not worn and used properly. The level of protection required is determined by the scene and the hazards that can be anticipated during the investigation. A minimum level of personal protective equipment must always be worn: protection for the head, eyes, hands, feet, and respiratory system.

When selecting PPE, monitoring and sample collection must also be considered. For example, when an air purifying respirator is selected, OSHA requires continuous air monitoring. In some instances, a risk of chemical, biological, or radiological exposure may require a higher and more specialized level of personal protection. Investigators must select the appropriate protective clothing designed to provide protection against the present hazards, and to accommodate the intended actions. Training in the use of these protective ensembles is essential to ensure that this equipment is used and maintained to provide optimal performance.

CAUTION: All protective equipment has inherent limitations. Investigators must understand the uses and limitations of selected equipment.

- Conduct gross decon of PPE while still at the fire scene.

- Use approved wipes at the scene to clean your hands and arms, face and neck, and any exposed areas **(Figure 2.3)**.

- Change out of PPE and work clothes before leaving the scene.

- Do not store contaminated gear or clothing in your home, living area, or vehicle.

- Shower after the investigation.

- Properly clean all gear after an investigation.

Figure 2.3 Using wet wipes on the scene can help minimize contamination.

Personal Protective Equipment

NFPA 1033 (2022): 4.1.3, 4.2.1

Interior origin and cause investigation generally does not begin until a fire has been extinguished. There may be situations in which investigators must begin their investigations while a fire scene is still active. In any situation, investigators should wear the appropriate **personal protective equipment (PPE)** for the situations they are entering. **Skill Sheet 2-1** shows the steps for donning and doffing personal protective equipment.

Fire investigators must be trained to recognize the level of hazard present and use safety equipment properly. Although no protective ensemble can totally protect the wearer from all risks present at the incident scene, the use of well-maintained PPE conforming to national standards may reduce the possibility of injury to an investigator.

The sections that follow describe some basic PPE. Investigators should remain aware that more specialized gear may be required at complicated scenes. Investigators should rely upon the expertise of incident commanders and safety officers when there is any doubt about what protective equipment they will need.

 **Cancer and Other Serious Health Hazards**

Much less research has been conducted with respect to fire investigators and cancer than has been completed on firefighters and the link to cancer. At the same time, investigators may be exposed to many of the same hazards as firefighters, for longer and more frequent periods.

Fire investigators may be exposed to respiratory hazards including asbestos, atmospheric toxins, gases and dusts during the course of the investigation. Respiratory protection should be used while conducting the fire investigation. The level of respiratory protection should match the level of hazard encountered. For example, the investigator may be able to work with a half face respirator with cartridges in a specific environment where no hazards exist for absorption through the eyes is present or the use of safety glasses is appropriate. A full face respirator may be more appropriate in other environments. An SCBA may be needed to enter active fire scenes or those scenes recently extinguished or where smoldering is still present.

The long term effects of inhalation of materials such as benzene, hydrogen cyanide and other hazardous substances has been found to cause illness and even death. Proper use of PPE and conducting a scene assessment prior to initiating the scene examination as well as having appropriate operating guidelines to ensure the health and safety of fire investigators is of the utmost importance.

Structural Fire Fighting Personal Protective Equipment

Structural fire fighting personal protective equipment (PPE) is generally suitable for postfire activities. Wearing this type of PPE protects the body from abrasions and cuts, minor burns, etc. All structural personal protective equipment should meet the standards set forth in NFPA 1971, *Standard on Protective Ensembles for Structural Fire Fighting and Proximity Fire Fighting*.

Any clothing used on the fireground should be considered contaminated and must be cleaned in the appropriate manner. For a discussion of cleaning contaminated clothing, see the IFSTA manual, **Essentials of Fire Fighting**.

Liquid Absorption on PPE

Remember that PPE may absorb liquids and become saturated. Some hazardous substances, such as benzene, for example, that are present at fire scenes can permeate material and become absorbed into the skin.

Coveralls

Coveralls are versatile and can be worn over other clothing while providing a limited measure of protection, uniformity, and comfort. Generally, coveralls do not provide the level of protection afforded by the structural fire fighting ensemble and they must be thought of as simply an additional layer of clothing. Some coveralls are also designated as fire resistant or *FR*. The type of coveralls selected should match the potential hazard **(Figure 2.4)**.

Special disposable coveralls may be worn that are liquid resistant and resistant to other irritants such as fiberglass. Some types of disposable coveralls may also be worn in fire scenes where a crime may have been committed and where additional special precautions may be necessary in order to reduce cross-contamination within the scene.

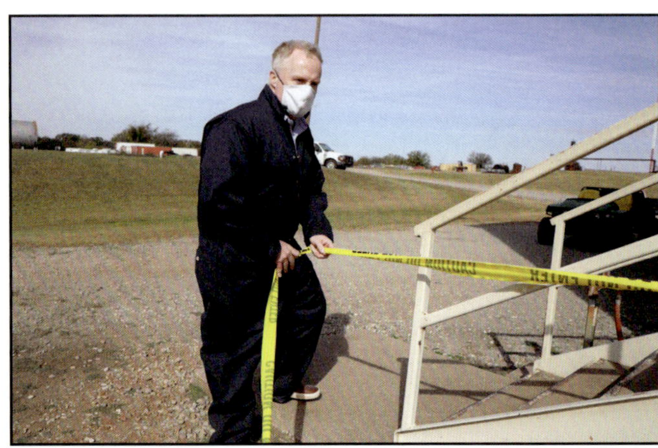

Figure 2.4 In any situation, investigators should wear the appropriate personal protective equipment (PPE) for the situations they are entering.

Helmets and Hard Hats

Damage caused by either the fire itself or by fire-fighting activities may weaken structural supports and other building materials. Therefore, head protection must be one of the first concerns of fire investigators. A fire investigator should always wear either a hard hat or a fire helmet for protection from falling debris. Hard hats should meet the requirements of ANSI Standard Z89.1, *Standard for Industrial Head Protection*, and Occupational Safety and Health Administration (OSHA) regulations in the U.S. or Canadian Centre for Occupational Health and Safety (CCOHS) regulations in Canada. If fire helmets are worn, they should meet the specifications set forth in NFPA 1971. Some investigators prefer the use of hard hats with chin straps to ensure that the helmet stays in place during activities such as bending, crawling and kneeling. Investigators must ensure that the helmet complies with the above regulations.

Face, Eye, and Hearing Protection

Proper protection for face, eyes, and hearing should be determined based on the scene requirements from the site safety assessment. When protection from ultra-bright light is needed, darkly tinted glasses may be required. A clear plastic face shield attached to the helmet can provide some face and eye protection; however, helmet face shields alone do not provide an adequate level of face and eye protection for most operations.

Potential face and eye hazards include the following, whether separately or in combination:

- Striking or scraping
- Penetration
- Chemical and thermal burns
- Harmful radiation
- Spraying liquids, including bodily fluids

When self-contained breathing apparatus (SCBA) is worn, the SCBA's facepiece provides face and eye protection. Proper and effective use of SCBA requires employee training, fit testing, and medical evaluation.

Hearing protection devices should conform to ANSI S3.19 or S12.6, and have a **noise reduction rating (NRR)** of at least 20 decibels (dB). Two general types of hearing protection devices used in most operations are ear plugs and headsets. For more detailed information on face, eye, and hearing protection, refer to the IFSTA manuals:

- **Fire and Emergency Services Safety Officer**
- **Occupational Safety Health and Wellness**
- **Fire Service Technical Search and Rescue**

> **CAUTION:** Some types of hearing protection may interfere with critical communications and may not be appropriate when communication is a high priority.

Gloves

The most important characteristics of gloves are the protection they provide against heat or cold penetration and their resistance to cuts, punctures, and liquid permeation. This protection can be accomplished by wearing latex gloves under strong, durable leather gloves **(Figure 2.5)**. Different types of gloves should be kept on hand to provide specific protection during the required task. Examples of different types of gloves include leather, vehicle extrication, mechanic's, and thick rubber gloves. During investigations, it is almost always necessary to sift through debris such as broken glass and charred furniture. In addition, some of the material may be too hot to handle without protection.

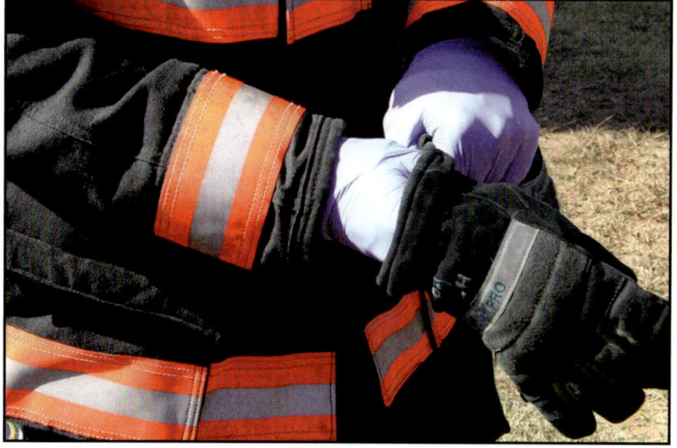

Figure 2.5 Investigators may wear latex gloves under protective gloves while excavating and reconstructing a scene to provide more protection against hazardous liquid permeation.

Gloves must fit properly and provide protection as well as affording dexterity, including enough tactile feel to perform the job effectively. If gloves are too cumbersome, the investigator may be unable to do fine, manipulative work. Contaminated gloves should be properly disposed of after each fire scene investigation to prevent contamination to other scenes. NFPA 1971 provides the requirements for gloves for fire investigators.

Boots

An investigator should select appropriate foot protection to ensure that the risk of injury from hazards is minimized. Normal turnout boots may be sufficient for postfire activities and should meet safety requirements such as NFPA 1971. Boots should be water resistant and provide adequate toe, mid-sole and ankle protection. The boots should protect against punctures and crushing injuries as well as provide extra ankle support to assist while working on uneven terrain.

Atmospheric Monitoring Devices

NFPA 1033 (2022): 4.1.3, 4.2.1

Monitors for testing environmental conditions should be used before anyone is allowed to enter an area where the **atmosphere** may be compromised. Several types of monitoring devices can be used to test the atmosphere. For more information on monitoring and detection devices, refer to the IFSTA manual, **Hazardous Materials Technician**.

Generally, detection and air monitoring should be performed in the following order:

1. Radiation

2. Corrosive gases and vapors (wet and dry pH paper)

3. Oxygen levels (multigas detectors including oxygen [O_2], carbon monoxide [CO], etc.) **(Figure 2.6)**

4. Flammable gases and vapors

5. Combustible gases

6. Toxic gases and vapors (colorimetric tubes)

This pattern allows the investigator to measure for hazards that can cause harm from the farthest distance working toward the scene. Also, some detectors may be damaged if they are used in some environments.

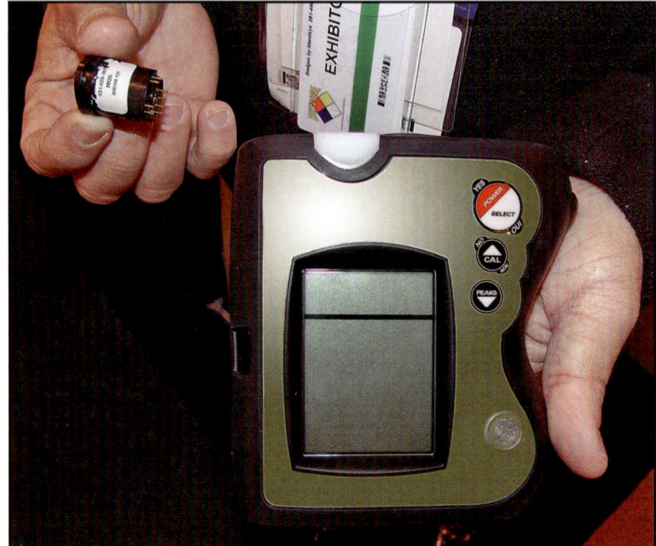

Figure 2.6 A multigas detector checks air quality against a wide range of harmful gases.

Radiation Monitoring

Radioactive materials may be present at fire scenes regardless of the expected hazards. All hazard response should consider all possibilities. There are three types of harmful radiation **(Figure 2.7)**:

- Alpha particles

- Beta particles

- Gamma rays

Alpha particle radiation usually does not penetrate beyond clothing or the outer layer of the skin. If it is prevented from entering the body, it will cause little damage. Beta particle radiation causes burning on the skin surface and damage to the subsurface skin circulatory system. Both alpha and beta particles can be extremely dangerous if they find their way into the body. Gamma radiation, however, passes through clothing and tissue causing severe, permanent injury to the body.

> **WARNING:** Structural fire fighting personal protective equipment and chemical protective clothing afford no protective barrier against gamma radiation.

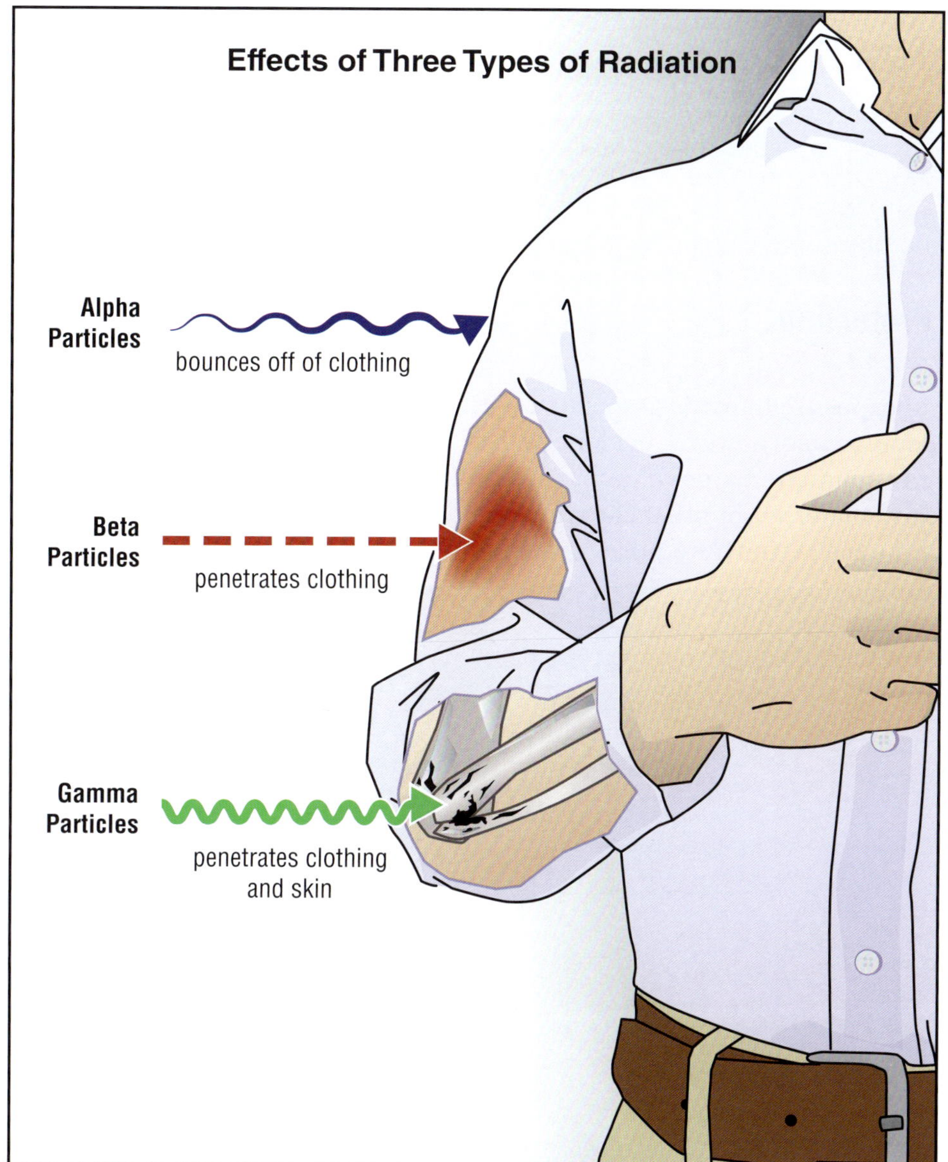

Figure 2.7 Radioactive materials may emit multiple types of radiation. *Source: Modified from U.S. Environmental Protection Agency.*

Where necessary, radiation detectors should be used to identify the types and levels of radiation at a scene. High levels of radiation require highly specialized PPE or may require the scene to remain secured and empty for long periods of time. Respirators reduce and may prevent radioactive materials from entering the body.

WARNING: A fire investigator must be thoroughly trained and knowledgeable about radiation emergencies before being involved in an investigation with this type of hazard. Under no circumstances should an investigator work at a scene with ionizing radiation without the proper PPE and training.

Air Monitoring

Air monitors provide fire investigators with some of the information they need to effectively protect themselves from the atmospheric hazards present at the scene. These versatile devices are designed to detect and/or measure multiple gases and hazardous atmospheres that may be encountered. They provide information such as the type and relative quantity of the hazard that is being monitored. From this information, the investigator can determine the type of personal protective clothing and equipment required to safely conduct the investigation until the hazards are eliminated.

Effects of Oxygen Concentration

Oxygen concentration in the atmosphere has a significant effect on both fire behavior and our ability to survive. Typically, an atmosphere having less than 19.5 percent oxygen is considered oxygen deficient and presents a hazard to persons not wearing respiratory protection, such as SCBA, to provide fresh air. When the oxygen concentration in the atmosphere exceeds 23.5 percent, the atmosphere is considered oxygen enriched and presents an increased fire risk.

Respiratory Protection

NFPA 1033 (2022): 4.1.3

The full range of exposure hazards associated with postfire scene examinations have not been fully determined or documented; however, evidence exists that indicates respiratory health hazards are severe in the postfire environment. Large volumes of toxic gases will be generated at an emergency involving chemical explosive materials or other chemical incidents. When an **air purifying respirator (APR)** is selected for use at a scene, OSHA requires continuous air monitoring **(Figure 2.8)**.

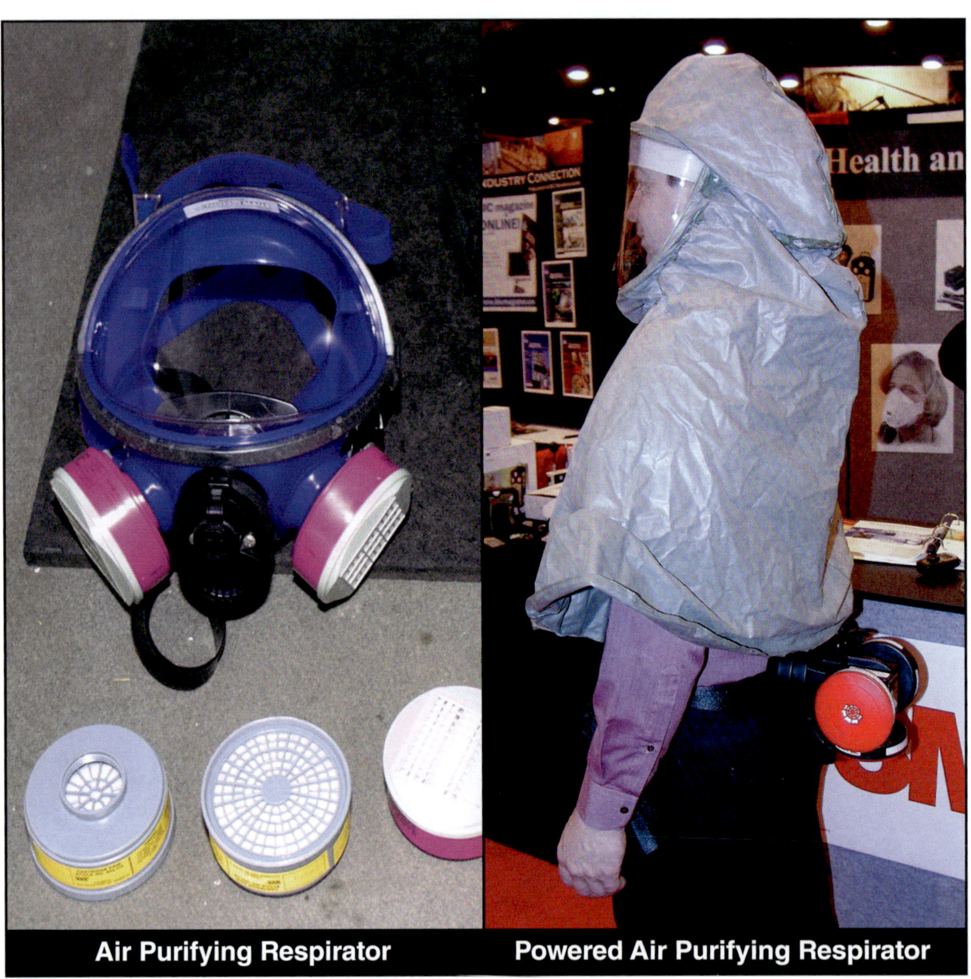

Figure 2.8 Investigators may choose between air purifying respirators (APR), left, and a powered air purifying respirator (PAPR), right.

Factors that Contribute to the Release of Toxic Gases

Information on the release of toxic gases is included later in this manual. The particular toxic gases that may be found at fire and explosion sites vary according to four common factors:

- Nature of the combustible material
- Temperature of the involved gases
- Rate of heating
- Oxygen concentration

As a result of the wide variety and volume of airborne contaminants at a fire scene, there should be constant vigilance in monitoring the atmosphere. Atmospheric and chemical monitoring devices should be used to verify or to establish levels of respiratory protection. A minimum level of protection such as a multipurpose respirator and eye protection should be worn in any work area where debris will be disturbed and dust created. Continuous monitoring may eventually indicate a lower level of respiratory protection. Work areas should be well-ventilated, naturally and/or mechanically.

Limits of Monitoring

The use of monitoring equipment beyond carbon monoxide and oxygen levels is usually not practical for most situations. No single meter can monitor for all respiratory hazards typically produced during a fire and are present during the overhaul phase. Research has shown that other hazardous gases, dusts, and metals are often present that exceed published short term exposure limit values. Investigators may consider wearing appropriate breathing apparatus or waiting until the atmosphere at the scene has been determined safe.

Investigators may need to have several different types of respirators or **self-contained breathing apparatus (SCBA)** available for the different hazards that may exist **(Figure 2.9, p. 32)**. Matching the correct respiratory equipment to the hazard is essential to protect against the harmful effects of atmospheric exposure. Annual face fit testing must be completed for each type of respirator that is part of the investigator's PPE.

Even with constant atmospheric monitoring of the emergency scene, an investigator must remain cautious and use prudence while working in these environments. Fire investigation scenes pose the same health and life threats as loss control and overhaul operations do during fire-suppression efforts. Smoldering fire(s) and other degrading chemicals that are common at fire scenes can quickly debilitate and incapacitate inadequately protected individuals, leading to major injury or death. Investigators must have available and be prepared to use respiratory protection equipment as a precaution against the harmful effects of atmospheric exposure to potential life and health safety risks. For example, N95 masks are not adequate respiratory protection for fire scenes.

WARNING: Potentially lethal atmospheres exist even after a fire is brought under control. All personnel entering a structure must wear appropriate respiratory protection.

Even fire scenes with no burning or smoldering fires present a considerable risk to investigators. When the debris from a fire scene is disturbed, some particles such as asbestos become airborne and susceptible to inhalation. Common particulates such as carbon (soot) are considered cancer risks in large quantities. These dust particulates can also be absorbed through the eye. Atmospheres with sufficient oxygen and low or undetectable gas hazards still require respiratory protection. An investigator in these conditions must use a full filter mask with maximum-protection filters or a half mask with nonvented goggles.

An air filter is the primary particulate filter and should be serviceable for several fires depending on the exposure time and particulate present. The primary means of evaluating its serviceability is the

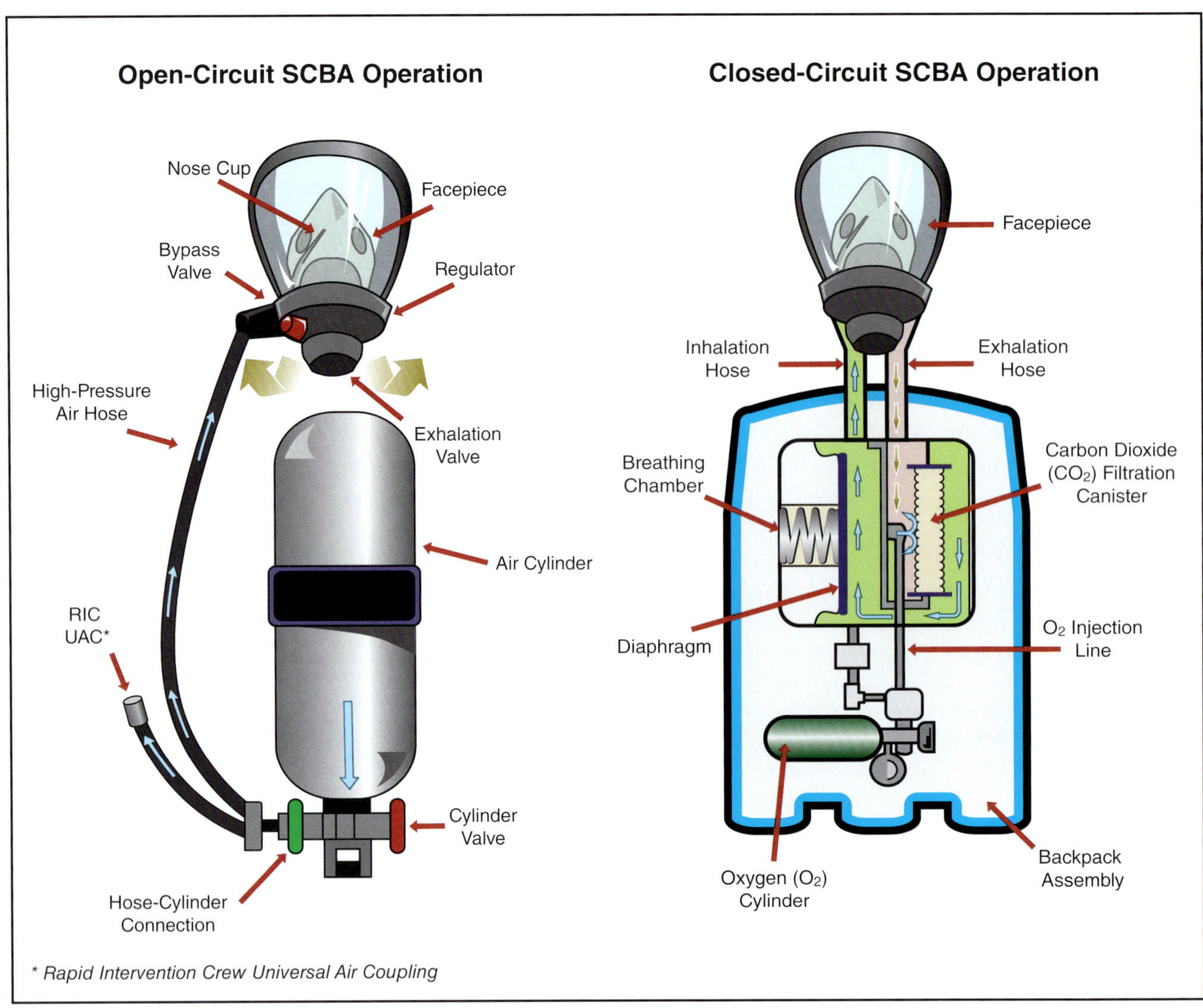

Figure 2.9 Investigators may need use of a self-contained breathing apparatus (SCBA) during an investigation, based on the hazards present at the scene.

ease or difficulty of inhaling. When the filter is new, breathing will be normal. As the filter removes particulates from the air, breathing becomes increasingly difficult.

A separate component, **organic vapor/acid gas cartridges** should be the primary filtering mediums. These cartridges are sealed and meant to be opened and placed in service at the scene. They are affected by humidity and begin to absorb vapors whether being used or not. For this reason, the cartridge is a *one-use* item.

The usable time depends upon the age and health of the investigator, time spent in the environment, and amount of particulate in the environment. Increased service life at a scene may be obtained by placing the respirator cartridges inside a respirator bag when not actively in use, to prevent passive filtering from the environment. Regardless of these factors, once the respirator cartridge is activated, it cannot be reused at a later incident.

The organic vapor/acid gas cartridge is effective for approximately 8 to 10 hours of continuous use. The indicator that it is no longer serviceable is the ability to detect a taste or smell from the environment that the cartridge would otherwise remove.

Skin Protection

NFPA 1033 (2022): 4.1.3

Investigators often evaluate a site after a fire is extinguished, and fire resistance may not be a critical factor in their choice of personal protective equipment. Depending on their AHJ or agency requirements and the scene conditions, fire investigators should be cautious about wearing clothing made of nonfire-resistant synthetic materials because these materials can melt when heated and stick to skin, causing serious burns.

NOTE: Refer to department policies concerning what can and cannot be worn while on duty or under PPE.

Examples of nonfire-resistant synthetic materials include:

- Nylon
- Polyester
- Iron-on patches
- Transfer decals

CAUTION: Do not wear clothing made of nonfire-resistant synthetic materials under PPE.

Fire investigators must wear appropriate personal protective equipment given their role and activities in any particular investigation. For more information and guidance on protective equipment and clothing, refer to NFPA 1975, *Standard on Emergency Services Work Apparel*; IAAI, *Fire Investigator Health and Safety Best Practices*; and the IFSTA manual, **Essentials of Fire Fighting**.

Factors that may influence an investigator's choice in personal protective equipment and clothing include:

- AHJ or agency
- Site safety assessment/evaluation
- Type of investigative activity

WARNING: Investigation work presents a number of serious hazards. Wear equipment/clothing suited to the site safety analysis.

Site Safety Assessment and Decontamination

NFPA 1033 (2022): 4.1.3

A site safety assessment must include consideration of the materials used to construct the buildings, their construction classifications, and the components that make up the structures. The assessment must also include reference to how the building's design and construction affect fire behavior and development inside the structure **(Figure 2.10)**.

An investigation may involve a structure that is hundreds of years old located near structures constructed in later

Figure 2.10 Construction standards change over time. An older structure may have more hidden spaces for fire to spread without warning.

decades **(Figure 2.11)**. Each may conform to a different building code intended to provide a certain level of fire safety and structural stability. Each will be constructed from the building materials and architectural design common to the period of construction. In addition, older structures may have been renovated with engineered lumber or modern insulation systems.

Decontamination methods can be divided into four broad categories: wet or dry methods and physical or chemical methods **(Figure 2.12)**. These methods vary in their effectiveness for removing different substances, and many factors may play a part in the selection decision such as weather conditions and the chemical and physical properties of the hazardous material(s). The response options may also affect which methods are used.

As their names imply, wet and dry methods are categorized by whether they use water or other resources as part of the decon process. Wet methods usually involve washing the contaminated surface with solutions or flushing with a hose stream or safety shower. Dry methods include scraping, brushing, and absorption.

Physical methods of decontamination remove the contaminant from a contaminated person without changing the material chemically (although wet methods may dilute the chemical). The contaminant is then contained (when practical) for disposal.

Chemical decontamination is beyond the scope of this manual. For more information on all types of decontamination methods, refer to the IFSTA manual, **Hazardous Materials for First Responders**.

Figure 2.11 The building in the foreground, constructed in 2005, was designed to coordinate with the aesthetic of the building in the background, constructed in 1911.

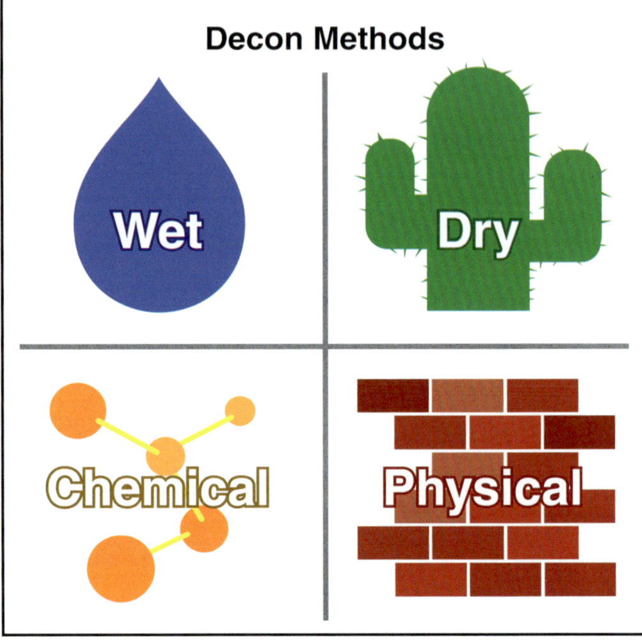

Figure 2.12 Four decontamination methods may be used to remove and contain hazardous materials.

 **Contaminants from Asbestos Insulation**

Asbestos may still be found in all types of structures. Asbestos can be found in a wide variety of materials including:

- Siding **(Figure 2.13)**
- Floor tile
- Pipe insulation
- Roofing materials

- Heating equipment
- Dry wall compound
- Structural insulation
- Electrical equipment

- Construction adhesives
- Contaminated soil around a structure
- Ceiling tile and stucco treatments on ceilings

Investigators must remember safe response to materials that may include asbestos:

- Any friable (crumbly) materials may contain asbestos, a respiratory hazard within a confined space.

Figure 2.13 Asbestos is highly effective at preventing fire spread when components are intact. When asbestos-containing materials are broken or disturbed, their component materials are a significant respiratory hazard.

- Any clothing exposed to asbestos containing materials must be cleaned immediately without contaminating residential areas.

- Disruption of asbestos materials may contaminate the wider environment, which can be a significant concern in a densely populated area.

Asbestos sidings may be covered with another siding or remain exposed. Asbestos inhalation risk is increased during overhaul, and scene excavation and reconstruction. While environmental tests are in progress, the scene may be unavailable.

Asbestos fibers are difficult to remove from contaminated clothing and other resources. Any investigation activities near or affecting asbestos products must include proper PPE and decontamination procedures. All appropriate precautions must be exercised while in proximity to components that may potentially contain asbestos. The AHJ or agency may require contaminated PPE and resources to be discarded after use. For additional information, refer to NFPA 1851, *Standard of Selection, Care, and Maintenance of Protective Ensembles, Structural Fire Fighting and Proximity Fire Fighting.*

Gross decontamination of asbestos at an investigation scene should include wet removal methods to prevent materials from becoming airborne. PPE must be removed before leaving the scene, double-bagged, transported safely, and cleaned per AHJ or agency requirements.

The hazards presented by asbestos can be too severe for conventional abatement mitigation work. For example, Libby, Montana was designated a Superfund site because of its significant asbestos contamination ("Asbestos at Superfund Sites" 2019).

CAUTION: Assume that a structure of any age includes asbestos components until you can prove otherwise.

Vehicle Contamination

Transporting equipment and clothing after an investigation must be handled strategically to minimize the risks of cross-contamination and exposure from off-gassing within a closed space ("Fire Investigator Health and Safety Best Practices." 2018) **(Figure 2.14)**. These risks can be minimized through proper hygiene habits as described below.

Storage hygiene:

- Wipe down vehicle interiors every day.

- Separate PPE storage from living areas.

- Vacuum and mop; sweeping can disturb dust.

- Store PPE in closed lockers, gear bags, and containers with a gas seal.

Reducing exposure to contaminants:

- Clean all PPE and SCBA.

- Use wet wipes to clean hands, neck, jaw, face.

- Shower immediately upon returning to the station.

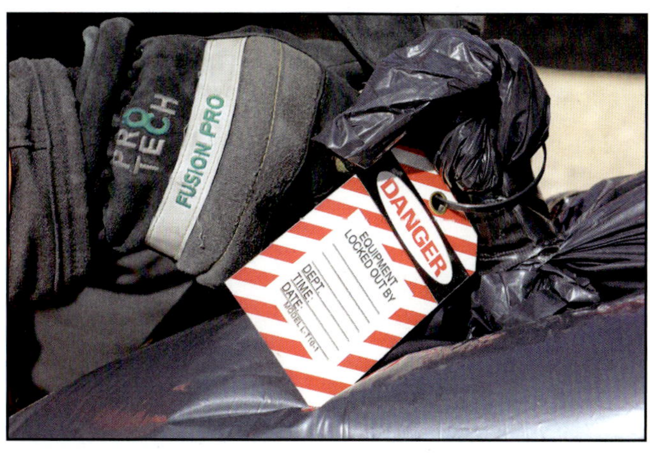

Figure 2.14 PPE should be thoroughly contained for transport before it is cleaned or disposed of.

- Wear appropriate PPE and respiratory protection for all on scene operations.
- Gross decon of PPE, transport in sealed bag outside of the vehicle's passenger compartment. Change clothes before entering the vehicle.

Rehabilitation Facilities

NFPA 1033 (2022): 4.1.3

Rehabilitation (rehab) facilities should be established that offer relief from conditions at the emergency scene. For example, in frigid conditions, the rehab center should be a warmer environment. Similarly, in warm weather, the rehab area should be cooler than the scene conditions. In addition, rehab areas should be stocked with water and food so that investigators and other emergency workers can replenish fluids and energy.

This place of rest must be remote enough to ensure a safe atmosphere as well as provide facilities for fire investigators to decontaminate their face, hands, and any exposed skin. Investigators may wish to establish a decontamination protocol to be used at the scene to ensure that any particulate is removed prior to eating or drinking. It is recognized that fresh water and cleaning or decontamination liquids are not always available at all fire scenes. An example of a decontamination protocol to prevent contamination during the ingestion of any liquids or foods is as follows:

- Have a supply of disposable wipes suitable for removing contaminants from the skin
- Use multiple wipes to decontaminate the face, neck, hands and any exposed skin or hair

Generally, rehab areas are also places where fire investigators can take some relief from the heat of wearing additional clothing at the fire scene particularly in warm weather. Frequent breaks outside the fire scene are necessary to ensure adequate hydration, rest, and occasional snacks to provide energy. Cleanliness is essential to ensure that there is no contamination of drinks or foods either in handling or during ingestion. Private sector investigators should also plan for basic rehabilitation resources.

Scene Safety Hazards

NFPA 1033 (2022): 4.1.7, 4.2.1, 4.2.2

The potential for injury to fire investigators on the scene of an emergency is extremely high. Investigators must exercise situational awareness at all times, from the moment of notification until the incident has been demobilized. For safety reasons, it is preferable to operate at a scene with at least two investigators familiar with fire investigation scene safety. If it becomes necessary to operate alone, investigators should make their location and status known to their employer. This notification can be accomplished by contacting the agency's telecommunications/dispatch center upon both scene arrival and departure. Investigators may also communicate their plans with a local law enforcement office or fire department, especially in areas where telephone communications are unreliable. An investigator may consider establishing a system of regular check in times so that in the event an investigator becomes injured/incapacitated, help can be sent if the investigator does not check in within an established period of time.

The sections that follow discuss some hazards common to a fire investigation. The IAAI *Fire Investigator Health and Safety Best Practices* includes guidelines and considerations for investigators en route to and at a scene.

Hazardous Materials

NFPA 1033 (2022): 4.1.7, 4.2.1, 4.2.2

All fire investigators must be aware of the possibility that **hazardous materials** may be present at an investigation scene. A fire investigator should have a basic understanding of ways to identify potential hazardous materials and should determine scene threats and safety requirements.

NOTE: In Canada, hazardous materials are frequently referred to as **dangerous goods**.

Hazardous Materials Awareness

The investigator must also determine whether the area is safe or whether specialized assistance is necessary before entering the area. The priority is self-protection. For example, binoculars can be used to evaluate the safety of the situation while maintaining a safe distance from the scene (**Figure 2.15**). For information on identifying indications of unsafe labs and activities, refer to the IFSTA manual, **Hazardous Materials First Responder**.

If the incident involves a release or spill of a hazardous chemical, the investigator should observe the scene from a safe distance and wait for the area to be declared safe to enter. For more information on hazardous materials identification, refer to the IFSTA manual, **Hazardous Materials for First Responders**.

Unsafe Labs and Activities

Investigators must be aware of many types of unsafe labs and activities that could be encountered during the course of an investigation. Some of these labs and activities may be legal in some states/provinces, but still present a significant hazard. Regardless of legality, safety should be the investigator's primary concern.

NOTE: Some jurisdictions may offer continuing education on hazards as they develop in an area.

Figure 2.15 Investigators should make their own determination about the safe operations they can perform at a each incident site.

These types of labs are increasingly difficult to identify because they now use smaller spaces and fewer resources than has been historically possible. The hazard from unsafe labs may not be significantly different from chemicals collected with benign intent, such as chemicals used for maintaining a residential swimming pool. An unsafe lab, however, will likely include a wider range of materials including a high quantity of oxidizers, and production at various stages such as crystals with high sensitivity to heat, friction, or shock.

Chemicals and precursors used in the production of methamphetamine are often found in explosives labs. Should an investigator discover apparent evidence of an illicit lab, whether in direct evidence or plausible precursors, they should call an appropriate AHJ or agency, such as a bomb squad or law enforcement agency, to address the concern. The Bureau of Alcohol, Tobacco, Firearms and Explosives (ATF) distributes information cards to the public sector emergency services indicating common indicators. For more information on identifying illicit labs, refer to the IFSTA manual, **Hazardous Materials First Responder**.

Hazmat Facilities and Transportation

Hazardous materials that are manufactured, stored, processed, or used at a particular site are not subject to the same regulations affecting transported materials. Local agencies or state/provincial governments may adopt their own identification system. Alternatively, they may decide to use a **placard** as recommended by NFPA 704, *Standard System for the Identification of the Hazards of Materials for*

Emergency Response **(Figure 2.16)**. The ratings for each hazard (health, flammability, and instability) are described in the standard itself. Some private companies and the military have their own policies that regulate the marking of these materials.

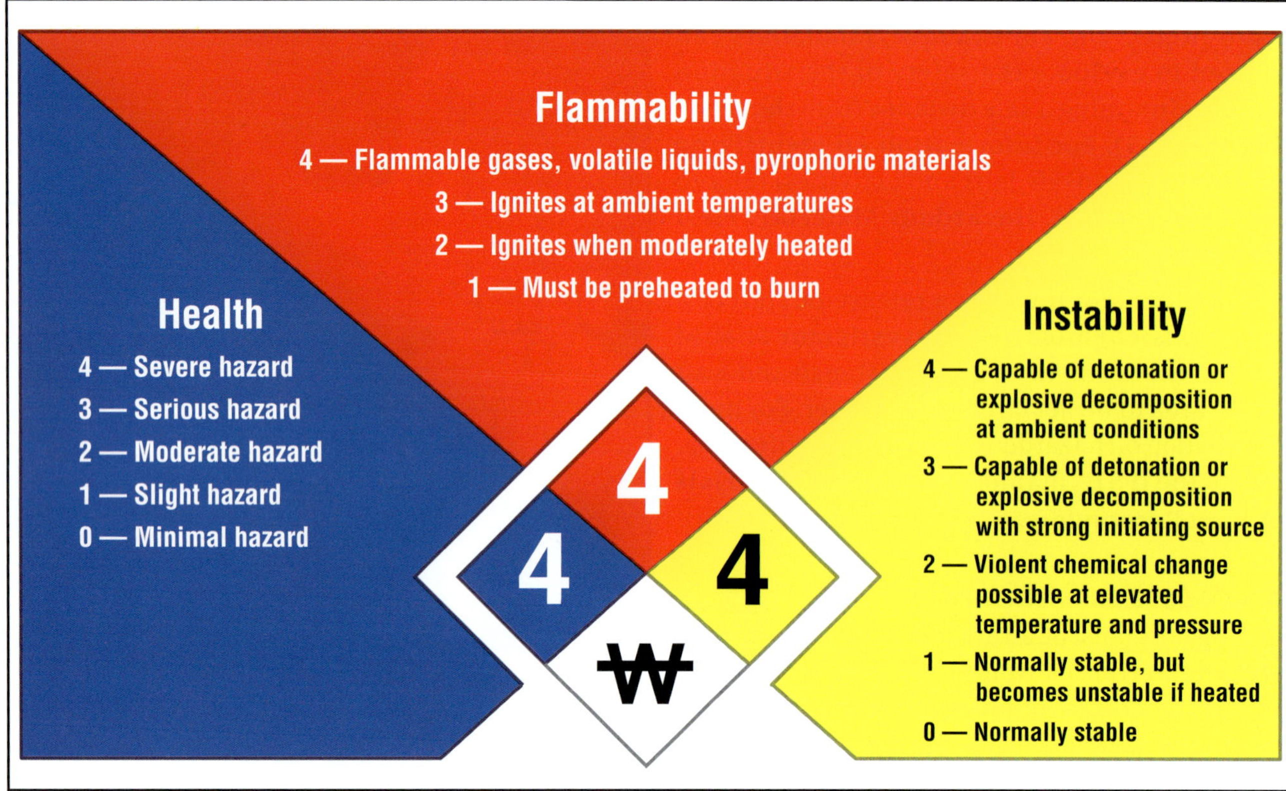

Figure 2.16 Investigators should recognize the significance of NFPA hazmat placards.

NOTE: For more information on the NFPA 704 marking system, consult the IFSTA manual, **Hazardous Materials for First Responders**. Response at facilities is discussed extensively in the IFSTA manual, **Facility Fire Brigades**.

For incidents involving vehicles that are transporting hazardous materials, the investigator can look for placards to help identify the material(s) in question. Placards are diamond-shaped signs that are affixed to each side of a commercial vehicle transporting hazardous materials **(Figure 2.17)**. The placard indicates the primary class of the material and, in some cases, the exact material transported.

Figure 2.17 Investigators may encounter vehicles transporting hazardous materials. *Courtesy of Rich Mahaney.*

Safety Data Sheets

All flammable and combustible materials maintained in a facility must be inventoried and have safety data sheets (SDSs) collected on each item. This information must be kept in an SDS book on the site, and it must contain health, safety, and injury prevention information. Personnel must be trained in the use of the flammable and combustible liquids and the SDS book.

Biohazards

NFPA 1033 (2022): 4.2.1

Fire investigators face potential exposure to biological hazards from blood, body fluids, and other potentially infectious materials that may be present at the scene of a fire or explosion. Acquired immune deficiency syndrome (AIDS) caused by the human immunodeficiency virus (HIV), and hepatitis B, pose significant threats. The hepatitis virus is particularly threatening because it has been found to live several days in dried blood spills. Fire investigators can be exposed wherever blood, body fluids, or contaminated equipment are present.

> **CAUTION:** Investigators must be alert to hazards such as discarded used syringes.

Infection-control procedures and **body substance isolation (BSI)** practices recognize that *all* body fluids must be treated as if known to be infectious. The best method for preventing infection by biological agents is to provide a barrier between the person and the infectious agent by wearing disposable gloves, masks, gowns, and eye protection **(Figure 2.18)**. Where exposure is possible such as scenes involving fatalities, OSHA (Title 29 CFR 1910.1030, Bloodborne Pathogens) mandates establishing a comprehensive protection program.

Guidelines for working safely near biohazards are as follows:

- Decontaminate all equipment after use.

- Wear eye and face shields to protect against exposure from infectious material splashes or sprays.

- Do not eat, drink, or smoke where blood, body fluid, or other potentially infectious materials are found.

- Always use disposable protective gloves, coveralls, and shoe covers when exposed to potentially infectious materials.

- Dispose of items not intended for reuse in an approved manner that eliminates the possibility of remote contamination.

- Ensure that exposed skin does not come in contact with contaminated exterior equipment surfaces when removing protective equipment.

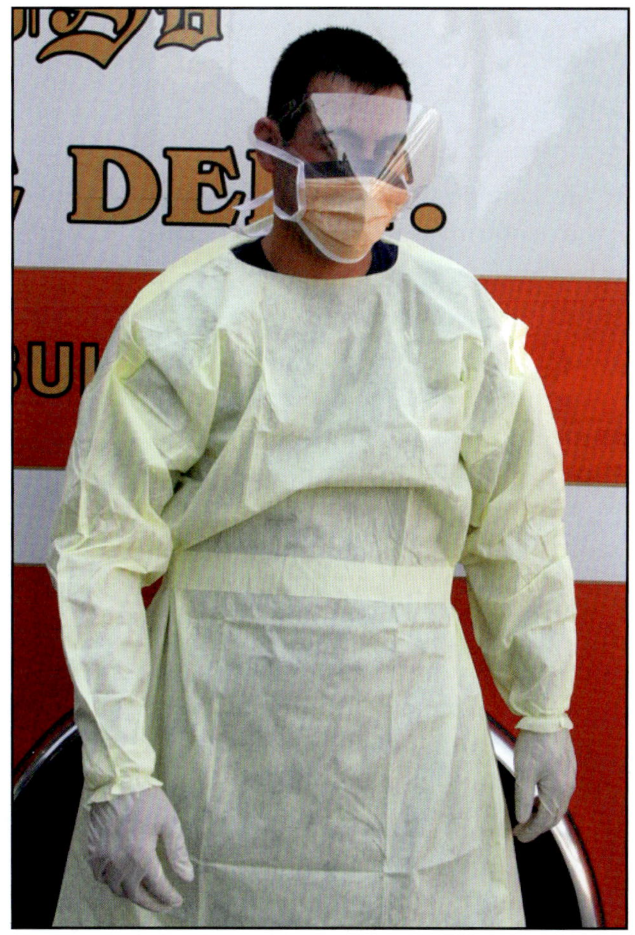

Figure 2.18 Disposable combination mask and eye shields are widely used by emergency response agencies.

Structural Hazards

NFPA 1033 (2022): 4.2.1

Structural stability is critical to the safety of all fire investigators. The investigator should reassess the scene before entering a structure or attempting any investigation to determine fire origin or cause. Investigators should conduct an examination of all exterior surfaces in order to determine whether the building is sound before entering.

Structural Instability

Observing load-bearing building components will aid in determining whether the structure is safe to enter. For example, the load-bearing walls of a structure will usually be the ones that have the longest length. Observing the condition of these walls as well as the beams, joists, and columns associated with a particular assembly, such as floors and roofs, will provide guidance concerning their stability as well as for the possible need for shoring and bracing. Some investigators have the specialized training to evaluate these risks; however, additional resources such as a structural engineer may be required to further evaluate the risks and recommend solutions.

NOTE: A potential **collapse zone** for walls and structurally damaged walls and/or chimneys should be 1.5 times the height of the object.

Always consider the following factors before entering structures that have experienced a fire or explosion:

- **Building construction** — Wood frame construction may fail quickly under fire conditions. Investigators should delay entry into any damaged structures until the specific type of construction is determined and the appropriate safety precautions are implemented.

- **Content load** — Building contents such as paper, carpet, and furniture that absorb water will readily increase weight and contribute to the probability of collapse.

- **Lightweight building materials** — Buildings with truss construction and structural members that have less mass will generally deteriorate and fail more readily without warning.

- **Length of burning time** — The longer a building is subjected to extreme fire conditions and environmental factors, the more probable it will collapse.

- **Accumulating water or ice** — Buildings with large quantities of standing water or ice tend to collapse more readily. Extended fire-suppression operations may cause heavy loads to be placed on weakened structural members that may eventually fail.

- **Parallel chord lightweight truss beams** — These beams (usually either wood or steel) are commonly used in many residential, commercial, and industrial applications and are more probable to fail after a fire or explosion than heavier wood or steel joists.

- **Special conditions** — Structures where equipment such as heating and air conditioning units are roof-mounted are susceptible to postfire collapse due to the static loads placed on weakened or damaged structural members. Freestanding chimneys and any signs of building braces or supports such as corner angle braces, "star" supports, and shoring should be noted, and these areas should be avoided until qualified personnel determine that they are safe.

- **Exposed steel** — Exposed steel can begin to lose strength with prolonged exposure to temperatures common in virtually all fires **(Figure 2.19)**. The result of exposure can be an expansion that may compromise structural integrity. Investigators should proceed with caution and look for any visual changes in exposed steel that may indicate the probability of failure.

Water and Wind Effects

Remember that water adds weight to a structure. Steps must be taken to reduce water accumulation and prevent additional water from collecting. Freezing weather may add an additional hazard in the accumulation of ice and snow. Ice might have stabilized the structure, which may become unstable when it thaws. Winds can cause collapse of structural elements, and even moderate winds are capable of toppling fire-damaged chimneys.

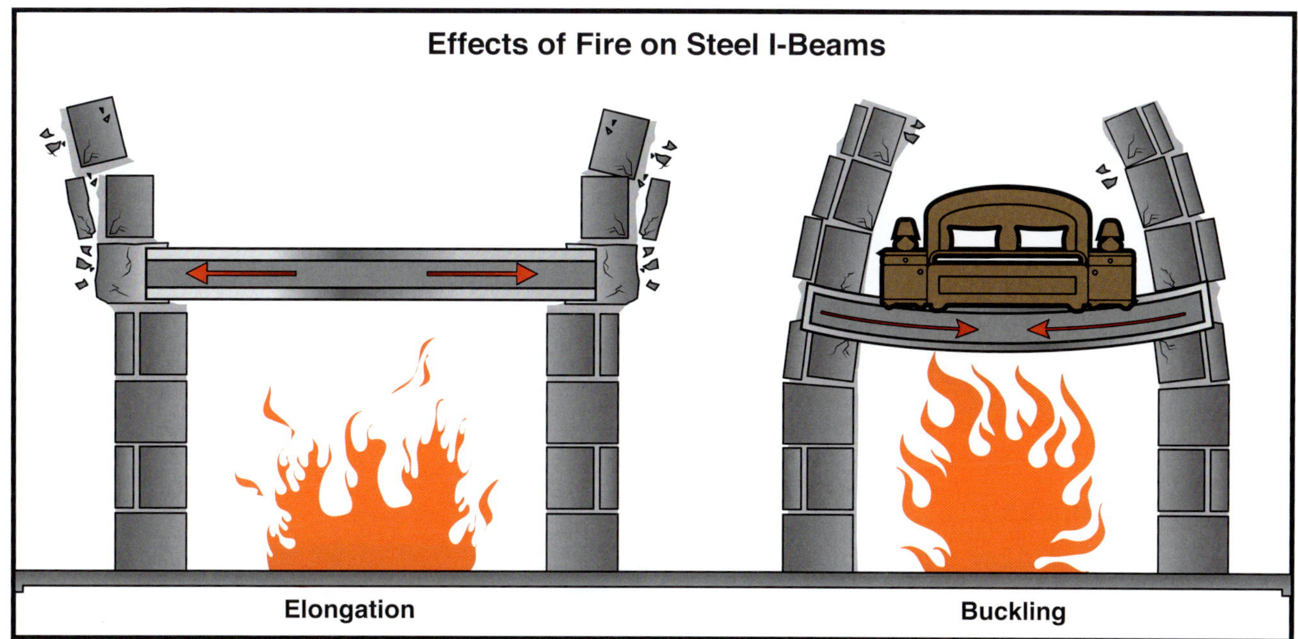

Figure 2.19 Heat can cause steel structural members to lengthen or buckle depending upon fire conditions, building construction, and content placement.

Other Structure-Related Hazards

Fire investigators must recognize and consider hazards at a structure that has been damaged by fire and suppression activities. Some common hazards at a fire investigation scene are:

- Holes in floors
- Ignition sources
- Broken glass, nails, or torn metal
- Damaged natural gas/propane lines
- Damaged electrical appliances or equipment
- Confined spaces including wells, cisterns, or septic tanks
- Loose flooring or steps, slippery surfaces, and protruding objects
- Improper space for body mechanics including insufficient space for safe lifting practices
- Exposed, possibly energized, electrical wiring (underground, overhead, and residential services)
- Environmental factors including weather conditions such as extreme cold or heat, or lightning storm
- High noise levels, particularly during ongoing response or when propane generators are used to power tools during investigation
- Damaged, hanging, or unsecured objects that can fall and/or swing from elevated areas, such as structural beams, light fixtures, and debris

Continual Evaluation

Once an investigator is inside a structure, all walls, ceilings, chimneys, and floors should be visually assessed **(Figure 2.20a-c, p. 42)**. Investigators should remember that chimneys may be visible or hidden. As the scene is examined and debris is moved, structural stability must be continually reassessed. For example, debris may hide or support damaged flooring.

Figure 2.20a-c Common hazards that may be encountered at an investigation scene include shattered glass (a), burnthroughs in floors (b), and fallen structural members (c). *Photo (a) courtesy of Donny Howard, Agent Oklahoma State Fire Marshal's office. Photo (b) courtesy of Yates & Associates. Photo (c) courtesy of Wayne Chapdelaine, Metro-Rural Fire Forensics.*

Poor Lighting

NFPA 1033 (2022): 4.2.1

Many incidents occur in low- or poor-light conditions because of the time of day or location inside a large, windowless building. Providing artificial light under these circumstances is essential both to enable investigators to perform their jobs and to reduce the risk of injury **(Figure 2.21)**. Many departments use floodlight units to provide exterior lighting. Although these units serve a significant benefit, the area they illuminate is generally limited to areas close to the power supply such as apparatus. Applying intensely concentrated lighting to the exterior of a building may disorient responders as they move between dark smoky areas and brightly lit areas. Investigators should be aware of this effect and exercise caution to avoid tripping over obstructions such as hoselines and debris.

When interior lighting is necessary, it is usually best provided by portable lights. Caution should be used when attaching power cords to floodlights and other electric-powered equipment. Power cords for these units must have proper insulation and highly visible colors such as yellow or orange. Safety considerations with power cords include both trip hazards and shock hazards.

Investigators should also carry battery-operated flashlights for interior illumination. Flashlights increase safety when working inside a dark or smoky facility. Even daylight operations may require a flashlight for efficiency and safety.

Figure 2.21 Portable lighting is necessary if an investigation must take place at night. *Courtesy of Jocelyn Augustino, FEMA News.*

> **WARNING:** Use intrinsically safe equipment, including flashlights, when working in a hazardous atmosphere.

Electrical generators may be needed at a scene to provide electricity for portable lights. These generators are often gasoline-powered and should be operated outside the fire scene, in open areas, or in specially designed apparatus mounts to ensure safety as well as avoid potential contamination of the scene. Because gasoline-powered generators produce CO and other exhaust gases, using them indoors at a fire scene could contaminate the scene and add an additional safety hazard. Battery powered lighting is an alternative if used in a proper environment.

Improper Ventilation

NFPA 1033 (2022): 4.2.1

Ventilation is the process of removing contaminants from the work environment either naturally or mechanically. In the case of a hazardous material incident, ventilation may serve to dilute and eventually displace toxic gases with fresh air. Dusts and fumes also pose a risk to investigators. Proper ventilation can direct these irritants away from personnel, allowing them to safely perform their assigned tasks.

Insufficient Scene Security

NFPA 1033 (2022): 4.2.1

Perimeters are established to ensure scene security and preserve evidence. As the size and scope of an incident widens, criminal activity may be uncovered; and the need to establish a perimeter for personal security, animal control, and crowd control becomes more essential. Upon arrival at the scene, the fire investigator may request the IC to facilitate the installation of a visible perimeter when it is practical to do so. Fire line or law enforcement line tape may already be in place from the emergency response.

Whenever possible, the perimeter should surround the entire scene. At an explosion scene, the perimeter should be a minimum of 1.5 times the distance of the farthest piece of debris found.

Fuel Spill Contamination

Exercise caution to avoid contaminating the fire scene with spilled or leaking fuel **(Figure 2.22)**. The spill area should be covered or barricaded, and personnel who do step in the spill should cover or remove contaminated boots.

Figure 2.22 When moving through a scene, investigators should avoid any spilled liquid. *Courtesy of Donny Howard, Agent, Oklahoma State Fire Marshal's Office.*

CAUTION: Fire line tape does not prevent people or animals from entering the scene. When working alone, the investigator must maintain situational awareness.

Personal Security

In cases where a threat possibly exists from suspects or individuals associated with a fire scene, the perimeter should be constantly monitored by the appropriate law enforcement entity. A monitored perimeter allows fire investigators to concentrate on debris removal as well as maintain a safe work environment **(Figure 2.23)**. If the perimeter is breached, investigators may have to resort to defusing the situation with an appropriate use of force. Proper perimeter security minimizes the possibility that an investigator will have to use force in self-defense, or pursue someone bypassing security.

Figure 2.23 An investigator must concentrate on safe practices at an incident scene. *Courtesy of Yates & Associates.*

There are also personal safety precautions when conducting witness and suspect interviews. Postfire scene interviews frequently occur at a person's residence, place of employment, or other public facility. The interview function becomes inherently safer when conducted in teams of two. Conducting interviews alone could leave one person vulnerable to accusations of impropriety. Special consideration should be given when dealing with minors and persons of the opposite sex. Interviews are further discussed in Chapter 8 of this manual.

Crowd Control

As the size and scope of an incident widens from fire suppression and emergency medical attention to operations that may uncover criminal activity, crowd control becomes more essential. Crowds may gather for more than simple curiosity. Some people may want to witness the results of their crimes or to intentionally disrupt the investigation or present a danger to investigators. Investigators may seek support from law enforcement personnel or leave the scene if they think their safety is, or may become, compromised.

Uncontrolled Animals

NFPA 1033 (2022): 4.2.1

Animals of almost any type can be encountered in residential areas, including domesticated animals, wildlife, and insects. Although pets may be gentle under normal circumstances, any fire and the resulting confusion and noise may make them fearful and difficult to handle. Investigators should try to be as calm as possible because animals can sense fear. Wait until the animals have been controlled before entering the area to continue the investigation.

When entering any structure, consider all animals as potential hazards. If an animal bite punctures the skin, immediate medical treatment is necessary; and the animal must be captured and impounded for rabies observation. Local municipalities or counties usually retain animal control personnel who are trained in the proper handling of animals.

In addition to animals, insects may be a safety factor during an investigation. Wasps or bees may build hives in breaker boxes or in eaves **(Figure 2.24)**. Latent insects, such as bedbugs and fleas, may become embedded in PPE. In certain regions, fire ants may be present at the investigation scene. Proper handling of PPE after an investigation will help minimize the risk of bringing insects out of the investigation areas.

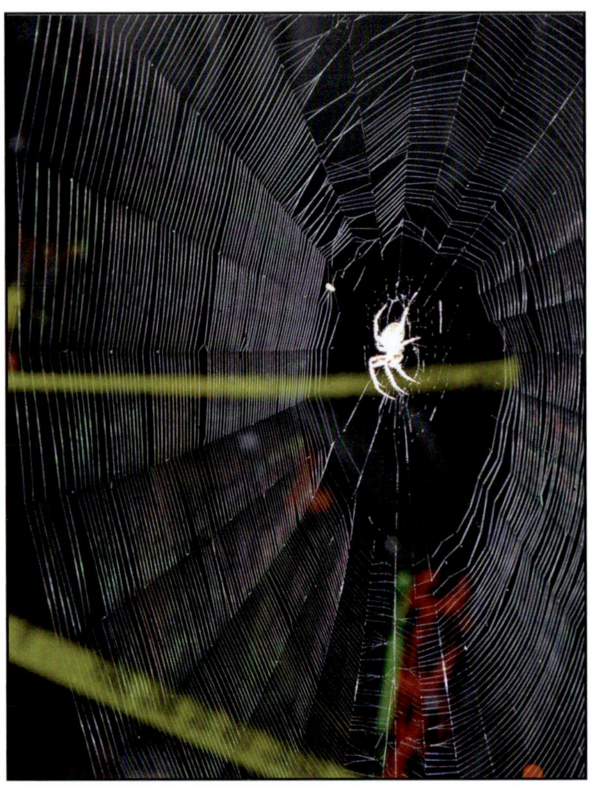

Figure 2.24 Insects are common at fire investigation scenes. *Courtesy of Donny Howard, Agent, Oklahoma State Fire Marshal's Office.*

Insects at the Investigation

While preparing for an investigation, investigators should consider the risk of allergic reaction should they be stung by an insect. Preparation for an allergic reaction can include preemptive allergy testing, and the inclusion of an epinephrine dose for quick emergency response to severe allergic symptoms. Use caution when using insect repellent at an investigation scene because it may contaminate a scene with introduced ignitable liquids such as butane or propane often as a propellant in aerosol sprays.

Utility Systems

NFPA 1033 (2022): 4.1.7, 4.2.1, 4.2.8

Utility systems can be a safety concern for investigators. Electricity left on or not de-energized properly at a scene, for example, could injure an investigator attempting to examine evidence from an electrical socket. As a result, investigators should be very familiar with utility systems and understand how to verify their safety before entering a scene. The sections that follow discuss the most common utility systems: electricity, fuel gas, water, and sewage/drainage. More information about utilities can be found later in this manual.

Building Fire Detection and Suppression Systems

NFPA 1033 (2022): 4.2.1, 4.2.8

Investigators will often encounter different types of systems within the structure that are designed to detect and suppress fire, and prevent fire spread. These systems should be examined to ensure they functioned properly and to also ensure they have not been tampered with or disabled prior to the incident. Building systems are further discussed in Chapter 6 of this manual.

Electricity

NFPA 1033 (2022): 4.1.7, 4.2.1, 4.2.8

Fire investigators must be cautious at all times for the possibility of electrical hazards. Electricity is discussed extensively in Chapter 7 of this manual. Consider the following precautions when dealing with electrical hazards:

- Treat all wires as "live" wires.

- Do not remove electrical meters **(Figure 2.25)**.

- Do not rely on rubber footwear as an insulator.

- Let power company personnel cut electrical supply wiring.

- Do not walk in "standing water" when the electrical system is energized.

- Exercise extreme caution when using all ladders around electrical hazards.

- Use only explosion-proof equipment where a potential explosion hazard exists.

- Do not switch energized electrical equipment on or off while standing in water.

- Switch off electrical power at a point remote from any potential explosive atmosphere.

- Be alert to fallen electrical wires on the ground because they can energize the entire area.

- Be aware that there may be more than one electrical service line or electrical power source into a structure that may not be disconnected.

- Remember that structures that rely on renewable energy may also have energy stored within batteries. This energy storage is not really electrical energy but

Figure 2.25 Electrical equipment intended for significant power output should be left to the relevant utility company to manage should the equipment need to be removed.

stored chemical energy that may still be converted to electrical energy. Investigators should check for the presence of batteries.

- Be aware of the potential for electrical shock from energy stored in capacitors or suppressors within appliances. These appliances include many appliances such as microwaves, washing machines and clothes dryers. The amount of time the energy is stored prior to dissipation varies widely between appliances.

- Consider the possibility that back-up generators may automatically turn on and reenergize the structure after electrical service had been cut or turned off. This presents an additional hazard to investigators and they should ensure that no back up system is present or that it has been disabled/locked out prior to entering the structure.

Energized Enclosures

Do not enter an electrical vault, substation, generator station, or transfer station before the power has been de-energized (circuit opened). Always consider everything within these enclosures as being energized until confirmed by a qualified representative of the electric utility company.

Gas Utilities

NFPA 1033 (2022): 4.2.1, 4.2.8

Many houses, manufactured homes, businesses, and industrial properties use natural gas or **liquefied petroleum gas (LPG)** for cooking, heating, or industrial processes. To control gas utilities, the investigator must have a working knowledge of the hazards and correct procedures for handling incidents involving natural gas and LPG.

Natural Gas

Natural gas is lighter than air, so it tends to rise and diffuse in the open. Although natural gas has no odor of its own, the utility companies add an odorant, such as ethyl mercaptan, which has a distinctive sulphur-like odor, similar to rotten eggs.

Natural gas is often present in structures as a fuel for cooking and heating. The presence of natural gas service is often indicated by a meter outside the structure. Other sources of natural gas may not be apparent to an investigator without asking the owner. In cases where the owner has accessed a natural gas well, the source will be adapted with a series of regulators. These alternative supplies may not be treated with an odorant.

When ordered, the natural gas supply to a structure must be shut off at the meter. The meter is usually located outside the structure near the foundation or on the easement near the property line. However, it may be inside the structure in a basement or mechanical space. In some industrial and institutional occupancies, critical equipment and processes depend upon an uninterrupted supply of natural gas; for example, natural gas fuels emergency generators in some hospitals.

The gas shutoff is an inline valve located on the owner supply side of the meter between the distribution system and the meter. When the valve is open, the tang (a rectangular bar) is in line with the pipe. To close the valve, use a spanner wrench, pipe wrench, or similar tool to turn the tang until it is 90 degrees to the pipe. If shutting off the inline valve does not stop the gas supply to the structure, take no further action and contact the gas utility. The local utility will provide:

- Distribution system maps

- An emergency response crew equipped with nonsparking tools

- Training and experience needed to help control the flow of gas

Contact the local utility company when the gas has been shut off or when any emergency involving natural gas occurs in its service area. The utility company, not the fire department, is responsible for turning gas utilities back on after they have been shut off.

Methane. Natural gas in its pure form is methane. Methane has a flammability range of 5 percent to 15 percent. While it is nontoxic, it is classified as an asphyxiant because it may displace oxygen in a confined space and lead to suffocation.

Compressed Natural Gas (CNG). Compressed natural gas (CNG) is increasingly used to power motor vehicles and other mechanized equipment. Some fleets, including municipal vehicles such as busses and waste management vehicles, use this resource. CNG pumps may be found at gasoline service stations.

Liquefied Petroleum Gas (LPG)

Many houses, mobile homes, and businesses use natural gas or LPG for cooking, heating, or industrial processes. Investigators must have a working knowledge of the physical properties, hazards, and correct procedures for handling incidents related to natural gas and liquefied petroleum gas (LPG). Investigators must be familiar with gas distribution and usage. This knowledge helps prevent personal injury or additional damage caused by these gases. Consult NFPA 54, *National Fuel Gas Code* and *The NFPA National Fuel Gas Code Handbook* for additional information.

Also known as bottled gas, liquefied petroleum gas (LPG) refers to fuel gases stored in a liquid state under pressure. While there are two main gases in this category — butane and propane — propane is the most widely used. Propane is used primarily as a fuel gas in:

- Businesses
- Rural homes
- Manufactured homes
- Agricultural applications
- Campers and recreational vehicles
- Motor vehicles and mechanized equipment

Propane gas has no natural odor, but added ethyl mercaptan gives it a very distinctive smell. The gas is nontoxic, but it is classified as an **asphyxiant** because it may displace normal breathing air in a confined space and lead to suffocation.

LPG is about 1½ times as heavy as air, so it will sink to the lowest point possible. The gas is explosive in concentrations between 2 percent and 9.5 percent. LPG is shipped from its distribution point to its point of usage in cylinders and in tanks on cargo trucks. It is stored in cylinders and tanks near its point of use. Steel piping and copper tubing then connect the tank or cylinder to gas powered appliances. Shutting the tank valve may stop the supply of gas going into a structure. An LPG leak will produce a visible cloud of vapor that hugs the ground. A fog stream of at least 100 gpm (400 L/min) can be used to dissipate this cloud of unburned gas.

A shutoff valve should be located at the point where the supply line from the LPG tank enters the structure **(Figure 2.26)**. This valve may be similar to the type used for natural gas or water supply lines. There will also be a control valve located on the LPG tank **(Figure 2.27)**. Do not attempt to open the valve after the fire or emergency has been terminated. Turning on the gas is the responsibility of the owner and the LPG supplier.

Water

NFPA 1033 (2022): 4.2.1, 4.2.8

Although a water system is not usually considered a safety problem for investigators, water leaking from broken pipes, sprinkler systems, or disrupted underground mains can create major safety concerns. The addition of the resulting **dead load** to an already-weakened structure can cause an unexpected collapse resulting in injury or death even hours after the initial fire or explosion. Another major concern is the collection of water in basements or low spaces in and near the incident site. Open wells and cisterns can also be a safety concern at some incidents.

Because these hazards can be difficult to detect, they represent a significant falling or drowning danger. To guard against such a situation, the water utility should be notified and requested to turn off service to the incident site. Drain chutes and catchalls can be constructed with salvage covers to divert or retain water as needed **(Figure 2.28)**. Hazardous areas should be marked or barricaded to prevent investigators (or anyone else) from falling and becoming injured.

Sewage/Drainage

NFPA 1033 (2022): 4.2.1, 4.2.8

Some incidents may compromise the sanitary or storm sewer systems. Large quantities of wastewater can contribute to the amount of dead load being supported by a structure. As with water utilities, this situation can lead to a structural failure and collapse. Aside from the possibility of collapse and drowning, the added concern of contamination and disease must be considered. The sewer utility and the local public health officer must be notified when situations of this type are found. Low areas, and confined spaces in and near the incident area may collect wastewater and contaminants.

Care must also be taken when working in a rural environment, and it is necessary to take precautions when working around cesspools, farm manure, or drainage ditches. These locations should be cordoned using scene or biological hazard tape.

Figure 2.26 Gas supply lines feature shutoff valves located for the ease of use when needed.

Figure 2.27 The LPG supply tank will also feature a shutoff valve location.

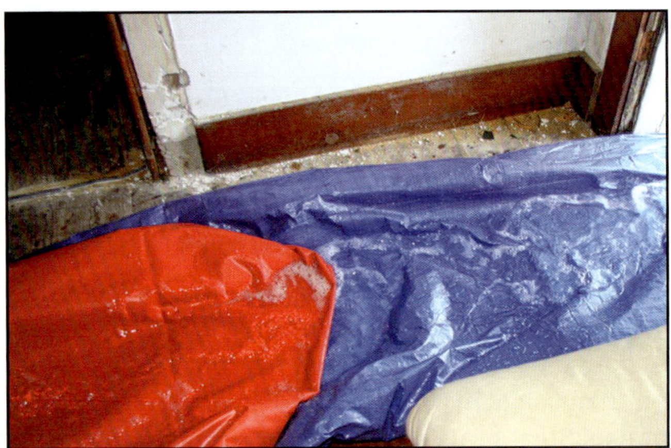

Figure 2.28 Fire suppression activities may leave significant water collection in low areas. Splicing a chute to a catchall can help divert water from a catchall to the building's exterior.

Review Questions

1. Why is it important for investigators to function within the incident management system?

2. What topics and conditions should be evaluated during a site safety assessment?

3. How do investigators determine what type of PPE is appropriate for the situation?

4. What types of substances are generally monitored for prior to beginning an investigation?

5. Why may respiratory protection be necessary even after the fire has been extinguished?

6. Why are rehabilitation and decontamination both important parts of staying safe at an incident?

7. What are five common hazards that an investigator may encounter at an incident?

8. Why must investigators be aware of hazardous materials at an incident?

9. What types of biohazards should investigators be concerned about coming in contact with at an incident?

10. Why are perimeter control and crowd control such a major concern for fire investigators?

11. What actions should an investigator be prepared to take if utilities become a concern at a scene?

Discussion Question

1. Give an example of an incident that might require the use of unified command.

Chapter References

"Hazard Prevention and Control." Recommended Practices for Safety and Health Programs. United States Department of Labor. Accessed online.

"Asbestos at Superfund Sites: Cleanup Examples." Superfund. United States Environmental Protection Agency, 2019. Accessed online.

"Fire Investigator Health and Safety Best Practices." International Association of Arson Investigators, Inc., Health and Safety Committee; June 15, 2018. Accessed online.

Chapter Key Terms

Air Purifying Respirator (APR) — Respirator that removes contaminants by passing ambient air through a filter, cartridge, or canister; may have a full or partial facepiece.

Asphyxiant — Any substance that prevents oxygen from combining in sufficient quantities with the blood or from being used by body tissues.

Atmosphere — The ambient air in a given place.

Body Substance Isolation (BSI) — (1) The practice of taking proactive, protective measures to isolate body substances in order to prevent the spread of infectious disease. (2) Comprehensive method of infection control in which every patient is assumed to be infected; personal protective equipment is worn to prevent exposure to bodily fluids and bloodborne and airborne pathogens.

Collapse Zone — Area beneath a wall in which the wall is likely to land if it loses structural integrity.

Dangerous Goods — (1) Any product, substance, or organism included by its nature or by regulation in any of the nine United Nations classifications of hazardous materials. (2) Alternate term used in Canada and other countries for hazardous materials. (3) Term used in the U.S. and Canada for hazardous materials aboard aircraft.

Dead Load — Weight of the structure, structural members, building components, and any other features permanently attached to the building that are constant and immobile.

Decontamination — Process of removing a hazardous foreign substance from a person, clothing, or area. *Also known as* Decon.

Gross Decontamination — Quickly removing the worst surface contamination, usually by rinsing with water from handheld hoselines, emergency showers, or other water sources.

Hazardous Material — Any substance or material that poses an unreasonable risk to health, safety, property, and/or the environment if it is not properly controlled during handling, storage, manufacture, processing, packaging, use, disposal, or transportation.

Incident Commander (IC) — Person in charge of the incident command system and responsible for the management of all incident operations during an emergency.

Incident Management System (IMS) — (1) System described in NFPA® 1561, *Standard on Emergency Services Incident Management System*, that defines the roles, responsibilities, and standard operating procedures used to manage emergency operations. Such systems may also be referred to as Incident Command Systems (ICS). (2) Management system developed by the National Fire Service Incident Management System Consortium, combining pre-existing command systems into one.

Liquefied Petroleum Gas (LPG) — Any of several petroleum products, such as propane or butane, stored under pressure as a liquid.

Noise Reduction Rating (NRR) — The measurement of how well a hearing protection device will block noise. The rating is stated in decibels (dB) and established by the U.S. Environmental Protection Agency (EPA).

Organic Vapor/Acid Gas Cartridge — Device in respiratory protection equipment that absorbs harmful vapors and gases from breathable air, preventing inhalation. Each cartridge is a one-time use item that lasts 8 to 10 hours.

Personal Protective Equipment (PPE) — General term for the equipment worn by fire and emergency services responders; includes helmets, coats, trousers, boots, eye protection, hearing protection, protective gloves, protective hoods, self-contained breathing apparatus (SCBA), personal alert safety system (PASS) devices, and chemical protective clothing. When working with hazardous materials, bands or tape are added around the legs, arms, and waist. *Also known as* Bunker Clothes, Chemical Protective Clothing, Full Structural Protective Clothing, Protective Clothing, Turnout Clothing, or Turnout Gear.

Placard — Diamond-shaped sign that is affixed to each side of a structure or a vehicle transporting hazardous materials to inform responders of fire hazards, life hazards, special hazards, and reactivity potential. The placard indicates the primary class of the material and, in some cases, the exact material being transported; required on containers that are 640 cubic feet (18 m^3) or larger.

Rehabilitation — (1) Activities necessary to repair environmental damage or disturbance caused by wildland fire or the fire-suppression activity. (2) Allowing firefighters or rescuers to rest, rehydrate, and recover during an incident; also refers to a station at an incident where personnel can rest, rehydrate, and recover. *Also known as* Rehab.

Safety Officer (SO) — (1) Fire officer whose primary function is to administrate safety within the entire scope of fire department operations. *Also known as* Health and Safety Officer. (2) Member of the IMS command staff responsible to the incident commander for monitoring and assessing hazardous and unsafe conditions and developing measures for assessing personnel safety on an incident. *Also known as* Incident Safety Officer.

Self-Contained Breathing Apparatus (SCBA) — Respirator worn by the user that supplies a breathable atmosphere that is either carried in or generated by the apparatus and is independent of the ambient atmosphere. Respiratory protection is worn in all atmospheres that are considered to be Immediately Dangerous to Life and Health (IDLH). *Also known as* Air Mask *or* Air Pack.

Site Safety Assessment — Initial observation and evaluation of an investigation site; related more to situational awareness than to problem mitigation.

Situational Awareness — Perception of the surrounding environment and the ability to anticipate future events.

Unified Command (UC) — In the Incident Command System, a shared command role in which all agencies with geographical or functional responsibility establish a common set of incident objectives and strategies. In unified command there is a single incident command post and a single operations chief at any given time.

NOTE: This skill provides basic steps for donning and doffing PPE. Certain scenes may require other specialized protection. If structural fire fighting coats and trousers are worn, they must be compliant with NFPA 1971. Always follow AHJ, SOPs, and OSHA safety requirements.

Donning PPE

Step 1: Don appropriate PPE **(Figure 2.29)**.

Step 2: Don respiratory protection and eye protection.

NOTE: Once the donning is complete, perform positive and negative pressure testing to ensure that the respiratory protection is properly adjusted and functioning correctly. Follow instructions for respiratory protection equipment fit test described in local SOPs.

Step 3: Don helmet.

Step 4: Don gloves.

Doffing PPE

Step 1: Perform gross decontamination of PPE at the scene.

Step 2: Doff PPE in reverse order.

Step 3: Store PPE appropriately until it can be thoroughly cleaned and prepared for reuse **(Figure 2.30)**.

NOTE: Follow directions to inspect, clean and maintain personal protective equipment issued by AHJ and described in SOPs.

Figure 2.29

Figure 2.30

Photo courtesy of Donny Howard, Agent, Oklahoma State Fire Marshal's Office.

Chapter Contents

NFPA 1033 JPRs addressed in this chapter

This chapter provides information that addresses the following job performance requirements (JPRs) of NFPA 1033, *Standard for Professional Qualifications for Fire Investigator* (2022):

4.1.7 4.2.2 4.2.3 4.2.4 4.2.5 4.2.6

This chapter provides information that addresses the following course outcomes for the FESHE courses:

Fire Investigation I (C0283)

6. Explain how the basic elements fire dynamics, construction, and fire protection systems as to how they affect origin and cause determination.

Fire Investigation II (C0284)

3. Describe the nature and behavior of fire as it relates to fire dynamics.

Learning Objectives

1. Explain the basic principles of fire science. [NFPA 1033, 4.1.7, 4.2.2, 4.2.3, 4.2.4, 4.2.5, 4.2.6]

2. Describe how thermal energy impacts fire dynamics. [NFPA 1033, 4.1.7, 4.2.2, 4.2.3, 4.2.4, 4.2.5]

3. Explain the function of fuel within the combustion process. [NFPA 1033, 4.1.7, 4.2.2, 4.2.3, 4.2.4, 4.2.5]

4. Explain the function of oxygen within the combustion process. [NFPA 1033, 4.1.7, 4.2.2, 4.2.3, 4.2.4, 4.2.5]

5. Explain the self-sustained chemical reaction involved in flaming combustion. [NFPA 1033, 4.1.7, 4.2.2, 4.2.3, 4.2.4, 4.2.5]

6. Differentiate among the stages of fire development. [NFPA 1033, 4.1.7, 4.2.2, 4.2.3, 4.2.4, 4.2.5]

7. Explain how fire fighting operations can influence fire dynamics in a structure. [NFPA 1033, 4.1.7, 4.2.2, 4.2.3, 4.2.4, 4.2.5]

Chapter 3
Fire Dynamics:
Chemistry and Physics

Courtesy of Donny Howard, Agent, Oklahoma State Fire Marshal's Office.

Fire investigators often begin a scene examination after suppression is complete. To complete a credible examination, investigators need to understand how fire moves in given conditions without being able to see the incident in progress. This chapter mirrors the IFSTA manual, Essentials of Firefighting with some changes to focus on a fire investigator's work. The foundations for understanding fire dynamics include context for:

- Fire science
- Thermal energy (heat)
- Fuel
- Oxygen
- Self-sustained chemical reaction
- Compartment fire development
- Structure fire development

Fire Science

NFPA 1033 (2022): 4.1.7, 4.2.2, 4.2.3, 4.2.4, 4.2.5, 4.2.6

Fire investigators should have a scientific understanding of fire. This understanding should begin with the knowledge that **fire** can take various forms. For example, a smoldering fire that does not generate flames is considered condensed-phase **combustion**. In contrast, flaming fires are considered gas-phase combustion. All fires involve a heat-producing chemical reaction between some type of **fuel** and an **oxidizer**, most commonly oxygen in the air **(Table 3.1)**. Oxidizers are not combustible but will support or enhance combustion.

Table 3.1 Common Oxidizers	
Substance	**Common Use**
Calcium Hypochlorite (granular solid)	Chlorination of water in swimming pools
Chlorine (gas)	Water purification
Ammonium Nitrate (granular solid)	Fertilizer
Hydrogen Peroxide (liquid)	Industrial bleaching (pulp and paper and chemical manufacturing)
Methyl Ethyl Ketone Peroxide	Catalyst in plastics manufacturing

Courtesy of Ed Hartin.

Physical Science Terminology

Physical science is the study of **matter** and **energy**, which includes chemistry and physics. This chapter will concentrate on physical and chemical changes that occur in materials when exposed to fire.

A physical change occurs when a substance remains chemically the same but changes in size, shape, appearance, or state of matter (e.g. solid, liquid, or gas/**vapor**). Examples of physical change are:

- Charring as a result of a chemical reaction damage
- Water freezing (liquid to solid) and boiling (liquid to gas)
- Plastics melting to a liquid through the application of heat

A chemical change occurs when a substance changes from one form of matter into another, such as two or more substances combining to form compounds. More information on chemical changes related to fire is presented throughout this manual. Examples of chemical change include:

- Charring wood
- Melting plastic begins to burn and produce toxic vapors
- One of the byproducts of burning hydrocarbons is the generation of water vapor

Oxidation is a chemical reaction involving the combination of an oxidizer, such as oxygen in the air, with other materials. Oxidation can be slow, such as the combination of oxygen with iron to form rust, taking days, weeks, or months. In contrast, oxidation can be rapid, as in combustion, occurring in fractions of a second **(Figure 3.1)**.

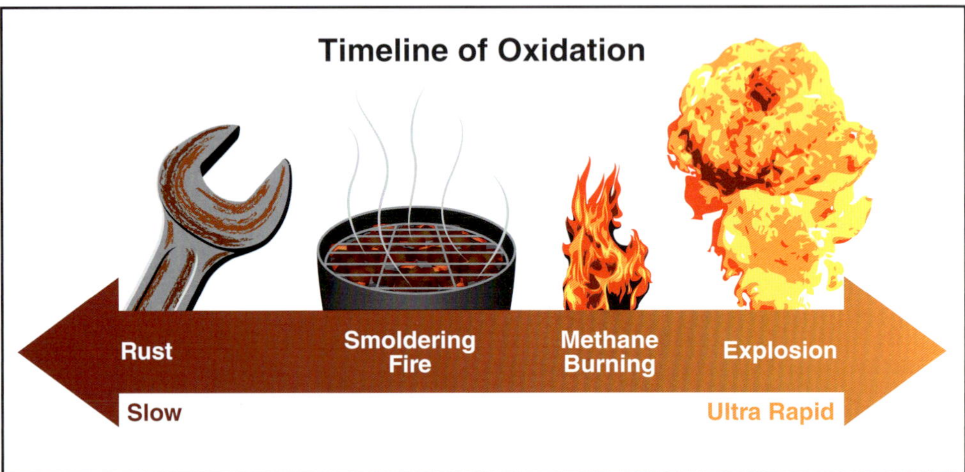

Figure 3.1 The oxidation timeline illustrates the speed differences among each oxidation type.

Energy

Energy is the capacity to perform work. Work occurs when a force is applied to an object over a distance or when a substance undergoes a chemical, biological, or physical change. In the case of **heat**, work means increasing a substance's **temperature**.

Forms of energy are classified as either potential or kinetic **(Figure 3.2)**. **Potential energy** represents the amount of stored energy that an object can release at some point in the future. Fuels have a certain amount of potential energy before they are ignited, based on their chemical composition. This potential energy available for release in the combustion process is known as the **heat of combustion**.

Figure 3.2 Potential energy is stored energy. In contrast, kinetic energy is actively being released. *Courtesy of Dan Madrzykowski, NIST.*

The rate at which a fuel releases energy over time depends on many variables including:

- Arrangement
- Density of the fuel
- Chemical composition
- Availability of oxygen for combustion

Kinetic energy is the energy that a moving object possesses. As the heat (**thermal energy**) increases, these molecules vibrate more and more rapidly. The fuel's kinetic energy is the result of these vibrations in the molecules.

There are many types of energy including:

- Light
- Sound
- Nuclear
- Thermal
- Electrical
- Chemical
- Mechanical

All energy can change from one type to another. For example, mechanical energy from a machine can convert to thermal energy when friction between moving parts generates heat. In terms of fire behavior, the potential chemical energy of a fuel converts into heat and light during combustion.

Energy is measured in **joules (J)** in the International System of Units (SI). The quantity of heat required to change the temperature of 1 gram of water by 1 degree Celsius is 4.2 joules. In the customary system, the unit of measurement for heat is the British thermal unit (Btu). A British thermal unit is the amount of heat required to raise the temperature of 1 pound of water by 1 degree Fahrenheit. While not used in scientific and engineering texts, the Btu is still frequently used in the fire service. When comparing joules and Btu, 1 055 J = 1 Btu. **Watts** are discussed later in this chapter in terms of **power**.

Chemical and physical changes almost always involve an exchange of energy. A fuel's potential energy releases during combustion and converts to kinetic energy. Reactions that emit energy as they occur are **exothermic reactions**. Fire is an exothermic *chemical* reaction that releases energy in the form of heat and sometimes light.

Reactions that absorb energy as they occur are **endothermic reactions** (**Figure 3.3**). For example, converting water from a liquid to a gas (steam) requires the input of energy resulting in an endothermic *physical* reaction. Converting water to steam is a tactic for controlling and extinguishing some types of fires.

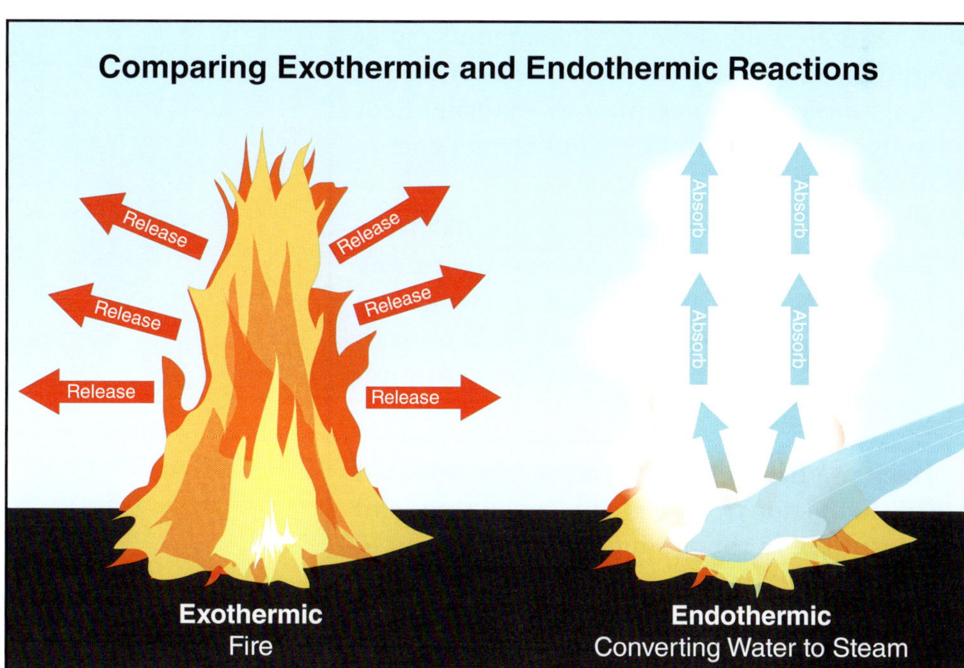

Figure 3.3 Exothermic reactions release energy; endothermic reactions absorb energy.

Fire Triangle and Tetrahedron

The fire triangle is the oldest and simplest model demonstrating the three elements necessary for combustion to occur: fuel, oxygen, and heat. Remove any one of these elements and the fire will be extinguished or prevented (**Figure 3.4**).

The fire triangle was used prior to the general adoption of the fire tetrahedron, which includes an uninhibited chemical chain reaction. An uninhibited chemical chain reaction must also be present for a fire to develop. The fire tetrahedron model includes the chemical chain reaction to explain flaming or gas-phase combustion.

Ignition

NFPA 1033 (2022): 4.2.6

The process of ignition begins with the state of the fuels. Fuels must be in a gaseous state to burn; therefore, solids and liquids must become gaseous in order for ignition to occur. When heat is transferred to a liquid or solid, the substance's temperature increases and the substance starts to convert to a gaseous state (off-gassing). In solids, off-gassing is a chemical change known as **pyrolysis**. When heat is applied to liquids, the physical change to a gas is called **vaporization** (**Figure 3.5**).

Ignition occurs when a heat source has sufficient time, energy, and temperature to raise some material to its **ignition temperature**. There are two types of ignition:

- **Piloted ignition** is the most common form of ignition and occurs when a mixture of gaseous fuel and oxygen encounter an external heat source with sufficient heat or thermal energy to start the combustion reaction.

- **Autoignition** occurs without any external flame or spark to ignite the fuel gases or vapors. The fuel's surface is heated to the point at which the combustion reaction occurs, such as occurs with ignition of a material by means of **radiation** (**Figure 3.6**).

Once the fuel is ignited, the energy released from combustion transfers to the remaining solid fuel resulting in the production and ignition of additional fuel vapors or gases. This exchange of energy from the burning gases to the fuel results in a sustained combustion reaction.

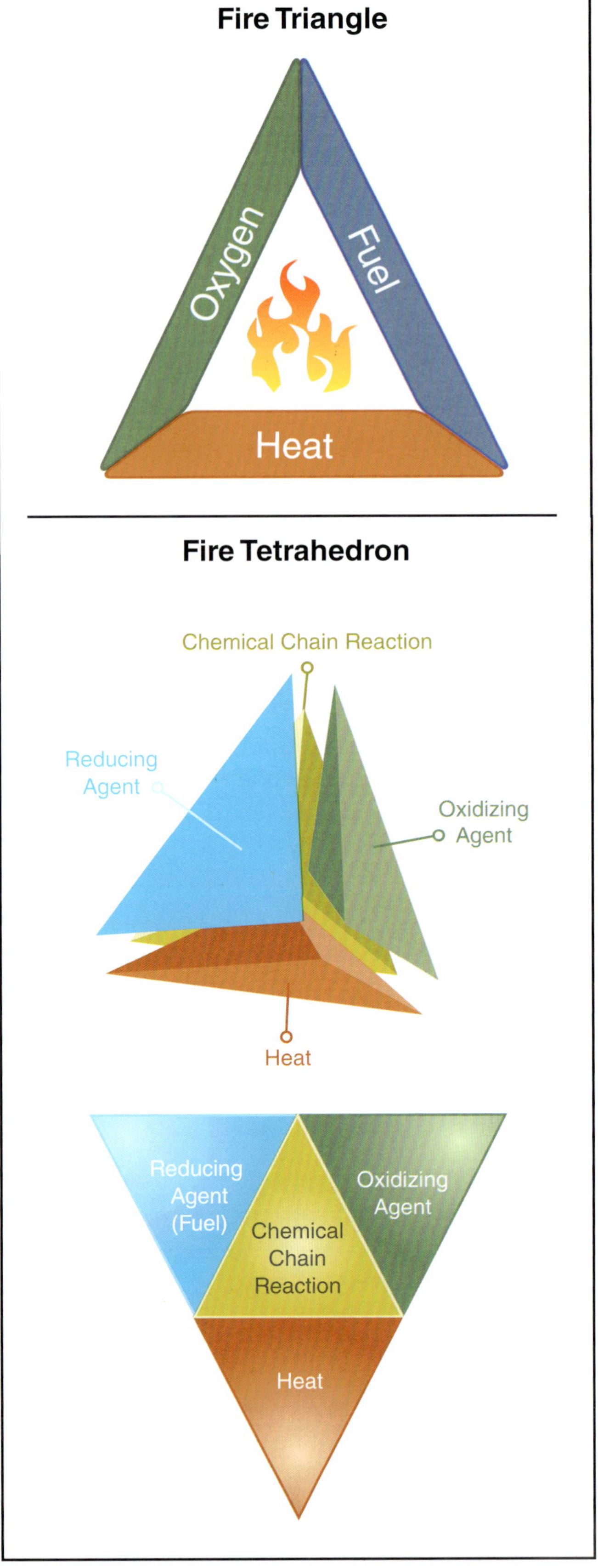

Figure 3.4 The fire triangle illustrates the three components needed for a fire. The fire tetrahedron demonstrates the four components needed for a self-sustaining fire.

Figure 3.5 Pyrolysis occurs when a solid fuel converts into a gaseous fuel. Vaporization is the conversion of a liquid to a vapor using heat energy. *Both courtesy of Dan Madrzykowski, NIST.*

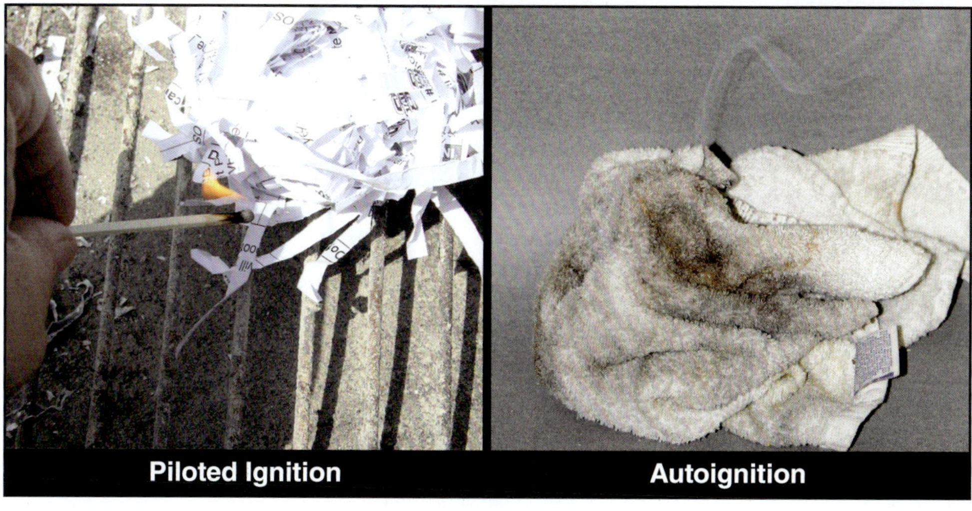

Figure 3.6 Piloted ignition involves the introduction of an external ignition source. Autoignition occurs under special conditions without the heat of a spark or other source.

Autoignition temperature (AIT) is the minimum temperature at which a gaseous fuel in the air must be heated in order to start self-sustained combustion. The autoignition temperature of a substance is always higher than its piloted ignition temperature.

Combustion

Fire and combustion are similar conditions, and the terms may be used interchangeably. At the same time, flaming combustion is associated with fire, and nonflaming combustion is associated with smoldering **(Figure 3.7)**.

Figure 3.7 Nonflaming combustion (left) features lower temperatures and smoldering conditions. Flaming combustion (right) displays visible flames above the burning fuel.

Flaming Combustion

The fire tetrahedron accurately reflects the conditions required for flaming combustion. Each element of the tetrahedron must be in the proper proportion and in close physical proximity for flaming combustion to occur. Removing any element of the tetrahedron interrupts the chemical chain reaction and stops flaming combustion. However, the fire may continue to smolder depending on the characteristics of the fuel.

Flaming combustion is considered gas-phase combustion, and produces a visible flame above the material's surface. Flaming combustion occurs when a gaseous fuel mixes with oxygen in the correct ratio and heats to ignition temperature.

Gases in flaming combustion can be generated from liquid or solid fuels through the addition of heat (vaporization or pyrolysis, respectively). When heated sufficiently above the ignition temperature of the fuel, the vapors will ignite, producing flames above the material's surface. The fire also generates heat. As the heat transfers to the gaseous **products of combustion**, they expand and begin to rise and move away from the fire due to buoyancy.

In other words, hot combustion products are less dense than the surrounding air. The combustion products "float" on the dense cool air surrounding the fuel, creating the layers of smoke and fuel gases that fill a compartment during a fire.

Discussion of competent ignition sources is included in Chapter 13. For the purposes of this section, ignition is simply defined as the beginning of the combustion process:

- A heat source **pyrolizes** a fuel, creating fuel gases.

- Those gases mix with oxygen and ignite, creating a fire.

Generally speaking, a fire can be compared to a pump. Fresh oxygen is "pumped in" and mixes with fuel gases. Then as it burns, the fire "pumps out" combustion products that have larger amounts of mass and a higher level of energy than the inlet air. When the fire has limitations in air intake and exhaust output, there may be well defined **flow paths (Figure 3.8)**. In the case of **open burning**, the "pump" does not have a well-defined inlet or outlet, as the air is being **entrained** (drawn in) from all around the burning fuel.

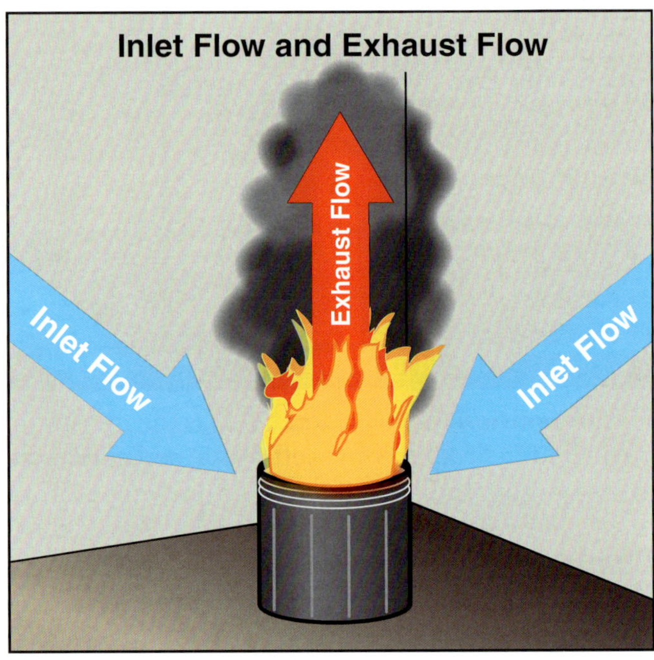

Figure 3.8 Inlet air flow (intake) consists of fresh oxygen. Exhaust flow (outlet) consists of products of combustion.

Smoldering Combustion

Smoldering, while a form of combustion, is not generally considered to be fire. Smoldering occurs more slowly than flaming combustion, and at a lower temperature, producing a glow in the material's surface. Smoldering combustion may be localized on or near the fuel's surface where it is in contact with oxygen. The fire triangle illustrates the elements/conditions required for this mode of combustion. This type of combustion is considered condensed-phase or solid-phase.

Examples of smoldering combustion include:

- Burning charcoal
- Smoldering wood or fabric
- Ignition sequence of linseed oil on rags, or Latex gloves prior to the flaming combustion phase

Products of Combustion

As a fuel burns, its chemical composition changes, which produces new substances. These products of combustion are often simply described as heat and smoke. While the heat from a fire is a danger to anyone directly exposed to it, exposure to toxic gases and/or lack of oxygen cause most fire deaths.

Complete Combustion. Complete combustion occurs when the balance between the available oxygen and fuel results in no visible soot, smoke, or ash. For example, combustion which occurs in a burner of a kitchen range is an example of complete combustion.

Incomplete Combustion. Smoke is the product of **incomplete combustion**. Simply stated, combustion is incomplete when any of the fuel is left after combustion has occurred. Smoke and ash are examples of leftover fuel from incomplete combustion. Smoke is comprised of gases, vapors, and solid particulates resulting from incomplete combustion, and **pyrolysates**.

Combustion is incomplete in a structure fire, meaning that some of the fuel does not burn, but instead gets entrained with hot gases, and rises. This unburned fuel is smoke, and it has the potential to burn **(Figure 3.9)**.

Figure 3.9 Smoke consists of wide-ranging toxic and flammable gases and particulates.

Most structure fires involve multiple types of fuels including natural materials such as wood and cotton, and synthetic materials such as polyurethanes and polyesters, synthetic fabrics. As the fire develops, the air supply becomes limited within an enclosed compartment. This results in a higher level of incomplete combustion, which produces more smoke. These factors result in complex chemical reactions that generate pyrolysates and a wide range of products of combustion including toxic and flammable gases, vapors, and particulates that comprise smoke.

Pressure Differences

Pressure is the force per unit of area applied perpendicular to a surface. For example, atmospheric pressure (1 atmosphere [app. 101 kPa]) at standard temperature (68° F [20° C]) indicates the amount of pressure the atmosphere applies to the surface of the earth. At standard temperature and atmospheric pressure, gases remain calm and move very little. Differences in pressure above or below standard pressure create movement in gases. Gases always move from areas of higher pressure to areas of lower pressure. Even small differences in pressure, such as the 0.1 kPa or less differences created in most compartment fires, create this movement.

Heat from a fire increases the pressure of the surrounding gases. This increased pressure will seek to expand and equalize with areas of lower pressure. Heated gases will rise, remain aloft (**buoyant**) and generally travel up and out. At the same time, cooler, fresh air will generally travel inward toward the fire. This exchange of air creates a convective flow. As the pressure difference between high and low pressure areas increases, the speed with which gases will move from high to low also increases. It is critical for investigators to understand how small changes to the gas pressure within a structure can dramatically affect fire behavior **(Figure 3.10, p. 62)**.

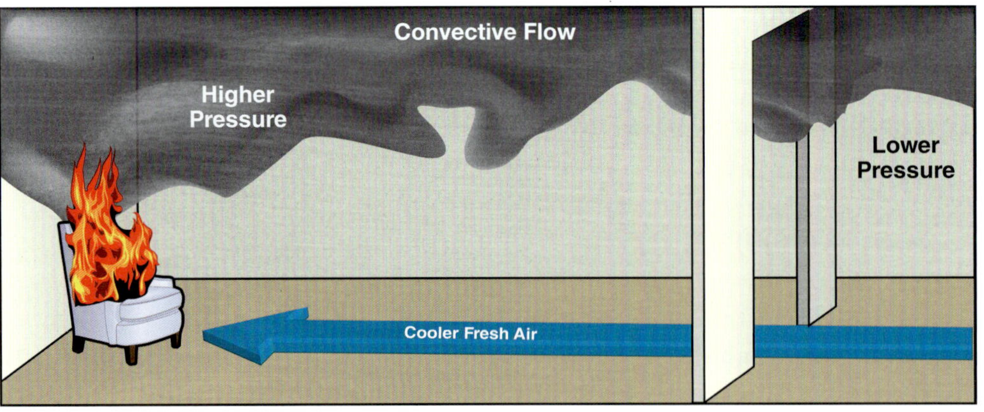

Figure 3.10 Heated gases will travel upward and outward from a fire, while cooler, fresher air draws toward the fire, creating a convective flow.

Thermal Energy (Heat)

NFPA 1033 (2022): 4.1.7, 4.2.2, 4.2.3, 4.2.4, 4.2.5

A working knowledge of **fire dynamics** requires an understanding of temperature, energy, and power or **heat release rate (HRR)**. These concepts are described in the following sections.

Difference between Heat Release Rate and Temperature

Temperature is the measurement of heat. More specifically, temperature is the measurement of the average kinetic energy in the particles of a sample of matter. A block of wood at room temperature has stable molecules and is in no danger of ignition. When thermal energy transfers to the wood, the wood is heated. The temperature of the wood rises because its molecules have begun to vibrate and move more freely and rapidly.

Temperature can be measured using several different scales. The most common are the Celsius scale, used in the International System of Units (SI) (metric system), and the Fahrenheit scale, used in the customary system. The freezing and boiling points of water provide a simple way to compare these two scales **(Figure 3.11)**.

A dangerous misconception is that temperature is an accurate predictor or measurement of heat transfer. It is not. Heat release rate is the total amount of heat released per unit time. For example, one candle burns at the same temperature as ten candles. However, the heat release rate of the ten candles is ten times greater than one candle at the same temperature. The increased heat release rate results in an increased heat transfer rate to an object. This energy flow to a unit area (**heat flux**) is measured in kilowatts per square meter.

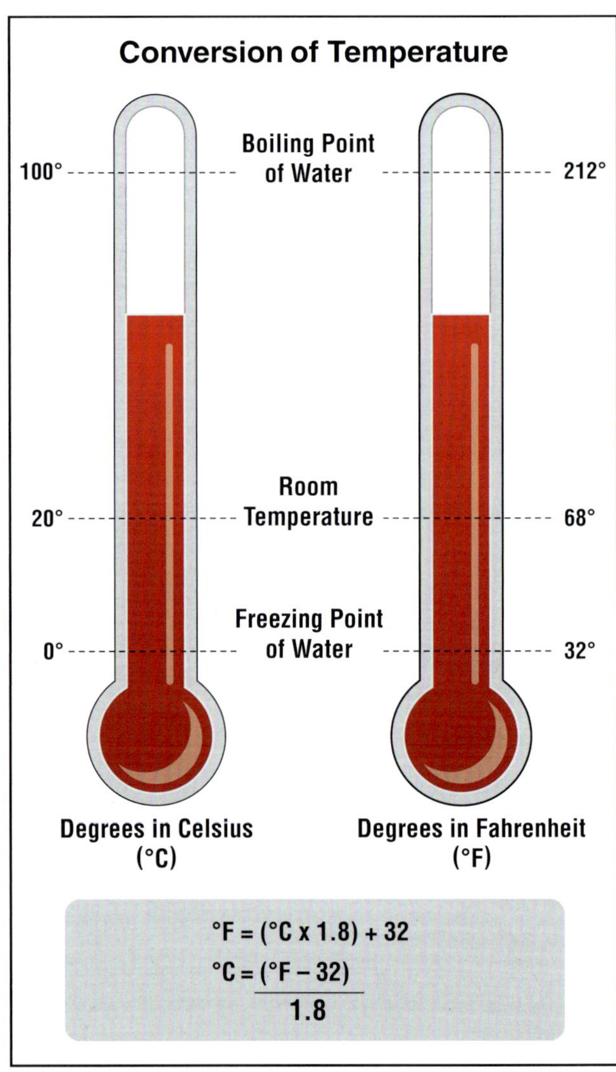

Figure 3.11 The two common scales for measuring temperature are the Celsius scale (International System of Units [SI or metric system]) and the Fahrenheit scale used in the Customary System.

Sources of Thermal Energy

To ignite a fuel, a thermal energy source must be able to transfer heat and cause the temperature of a substance to increase. The three most common ignition sources of structure fires are chemical, electrical, and mechanical energy.

Chemical Energy

Chemical energy is the most common source of heat in combustion reactions. The potential for oxidation exists when any combustible fuel is in contact with oxygen. The oxidation process almost always results in the production of thermal energy **(Figure 3.12)**.

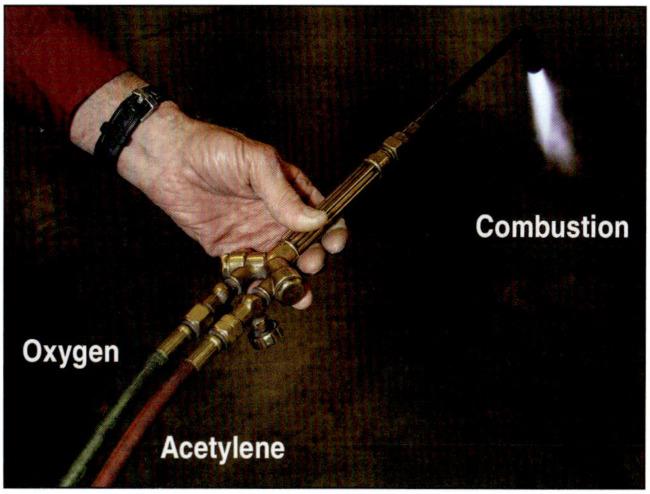

Self-heating, a form of oxidation, is a chemical reaction that increases the temperature of a material without the addition of external heat. Self-heating can lead to **spontaneous ignition** which is ignition without the addition of external heat.

Figure 3.12 A cutting torch flame is an example of heat generation from a chemical reaction.

Oxidation normally produces thermal energy slowly. The energy dissipates almost as fast as it is generated. An external heat source such as sunshine can initiate or accelerate the process. For self-heating to progress to spontaneous ignition, the following factors are required:

- The available air supply in and around the heated material must be adequate to support combustion.

- The rate of heat production must be great enough to raise the temperature of the material to its autoignition temperature.

- The insulation properties of the material immediately surrounding the fuel must be such that the heat cannot dissipate as fast as it is generated.

Rags soaked in linseed oil, rolled into a ball, and thrown into a corner have the potential for spontaneous ignition. The natural oxidation of this vegetable oil and the cloth will generate heat if some method of heat transfer such as air movement around the rags does not dissipate the heat. The cloth could eventually increase in temperature enough to cause ignition.

The rate of most chemical reactions increases as the temperature of the reacting materials increases. The oxidation reaction that causes heat generation accelerates as the fuel generates and absorbs more heat. When the heat generated exceeds the heat being lost, the material may reach its autoignition temperature and ignite spontaneously. A number of common materials are subject to self-heating **(Table 3.2)**.

Table 3.2 Spontaneous Heating Materials and Locations	
Type of Material	**Possible Locations**
Charcoal	Convenience stores Hardware stores Industrial plants Restaurants Residences
Linseed oil-soaked rags	Woodworking shops Lumber yards Furniture repair shops Picture frame shops Residential/Commercial Construction/Remodeling sites
Hay and manure	Farms Feed stores Arenas Feedlots

Electrical Energy

Electrical energy can generate temperatures high enough to ignite any combustible materials near the heated area. Electrical heating can occur in several ways, including **(Figure 3.13)**:

- **Resistance heating** — Electric current flowing through a conductor produces heat. Some electrical appliances, such as incandescent lamps, ranges, ovens, or portable heaters, are designed to make use of resistance heating. Other electrical equipment is designed to limit resistance heating under normal operating conditions.

- **Overcurrent or overload** — When the current flowing through a conductor exceeds its design limits, the conductor may overheat and present an ignition hazard. Overcurrent or overload is unintended resistance heating.

- **Arcing** — In general, an arc is a high-temperature luminous electric discharge across a gap or through a medium such as charred insulation. Arcs may be generated when there is a gap in a conductor (such as a cut or frayed wire) or when there is high voltage, static electricity, or lightning.

- **Sparking —** When an electric arc occurs, luminous (glowing) particles can form and splatter away from the point of arcing.

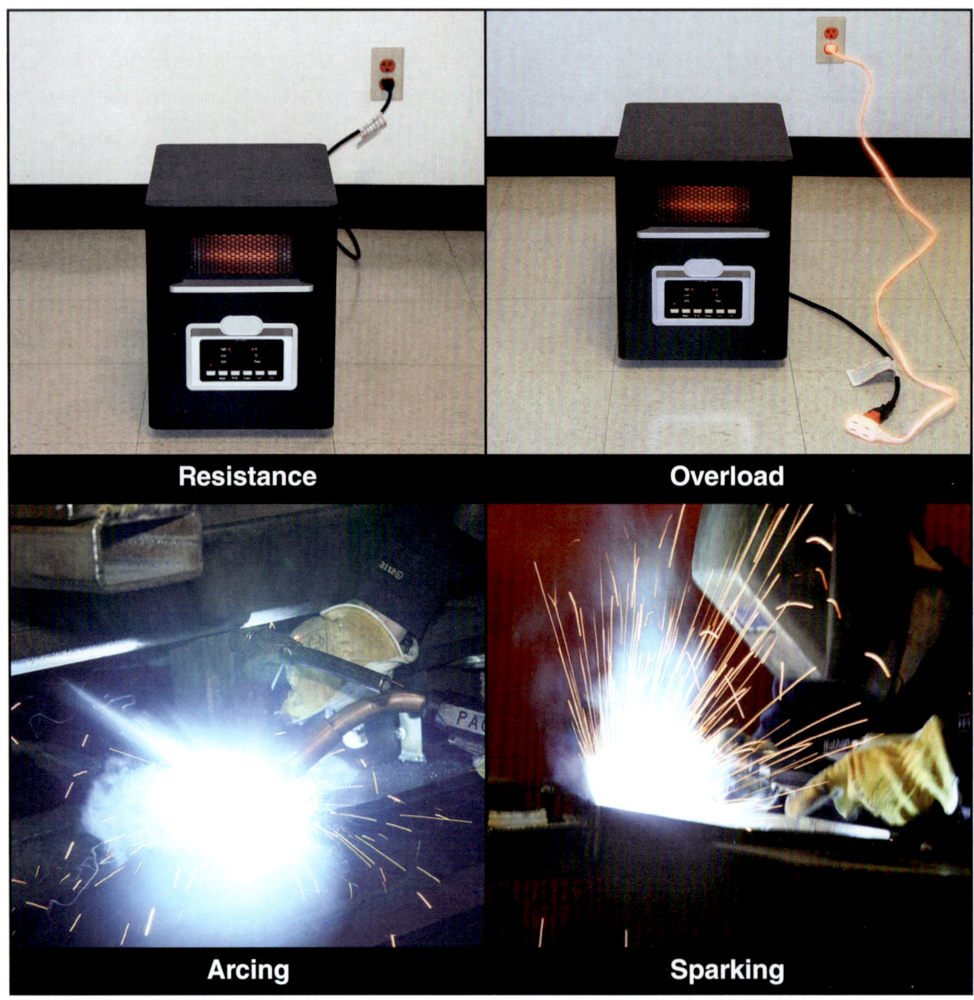

Figure 3.13 Examples of resistance heating, overcurrent or overload heating, arcing, and sparking.

Mechanical Energy

Mechanical energy is the result of friction or compression **(Figure 3.14)**. Friction, the movement of two surfaces against each other, generates heat and/or sparks. For example, a belt on a motor which ceases to turn, but gets heated by the spinning wheel, is an example of friction heating.

Heat is also generated when a gas is compressed. For example, diesel engines use compression to ignite fuel vapors without spark plugs.

Heat Transfer

The transfer of heat from one point or object to another is part of the study of thermodynamics. Heat transfer from the initial fuel package (burning object) to other fuels in and beyond the area of fire origin affects the growth of any fire. Heat transfers from warmer objects to cooler objects because heated materials will naturally return to a state of **thermal equilibrium** in which all areas of an object are a uniform temperature. Objects at the same temperature do not transfer heat.

The rate at which heat transfers is related to the temperature differential of the bodies and the **thermal conductivity** of the materials involved. The greater the temperature differences between the bodies, the greater the transfer rate. A material with higher thermal conductivity will transfer heat more quickly than other materials.

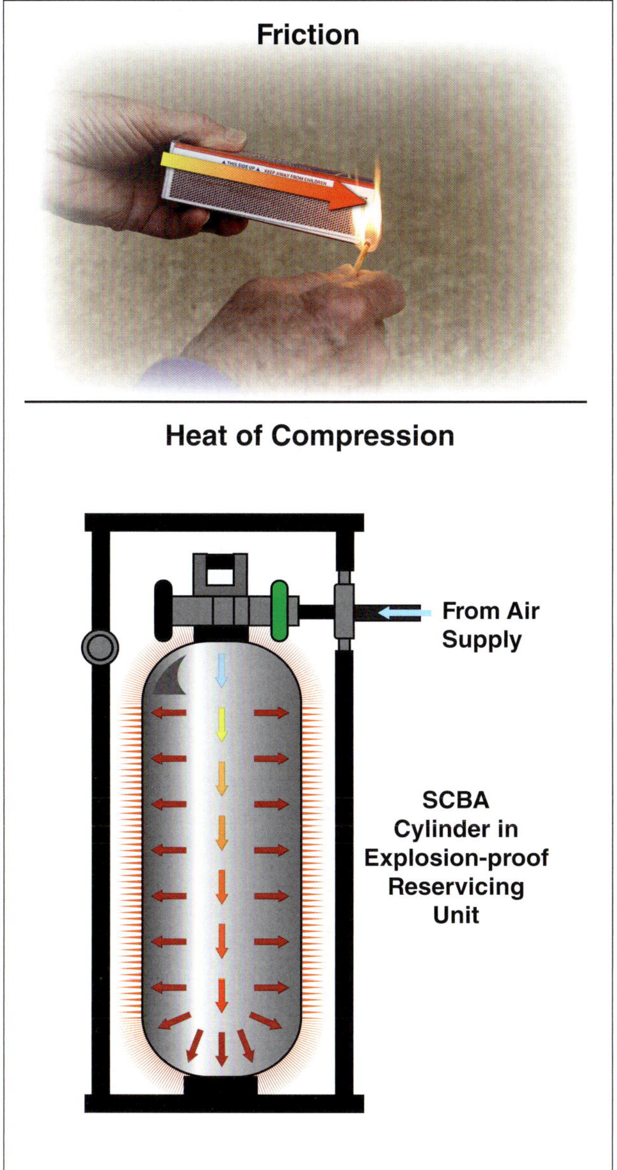

Figure 3.14 The friction of the match head scratching the box's striker generates the heat to ignite the match. When gas is compressed, it generates heat.

Thermal inertia is the product of three material properties of solid fuels: thermal conductivity, density, and heat capacity. Thermal inertia provides a measure of how rapidly the surface temperature of a material will increase when the material is exposed to heat. When the surface of a given material heats up faster relative to another material exposed to the same heat source, the material with the higher surface temperature is likely to ignite first. For example, you could have two items made of the same plastic material. While the two items are composed of the same material based on chemistry, if the two items have significantly different densities, that would cause the flow of heat into each of the items to be different as the thermal inertia would be different. If the two items were exposed to the same heat flux, the time to ignition for each item would be different.

Thermal inertia is often referred to as *k rho c*. In theory, the lower the value, the faster the material's surface temperature will increase, which could lead to a more rapid ignition relative to a material with a higher thermal inertia.

This formula is derived from the units:

- **Thermal conductivity** is measured in units of W/(m k) and is represented by k.
- **Material density** is measured in units of (kg/m^3) and is represented by the Greek letter rho.
- **Heat capacity** is measured in J/(kg K) and is represented by c.

The three methods of heat transfer from one body to another are:

- Conduction
- Convection
- Radiation

Conduction

Conduction is the transfer of heat through and between solids. Conduction occurs when a material is heated as a result of direct contact with a heat source **(Figure 3.15)**. Conduction results from increased molecular motion and collisions among a substance's molecules. The more closely packed the molecules of a substance are, the more readily it will conduct heat. For example, if a fire heats a metal pipe on one side of a wall, heat conducted through the pipe can ignite wooden framing components in the wall or nearby combustibles on the other side of the wall.

Heat transfer due to conduction is dependent upon three factors:

- Area being heated
- Thermal conductivity of the heated material
- Temperature difference between the heat source and the material being heated

Figure 3.15 Conduction occurs when heat transfers between solid objects, in this case, between the door and the firefighter's hand.

Common materials may have varying thermal conductivity at the same ambient temperature (68°F [20°C]) **(Table 3.3)**. For example, copper will conduct heat more than seven times faster than steel. Likewise, steel is nearly forty times as thermally conductive as concrete. Air is the least able to conduct heat of most substances, so it is a very good insulator.

Insulating materials slow the conduction of heat from one solid to another. Good insulators are materials that do not conduct heat well because their physical makeup disrupts the point-to-point transfer of heat or thermal energy. The best commercial insulators used in building construction are those made of fine particles or fibers with void spaces between them filled with a gas such as air. Gases do not conduct heat very well because their molecules are relatively far apart.

Table 3.3 Thermal Conductivity of Common Substances		
Substance	**Temperature**	**Thermal Conductivity (W/mK)**
Copper	68°F (20°C)	386.00
Steel	68°F (20°C)	36.00 – 54.00
Concrete	68°F (20°C)	0.8 – 1.28
Gypsum Wall Board	68°F (20°C)	0.50
Wood (pine)	68°F (20°C)	0.13
Air	68°F (20°C)	0.03

Convection

Convection is the transfer of thermal energy by the circulation or movement of a fluid (liquid or gas) **(Figure 3.16)**. In the fire environment, convection usually involves transfer of heat through the movement of hot smoke and fire gases. As with all heat transfer, the heat flows from the hot fire gases to the cooler structural surfaces, building contents, and air. Convection may occur in any direction.

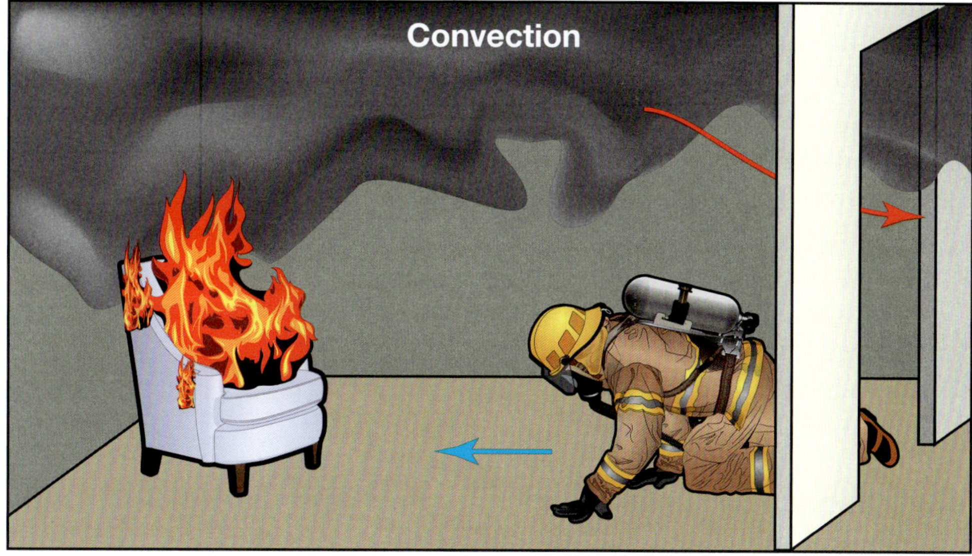

Figure 3.16 Convection is the transfer of heat by liquid or gas circulation.

Vertical movement is due to the buoyancy of smoke and fire gases. Lateral movement is usually the result of pressure differences (movement from high to low pressure).

Heat transfer due to convection is dependent upon three factors:

- Area being heated
- Turbulence and velocity of moving gases
- Temperature difference between the hot fluid or gas and the material being heated

Radiation

Radiation is the transmission of energy as electromagnetic waves, such as light waves, radio waves, or X-rays, without an intervening medium **(Figure 3.17)**. Radiant heat can become the dominant mode of heat transfer as the fire grows in size and can have a significant effect on the ignition of objects located some distance from the fire. Radiant heat transfer is also a significant factor in fire development and spread in compartments. For example, radiant energy from a **hot gas layer**, even before full involvement, can ignite fuels below the layer.

Numerous factors influence radiant heat transfer, including:

- **Nature of the exposed surfaces** — Dark-colored materials emit and absorb heat more effectively than light-colored materials; smooth or highly-polished surfaces reflect more

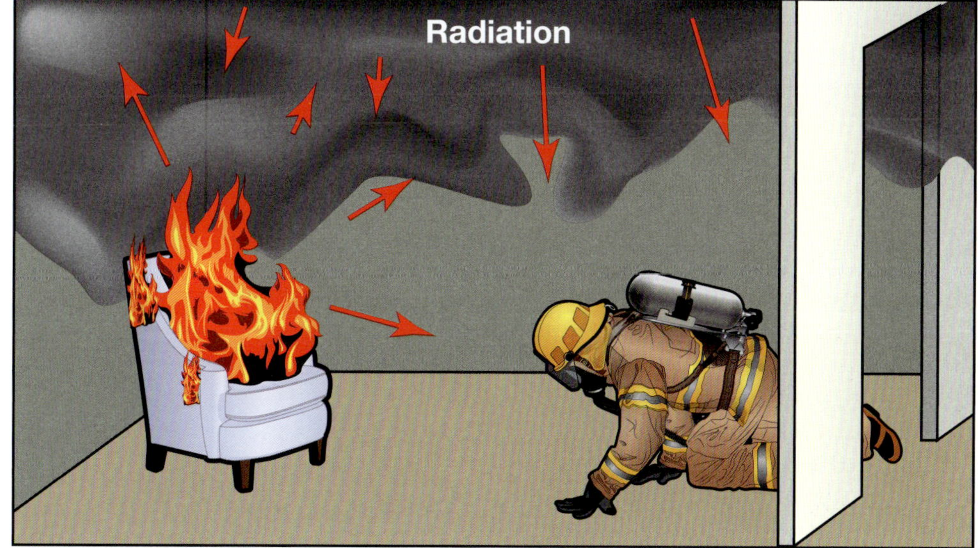

Figure 3.17 Radiation is the transfer of heat by electromagnetic waves without another medium to transfer the heat energy.

radiant heat than rough surfaces. This concept, known as **emissivity**, is used to protect some items, for example by wrapping vulnerable materials with foil.

- **Distance between the heat source and the exposed surfaces** — Increasing distance reduces the effect of radiant heat **(Figure 3.18)**.

- **Temperature of the heat source** — Unlike other methods of heat transfer that depend on the temperature of both the heat source and exposed surface, radiant heat transfer primarily depends on the energy released from the heat source. As the heat release rate of the heat source increases, the radiant energy also increases **(Figure 3.19)**.

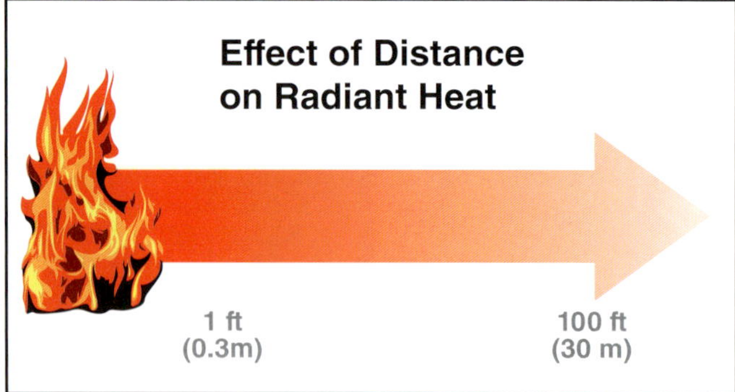

Figure 3.18 The effects of radiant heat diminish as the distance between the origin point and an exposure increases.

Figure 3.19 The fire on the right is giving off more thermal radiation than the fire on the left. *Courtesy of NIST.*

As an electromagnetic wave, radiated heat energy travels in a straight line. Examples of radiant heat transfer include:

- Heat from the sun
- Broiler in an oven
- Heating bread in a toaster

Radiation is a common cause of **exposure fires**. As a fire grows, it radiates more energy which other objects absorb as heat. In large fires, it is possible for the radiated heat to ignite buildings or other fuel packages a considerable distance away. Radiated heat travels through vacuums and air spaces that would normally disrupt conduction or convection. However, materials that reflect, absorb, or scatter radiated energy will disrupt the heat transmission. While flames have high temperature resulting in significant radiant energy emission, hot smoke or flames in the upper layer can also radiate significant heat.

The Importance of Understanding Temperature and Heat Transfer Rate

Heat flux (kW/m^2) from radiated heat emitted from flames or hot surfaces such as the walls and ceiling may cause ignition. This can occur even when the temperature of the gases adjacent to the target fuel are at temperatures less than the ignition temperature of the exposed fuel. As the fuel absorbs the radiant heat, the temperature of the fuel increases. If the fuel stores more heat than it transfers away, the temperature of the fuel increases. As the heating process continues, the fuel will reach its autoignition temperature.

Interaction among the Methods of Heat Transfer

The methods of heat transfer rarely occur individually during a fire. The fire radiates heat, causes convection of heat through hot fuel gases, and conducts heat through burning materials or metals that are involved in the fire.

Convected heat and radiated heat that reaches walls and ceilings will heat those surfaces. The surface materials absorb (conduct) heat within the heated material, at a rate based on the material's thermal conductivity. Basically, one side of the object is warm and slowly warms through the object until the opposite side is of equal temperature with the heated side.

A heated surface will then, in turn, begin to radiate heat which could lead to ignition, combustion, convection, and so on. This cycle continues until interrupted.

Fuel

NFPA 1033 (2022): 4.1.7, 4.2.2, 4.2.3, 4.2.4, 4.2.5

Fuel is the oxidized or burned material or substance in the combustion process. The fuel in a combustion reaction is known as the **reducing agent**. Fuels may be inorganic or organic. Inorganic fuels, such as hydrogen or magnesium, do not contain carbon. Most common fuels are organic, containing carbon and other elements. Organic fuels can be divided into hydrocarbon-based fuels, such as:

- Fuel oil
- Plastics
- Gasoline
- Cellulose-based materials (wood and paper)

A fuel's chemical content influences both its heat of combustion and heat release rate. The fuel's heat of combustion is the total amount of thermal energy released when a specific amount of that fuel burns. In other words, different materials release more or less heat than others based on their chemical makeup. Many plastics, flammable liquids, and flammable gases contain more potential thermal energy than wood **(Table 3.4)**.

Table 3.4 Representative Peak Heat Release Rates (HRR) During Unconfined Burning		
Fuel Material	**Peak HRR in kilowatts**	**Common Locations for Material**
Small wastebasket	4-50	Homes, businesses, shops
Cotton mattress	40-970	Homes, furniture stores, motels
Cotton easy chair	290-370	Homes, furniture stores, office buildings
Small pool of gasoline	400	Traffic crash, fuel stations
Dry Christmas tree	3000-5000	Homes, trash facilities, dumpsters, recycling sites
Polyurethane mattress	810-2630	Homes, furniture stores, motels, dormitories, jails
Polyurethane easy chair	1350-1990	Homes, furniture stores, motels
Polyurethane sofa	3120	Homes, furniture stores, motels, dormitories, office buildings

Adapted from NFPA® 921, 2017 edition

Synthetic materials are common in modern construction and furnishings. These materials are synthesized from petroleum products, and as a result, they have higher heats of combustion and may generate higher heat release rates than wood on a per-mass basis.

Power is the rate at which energy transfers. Another way to describe power is the rate at which energy converts from one form to another. The standard international (SI) unit for power is the watt (W). One watt is 1 joule per second (J/s).

In terms of fire behavior, power is the heat release rate during combustion. When a fuel is heated, work is being performed (energy is being transferred). The speed with which this work occurs, heat release rate, is the amount of generated power. Heat release rate is the energy released per unit of time as a fuel burns and is usually expressed in kilowatts (kW) or megawatts (MW). Heat release rate depends on the type, quantity, and orientation of the fuel **(Figure 3.20)**. Heat release rate directly relates to oxygen consumption because the combustion process requires a continuous supply of oxygen to continue. Typically, the more oxygen is available, the higher the heat release rate. Similarly, the heat release rate decreases if all available oxygen is consumed and not replenished.

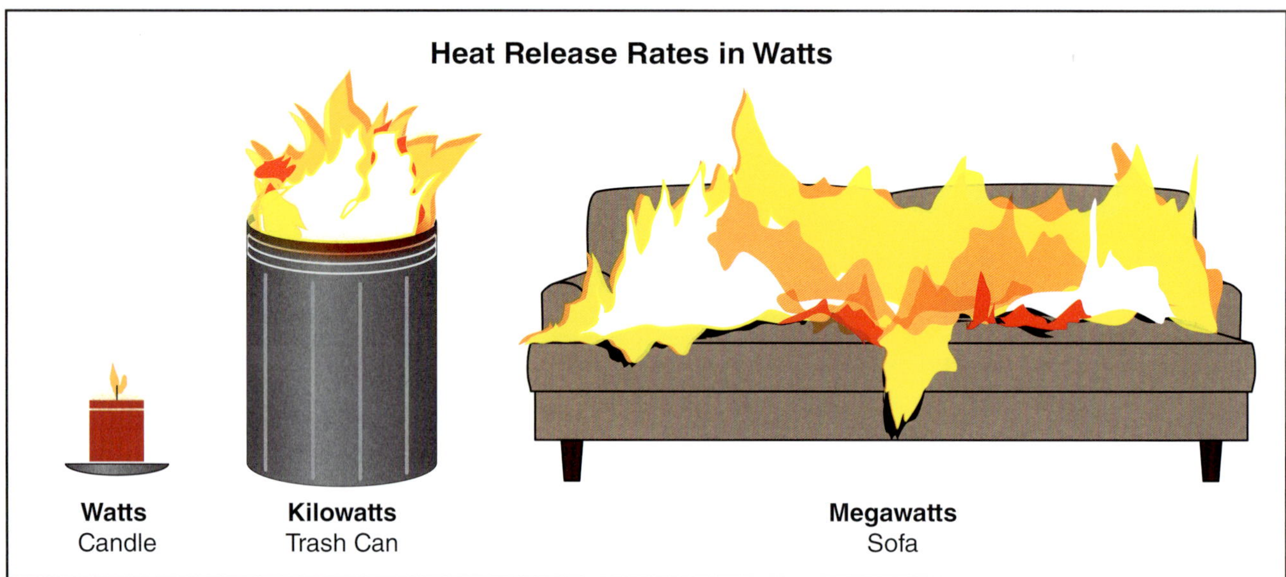

Figure 3.20 Examples of heat release rate conditions that may be measured in watts, kilowatts, or megawatts.

Prefixes for Units of Measure: Kilo and Mega

The International System of Units (SI) specifies a set of prefixes that precede units of measure to indicate a multiple or fraction of that unit. Two common prefixes encountered when discussing energy (joules) and heat release rate (watts) are *kilo* (one thousand) and *Mega* (one million).

Gases

As previously described, for flaming combustion to occur, fuels must be in the gaseous state. Thermal energy is required to change solids and liquids into the gaseous state. Gaseous fuels can be the most dangerous of all fuel types because they are already in the physical state required for ignition. When wood burns inefficiently, the combustion products may contain methane, acetylene and other fuel gases **(Table 3.5)**. Vapor is the common term used to describe the gaseous state of a fuel that would normally exist as a liquid or a solid at standard temperature and pressure. More information about gases and their reactions is included in the IFSTA manual, **Hazardous Materials for First Responders**.

Material	Vapor Density	Ignition Temperature
Methane (Natural Gas)	0.55	(1004°F) 540°C
Propane (Liquefied Petroleum Gas)	1.52	(842°F) 450°C
Carbon Monoxide	0.96	(1,128°F) 620°C

Source: *Computer Aided Management of Emergency Operations* (CAMEO)

Vapor Density

Vapor density is the weight of a given volume of pure vapor or gas compared to the weight of an equal volume of dry air at the same temperature and pressure. For more information on vapor density, refer to the IFSTA manual, **Hazardous Materials for First Responders**.

Few gases are lighter than air. Any gas with a vapor density less than one will be lighter than air and will rise into the atmosphere. These gases can spread to a wide geographical area. Depending on the product, this could create a hazardous area downwind of a release. Examples of materials with a vapor density less than one include helium, neon, acetylene, and hydrogen.

Most gases have a vapor density greater than one; they will sink in relation to ambient air and will displace oxygen at low elevations **(Figure 3.21)**. Heavier vapors and gases are likely to concentrate in low places along or under floors; in sumps, sewers, and manholes; and in trenches and ditches where they may create fire or health hazards. Examples of common materials with a vapor density greater than one include: Propane, Butane, Chlorine, and Sulfur dioxide.

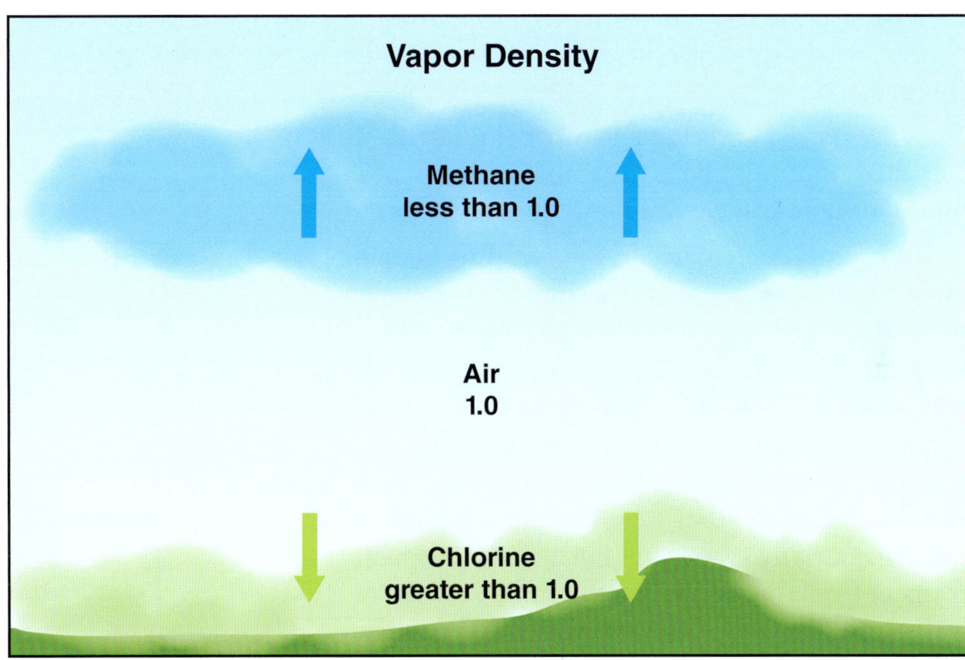

Figure 3.21 The vapor density of a gas indicates where a gas will collect at an incident.

Vapor density varies with the temperature of the vapor or gas. Hot vapors will rise, but unless totally dispersed, they will sink once they have cooled. Cold vapors are dense and will stay low but will rise when they warm.

Personnel cannot precisely predict the spread of vapors from the vapor density because topography, weather conditions, and the vapor mixture with air affect vapors. However, knowing the vapor density gives a general idea of what to expect from a specific gas or vapor.

NOTE: All vapors and gases will mix with air, but the lighter materials tend to rise and dissipate (unless confined).

Specific Gravity

Specific gravity is the ratio of the density (mass per volume) of a material to the density of a standard material, usually an equal volume of water, at standard conditions of pressure and temperature. If a volume of a material weighs 8 pounds (3.6 kg), and an equal volume of water weighs 10 pounds (4.5 kg), the material is said to have a specific gravity of 0.8. Materials with specific gravities less than one will float in (or on) water. Materials with specific gravities greater than one will sink in water.

Solubility plays an important role in specific gravity. Highly soluble materials will mix or dissolve more completely in water (distributing themselves more evenly throughout), rather than sinking or floating (without dissolving) according to their specific gravities. Most (but not all) flammable liquids have specific gravities less than one and will float on water. An important consideration for fire-suppression activities is that flammable liquids will float on water.

Liquids

Liquids have mass and volume but no definite shape except for a flat surface or the shape of their container. Unlike solids, liquids retain their state of matter partly due to standard atmospheric pressure. Unlike gases, liquids will not expand to fill all of a container. When released on the ground, liquids will flow downhill and pool in low areas.

Just as gas density is compared to air, liquid density is compared to water. Specific gravity is the ratio of the mass of a given volume of a liquid compared to the mass of an equal volume of water at the same temperature. Water is assigned a specific gravity of 1. Liquids with a specific gravity less than 1, such as gasoline and most other flammable liquids, are lighter than water and will float on its surface. Liquids with a specific gravity greater than 1, such as corn syrup, are heavier than water and will sink **(Figure 3.22)**.

As previously described, to burn, liquids must vaporize. Vaporization is the transformation of a liquid to vapor or a gaseous state.

For vaporization to occur, the escaping vapors must be at a greater pressure than atmospheric pressure. **Vapor pressure** indicates how easily a substance will evaporate into air. Flammable liquids with a high vapor pressure present a special hazard.

The vapor pressure of the substance and the amount of thermal energy applied to it determines the rate of vaporization. For example, a puddle of water eventually evaporates because of slow heat transfer from the sun. When the same amount of water is heated on a stove, however, it vaporizes much more rapidly because there is more thermal energy applied. The volatility or ease with which a liquid gives off vapor influences how easily it can ignite. The size of a liquid's surface area also influences the extent to which the liquid will give off vapor. In many open containers, the surface area of liquid exposed to the atmosphere is limited.

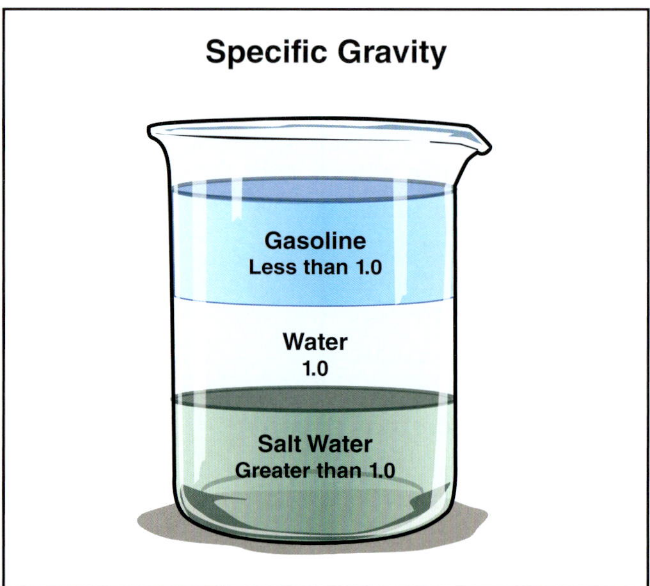

Figure 3.22 The specific gravity of a liquid indicates whether the liquid will float or sink relative to the water's surface.

Flash point is the minimum temperature at which a liquid gives off sufficient vapors to ignite, but not sustain combustion, in the presence of a piloted ignition source. **Fire point** is the temperature at which a piloted ignition of sufficient vapors will begin a sustained combustion reaction **(Figure 3.23)**. Flash point is commonly used to indicate the flammability hazard of liquid fuels. For example, liquid fuels

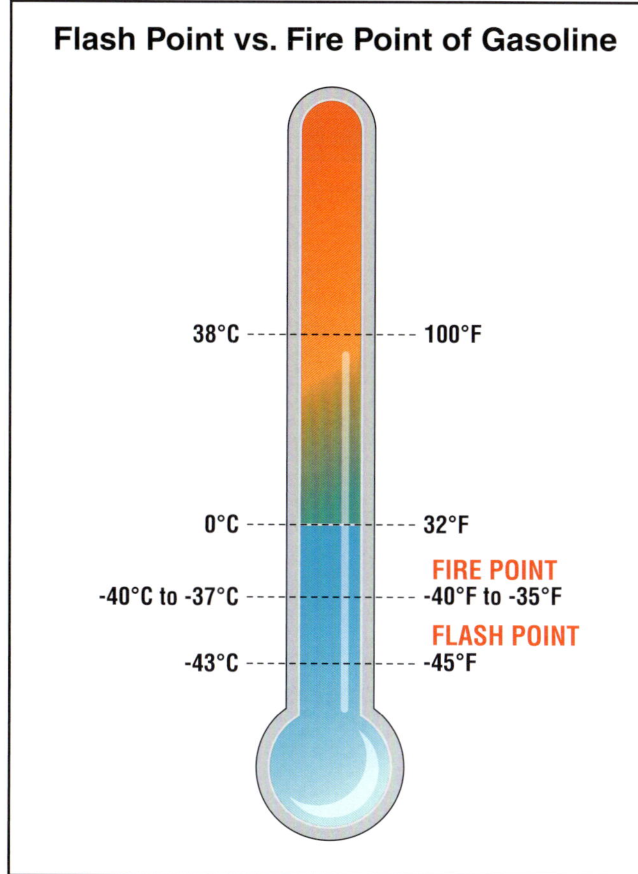

Figure 3.23 A liquid's flash point indicates the temperature at which the liquid will ignite temporarily. The fire point indicates the temperature at which the liquid, once ignited, will continue to burn.

that vaporize sufficiently to burn at temperatures under 100°F (38°C) present a significant flammability hazard.

Solids

Solids have definite size and shape. Different solids react differently when exposed to heat. Some solids such as wax and metals will change their state from a solid to a liquid, and melt, while others such as wood will not. Some plastics will melt, or sublimate, or char, depending on their chemical composition.

Process of Pyrolysis

As previously described, when solid fuels are heated, they begin to pyrolize (off-gas). The solid fuels begin to decompose and emit combustible vapors. If there is enough fuel and heat, the process of pyrolysis generates sufficient flammable vapors to ignite in the presence of sufficient oxygen or another oxidizer.

When wood first heats, it begins to pyrolize and decompose into its volatile components and carbon. Pyrolysis of wood begins at temperatures below 400°F (204°C), which is lower than the temperature required for ignition of the released vapors. The pyrolysis process of materials such as wood and polyurethane foam have significant differences **(Table 3.6, p. 74)**.

Surface-to-Mass Ratio

Solid fuels have a definite shape and size which significantly affects how easily they ignite. The primary consideration is the surface area of the fuel in proportion to its mass, called the **surface-to-mass ratio**. Consider a large tree:

1. To produce lumber, the tree must be felled and cut into a log. The surface area of this log is low compared to its mass; therefore, the surface-to-mass ratio is low.

2. The log is then cut into planks. This reduces the mass of the individual planks compared to the log. The resulting surface area increases, thus increasing the surface-to-mass ratio.

3. The chips and sawdust produced as the planks are cut into boards have an even higher surface-to-mass ratio.

4. If the boards are milled or sanded, the shavings or sawdust have the highest surface-to-mass ratio of any of the examples.

As this ratio increases, the fuel particles become more finely divided like shavings or sawdust. Therefore, the particles' ability to ignite increases tremendously. As the surface area increases, more of the material is exposed to the heat and generates combustible pyrolysis products more quickly **(Figure 3.24, p. 75)**.

The proximity and orientation of a solid fuel relative to the source of heat also affects the way the fuel burns **(Figure 3.25, p. 75)**. For example, if you ignite one corner of a sheet of ⅛-inch (3 mm) plywood

Wood	Polyurethane Foam (PUF)
Stage 1	**Stage 1**
Temperature: Less than 392° F (200° C)	**Temperature:** Less than 392° F (200° C)
Physical and Chemical Changes: Moisture is released as the wood begins to dry; combustible and noncombustible materials are released to the atmosphere although there is insufficient heat to ignite them.	**Physical and Chemical Changes:** As flexible polyurethane foam (PUF) thermally degrades (pyrolyzes), it transforms into combustible gases and liquid.
Stage 2	**Stage 2**
Temperature: 392° F – 536° F (200° C) – (280° C)	**Temperature:** 392° F – 536° F (200° C) – (280° C)
Physical and Chemical Changes: The majority of the moisture has been released; charring has begun; the primary compound being released is carbon monoxide (CO); ignition has yet to occur.	**Physical and Chemical Changes:** As the liquid polyols continue to be heated, they will vaporize into combustible gases, as well. Ignition of these gases may occur in this stage.
Stage 3	**Stage 3**
Temperature: 536° F – 932° F (280° C) – (500° C)	**Temperature:** 536° F – 932° F (280° C) – (500° C)
Physical and Chemical Changes: Rapid pyrolysis takes place; combustible compounds are released and ignition can occur; charcoal is formed by the burning process.	**Physical and Chemical Changes:** Pyrolysis continues at an increased rate. Ignition of PUF occurs 698° F (370° C). Auto-ignition of PUF can occur at temperatures in the range of 797° F to 833° F (425° to 445° C). No char layer is formed.
Stage 4	
Temperature: Greater than 932° F (500° C)	
Physical and Chemical Changes: Free burning exists as the wood material is converted to flammable gases.	

Sources: Wood data adapted from NFPA *Fire Protection Handbook*®, 19ᵗʰ edition, Volume II, pages 8-35 and 36. Polyurethane foam data from UL-FSRI, ASTM 1929 test (NIST NCSTAR 2); SFPE *Handbook of Fire Protection Engineering*, 4ᵗʰ edition; and *The Ignition Handbook* (V. Babrauskas).

paneling that is lying horizontally (flat), the fire will consume the fuel at a relatively slow rate. The same type of paneling in a vertical position (standing on edge) burns much more rapidly because the heated vapors rise over more surface area and transfer more heat to the paneling.

Oxygen

NFPA 1033 (2022): 4.1.7, 4.2.2, 4.2.3, 4.2.4, 4.2.5

Oxygen in the air is the primary oxidizing agent in most fires. Normally, air consists of about 21 percent oxygen. The energy release in fire is directly proportional to the amount of oxygen available for combustion. When a fire ignites in an open area where air is plentiful, the fire will release energy based on the given surface area of the fuel. In contrast, when a fire ignites within a compartment with limited air supply

Figure 3.24 As the surface-to-mass ratio of a fuel increases, the energy required for ignition reduces.

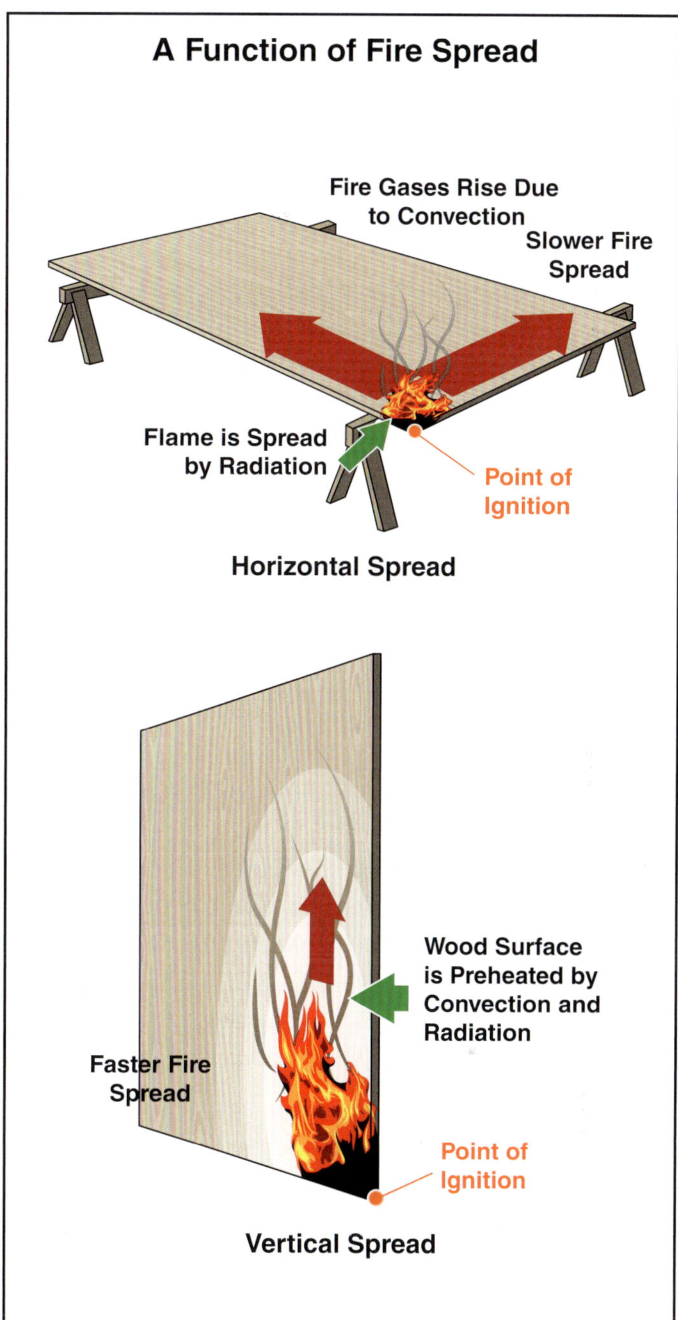

Figure 3.25 This illustration demonstrates how the position (orientation) of a fuel impacts fire spread.

the fire can only react with oxygen from the compartment's air and any additional oxygen supplied through openings. Thus, in most compartment fires, the energy released is proportional to the limited amount of oxygen available, not the amount of fuel available to burn.

Oxygen Concentration

At normal ambient temperatures (68°F [20°C]), materials can ignite and burn at oxygen concentrations as low as 15 percent. When oxygen concentration is limited, the flaming combustion will diminish, causing combustion to continue in the nonflaming mode. Nonflaming or smoldering combustion can continue at extremely low oxygen concentrations even when the surrounding environment's temperature is relatively low. However, at high ambient temperatures, flaming combustion may continue at considerably lower oxygen concentrations.

Vapor-to-Air Ratio

For combustion to occur after a fuel converts into a gaseous state, the fuel must be mixed with air (an oxidizer) in the proper ratio (fuel-to-air ratio). The range of concentrations of the fuel vapor and air is called the **flammable (explosive) range (Table 3.7)**. The fuel's flammable range is reported using the percent by volume of gas or vapor in air for the **lower explosive (flammable) limit (LEL)** and for the **upper explosive (flammable) limit (UEL)**. Atmospheres within the flammable range are particularly dangerous **(Figure 3.26)**.

Table 3.7 Flammable Ranges of Common Flammable Gases and Liquids (Vapor)	
Substance	**Flammable Range**
Methane	5%–15%
Propane	2.1%–9.5%
Carbon Monoxide	12%–75%
Gasoline	1.4%–7.4%
Diesel	1.3%–6%
Ethanol	3.3%–19%
Methanol	6%–35.5%

Source: *Computer Aided Management of Emergency Operations* (CAMEO)

The LEL is the minimum concentration of fuel vapor and air that supports combustion. Concentrations below the LEL are said to be *too lean* to burn.

The UEL is the concentration above which combustion cannot take place. Concentrations above the UEL are said to be *too rich* to burn.

Chemical handbooks and documents such as the National Fire Protection Association (NFPA) *Fire Protection Guide to Hazardous Materials* present the flammable limits for combustible gases. *The Guide* and other sources normally reports the limits at standard temperature and atmospheric pressures. Variations in temperature and pressure can cause the flammable range to vary considerably.

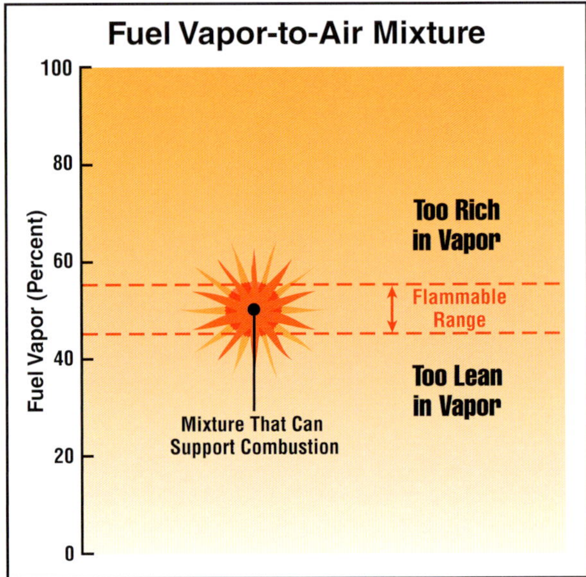

Figure 3.26 The flammable range is a relatively narrow band of conditions at which a mixture of fuel vapors and air will burn.

Self-Sustained Chemical Reaction

NFPA 1033 (2022): 4.1.7, 4.2.2, 4.2.3, 4.2.4, 4.2.5

The self-sustained chemical reaction involved in flaming combustion is complex. As flaming combustion occurs, the molecules of a fuel gas and oxygen (O_2) break apart to form **free radicals** (electrically charged, highly reactive parts of molecules). Free radicals combine with oxygen or with the elements released from the fuel gas to form new substances (molecules) and even more free radicals. The process also increases the speed of the oxidation reaction.

The combustion of a simple fuel such as methane and oxygen provides a good example. Complete oxidation of methane releases the elements needed to create carbon dioxide and water as well as release energy in the form of heat and light. The elements released when methane molecules break down (carbon and hydrogen) recombine with oxygen in the air to form CO_2 and H_2O (carbon dioxide and water) **(Figure 3.27)**.

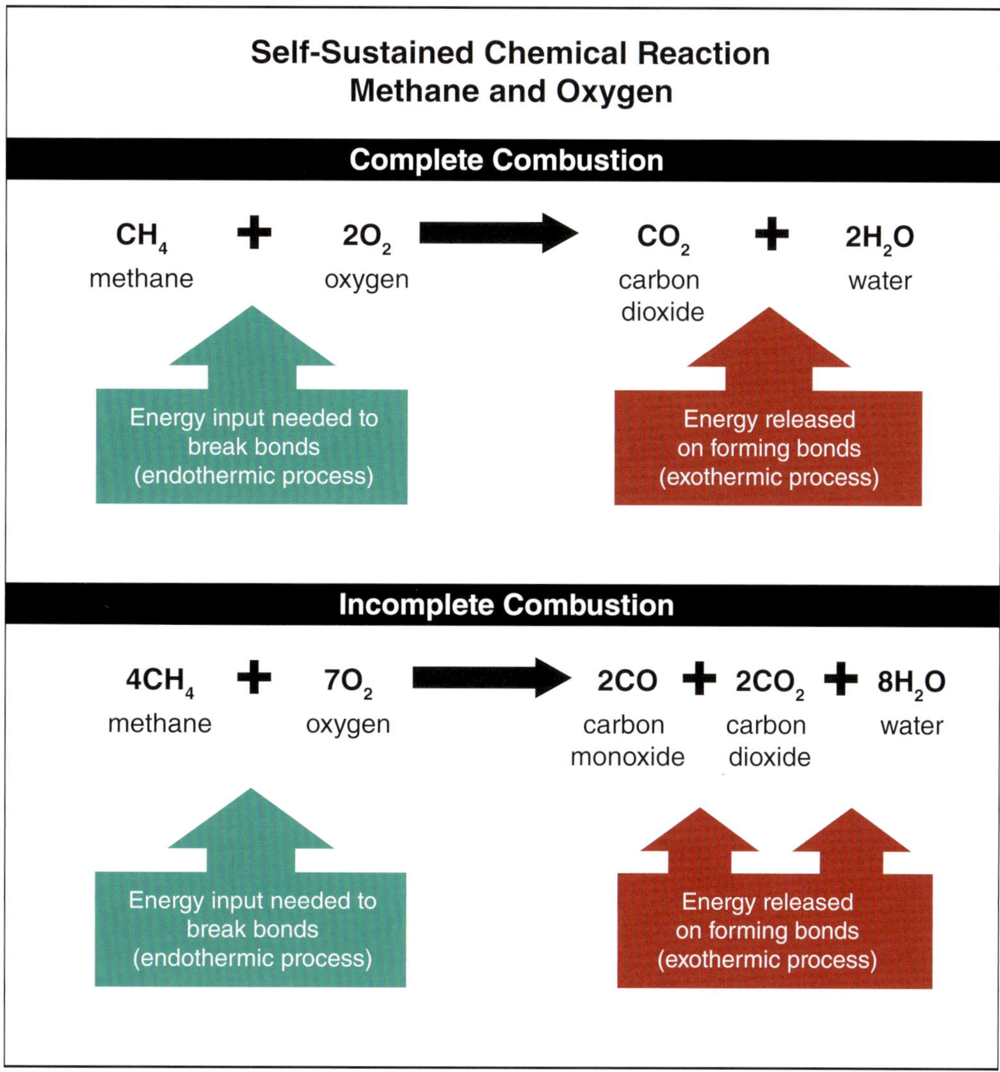

Figure 3.27 This illustration depicts the concepts of complete and incomplete combustion of methane.

At various points in the combustion of methane, this process results in production of carbon monoxide and formaldehyde, which are both flammable and toxic. When more chemically complex fuels burn, their combustion creates different types of free radicals and intermediate combustion products, many of which are also flammable and toxic.

Flaming combustion is one example of a chemical chain reaction. Sufficient heat will cause fuel and oxygen to form free radicals and initiate the self-sustained chemical reaction. The fire will continue to burn until it consumes the fuel or oxygen or an extinguishing agent, applied in sufficient quantity, interferes with the ongoing reaction. **Chemical flame inhibition** occurs when an extinguishing agent, such as dry chemical or Halon-replacement agent, interferes with this chemical reaction, forms a stable product, and terminates the combustion reaction.

Compartment Fire Development

NFPA 1033 (2022): 4.1.7, 4.2.2, 4.2.3, 4.2.4, 4.2.5

A fire is often defined by more than the burning fuel itself. The compartment surrounding that burning fuel has a significant impact on the available ventilation, access to additional fuel, and heat losses or gains.

Compartment fire development depends upon whether the fire is **fuel-limited** (fuel-controlled) or **ventilation-limited** (ventilation-controlled). When sufficient oxygen is available for flaming combustion, the fire is said to be fuel-limited. Under fuel-limited conditions, the fuel's characteristics such as heat release rate and configuration control fire development. As long as the fire can reach more ignitable fuel, it will continue to burn.

Conversely, ventilation-limited fires have access to all of the fuel needed to maintain combustion. However, the fire does not have access to enough oxygen to continue to burn or to spread to all available fuels.

All compartment fires begin in the incipient stage as fuel-limited fires. Once the fire reaches the growth stage, the fire will either remain fuel-limited if there is enough oxygen to support continued growth, or the fire will consume all available oxygen and become ventilation-limited. A fuel-limited fire will usually progress through the stages of fire development in order. Ventilation-limited fires tend to enter an early state of decay at the end of the growth stage because there is no longer enough available oxygen for the fire to become fully developed.

The remainder of this section will define the stages of fire development and then describe the progression of a fire in a compartment. The examples in the information boxes describe fire behavior in a room with one exterior window, an exterior doorway, and typical modern furnishings found in a residential living room.

Stages of Fire Development

Fires develop through four stages: incipient, growth, fully developed, and decay. These stages can occur with any fire; however, three key factors control how the fire develops:

- Fuel properties
- Heat conservation
- Ventilation available

Open burning or a *free burn* condition provides the most basic fire growth curve

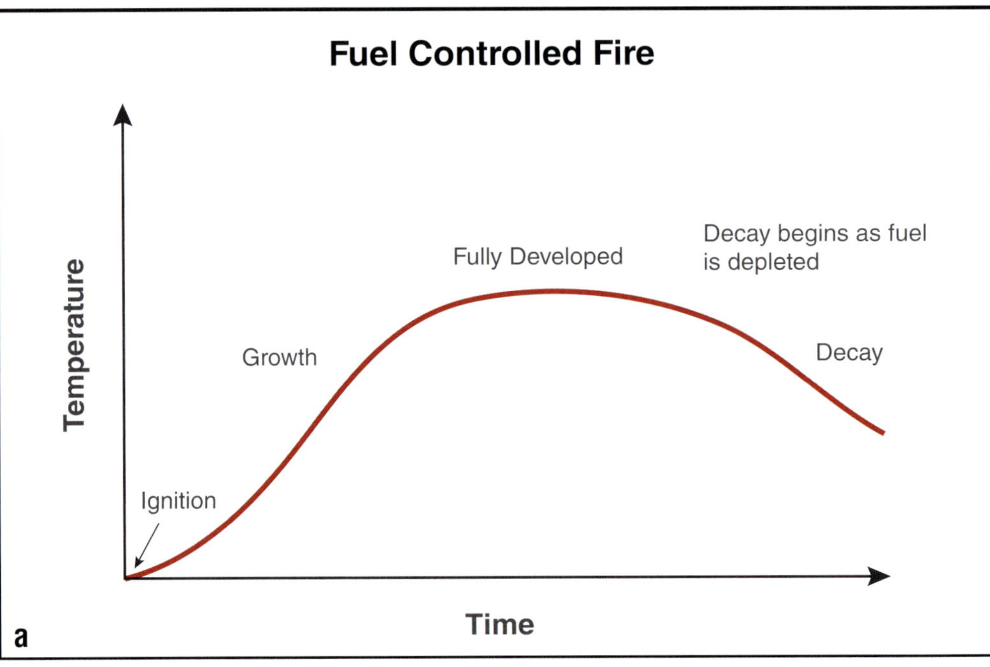

Figure 3.28a This line graph shows the progression of a fuel-controlled fire. *Courtesy of Dan Madrzykowski, NIST.*

(Figure 3.28 a and b). The fuel properties of open burning are representative of a fuel-limited fire, such as a campfire, a pile of wood pallets, or a sofa in a large, open, empty warehouse. The ventilation available for a single item/area burning either outside or in a large, well-ventilated space means there is sufficient oxygen available to burn the fuel until it can no longer sustain combustion. As heat and fire gases are produced, they move away from the fuel and disperse throughout the environment remote from the burning fuel.

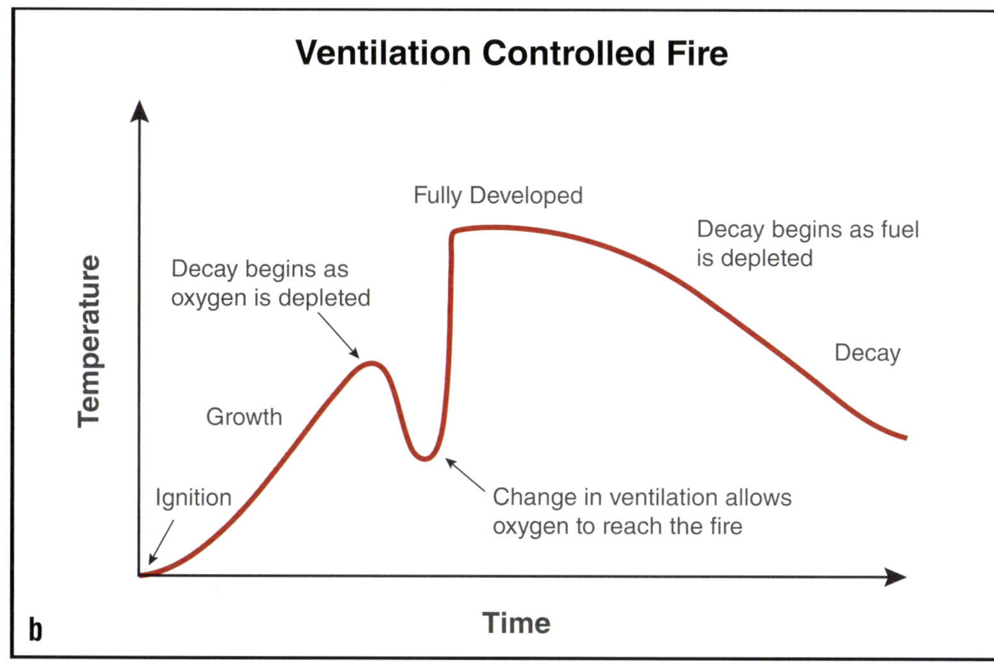

Figure 3.28b This line graph shows the progression of a ventilation-controlled fire. *Courtesy of Dan Madrzykowski, NIST.*

Heat is not conserved. The only limit or control on the heat release rate of a fire burning out in the open is the fuel itself.

The four fire development stages described in the following sections are often defined in terms of a confined space or compartment. Each stage may exhibit different characteristics or occur in a different sequence.

Incipient Stage

The incipient stage is the beginning of the fire (**Figure 3.29**). Once ignition occurs and the combustion process begins, development in the incipient stage depends largely upon the characteristics and configuration of the fuel involved (fuel-limited fire). Air in the compartment provides adequate oxygen to continue fire development. When a compartment fire enters the incipient stage:

- Hot gases in contact with the surfaces of the compartment and its contents transfer heat to other materials.

- The hot gases in the plume rise until they encounter the ceiling and then begin to spread horizontally. This flow of fire gases is called the **ceiling jet**.

Figure 3.29 An incipient fire on a couch will have sufficient fuel to grow. *Courtesy of Dan Madrzykowski, NIST.*

- Radiant heat warms the adjacent fuel and continues the process of pyrolysis. A thin plume of hot gases and flame rises from the fire and mixes with the cooler air in the compartment.

In this early stage of fire development, the fire has not yet influenced the environment within the compartment to a significant extent. The temperature, while increasing, is only slightly above ambient in areas that the fire, plume, and ceiling jet directly affect. During the incipient stage, occupants can safely escape from the compartment, and a portable extinguisher or small hoseline can safely extinguish the fire.

The transition from incipient to growth stage can occur quickly (in some cases in seconds), depending on the type and configuration of fuel involved. A visual indicator that a fire is leaving the incipient stage is flame height. When flames reach 2.5 feet (750 mm) high, radiated heat begins to transfer more heat than convection. The fire will then enter the growth stage.

Example Compartment: Incipient Stage

To begin our example of a compartment fire's development, let's assume that a fire started in the cushions of a chair in the corner of a room. The window and the door are closed. Within the room, there is enough oxygen for the incipient fire to entrain ("pump in") air and create ("pump out") fuel gases and smoke. The fire begins to spread from the cushions to the rest of the chair. The polyurethane foam of the cushions burns quickly, creating black, fuel-rich smoke which begins to form a plume above the chair **(Figure 3.30)**.

Growth Stage

Within the growth stage, a variety of fire behaviors can occur, depending upon the number of ventilation sources. The fire may consume all of its available oxygen and enter a ventilation-limited state of decay or ventilation may provide enough oxygen for rapid fire development and/or growth to full development. Rapid fire development usually occurs during the growth stage. Understanding fire dynamics is largely an understanding of everything that can happen during the growth stage.

NOTE: Keep in mind that if the fire enters ventilation-limited decay, that does not indicate that the fire is in its final stage of development.

Figure 3.30 The graph in the lower left of the photo shows the progression of the fire in its incipient stage. *Courtesy of Dan Madrzykowski, NIST.*

As the fire transitions from incipient to growth stage, it begins to influence more of the compartment's environment and has grown large enough for the compartment configuration and amount of ventilation to influence it. The first effect is the amount of air that is entrained into the fire.

Entrainment

Unconfined fires draw air from all sides and the entrainment (drawing in) of air cools the plume of hot gases, reducing flame length and vertical extension **(Figure 3.31)**. In a compartment fire, the location of the fuel package in relation to the compartment walls affects the amount of air that is entrained and thus the amount of cooling that takes place. The following tenets describe entrainment based on the positioning of fuel packages:

- Fires in fuel packages in corners can only entrain air from two sides.
- Fires in fuel packages near walls can only entrain air from three sides.
- Fires in fuel packages in the middle of the room can entrain air from all sides.

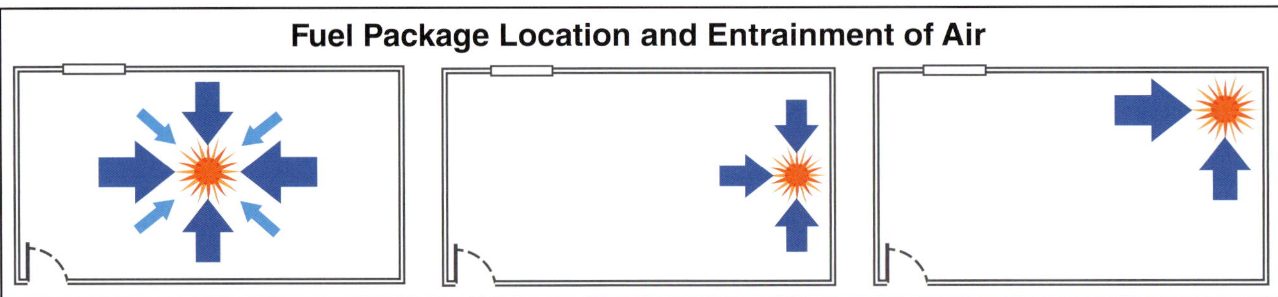

Figure 3.31 A fire's location in a compartment influences the entrainment of air into the fire.

Therefore, when the fuel package is not in the middle of the room, the **combustion zone** (the area where sufficient air is available to feed the fire) expands vertically and a higher plume results. A higher plume increases the temperatures in the developing hot gas layer at ceiling level and increases the speed of fire development. In addition, heated surfaces around the fire radiate heat back toward the burning fuel which further increases the speed of fire development.

Transition to Full Development

A fire is said to be in the growth stage until the fire's heat release rate has reached its peak, either because of a lack of fuel or a lack of oxygen. In other words, when a fire cannot grow without the introduction of a new fuel source or a new oxygen source, it has left the growth stage and become fully developed. Two common routes to full development are as follows:

- Fires that consume all available oxygen and transition to a state of ventilation-limited decay.
- Fires that have enough oxygen and move to the growth phase and possibly through rapid fire development.

Example Compartment: Growth Stage – Transition from Incipient Stage

As the fire spreads to the chair, the entire chair becomes involved. Air entrains from only two sides, so the fire releases more energy than it would in the middle of the room and thus accelerates the fire's growth. The side table next to the chair begins to heat and pyrolize. The ceiling above the chair begins to blacken as the plume grows taller and transfers heat to the ceiling and surrounding walls. A ceiling jet begins to form **(Figure 3.32)**.

Development of the Hot Gas Layer

The heat, smoke, and fire gases rise in the plume to the ceiling. As the plume strikes the ceiling the smoke and gases move away in all directions from the centerline of the plume until stopped by a vertical surface, such as walls. As more smoke rises in the plume and the layer of hot gases and smoke begins to thicken, the layer moves downward from the ceiling.

As the ceiling layer thickens and become deeper, it also gets hotter. Items now in the hot gas layer, such as bookshelves or curtains attached to curtain rods, will begin to **pyrolyze**. This adds smoke and fire gases to the hot gas layer, and these materials will ignite if and when they reach their ignition temperature. As additional items pyrolyze, the pyrolysis products of smoke, heat, and fire gases will add to the development of the hot gas layer, thus speeding up the fire development process in the compartment. This process can lead to full involvement.

Figure 3.32 The graph in this image shows the transition of a fire from the incipient stage into the growth stage. *Courtesy of Dan Madrzykowski, NIST.*

Development of the Hot Gas Layer. Heat from the hot gas layer will radiate in all directions, including downward and can begin to heat items below the hot gas layer. Exposed to radiant heat, those items will begin to pyrolyze, adding to the development of the hot gas layer, and ignite if and when they reach their ignition temperature. This process will continue, provided there is sufficient oxygen. As the hot gas layer continues to thicken and move downward from the ceiling, more fuels are pyrolyzed and ignited both in and below the hot gas layer. Depending on the ventilation of hot gases from the compartment, the hot gas layer can get very close to the floor.

Isolated or *intermittent* **flames** may move through the hot gas layer. Combustion of these hot gases indicates that portions of the hot gas layer are within their flammable range, and that there is sufficient heat to cause ignition. As these hot gases circulate to the outer edges of the plume or the lower edges of the hot gas layer, they find sufficient oxygen to ignite. This phenomenon frequently occurs before more substantial involvement of flammable products of combustion in the hot gas layer. The appearance of isolated flames is sometimes an immediate indicator of **flashover**.

Flow Path and Thermal Layering. In addition to the effects of heat transfer through radiation and convection described earlier, radiation from the hot gas layer also acts to heat the interior surfaces of the compartment and its contents. Changes in ventilation and flow path can significantly alter thermal layering. The flow path is defined as the space between the air intake and the exhaust outlet. Multiple openings (intakes and exhausts) create multiple flow paths.

The products of combustion from the fire begin to affect the environment within the compartment. As the fire continues to grow, the hot gas layer within the fire compartment gains mass and energy. As the mass and energy of the hot gas layer increases, so does the pressure. Higher pressure causes the hot gas layer to spread downward within the compartment and laterally through any openings such as doors or windows. If there are no openings for lateral movement, the higher pressure gases have no lateral path to follow to an area of lower pressure. As a result, the hot gases will begin to fill the compartment starting at the ceiling and filling down.

The interface between the hot gas layers and cooler layer of air is commonly referred to as the **neutral plane** because the net pressure is zero, or neutral, where the layers meet. The neutral plane exists at openings where hot gases exit and cooler air enters the compartment. At these openings, hot gases at higher than ambient pressure exit through the top of the opening above the neutral plane. Lower pressure air from outside the compartment entrains into the opening below the neutral plane **(Figure 3.33)**.

Figure 3.33 This illustrates the location of a compartment fire's neutral plane. Hot gases are exiting through the upper part of the doorway while cooler air enters through the lower part of the doorway. *Courtesy of Dan Madrzykowski, NIST.*

Example Compartment: Growth Stage

The plume coming from the chair has now reached the ceiling and become a ceiling jet. Hot fire gases and fuel-rich smoke begin to spread horizontally across the ceiling. Both the door and the window are closed, so the smoke has no way to leave the compartment. The compartment begins to fill with hot gases and smoke. The dividing line between the dwindling air in the compartment and the increasing amount of smoke in the compartment steadily lowers toward the floor. The end table is completely involved with flames now. The walls and ceiling have also heated and are radiating heat back into the room. The coffee table has begun to pyrolize and the flat-screen television has begun to melt. The hydrocarbon materials in the compartment burn quickly and inefficiently, creating fuel-rich, black smoke **(Figure 3.34)**.

Transition to Ventilation-Limited Decay

Most residential fires that develop beyond the incipient stage become ventilation-limited (ventilation controlled). Even when doors and windows are open, insufficient air entrainment may prohibit the fire from developing based on the available fuel. When windows are intact and doors are closed, the fire may move into a ventilation-limited state of decay even more quickly. While a closed compartment reduces the heat release rate, fuel may continue to pyrolize, creating fuel-rich smoke.

As the interface height of the hot gas layer descends toward the floor, the greater volume of smoke begins to interrupt the entrainment of fresh air and oxygen to the seat of the fire and into the plume. This interruption causes the fire within the compartment to burn less efficiently. As the efficiency of combustion decreases (incomplete combustion), the heat release rate decreases and the amount of unburned fuel within the hot gas layer increases.

Figure 3.34 This image shows a compartment fire's growth stage as thermal layering begins to occur within the compartment. *Courtesy of Dan Madrzykowski, NIST.*

The fire is now in a state of ventilation-limited decay because:

- There is not enough oxygen to maintain combustion.

- The heat release rate has decreased to the point that fuel gases will not ignite.

During ventilation-limited decay, the environment stays untenable despite a decreased heat release rate because:

- The temperature in the room may remain high.

- Oxygen is present in levels too low to maintain combustion.

- The compartment fills with fuel-rich gases that may ignite with the addition of oxygen.

Even if temperatures decrease, pyrolysis can continue. Under these conditions, a large volume of flammable products of combustion can accumulate within the compartment. These gases are fuel that can ignite if they are provided a new source of oxygen.

If no other source of oxygen exists, the compartment will fill with black smoke and slowly cooling fuel gases. The compartment will show no visible flames. The characteristics of the fuel and fuel load in today's typical fires will cause fires to quickly become ventilation-limited.

In order for a ventilation-limited fire to grow, it needs a new supply of oxygen. Ventilation introduces outside air to the fire as this new source of oxygen. If windows or doors fail, the sudden introduction of fresh air creates a rapid increase in the heat release rate and growth of the fire. This rapid increase can also occur when firefighters open a door or window to enter the compartment for extinguishment, which creates a • new flow path **(Figure 3.35)**.

Figure 3.35 By breaking the window, the firefighter introduces new oxygen into the compartment, increasing the fire's heat release rate and growth. *Courtesy of Dan Madrzykowski, NIST.*

The pressure outside the compartment is lower than the pressure inside the compartment **(Figure 3.36)**. Because of these pressure differences, any ventilation to the outside – opening an interior or exterior door, or breaking or opening a window – provides a flow path along which the hot gases can now move from the high pressure area inside to the low pressure area outside.

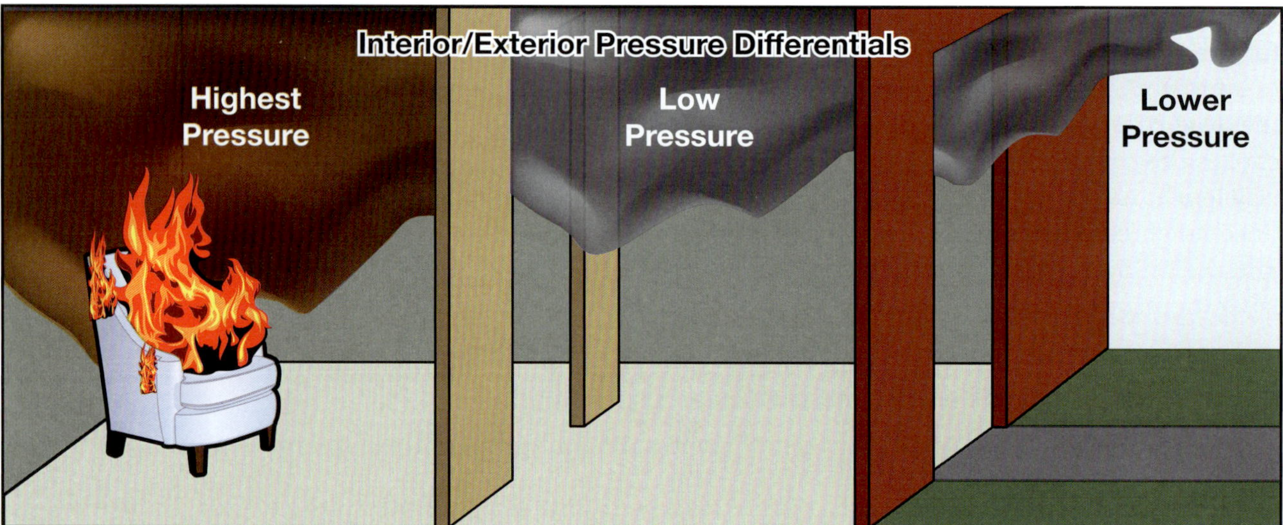

Figure 3.36 The pressure differences between the interior of a compartment (higher) and the exterior (lower) provide a flow path that hot gases may move along.

Example Compartment: Growth Stage – Transition to Ventilation-Limited Decay

The window in the compartment has not failed and the door is still closed. The hot gas layer in the room has lowered to about 2 feet (600 mm) off the floor. The small amount of oxygen left below the gas layer is no longer sufficient to sustain flaming combustion **(Figure 3.37)**. No flames are visible, but the remainder of the furnishings in the room slowly continue to pyrolize and add fuel gases to the compartment. The walls and ceiling still radiate heat. Though the heat release rate is low, use of a thermal imager from outside the compartment shows temperatures hot enough to ignite flammable gases. From outside the window, only smoke is visible, and there are pulses of smoke in the cracks around the door.

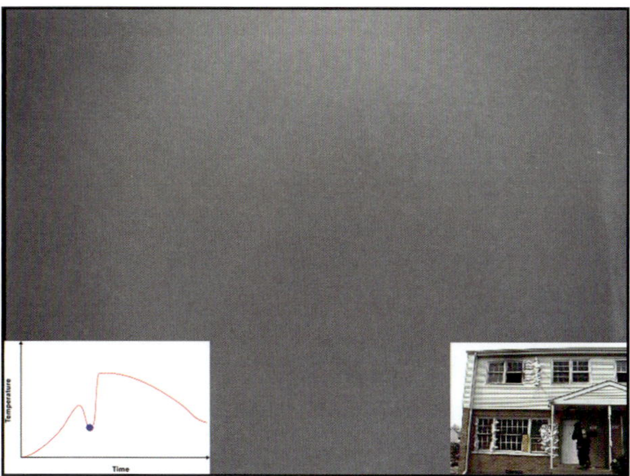

Figure 3.37 This image shows a fire's transition to ventilation-limited decay. *Courtesy of Dan Madrzykowski, NIST.*

Rapid Fire Development

Rapid fire development refers to the rapid transition from the growth stage or early decay stage to a ventilation-limited, fully developed stage **(Figure 3.38)**. Among these events are flashover and **backdraft**. In this section, rapid fire development conditions are described along with their indicators.

NOTE: Smoke explosions are also incidents of rapid fire development, but they involve more than just one compartment of a structure. Smoke explosions will be described later in this chapter.

Flashover. Rapid transition from the growth stage to the fully developed stage is known as flashover. When flashover occurs, the combustible materials and fuel gases in the compartment ignite almost simultaneously; the result is full-room fire involvement. Flashover typically occurs during the fire's growth stage, but may occur during the fully developed stage as the result of a change in ventilation.

Flashover conditions are defined in various ways; however, during flashover, the environment of the room changes from a two-layer condition (hot on top, cooler on the bottom) to a single, well mixed hot gas condition from floor to ceiling. The environment is untenable. As flashover occurs, the gas temperatures in the room reach 1,100 °F (600°C) or higher.

A significant indicator of flashover is **rollover**. Rollover describes a condition where the unburned fire gases that have accumulated at the top of a compartment ignite and flames propagate through the hot gas layer or across the ceiling.

Figure 3.38 This series of images shows a fire's evolution from the growth stage through decay to rapid fire development that occurs once a door is opened. *Courtesy of Dan Madrzykowski, NIST.*

Rollover may occur during the growth stage as the hot gas layer forms at the ceiling of the compartment. Flames may appear in the layer when the combustible gases reach their ignition temperature. While the flames add to the total heat generated in the compartment, this condition is not flashover. Rollover will generally precede flashover, but it may not always result in flashover. Rollover contributes to flashover conditions because the burning gases at the upper levels of the room generate tremendous amounts of radiant heat which transfers to other fuels in the room. The new fuels begin pyrolysis and release the additional gases necessary for flashover.

The transition period between pre-flashover fire conditions (growth stage/ventilation-limited decay) to post-flashover (fully developed stage) can occur rapidly. Radiation from the compartment's upper layer heats the compartment's contents until they reach their ignition temperature simultaneously. When the upper layer ignites, the amount of radiation increases to levels which rapidly ignite contents in the room, even if they are remote from the fire. During flashover, the volume of burning gases can increase from approximately ¼ to ½ of the room's upper volume to fill the room's entire volume and extend out of any openings from the room. When flashover occurs, burning gases push out of compartment openings (such as a door to another room) at a substantial velocity.

Four common elements of flashover are:

- **Transition in fire development** — Flashover represents a transition from the growth stage to the fully developed stage.

- **Rapidity** — Although it is not an instantaneous event, flashover happens rapidly, often in a matter of seconds, to spread fire completely throughout the compartment.

- **Compartment** — There must be sufficient enclosure to generate a hot gas layer sufficient for rollover conditions.

- **Pyrolysis of all exposed fuel surfaces** — Fire gases from all of the combustible surfaces in the enclosed space ignite, provided that there is sufficient oxygen to support flaming combustion.

Two interrelated factors determine whether a fire within a compartment will progress to flashover. First, there must be sufficient fuel and the heat release rate must be sufficient for flashover conditions to develop. For example, ignition of discarded paper in a small metal wastebasket may not have sufficient heat to develop flashover conditions in a large room lined with gypsum drywall. On the other hand, ignition of a sofa with polyurethane foam cushions placed in the same room will likely result in flashover provided the fire has sufficient oxygen.

The second factor is ventilation. Regardless of the type, quantity, or configuration of fuel, heat release depends on oxygen. A developing fire must have sufficient oxygen to reach flashover, an amount that a sealed room may not provide. The available air supply limits the heat release. If there is insufficient natural ventilation, the fire may enter the growth stage but not reach the heat release rate or gaseous fuel production to transition through flashover to a fully involved fire.

NOTE: The autoignition temperature of CO, the most abundant fuel gas created in most fires, is approximately 1,100° F (600°C).

When a fire is in ventilation-limited decay, the introduction of new oxygen can trigger flashover quickly. Flashover may occur whenever sufficient oxygen and ventilation are available for fire growth. However, in an uncontrolled situation, it may be difficult to identify what stage a fire is in.

Flashover may not occur in every compartment fire, such as in large-area compartments or compartments with high ceilings. Fire development may take an alternative path in a compartment that quickly becomes ventilation-limited, before the thermal energy can build within the compartment. The fire may not progress to flashover but instead become ventilation-limited, limiting heat release rate and causing the fire to enter the decay stage while continuing the process of pyrolysis and increasing the fuel content of the smoke.

Example Compartment: Growth Stage – Flashover

Let's back up a moment and change the conditions of our example. When the hot gas layer reaches about halfway down the compartment, a firefighter opens the door to ventilate the room. When the door was closed, the fire was on its way to ventilation-limited decay. With the introduction of fresh air from outside, the fire grows rapidly **(Figure 3.39)**. Firefighters can observe the neutral plane in the doorway with smoke exiting the top half of the door. The amount of smoke increases, and the neutral plane lowers in the doorway as more of the surfaces in the room pyrolize. Firefighters observe flames moving through the top of the hot gas layer. These flames radiate a large amount of energy to the compartment contents, causing the contents to pyrolize and release more fuel while rapidly heating to their ignition temperature. Suddenly, the hot gases and incoming oxygen reach the correct mix, and the isolated flames become a room full of flames. All of the hot gases ignite at once. The heat release rate rises dramatically, igniting all of the flammable fuels within the compartment. Flames extend up and around the door frame of the open door.

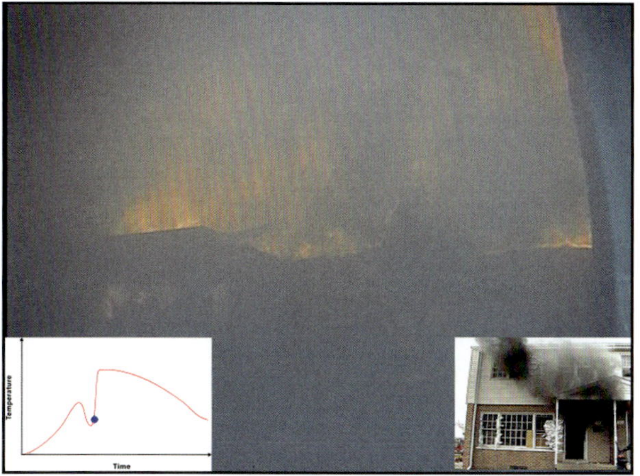

Figure 3.39 In this image, the fire grows rapidly as oxygen is introduced into the compartment. *Courtesy of Dan Madrzykowski, NIST.*

Backdraft. A ventilation-limited compartment fire can produce a large volume of flammable smoke and other gases due to incomplete combustion. While the heat release rate from a ventilation-limited fire decreases, elevated temperatures may still be present within the compartment. Backdraft occurs in a space containing a high concentration of heated flammable gases that lack sufficient oxygen for flaming combustion.

When potential backdraft conditions exist in a compartment, the introduction of a new source of oxygen will return the fire to a fully involved state rapidly (often explosively). A backdraft can occur with the creation of a horizontal or vertical opening. All that is required is the mixing of hot, fuel-rich smoke with air. Backdraft conditions can develop within a room, a void space, or an entire building.

Example Compartment: Growth Stage – Backdraft

Let's again change the conditions in our compartment. The compartment is ventilation-limited, and all of the petroleum products in the room (chair cushions, plastic in the flat-screen TV) have produced large amounts of fuel-rich, black smoke. The bottom of the hot gas layer is barely off the floor. There is nothing visible but dense, black smoke through the window. At this moment, the window fails. The high volume of confined smoke rushes to the new area of low pressure and billows out the window. Air from the outside rushes in at the same speed. The hot gases and air mix so quickly that they reach their explosive limit almost immediately. Flames propagate through the hot gases seemingly all at once, and an explosion of fire erupts out of the open window.

Smoke Explosions. A **smoke explosion** occurs when a mixture of unburned fuel gases and oxygen comes in contact with an ignition source. When smoke travels away from the fire it can accumulate in other areas and mix with air. When the fuel and oxygen are within the flammable range and contact an ignition source, the result will be explosive, rapid combustion. Smoke explosions are violent because they involve premixed fuel and oxygen (Fleishmann, 2013).

Fully Developed Stage

The fully developed stage occurs when the heat release rate of the fire has reached its peak, because of lack of either fuel or oxygen. The two main types of fully developed fires are ventilation-limited and fuel-limited fires. The factor limiting the peak heat release rate is used to identify which type of fully developed fire exists.

The term "fully developed" is often misinterpreted to mean that the fire can no longer grow. A more accurate description would be that the fire has grown *as much as it can*. New sources of fuel introduced after full development will allow fuel-limited fires to grow. Likewise, new sources of oxygen introduced after full development will allow ventilation-limited fires to grow.

Fuel-Limited Conditions

The available fuel limits the peak heat release in a fully developed, fuel-limited fire. The most effective method of increasing the heat release rate is to provide more fuel. A campfire located in a fire ring is a good example of fuel-limited conditions. The fire reaches its peak when all the fuel becomes involved. The fire ring separates the burning fuel from other potential fuel resulting in a fuel-limited, fully developed fire. Adding additional fuel or firewood would increase the energy release of the fire to a new peak heat release rate. Fuel-limited full development usually occurs when fires are not contained within compartments such as wildland fires, vehicular fires, or fires burning in collapsed structures.

Ventilation-Limited Conditions

In contrast to fuel-limited conditions, a fully developed ventilation-limited fire lacks sufficient oxygen to grow. The fire reaches a peak when it consumes all the available oxygen from the air intake, typically with incomplete combustion. Additional fuel is available and gaseous fuel is leaving the compartment in the smoke; however, the fire cannot release any more energy. Allowing additional air into the compartment via an additional opening or enlarging the existing opening will provide more oxygen, resulting in a higher peak heat release rate.

Ventilation-limited, fully developed fires present a hazardous situation. The potential for a window failure to provide fresh oxygen and increase the peak heat release rate can endanger anyone inside the structure. To reduce the risk of the unpredictable window failure, firefighters must transition the fire from ventilation-limited to fuel-limited. With the high heat of combustion found in modern furnishings, the only mechanism to transition the fire is to extinguish some of the burning fuel. It is not possible to make enough openings in a compartment to transition a fire from ventilation-limited to fuel-limited conditions.

Decay Stage

A fire is said to be in the decay stage when it runs out of either available fuel or available oxygen. Either fuel or oxygen is an integral part of the fire triangle introduced earlier. Without all three components of the triangle, the fire will decay and extinguish.

In fuel-limited fires, the decay stage is usually the fire's final stage, leading to the fire's self-extinguishment when it runs out of available fuel. Ventilation-limited fires can also self-extinguish due to lack of oxygen. Both of these situations can result in the termination of the combustion reaction. However, just like throwing another log on top of a smoldering campfire, introducing new oxygen to a ventilation-limited fire can cause it to reenter the growth stage.

Fuel-Limited Decay

After a fuel-limited fire reaches the fully developed stage the fire will decay as the fuel is consumed. As the fire consumes the available fuel and the heat release rate begins to decline, the fire enters the decay stage. The heat release rate will decrease, but the temperature of surrounding objects may remain high for some time due to absorbed heat.

Ventilation-Limited Decay

When a fire enters a ventilation-limited state of decay, this stage is not necessarily the last stage of the fire's development. As stated earlier, the fire awaits a new supply of oxygen to return to the growth stage. This statement is true even if compartment ventilation has already occurred.

To ensure that the decay stage of a ventilation-limited fire is the fire's final stage, a controlled transition from ventilation-limited to fuel-limited must take place. If the compartment has no ventilation openings, the heat release rate will eventually decrease to the point that the heat in the compartment naturally transfers through the compartment itself to the outside.

Structure Fire Development

NFPA 1033 (2022): 4.1.7, 4.2.2, 4.2.3, 4.2.4, 4.2.5

Structures are essentially composed of individual compartments connected by hallways, stairways, or openings such as doorways. If a fire starts within one of the compartments, how it grows or decays is based on the model growth curves presented earlier. However, a fire within a structure has the potential to involve more than one compartment or spread beyond the contents of a compartment and involve the structural members of the building itself. The fire investigator will need to identify the area of origin and be able to articulate how the fire moved through the structure.

Flow Path

In a structure fire, the method by which the fire receives the needed oxygen to sustain the combustion reaction occurs through one or many flow paths. The flow path is composed of two regions: the ambient air flow in and the hot exhaust flow out **(Figure 3.40)**. The flow is always unidirectional due to pressure differences where the ambient air flows toward the seat of the fire and reacts with the fuel. The products of combustion flow away from the fire toward the exhaust or low pressure outlet.

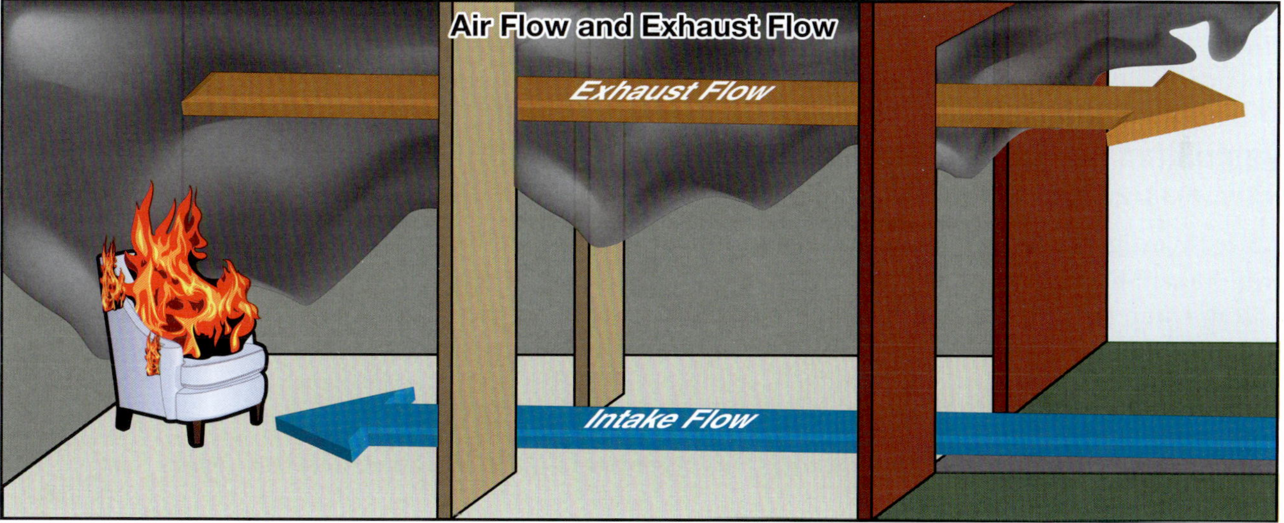

Figure 3.40 Exhaust flow moves toward an area of lower pressure. Air intake flow moves toward the seat of a fire.

In a structure fire, the floor plan and openings within the structure determine the available flow path. For example, hot gases from a fire in a bedroom will travel out of the doorway and into the hallway if the door is open. If other doors in the structure are also open, the adjoining rooms also become possible parts of the flow path. The pressure in these other rooms is lower than the pressure in the fire room; therefore, hot fire gases and smoke will travel toward those areas unless the direction of flow is altered, for example, through tactical ventilation or door control **(Figure 3.41, p. 90)**. Air in those rooms will entrain toward the fire as the structure fills with fuel gases and the fire grows and spreads.

A flow path's effectiveness to transport ambient air to the seat of the fire is based on:

- Number of obstructions
- Length of the path traveled
- Size of the ventilation opening
- Elevation differences between the base of the fire and the opening

Figure 3.41 The floor plan and openings within a structure determine the available flow path of hot gases.

Investigators must be aware that when firefighters advance a hoseline or ventilate a building, they establish new flow paths between the fire compartment and exterior vents of the building. These new paths may affect fire pattern development.

When hot gases follow the flow path from areas of high to low pressure, they convect heat to a larger portion of the structure. They also carry the products of combustion into new areas of the structure. Because these gases are also fuel, fire can propagate through them, out of the fire room.

A structure fire that extends beyond the room of origin may have two or more compartments involved, each in different stages of development. The room of origin may be in a fully developed, ventilation-limited stage while the adjacent compartment may be in the growth stage and nearing flashover. Understanding the model growth curves and what conditions to expect based on fire dynamics will aid fire investigators in finding or establishing the area of origin.

Ventilation and Wind Considerations

NFPA 1033 (2022): 4.2.3

A fire investigator should communicate with the fire suppression team to determine the actions that were taken, how they changed conditions, and the resulting patterns and effects. Ventilation alone will not stop fire growth and spread, once the fire has filled the structure's compartments with hot, unburned, gaseous fuel **(Figure 3.42)**.

Unplanned Ventilation

Unplanned ventilation occurs when a structural member fails – usually because of exposure to heat – and introduces a new source of oxygen to the fire. This new oxygen source could result from the failure of a:

- Roof
- Wall
- Window **(Figure 3.43)**
- Doorway

The source of new oxygen does not have to originate from outside the building. When floors fail above basement fires, the interior air in the structure becomes a new oxygen source.

Unplanned ventilation, by definition, is unexpected. Unplanned ventilation is often the result of:

- Occupant action
- Fire effects on the building (such as window glazing)
- Actions other than planned, systematic, and coordinated tactical ventilation

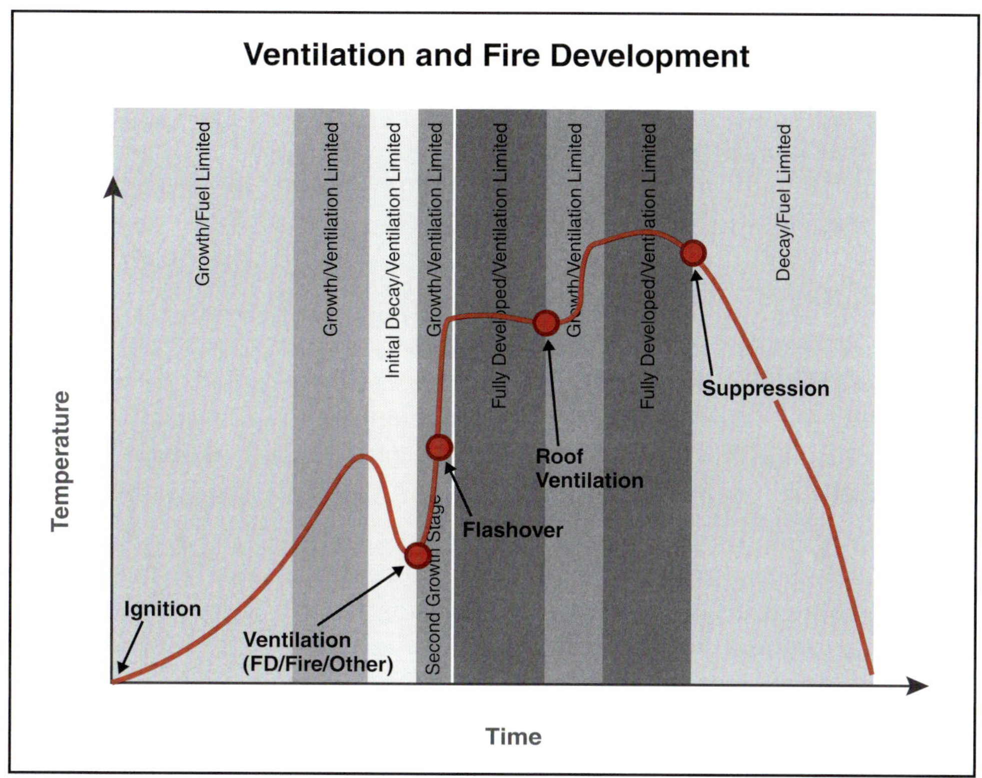

Figure 3.42 This model shows the evolution of a compartment fire. *Courtesy of UL FSRI.*

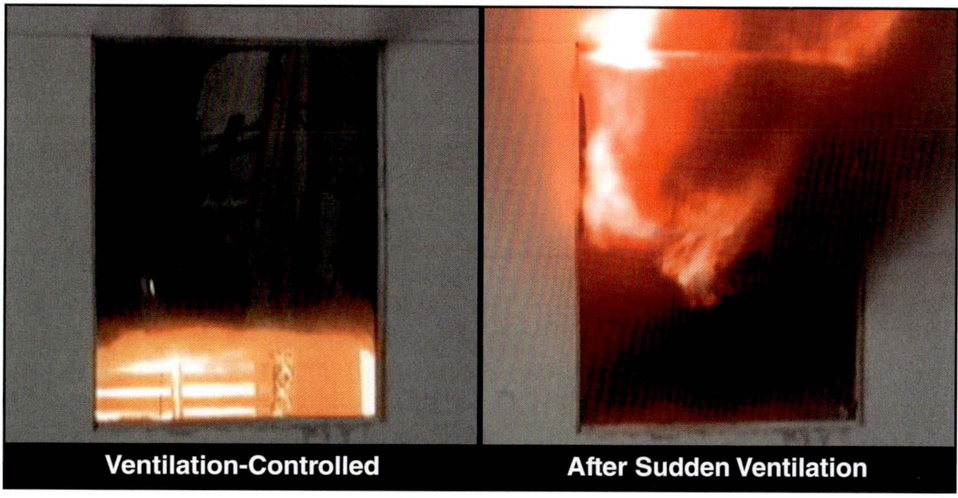

Figure 3.43 The sudden ventilation of a compartment fire can lead to rapid fire development. *Courtesy of Dan Madrzykowski, NIST.*

Wind Conditions

Wind can increase the pressure inside the structure, drive smoke and flames into unburned portions of the structure, and/or upset tactical ventilation efforts. Wind conditions can also create differences in pressure that can cause windows to fail. The exterior pressure on the upwind side of a structure will be higher than the pressure on the downwind side of the structure. As a result, the ambient air on the outside of the structure is constantly trying to move through the structure along the path from high to low pressure. If heat exposure weakens windows in this path, wind pressure could cause them to fail which introduces a new flow path for oxygen and hot fire gases.

> **WARNING:** Wind-driven conditions can occur in any type of structure. Wind speeds as low as 10 mph (16 km/h) can create wind-driven fire conditions.

Building Materials and Contents (Fuel Load)

Building materials are discussed in Chapter 5. Stated simply, the fire investigator must consider how the following elements affect fire spread within a structure:

- Exterior siding
- Interior finishes
- Types and contents
- Configurations and compartmentation

Effects of Fire Fighting Operations

Limiting or interrupting one or more of the essential elements in the combustion process depicted in the fire tetrahedron controls and extinguishes fire. Investigators should be aware that fire suppression operations can influence fire dynamics (development and spread) and resulting fire patterns in a number of ways:

- **Temperature change** — Successful suppression will reduce temperatures using water or a foam agent to cool fire gases and hot surfaces for the purposes of extinguishment. Uncoordinated or unsuccessful suppression may cause other conditions to deteriorate.

- **Fuel removal** — Eliminating sources of fuel in the path of the fire's spread that could provide a new source of fuel; typically, a tactic in wildland fires or liquid and gas fires.

- **Oxygen exclusion/flow path control** — Using door control and tactical ventilation techniques to control the amount of air available to the fire.

- **Chemical flame inhibition** — Using extinguishing agents other than water and foam, such as some dry chemicals, halogenated agents (Halons), and Halon-replacement "clean" agents, to inhibit or interrupt the combustion reaction and stop flame production.

Review Questions

1. What is the difference between a physical change and a chemical reaction?

2. What is the difference between the fire triangle and the fire tetrahedron?

3. What is the difference between piloted ignition and autoignition?

4. What is the difference between complete combustion and incomplete combustion?

5. What is the difference between heat and temperature?

6. Contrast the three methods of heat transfer.

7. How are gaseous, liquid, and solid fuels different?

8. How does oxygen concentration relate to flammability?

9. After flaming combustion occurs, what is necessary to extinguish the fire?

10. How do the four stages of fire development differ?

11. In what ways can firefighters' actions impact structure fire development?

Discussion Questions

1. Why is it important for fire investigators to understand the principles of fire science?

2. Provide an example of each of the methods of heat transfer: conduction, convection, and radiation.

3. How can understanding the stages of fire development help an investigator?

Chapter References

Fleishmann, Charles M., Chen, ZhiJian. 2013. "Defining the difference between backdraft and smoke explosions." Procedia Engineering 62, 324-330. Accessed online.

Kerber, Stephen. 2012. "Analysis of Changing Residential Fire Dynamics and Its Implications on Firefighter Operational Timeframes," Underwriters Laboratories, Inc.

Kerber, Stephen; Daniel Madrzykowski; James Dalton; Bob Backstrom. March 2012. "Improving Fire Safety by Understanding the Fire Performance of Engineered Floor Systems and Providing the Fire Service with Information for Tactical Decision Making," Underwriters Laboratories, Inc.

Putorti, Anthony Jr.; Amy Mensch; Nelson Bryner; George Braga. February 2013. "Thermal Performance of Self-Contained Breathing Apparatus Facepiece Lenses Exposed to Radiant Heat Flux," NIST Technical Note 1785 (NIST.TN 1785).

SFPE Handbook of Fire Protection Engineering. 2016. Hurley, M.J. et al. editors, Springer.

Chapter Key Terms

Autoignition — Initiation of combustion by heat but without a spark or flame. (Reproduced with permission from NFPA 921-2017, *Guide for Fire and Explosion Investigations*, Copyright 2018, National Fire Protection Association.)

Autoignition Temperature (AIT) — The lowest temperature at which a combustible material ignites in air without a spark or flame. (Reproduced with permission from NFPA 921-2017, *Guide for Fire and Explosion Investigations*, Copyright 2018, National Fire Protection Association.)

Backdraft — Instantaneous explosion or rapid burning of superheated gases that occurs when oxygen is introduced into an oxygen-depleted confined space. The stalled combustion resumes with explosive force; may occur because of inadequate or improper ventilation procedures.

Buoyant — The tendency or capacity of a liquid or gas to remain afloat or rise.

Ceiling Jet — Horizontal movement of a layer of hot gases and combustion by-products from the center point of the plume, when a horizontal surface such as a ceiling redirects the vertical development of the rising plume.

Chemical Flame Inhibition — Extinguishment of a fire by interruption of the chemical chain reaction.

Combustion — A chemical process of oxidation that occurs at a rate fast enough to produce heat and usually light in the form of either a glow or flame. (Reproduced with permission from NFPA 921-2017, *Guide for Fire and Explosion Investigations*, Copyright 2018, National Fire Protection Association.)

Combustion Zone — Area surrounding a heat source in which there is sufficient air available to feed a fire.

Conduction — Physical flow or transfer of heat energy from one body to another, through direct contact or an intervening medium, from the point where the heat is produced to another location, or from a region of high temperature to a region of low temperature.

Convection — Transfer of heat by the movement of heated fluids or gases, usually in an upward direction.

Emissivity — Measure of an object's ability to emit infrared energy. Emitted energy indicates the temperature of the object. Most organic, painted, or oxidized surfaces have emissivity values close to 0.95.

Endothermic Reaction — A reaction or process that absorbs thermal energy or heat.

Energy — Capacity to perform work; occurs when a force is applied to an object over a distance, or when a substance undergoes a chemical, biological, or physical transformation.

Entrain — To draw in and transport solid particles, liquids, or gases by the flow of a fluid.

Exothermic Reaction — A reaction or process that releases thermal energy or heat.

Exposure Fire — A fire ignited in fuel packages or buildings that are remote from the initial fuel package or building of origin.

Fire — A rapid oxidation process, which is a gas phase chemical reaction resulting in the evolution of light and heat in varying intensities.

Fire Dynamics — The detailed study of how chemistry, fire science, and the engineering disciplines of fluid mechanics and heat transfer interact to influence fire behavior. (Reproduced with permission from NFPA 921-2021, *Guide for Fire and Explosion Investigations*, Copyright 2020, National Fire Protection Association.)

Fire Point — Temperature at which a liquid fuel produces sufficient vapors to support combustion once the fuel ignites. The fire point is usually a few degrees above the flash point.

Flammable (Explosive) Range — Range between the upper flammable limit and lower flammable limit in which a substance can ignite.

Flash Point — Minimum temperature at which a liquid gives off enough vapors to form an ignitable mixture with air near the surface of the liquid.

Flashover — Rapid transition from the growth stage to the fully developed stage.

Flow Path — Composed of at least one intake opening, one exhaust opening, and the connecting volume between the openings. The difference in pressure determines the direction of the flow. Heat and smoke in a high pressure area will flow toward areas of lower pressure.

Free Radical — Electrically charged, highly reactive parts of molecules released during combustion reactions.

Fuel — A material that will maintain combustion under specified environmental conditions. (Reproduced with permission from NFPA 921-2021, *Guide for Fire and Explosion Investigations*, Copyright 2020, National Fire Protection Association.)

Fuel-Limited — A fire in which the heat release rate and growth rate are controlled by the characteristics of the fuel, such as quantity and geometry, and in which adequate air for combustion is available. *Also known as* Fuel-controlled. (Reproduced with permission from NFPA 921-2021, *Guide for Fire and Explosion Investigations*, Copyright 2020, National Fire Protection Association.)

Heat — Form of energy associated with the motion of atoms or molecules in solids or liquids that is transferred from one body to another as a result of a temperature difference between the bodies, such as from the sun to the earth. To signify its intensity, it is measured in degrees of temperature.

Heat Flux — The measure of the rate of heat transfer to or from a surface, typically expressed in kilowatts per square meter (kW/m^2).

Heat of Combustion — Total amount of thermal energy (heat) that could be generated by the combustion (oxidation) reaction if a fuel were completely burned. The heat of combustion is typically measured in kilojoules per gram (kJ/g) or megajoules per kilogram (MJ/kg).

Heat Release Rate (HRR) — Total amount of heat released per unit time. The heat release rate is typically measured in kilowatts (kW) or Megawatts (MW) of output.

Hot Gas Layer — A buoyant layer of hot gases and smoke produced by a fire in a compartment. (Reproduced with permission from NFPA 921-2021, *Guide for Fire and Explosion Investigations*, Copyright 2020, National Fire Protection Association)

Ignition Temperature — Minimum temperature to which a fuel (other than a liquid) in air must be heated in order to start self-sustained combustion independent of the heating source.

Incomplete Combustion — Result of inefficient combustion of a fuel; the less efficient the combustion, the more products of combustion are produced rather than burned during the combustion process.

Isolated Flame — Flame in the hot gas layer that indicate the gas layer is within its flammable range and has begun to ignite; often observed immediately before a flashover.

Joule (J) — Unit of work or energy in the International System of Units (SI); the energy (or work) when a unit force (1 newton) moves a body through a unit distance (1 meter). Joules are defined in terms of mechanical energy. In terms of thermal energy, joules refer to the amount of additional heat needed to raise the temperature of a substance, such as the 4.2 Joules needed to raise the temperature of 1 gram of water 1 degree Celsius. Takes the place of calorie for heat measurement (1 calorie = 4.19 J).

Kinetic Energy — Energy possessed by a moving object because of its motion.

Lower Explosive (Flammable) Limit (LEL) — Lower limit at which a flammable gas or vapor will ignite and support combustion; below this limit the gas or vapor is too lean or thin to burn (too much oxygen and not enough gas, so lacks the proper quantity of fuel). Also *known as* Lower Flammable Limit (LFL).

Matter — Anything that occupies space and has mass.

Neutral Plane — Level at a compartment opening where there is an equal difference in pressure exerted by expansion and buoyancy of hot smoke flowing out of the opening and the inward pressure of cooler, ambient temperature air flowing in through the opening.

Open Burning — Description of a fire burning completely in the open with no restrictions to its oxygen supply.

Oxidation — Chemical process that occurs when a substance combines with an oxidizer such as oxygen in the air; a common example is the formation of rust on metal.

Oxidizer — Any material that readily yields oxygen or other oxidizing gas, or that readily reacts to promote or initiate combustion of combustible materials. (Reproduced with permission from NFPA 400-2010, *Hazardous Materials Code,* Copyright 2010, National Fire Protection Association.)

Piloted Ignition — Moment when a mixture of fuel and oxygen encounters an external heat (ignition) source with sufficient heat or thermal energy to start the combustion reaction.

Potential Energy — Stored energy possessed by an object that can be released in the future to perform work once released.

Power — Amount of energy delivered over a given period of time.

Pressure — Force per unit area exerted by a liquid or gas measured in pounds per square inch (psi) or kilopascals (kPa).

Products of Combustion — Materials produced and released during burning.

Pyrolize — Description of the process of a solid beginning to emit gases due to heat exposure.

Pyrolysate — Product of decomposition through heat; a product of a chemical change caused by heating. (Reproduced with permission from NFPA 921-2017, *Guide for Fire and Explosion Investigations*, Copyright 2018, National Fire Protection Association.)

Pyrolysis — The chemical decomposition of a solid material by heating. Pyrolysis precedes combustion of a solid fuel.

Radiation — Transmission or transfer of heat energy from one body to another body at a lower temperature through intervening space by electromagnetic waves.

Reducing Agent — Fuel that is being oxidized or burned during combustion.

Rollover — Condition in which the unburned fire gases that have accumulated at the top of a compartment ignite and flames propagate through the hot-gas layer or across the ceiling. These superheated gases are pushed, under pressure, away from the fire area and into uninvolved areas where they mix with oxygen. When their flammable range is reached and additional oxygen is supplied by opening doors and/or applying fog streams, they ignite and a fire front develops, expanding very rapidly in a rolling action across the ceiling.

Self-Heating — The result of exothermic reactions, occurring spontaneously in some materials under certain conditions, whereby heat is generated at a rate sufficient to raise the temperature of the material. (Reproduced with permission from NFPA 921-2021, *Guide for Fire and Explosion Investigations*, Copyright 2020, National Fire Protection Association.)

Smoke Explosion — Form of fire gas ignition; the ignition of accumulated flammable products of combustion and air that are within their flammable range.

Specific Gravity — Mass (weight) of a substance compared to the weight of an equal volume of water at a given temperature. A specific gravity less than 1 indicates a substance lighter than water; a specific gravity greater than 1 indicates a substance heavier than water.

Spontaneous Ignition — Initiation of combustion of a material by an internal chemical or biological reaction that has produced sufficient heat to ignite the material. (Reproduced with permission from NFPA 921-2021, *Guide for Fire and Explosion Investigations*, Copyright 2020, National Fire Protection Association.)

Surface-To-Mass Ratio — Ratio of the surface area of the fuel to the mass of the fuel.

Temperature — Measure of the average kinetic energy of the particles in a sample of matter, expressed in terms of units or degrees designated on a standard scale.

Thermal Conductivity — The propensity of a material to conduct heat within its volume. Measured in energy transfer over distance per degree of temperature.

Thermal Energy — Kinetic energy associated with the random motions of the molecules of a material or object; often used interchangeably with the terms *heat* and *heat energy*. Measured in joules or Btu.

Thermal Equilibrium — The point at which two regions that are in thermal contact no longer transfer heat between them because they have reached the same temperature.

Thermal Inertia — The properties of a material that characterize its rate of surface temperature rise when exposed to heat; related to the product of the material's thermal conductivity (k), its density {p}, and its heat capacity (c). (Reproduced with permission from NFPA 921-2021, *Guide for Fire and Explosion Investigations*, Copyright 2020, National Fire Protection Association.)

Upper Explosive (Flammable) Limit (UEL) — Upper limit at which a flammable gas or vapor will ignite. Above this limit the gas or vapor is too *rich* to burn (lacks the proper quantity of oxygen).

Vapor Pressure — The pressure at which a vapor is in equilibrium with its liquid phase at a given temperature; liquids that have a greater tendency to evaporate have higher vapor pressures at a given temperature.

Vaporization — Physical process that changes a liquid into a gaseous state; the rate of vaporization depends on the substance involved, heat, pressure, and exposed surface area.

Ventilation-Limited — A fire in which the heat release rate or growth is limited or controlled by the amount of oxygen available to the fire. *Also known as* Ventilation-controlled.

Watt (W) — The SI unit of power or rate of work equal to one joule per second (J/s).

Photo courtesy of Donny Howard, Agent, Oklahoma State Fire Marshal's Office.

Chapter Contents

NFPA 1033 JPRs addressed in this chapter

This chapter provides information that addresses the following job performance requirements (JPRs) of NFPA 1033, *Standard for Professional Qualifications for Fire Investigator* (2022):

4.2.2   4.2.3   4.2.4   4.2.5   4.2.8

This chapter provides information that addresses the following course outcomes for the FESHE courses:

Fire Investigation I (C0283)

6. Explain how the basic elements fire dynamics, construction, and fire protection systems as to how they affect origin and cause determination.

Learning Objectives

1. Describe the fire resistance of building construction materials. [NFPA 1033, 4.2.2, 4.2.3, 4.2.5]

2. Describe common types of building materials. [NFPA 1033, 4.2.2, 4.2.3, 4.2.5]

3. Describe the five basic building classifications. [NFPA 1033, 4.2.2, 4.2.3, 4.2.5]

4. Describe unique or unusual construction types that an investigator may encounter. [NFPA 1033, 4.2.2, 4.2.3, 4.2.5]

5. Describe various structural systems and features. [NFPA 1033, 4.2.2, 4.2.3, 4.2.5, 4.2.8]

6. Describe how various types of construction features will be affected by exposure to fire. [NFPA 1033, 4.2.2, 4.2.3, 4.2.4, 4.2.5]

Chapter 4

Basics of Building Construction as It Relates to the Fire Investigator

A fire investigator must have a working knowledge of building construction techniques and, in particular, how building construction affects fire confinement or spread in a structure. This chapter focuses on providing the fire investigator an overview of:

- Different types of construction
- Construction and finish materials
- Various components found in building construction

In addition, the chapter addresses unique construction considerations and fire resistance in building construction. This overview of construction knowledge will become the basis for many of the judgments and decisions that investigators make in the field. For more information about this topic, refer to the IFSTA manual, **Building Construction Related to the Fire Service**. This chapter addresses the following aspects of building construction:

- Fire resistance
- Building materials
- Basic building classifications
- Unique construction considerations
- Structural systems and features
- Reaction of building construction to fire

Fire Resistance

NFPA 1033 (2022) 4.2.2, 4.2.3, 4.2.5

Fire resistance is the ability of a structural assembly to maintain its load-bearing capacity and structural integrity under fire conditions. Fire resistance is one of the most basic properties of building materials. Fire-resistive materials, including gypsum wallboard, are used to protect the structural components, such as studs, joist, beams, and columns of a building from fire exposure. As a result, fire-resistant assemblies are intended to prevent or delay structural failure under fire conditions. In the case of walls, partitions, and ceilings, fire resistance could also mean the ability to act as a barrier to keep a fire contained to a defined volume of space, often referred to as a *fire area*. An assembly's fire resistance is a function of the properties of the materials used, including:

- Density
- Dimensions
- Combustibility
- Thermal inertia
- Thermal conductivity
- Chemical composition

NOTE: In the case of walls, partitions, and ceilings, fire resistance indicates the ability to act as a barrier to fire.

The fire resistance of structural assemblies can be evaluated quantitatively and is known as the *fire-resistance rating*. Fire-resistance ratings are expressed in time units including hours and fractions of hours **(Figure 4.1)**. The fire-resistance ratings for structural components are incorporated into the construction classification of buildings in building codes. Building codes have requirements for the fire resistance of structural elements such as:

- Beams
- Columns
- Walls and partitions
- Floor and ceiling assemblies
- Roof and ceiling assemblies

Fire-resistance ratings are determined by laboratory testing from qualified testing laboratories for a specific construction design, such as the number and location of nails or screws used in the construction of a wall. However, these tests do not provide specific information on how the item will perform during actual fire conditions because factors, such as heat release rate of fuels, rate of temperature rise, exposure time, and ventilation, may differ widely from the controlled conditions. Certain construction variables, such as not building to the exact construction criteria of the assembly, can void the validity of tested ratings. As a result, the assembly may fail sooner than the rating implies and not prevent failure of the barrier for the listed rating.

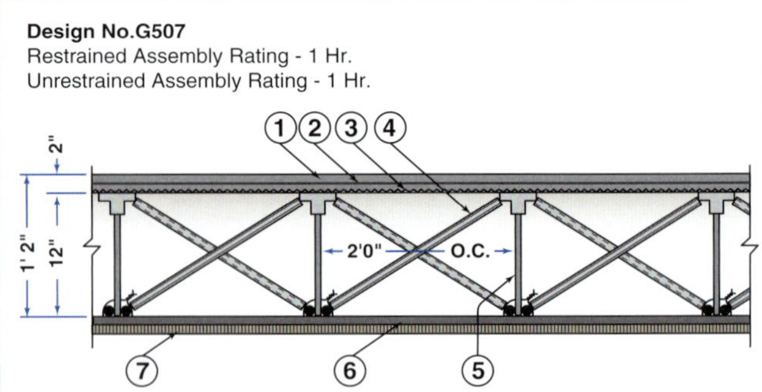

1. **Normal Weight Concrete** - Siliceous or carbonate aggregate, 150 (+or-) 3 pcf unit weight, 4000 psi compressive strength.
2. **Welded Wire Fabric** - 6 x 6, 8/8 SWG.
3. **Metal Lath** - 3/8 in. rib, 3.4 lb/sq yd expanded steel; tied to each joist at every other rib, and midway between joists at side lap with 18 SWG galv steel wire.
4. **Bridging** - 3/4 in., 16 USS gauge box channels or min 1/2 in. diam steel bars.
5. **Steel Joists** - Type 12J4 min size; spaced 24 in. OC and welded to end supports.
6. **Furring Channel** - 3/4 in. 0.30 lb furring channel or 7/8 in. 24 MSG nailing channels, 16 in. OC, fastened to each joist with double tie of galv 18 SWG wire, double furring at each butt joint of wallboard.
7. **Wallboard, Gypsum*** - 5/8 in. thick, secured to furring channels with No.6 flathead sheet-metal screws spaced 8 in. OC or to nailing channels with fetter ring barbed nails 1-1/4 in. long with 11 SWG shanks and 3/8 in. heads, spaced 6 in. OC. Joint treatment not required for this rating, except for tapered, rounded-edge wallboard where edge joints are covered with paper tape and joint compound.

> **Celotex Corp.** - Type B, C or FRP.
> **Continental Gypsum Company** - Type CG5-5.
> **G-P Gypsum Corp.** - Type 5 or C.
> **James Hardie gypsum Inc.** - Types Fire X, Max"C".
> **Republic Gypsum Co.** - Type RG-1 or Rg-3.

*Bearing the UL Classification Marking.

Figure 4.1 A fire investigator may be able to access structural assembly plans to determine whether the scene conditions matched the structural specifications at construction. This floor and ceiling assembly shows a 1-hour rating from the UL Fire Resistance Directory.

NOTE: To help better understand this chapter and typical construction topics in general, the most common construction terminology is presented in **Appendix C,** Common Construction Terminology. Any terms that the investigator is unfamiliar with in this chapter can be found in that appendix.

CAUTION: Investigators should not use fire-resistance ratings to establish a timeline for the duration of a fire.

Building Materials

NFPA 1033 (2022) 4.2.2, 4.2.3, 4.2.5

Building materials have a variety of properties that determine their usefulness in various architectural applications, including:

- Density
- Strength
- Durability
- Appearance
- Thermal conductivity
- Resistance to corrosion and insects

A fire investigator should understand the basic properties of these materials and the effect that exposure to heat and flame has on them. Some of the basic properties of materials that the investigator should know include the following:

- Combustibility
- Thermal inertia
- Thermal expansion
- Thermal conductivity
- Thermal effects on the material

Knowledge of the building involved can be a great asset to investigators. This information can come from preincident plans, inspection reports, or observations of similar types of structures. Building characteristics to consider include the following:

- Contents
- Ceiling height
- Construction type
- External exposures
- Occupancy classification
- Type and design of roof construction
- Square footage and **compartmentation**
- Type and location of fire protection systems
- Number of stories above and below ground level
- Heating, ventilation, and air conditioning (HVAC) system
- Extent to which a building connects to adjoining structures
- Number and size of exterior windows, doors, and other wall openings
- Number and location of staircases, elevator shafts, **dumbwaiters**, ducts, and roof openings

For additional details and greater discussion on building materials, see the IFSTA manual, **Building Construction Related to the Fire Service**. The following sections describe the materials most commonly used for the major structural components of buildings:

- Wood
- Masonry
- Steel
- Concrete
- Glass
- Aluminum
- Gypsum
- Plastic Construction Materials
- Artificial Laminates

Wood

Wooden structural members are found in Type III, Type IV, and Type V construction (see Basic Building Classifications section). Wood is a good insulator and therefore is not a good conductor of heat. Thus, a fire investigator must be concerned with the two primary properties of wood:

- Combustibility
- Effects from exposure to fire

These factors are discussed later in this chapter. This section presents an overview of the material properties. Wood in its normal state is combustible, but its mass, surface area, and moisture content affect how rapidly it burns.

Dimensions

The smaller the dimensions of the wood, the easier it is to ignite and the faster it will lose structural integrity. Ignition is introduced in Chapter 3, Fire Dynamics.

Large wooden beams, such as those used in heavy-timber construction, are difficult to ignite and retain their structural integrity even after prolonged exposure to direct flame impingement. From a structural-integrity perspective, a beam that is 6 by 10 inch (150 mm by 250 mm) will support its load for a significantly longer time than one with dimensions of 2 by 6 inches (50 mm by 150 mm). Lumber of smaller dimensions needs to be protected by gypsum wallboard or other insulation to increase its resistance to heat or fire. Fire effects on building materials based on dimensions, orientation, and other factors are discussed at the end of this chapter.

Laminated wood products are made of flat strips of wood that are joined with glue **(Figure 4.2)**. These products are used in many applications in building construction including beams and arches. While these products may behave similarly to solid wood under fire conditions when properly protected, they may fail more quickly without such protection.

Figure 4.2 Laminated wood products take many different sizes and shapes.

Age, Condition, and Type

The fire investigator must also consider the age, condition, and type of wood used in the construction of a building. The age and condition of wood will affect the moisture content of the wood. Wood with a high moisture content (sometimes referred to as *green wood*) does not ignite as readily nor burn as fast as wood that has been kiln-dried or dehydrated by exposure to air over a long period of time. In some cases, wood is pressure-treated with fire retardants to reduce the speed at which it ignites or burns. However, fire retardants are not always effective in reducing fire spread.

Organizations such as the American Wood Council may offer resources helpful in determining the features of specific types of wood products. For example, given comparable treatments, hardwoods (such as oak) may not burn as rapidly as conifer species (such as pine).

Masonry

Masonry is a fundamental construction technique consisting of stacking individual units on top of one another and bonding them with mortar into a solid mass. Because masonry materials do not easily burn, they are commonly used in the construction of fire walls. The fire resistance of a masonry wall depends on the type of masonry units used and the thickness of the wall.

Masonry units can be made of several materials including:

- **Brick** — Bricks rarely show any signs of loss of integrity or serious deterioration **(Figure 4.3)**.

Figure 4.3 This 2009 renovation of a building constructed in 1919 includes original features in innovative ways.

- **Non-fire-rated concrete block** — Non-fire-rated hollow concrete blocks may have little fire resistance and may spall and crumble when exposed to a fire.

- **Fire-rated concrete block** — Walls constructed with fire-rated concrete masonry units or bricks can have fire-resistance ratings of 2 to 4 hours. Thicker walls will have a higher degree of fire resistance.

- **Stone** — Stones may spall or lose small portions of their surface when heated. The primary problem with this form of construction is the deterioration of the mortar used to bond the units together.

Types of masonry units include clay tile block, gypsum block, and glass block. In mortared masonry, exposure to water (including fire streams) can erode mortar and result in wall collapse **(Figure 4.4)**. Rapid cooling, for example during extinguishment, may cause masonry materials to crack. Cracking is a common problem when water is used to extinguish masonry chimney flue fires.

Despite the basic structural qualities of masonry construction, total collapse of a masonry building is possible if it becomes heavily involved in fire. Masonry walls may not be free-standing, but they may be connected directly to plywood sheathing or to wooden studs. In these circumstances, masonry walls are more likely to collapse if the supporting structural members burn or are fire damaged. Collapsing interior floor or roof members can exert horizontal forces against a wall and push the wall outward. The horizontal forces create tensile forces at the inner face of the wall that the mortar joints cannot resist. Because the collapse of interior framing also removes interior bracing for the wall, the wall is simply pushed out from the building. If a wall is of reinforced construction, the reinforcing steel can withstand the tensile force and collapse is less likely.

Some floor joists are designed with a **fire cut** to allow the joist to fail while leaving the masonry wall standing **(Figure 4.5)**. The purpose of the fire cut is to allow the beam to fall away freely from a wall in case of structural collapse without acting as a lever to push against the masonry. However, fire cuts in joists do not completely preclude the collapse of a masonry wall.

There is no sure way to predict how or at what point a masonry wall may collapse when a building is heavily involved in fire. A masonry wall may collapse partially or completely. A collapsing roof may dislodge only a **parapet** or an entire wall. When a masonry building is heavily involved in fire, it should be assumed that a collapsing wall will fall out from the building a distance at least equal to the height of the wall. A collapse exclusion zone that is 1.5 times the height of the wall should be respected **(Figure 4.6)**. Because intersecting masonry walls tend to support each other, the corners of the building or other points of intersection, such as stairwells or elevator shafts, will be the strongest points in a masonry structure.

Figure 4.4 Mortar used in masonry walls decays at a quicker rate than most structural blocks. *Courtesy of Ed Prendergast, PE.*

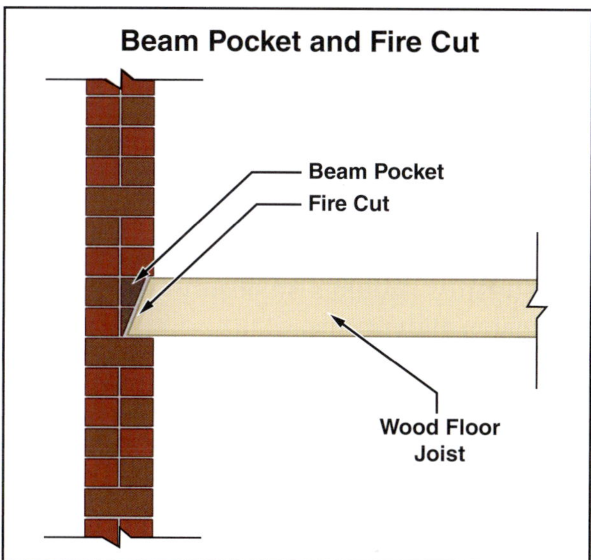

Figure 4.5 A fire cut in a horizontal support minimizes damage to a structural wall under fire conditions.

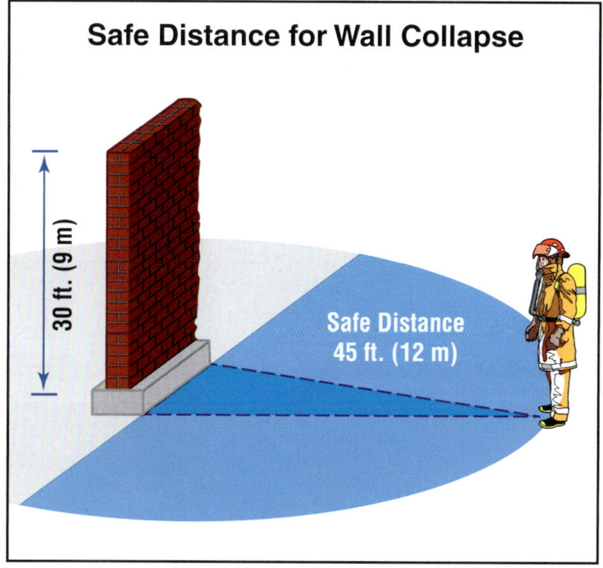

Figure 4.6 Safe distances to avoid wall collapse should be estimated as one and one-half the height of the wall.

Steel

For a fire investigator, the deterioration of the strength of steel at elevated temperatures is its most significant characteristic **(Figure 4.7)**. Fires normally encountered by the fire service do not create temperatures hot enough to melt steel. However, fires are hot enough to greatly weaken steel. Heat causes steel structural elements to elongate or expand (thermal expansion) significantly, which can cause the collapse of walls and roofs. Because temperatures in excess of 800° to 1,200°F (425°C to 650°C) are regularly encountered in fires, failure of unprotected steel to a greater or lesser degree can be anticipated. Steel members will weaken, depending on many variables including:

- Size and span
- Load and spacing
- Composition of the steel
- Duration of heat exposure
- Shape or configuration of the steel
- Amount of insulation or protection from heat

Steel is also a very good conductor of heat. Elevated temperatures at one end of a beam or column transmit through the member. This conduction of heat could result in the ignition of combustibles that are some distance from the point of initial exposure.

Investigators should consider if the steel components in the structure were protected as part of the building construction. Unprotected steel beams may require less heat to reach their failure temperature. Investigators should also observe the steel members for evidence of bowing or evidence of thermal expansion at the joints.

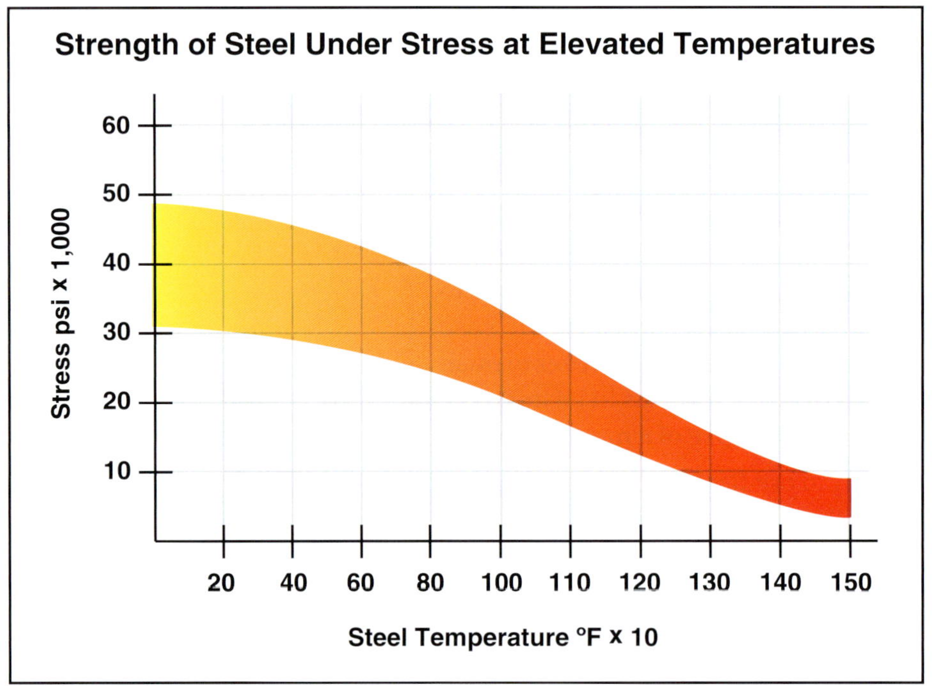

Figure 4.7 The strength of structural steel may change depending on the conditions of a fire-related temperature increase.

NOTE: For more information about the melting temperatures of steel and the other materials in this section, refer to NFPA 921 *Guide for Fire and Explosion Investigations*, and the IFSTA manual **Building Construction Related to the Fire Service**.

Concrete

Concrete is fire-resistive and **noncombustible** and contains good insulating properties. Concrete structural systems can have fire-resistance ratings from 1 to 4 hours. The fire-resistance of a concrete assembly is affected by variables such as the following:

- Concrete quality
- Concrete density
- Concrete thickness
- Load supported by the concrete
- Depth of concrete cover over the reinforcing bars

For example, structural lightweight concrete has a lower density than ordinary concrete and has a lower thermal conductivity. Therefore, it acts as a better insulator against the heat of a fire than ordinary concrete of comparable thickness.

The fire resistance of concrete assemblies can be compromised in several ways. If concrete floor slabs or wall panels are supported by non-fire-resistive members, the overall construction would not be fire resistive. Precast wall panels are horizontally braced by exposed steel roof beams. This very common design is used for one-story mercantile and industrial buildings. When openings exist in concrete slabs or walls, the ability of the concrete to act as a barrier is lost unless the opening is protected by an appropriately rated assembly such as a fire door or shutter.

One of the most significant effects fire and heat have on concrete is **spalling (Figure 4.8)**. The two causes of spalling are as follows (Sanderson, 1995):

Figure 4.8 Spalling is the result of an explosive build-up of steam within concrete.

1. Materials within concrete may expand at different rates from the concrete itself when exposed to a heat source, such as in the following situations:

    — Reinforcing rods/steel mesh and the surrounding concrete

    — Concrete mix and the aggregate (most common with silicon aggregates)

    — Fire-exposed surface and the interior of the slab

    — Rapid cooling of areas of concrete by application of water or a hose stream

2. Explosive buildup of steam from expanding water within the concrete that cannot migrate to the exposed surface. The expansion of the water creates tensile forces within the concrete. Because concrete has little resistance to tension, small pieces of the concrete break.

Reinforced Concrete

The steel used in reinforced concrete is not fire resistive and may expand under fire conditions, causing cracking or spalling of the surrounding concrete. This effect may result in earlier than expected failure of the concrete.

The concrete that surrounds the reinforcing steel acts as insulation to protect it from the heat of a fire. The overall fire resistance of reinforced concrete depends on the depth of cover of the concrete over the steel and the quality of the concrete.

Prestressed Concrete

Prestressed concrete systems may be somewhat more vulnerable to failure than ordinary reinforced concrete. The reason for this vulnerability is that the reinforcing cables and rods used in prestressed systems are made of high-strength steel that has lower yield-point temperatures. They can yield at a temperature of around 750°F (400°C). Therefore, for the same depth of cover, a prestressed assembly will fail sooner than a conventional reinforced assembly.

Relationship between Ignitable Liquids and Spalling

A fire investigator should not immediately interpret the presence of spalling as an indicator of the introduction of ignitable liquids (Sanderson, 1995). The presence of a burning ignitable liquid, such as a pool of gasoline, cannot cause spalling beneath the surface of the ignitable liquid due in part to the insulating effect of the liquid on the surface of the concrete. Secondly, when the pool of ignitable liquid is nearly consumed or no longer deep enough to "insulate" the surface of the concrete, there is insufficient liquid in the pool for burning to transfer sufficient heat over enough time to the surface to result in spalling, or there is insufficient liquid in the pool for burning to continue. In contrast, hot coals on a concrete floor are a good means to create spalling of concrete surfaces.

Glass

Glass is present in most buildings. Its common use is for windows, skylights, storefronts, and other applications where the transmission of light is desirable **(Figure 4.9)**. The architectural applications of glass extend to Gothic church windows, partition walls, and the exterior curtain walls of buildings.

As is the case with other building materials, several different types of glass are produced. The most commonly encountered glass types include:

Figure 4.9 Glass often serves as decoration in addition to transmitting light within a structure.

- Ordinary
- Tempered
- Laminated
- Glass block
- Heat strengthened
- Single-strength annealed

Glass Effects in Fire Conditions

Glass is noncombustible but is not fire resistive. When heated, internal thermal stresses cause glass to shatter and fall from its frame. However, wired glass or fired glass may be installed where fire resistance is required **(Figure 4.10)**.

Fire investigators often encounter cracked or broken glass at a fire scene. It is important to understand the mechanisms that cause breakage and incorporate that information into the investigation findings (Mowrer, 1998).

Glass rarely breaks as a result of excessive pressure on its surface. Research shows that the overpressures necessary for breakage to occur are approximately 0.3 psi to 1.0 psi (2.07 kPa to 6.89 kPa). The overpressures developed by a fire in a compartment are significantly lower than these figures and would not be sufficient to cause the glass to break. However, overpressures from explosions, including backdrafts, can be high enough to break glass.

Window glass can crack as a result of thermal stress. The portion of the glass exposed to the heat from the fire expands while the edges of the glass remain cool due to thermal shielding by the window frame. These cracks may be long, smooth, or wavy and may radiate across the pane. As cracks form, they can cause the pane to fail, allowing broken portions to fall from the frame **(Figure 4.11)**.

Figure 4.10 Wired glass prevents shattering when exposed to high temperatures.

Figure 4.11 Cracked glass indicates exposure to heat.

The crazing of glass is the formation of patterns of short cracks throughout the pane, which may or may not penetrate through the entire thickness of the glass. Crazing can occur when water is suddenly applied to one side of a hot pane of glass. Thus fire fighting operations could result in crazing when an attack stream is played on the surface of a hot window.

Analysis of Glass Effects

The appearance and types of cracks are data that should be noted and analyzed. Cracks are useful for conducting comparative analysis within a compartment or between multiple compartments and determining whether a window was broken through mechanical forces or thermal radiation. Analyses of glass should not be used to characterize a fire in terms of heat release rate or fuel type. In addition to differences in types of glass, the window frame itself may affect the failure rate. For example, vinyl frames melt at a lower temperature than many types of glazing and may allow the entire pane to drop out.

Aluminum

Aluminum is used in conventional residential construction, commercial construction, and manufactured homes (mobile homes). In conventional residential construction, aluminum is used primarily in window and door frames and as roof panels and siding **(Figure 4.12, p.108)**.

In commercial construction, aluminum is often used in window and door frames but is used most extensively in curtain walls on the exterior of high-rise buildings. It is also used as roof coverings, exterior wall coverings, and electrical wiring in some older units. With a relatively low melting temperature, the presence of melted aluminum at fire scenes is not uncommon.

Aluminum is also used extensively in the construction of mobile homes. Because these structures are designed to be transported from place to place, aluminum's high strength-to-weight ratio makes it a desirable construction material.

Gypsum

Gypsum is an inorganic product used in plaster and wallboards **(Figure 4.13)**. Gypsum has a high water content that is chemically bound into the material which gives it excellent heat-resistant and fire-retardant properties. The evaporation of this water (calcination) absorbs a great deal of heat (endothermic reaction). Gypsum wallboard after fire exposure is discussed further in Chapter 10, Scene Examination.

Because it decomposes gradually under fire conditions, gypsum is commonly used to provide insulation to steel and wood structural members that are less adaptable to high-heat situations. In areas where the gypsum has failed, the structural members behind it will be subjected to higher temperatures and could also fail as a result. Gypsum drywall is constructed for a number of specific applications and each may respond differently in fire conditions. For example, kitchens and utility rooms may include more protective varieties than other areas.

Plastic Construction Materials

The term *plastic* encompasses a large number of materials of high molecular weight that can be formed by pressure, heat, extrusion, and other methods. There are 20 to 30 major groups of plastics. In addition, variations can be produced within the major groups by varying the chemistry of individual materials. The large variety of plastics available permits their use in many different applications. In building construction, plastics are used for components such as the following:

Figure 4.12 Aluminum framing is lightweight and efficient for many applications.

Figure 4.13 Gypsum drywall protects structural components susceptible to damage under fire conditions. *Courtesy of McKinney (TX) Fire Department.*

- Siding
- Insulation
- Vapor barriers
- Floor coverings
- Sprinkler piping
- Lighting fixtures
- Pipe and pipe fittings
- Skylights and roof domes
- Tub and shower enclosures
- Lumber substitutes for decking materials
- Window frame assemblies using **thermoplastics** and **thermoset plastics**

Flammability and Melting of Plastics

The flammability and melting of plastics are of interest to a fire investigator. As with their other properties, the flammability of plastics varies widely. Some plastics, such as cellulose nitrate, burn so rapidly that they constitute a unique fire hazard. Other plastics may burn slowly and stop burning when the ignition source is removed. Fire retardants can be added to some plastics to reduce their flammability and ignition sensitivity. However, even plastics with low flammability are subject to pyrolization and may produce toxic gases at temperatures above 500°F (260°C).

The use of plastics in building construction increases the fire hazard to the extent that it increases the amount of fuel in a building and the toxicity of the products of combustion. Plastic materials frequently exhibit burning properties that are different from other materials. For example, nylon usually melts and drips when it burns. Foam plastics burn more intensely when tested on a large scale than when tested in small samples. Some plastics generate enormous quantities of heavy smoke. The products of combustion of some plastics are more toxic than nonplastic materials. The combustion of vinyl chloride, for example, produces hydrogen chloride — the gaseous form of hydrochloric acid. Hydrogen chloride is corrosive as well as toxic and increases the damage done to sensitive electrical equipment.

Figure 4.14 Thermoplastic composite lumber looks much like ordinary lumber but melts like plastic under fire conditions. *Courtesy of Donny Howard, Agent, Oklahoma State Fire Marshal's Office.*

Thermoplastic Composite Lumber

Thermoplastic composite lumber is a wood-like product made from wood fiber and **polyvinyl chloride (PVC)**, developed as an alternative to preservative-treated lumber **(Figure 4.14)**. Alternatively, plastic lumber material is based on High Density Poly Ethylene (HDPE).

These materials are manufactured as boards in sizes comparable to sawn lumber and in various shapes as architectural trim. Thermoplastic composite lumber is not intended to be used in the structural framing of a building. When heated, thermoplastic composite lumber will melt and flow.

Thermoset Plastics

Thermoset plastics are used for circuit breakers, sprinkler pipe, and appliance housing including duplex receptacle junction boxes. In contrast to thermoplastics, thermoset plastics will char in place under fire conditions; they do not drip or flow when heated **(Figure 4.15)**.

Artificial Laminates

Artificial laminates are made by adhering two different materials, such as wood and plastic, together to form a single composite material. In many cases, these laminates are used as interior finishes (for example, Formica® laminate), but they may also be used as

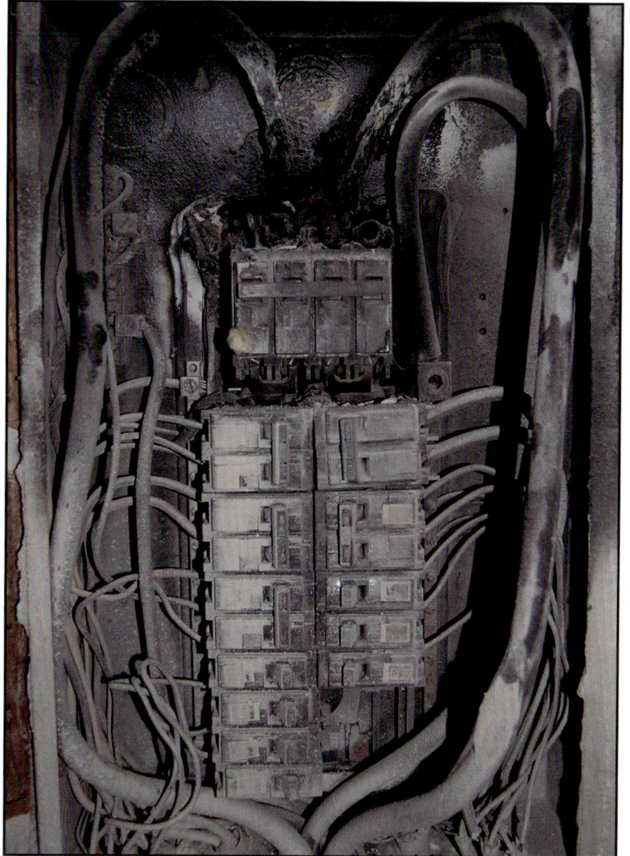

Figure 4.15 Thermoset plastics char, not melt, in high temperature conditions. *Courtesy of Mike Makela.*

structural components. For example, they may be used as laminate beams (engineered beams) and roof decking. Composite wood/plastic materials are increasingly popular for building decks and balconies as well. A fire investigator should be aware that the burning properties are different and most likely more aggressive than if each substance is burned separately. These products typically will fail earlier than regular wood resulting in loss of integrity and earlier collapse.

Basic Building Classifications

NFPA 1033 (2022) 4.2.2, 4.2.3, 4.2.5

An understanding of the different types and classifications of building construction common to most jurisdictions provides a fire investigator with some insight into how buildings are constructed and how they react when they are exposed to fire. In the fields of fire protection and building code enforcement, buildings are grouped into five major classifications. In the U.S., these classifications are as follows:

- **Type I Construction, Fire Resistive** — Construction type in which structural members including walls, columns, beams, floors, and roofs are made of noncombustible materials or limited combustible materials and have a specified degree of fire resistance **(Figure 4.16)**.

- **Type II Construction, Noncombustible or Protected Noncombustible** — Construction type that is similar to Type I except that the degree of fire resistance is lower **(Figure 4.17)**.

- **Type III Construction, Exterior Protected (masonry)** — Construction type (also referred to as *ordinary construction*) in which exterior walls and structural members are of noncombustible or limited-combustible materials. Interior structural members, including walls, columns, beams, floors, and roofs, are completely or partially constructed of wood **(Figure 4.18)**.

- **Type IV Construction, Heavy Timber** — Construction type in which exterior and interior walls and associated structural members are of noncombustible or limited-combustible materials. Interior structural framing is heavy timber that has minimum dimensions larger than those in Type III construction **(Figure 4.19)**.

- **Type V Construction, Wood Frame** — Construction type that has exterior walls, bearing walls, floors, roofs, and supports that are made completely or partially of wood or other approved materials of smaller dimensions than those used for Type IV construction **(Figure 4.20)**.

NOTE: The classifications are similar between the US IBC and the Canadian Building code.

Figure 4.16 Type I fire resistive construction uses combustible or noncombustible materials to meet a specified fire resistance rating.

Figure 4.17 Type II noncombustible construction is similar to Type I construction, but with two subclassifications to indicate the level of fire resistance reached.

Figure 4.18 Type III exterior protected construction uses external masonry components, and may include combustible internal structural components.

Figure 4.19 Type IV heavy timber construction uses structural members with larger nominal dimensions than other construction types to provide large open areas without concealed spaces.

Figure 4.20 Type V wood frame construction may use all combustible materials for structural purposes.

Figure 4.21 Mobile or manufactured homes are prevalent in most jurisdictions. *Courtesy of Donny Howard, Agent, Oklahoma State Fire Marshal's Office.*

Building construction techniques can vary widely within a jurisdiction, and use a wide array of different materials. As a result, the investigator should be aware that the fire may spread through the structure in a manner that is unanticipated. The investigator should evaluate the building materials as well as the construction type and technique.

Unique Construction Considerations

NFPA 1033 (2022) 4.2.2, 4.2.3, 4.2.5

Manufactured homes, prefabricated buildings, modular buildings, geodesic buildings, and log homes have unique construction considerations. Fire investigators must be familiar with these features when conducting investigations in these types of structures.

Manufactured and Mobile Homes

The typical manufactured home is an assembly of four major components: chassis and the floor, wall, and roof systems **(Figure 4.21)**. Although they are constructed of steel, wood, plywood, aluminum, gypsum wallboard, and other materials, they are basically frame construction.

The unique construction of manufactured homes does, however, affect the behavior of the fire within them because of the following factors:

- Small compartment sizes
- Low ceiling heights
- Composition of wall, floor, and ceiling assemblies
- Use of lightweight construction materials

Fire investigators should consider the different construction standards between a site built structure and manufactured construction as they analyze the fires in this type of unit. For example, the low ceiling heights generally found in manufactured housing allow for a more rapid fire spread and the onset of flashover conditions from the same fuel package burning in a site built structure with a higher ceiling. The construction features found in manufactured homes can influence fire development that can result in the relatively early failure of floors and roofs.

Requirements for Manufactured Homes

In the last two decades, changes in the design and construction of manufactured homes have reduced fire hazards related to these units. The driving force for change began in 1974 when the U.S. Department of Housing and Urban Development (HUD) released standards (Title 24 *CFR* Part 3280) that specified manufactured home construction and safety standards. HUD requirements are only valid in the United States. In Canada, the National Research Council (NRC) provides the national building and fire codes that would apply standards for these homes. In addition, provincial and territorial codes are often more restrictive than NRC building and fire code requirements.

Prefabricated

Prefabricated or panelized construction is often lightweight construction that is manufactured in a factory and then delivered to a site. This type of construction includes the walls, floors, and ceilings manufactured complete with plumbing, electrical wiring, and millwork (woodwork such as doors and trim). Once delivered to the site, the entire assembly is erected. Though structurally sound, this form of manufactured construction can result in rapid fire spread similar to other types of lightweight construction.

Lightweight beams and trusses may fail rapidly under fire conditions, especially if they are not protected. A fire investigator should consider this factor when conducting an investigation. The amount of damage may not appear to be consistent to the amount that a fire investigator would find with heavier framing members.

Modular

A modular building is built in two or more sections at the factory. All utilities, such as plumbing, heating, and electrical systems, are installed as an integral part of the construction phase. In addition, the manufacturer installs all the millwork, such as doors and windows. Some modular sections are designed to stack together to form multistory buildings.

Modular buildings are normally constructed to meet or exceed most of the model building codes. This fact, coupled with the need to construct a building to withstand transportation from the factory to the location where it will be erected, usually results in a building that is relatively strong. A modular method of construction may present unusual paths of fire travel. These paths can include **chases** and shafts that are not commonly constructed in buildings that are built on the site.

Geodesic Construction

A *geodesic structure* is defined as a dome or vault made of lightweight straight structural elements that are installed to form a tension load. The principle of geodesic construction is to reduce the weight at the tension points to make the structure economical to build **(Figure 4.22)**. However, being domed or hemispherical in shape promotes a greater rate of fire spread. These factors can lead to a rapid degradation of structural stability resulting in structural collapse.

Figure 4.22 Geodesic domes are constructed by strategically arranging triangle-shaped materials.

Log Homes

Log homes are another style of manufactured housing. These homes are constructed from solid logs ranging from 4 to 9 inches (100 mm to 228 mm) in diameter. Two sides and both ends of each log are machined to form a tight, structurally sound fit. Because of their mass and surface area, the walls of log homes are naturally well-insulated. Under fire conditions, this insulation can result in higher-than-expected temperatures in the building and less ventilation from the exterior through the walls. Varnish and other preservatives applied to the interior surfaces of the logs can also result in unusual fire patterns.

Structural Systems and Features

NFPA 1033 (2022) 4.2.2, 4.2.3, 4.2.5, 4.2.8

Up to now, this chapter has focused on the structural framing and types of materials used to form that frame. Should fire attack the frame of any building, serious damage can result as well as structural collapse. In addition to those structural components, fire investigators should understand:

- Foundations
- Floor systems
- Lightweight construction materials
- Wall assemblies
- Interior finishes
- Compartmentation
- Concealed and interstitial spaces
- Ventilation
- Ceiling assemblies
- Roof systems

Foundations

The function of a foundation is to transfer the structural load of a building to the ground, to support the dead load of a building, and to support the live load of its contents. The structural problems of foundations are normally of little interest to a fire investigator. However, the failure of a foundation can create or aggravate structural problems within the building.

Supports that shift or settle over time can alter the forces on the structural members in the upper part of a building. In severe cases, the following may occur:

- Floors may slope
- Walls and glass may crack
- Frame of a building may be distorted
- Doors and windows may not work properly
- Automatic sprinkler piping can be damaged

Under fire (or explosion) conditions, these altered load patterns can also hasten structural collapse. An investigator should consider whether to ask an engineer to evaluate the altered load patterns before proceeding with an investigation.

Many concrete foundation walls develop visible cracks for varying reasons. They usually do not significantly affect the ability of the wall to support or distribute the load that it is carrying. However, when inspecting a structure for stability, the investigator must closely observe any change in the size or extension of cracks or fissures. Any vertical or horizontal misalignment along the length of a crack in a foundation wall indicates a movement or shift in the structure, which may indicate a change in the way loads are being transmitted from structural members to the foundation.

Foundation walls may be constructed of concrete, stone, brick, or concrete block. Stone and brick foundation walls are found usually in old buildings. A distinguishing aspect of stone foundations is that they were often constructed without using any bonding mortar or cement. The stones were carefully quarried and meticulously assembled to form an amazingly tight-fitting and strong foundation.

Floor Systems

Floors support the building contents and transmit the resulting live load and dead load of the floor system to the structural frame. Building codes may require a specific fire-resistance rating for a floor assembly, depending on the construction classification of the building.

Wood Floors

A fire investigator should always be concerned with the integrity of timber connections under fire conditions and the likelihood of the collapse of roof or floor decks. In multistory buildings, floors are generally supported on **balloon-frame construction** or **platform-frame construction** using systems that can include:

- Solid joists
- Truss joists
- Wood I-beam joists

When truss joists are used in floor construction, it is possible for fire to spread in four directions: parallel to and perpendicular to the truss joists. Frequently, especially in new residential construction, the flooring consists of plywood or OSB subflooring with a carpet or tile floor surface. The thin subflooring may fail after a short period of exposure to fire. Particleboard may be used for flooring in manufactured homes. Particleboard can fail without warning from exposure to fire or just from the exposure of water.

A further area of concern is the possible existence of an unfinished floor over a basement space. Unprotected wood I-beams would be immediately exposed to any fire in the basement resulting in early failure and collapse **(Figure 4.23)**.

Concrete Floors

When a concrete beam or floor slab supports a load, the concrete in the part of the beam under tension is essentially not doing any work. Relatively thin floor slabs are sometimes reinforced with wire mesh rather than ordinary reinforcing bars **(Figure 4.24)**. This framing system is the lightest of the cast-in-place construction and is suitable for light floor loads in buildings with medium floor spans.

Figure 4.23 Exposed wood floor construction in this residential basement will not protect the sub-flooring, engineered I-beam, and small-diameter steel column under fire conditions. *Courtesy of Ed Prendergast, PE.*

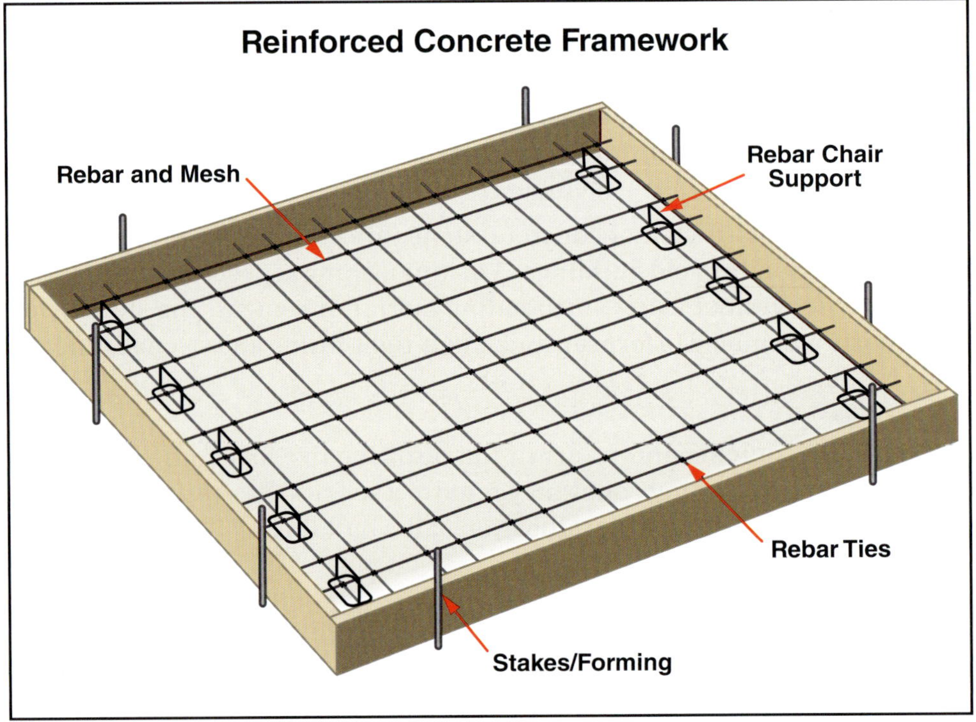

Figure 4.24 Reinforced concrete uses a planned grid of structural reinforcing materials to provide lateral and tensile support.

Steel Floor Systems

Steel structural members can support floors in multistory buildings in the following methods:

- **Steel beams** — Used sometimes where floor loads or spans dictate that steel beams should support the flooring instead of lighter open-web joists **(Figure 4.25)**.

- **Light-gauge steel joists** — Used to support metal decks or wood-panel flooring systems; similar to support given by open web-steel joists. Light-gauge steel joists are produced from cold-rolled steel and available in several cross-sectional varieties.

- **Open-web joists (bar joists) or trusses** — Use lightweight concrete having a minimum thickness of 2 inches (50 mm) supported by corrugated steel decking (a very common design of floors in steel frame buildings); can also be used to support precast concrete panels or wood decking. The corrugated steel is, in turn, supported by open web-steel joists. The steel joists can be supported by steel beams or directly supported on a masonry wall.

Figure 4.25 Open steel joists may support metal roofs and floors.

Flooring System Collapse

Weakened steel members are subject to collapse at any time. Always examine steel floor assemblies for warping or deformations of the assemblies themselves. Also examine where they come in contact with walls before working under or on any steel flooring system.

Lightweight Construction Materials

Lightweight construction materials are commonly used in construction as less-expensive alternatives to dimensional lumber, concrete, and heavier steel beams. Lightweight construction is most commonly found in homes, apartments, small commercial buildings, and warehouses.

Lightweight construction materials include:

- Engineered wood I-beams
- Glulam (glue-laminated) beams
- Plate-connected wood floor trusses
- Light-gauge cold-formed steel beams **(Figure 4.26)**

Figure 4.26 Unprotected, lightweight steel trusses can weaken and fail at elevated temperatures. *Courtesy of McKinney (TX) Fire Department.*

Lightweight Structural Members

All trusses are composed of one or more triangles (the strongest geometric shape known) and are designed to work as an integral unit. Some members are in tension (stresses that tend to pull things apart) and others are in **compression** (stresses that tend to press things together) **(Figure 4.27, p. 116)**. When a truss fails, the trusses next to it are likely to fail, and this domino effect can produce a total collapse almost instantaneously (Kerber, 2012). Regardless of the type of truss involved, if one member fails, the entire truss is likely to fail **(Table 4.1, p. 116)**.

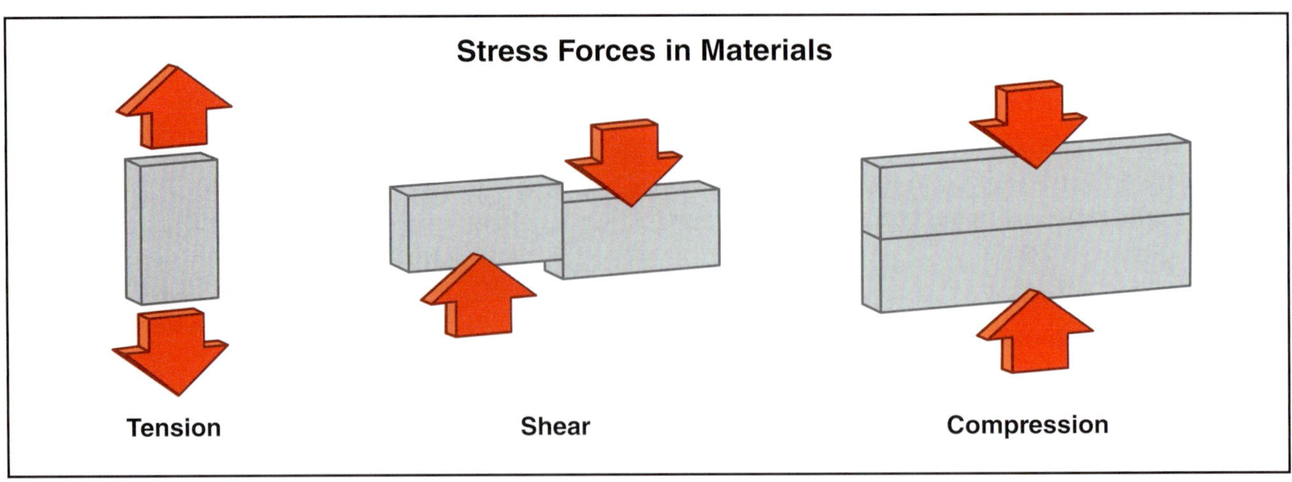

Figure 4.27 Differences between tension, compression, and shear forces are considered in structural analysis.

Table 4.1
Full Scale Floor System Laboratory and Field Fire Experiments

Laboratory Furnace Test, ASTM E119

Floor Support Elements	Time to Collapse (min:sec)
Solid Wood Joist (nominal 2 x 10) - Unprotected	18:35
Solid Wood Joist (nominal 2 x 10) – Covered with Gypsum Board (1/2 in)	44:40
Engineered I Joists (12 in.) – Unprotected	6:00
Engineered I Joists (12 in.) – Covered with Gypsum Wallboard (1/2 in.)	26:43

Note: Times to floor collapse after the ignition of the furnace.
Reference: Izydorek, M., Zeeveld, P., Samuels, M., Smyser, J., Structural Stability of Engineered Lumber in Fire Conditions, Underwriters Laboratories, Northbrook, Illinois, September 2008.

Field Townhouse Collapse Test

Floor Support Elements	Time to Collapse (min:sec)
Solid Wood Joist (nominal 2 x 12) - Unprotected, Vented Basement Fire	11:09
Engineered Wood I Joists (12 in.) – Unprotected, Vented Basement Fire	6:00
Parallel Chord Truss (12 in.) - Unprotected, Vented Basement Fire	3:28

Note: Times to floor collapse after the ignition of similar fires in the basement of a test structure. Each of the floors had a 65% design load.
Reference: Kerber, S., Madrzykowski, D., Backstrom, R., Dalton, J., Improving Fire Safety by Understanding the Fire Performance of Engineered Floor Systems, UL Firefighter Safety Research Institute, March 2012.

Exposed engineered steel and wooden trusses, particularly those without fire resistive protection of some type, can fail after 5 to 10 minutes of exposure to fire (Kerber, et al, "Improving ..." 2012). These trusses can fail from exposure to heat alone without flame contact. For steel trusses — 1,000°F (538°C) is the critical temperature of steel – the temperature at which steel begins to weaken (SFPE 2016). Metal

gusset plates in wooden trusses can fail quickly when exposed to heat. Although protective fire-retardant treatments can enhance the fire resistance of both steel and wooden trusses, most trusses lack this protection.

The traditional wood-joist roof uses solid wood joists that tend to lose their strength gradually when exposed to fire. This loss of strength causes a roof to become soft or "spongy" before failure, especially with a wood plank roof deck. Although a soft or sagging roof is an obvious indication of structural failure, it should not be considered the only sign of imminent collapse.

Fire Effects of Lightweight Construction

Experience has shown that unprotected lightweight steel and wooden trusses can fail after short periods of exposure to fire. These trusses can fail from exposure to heat alone without any flames. Unless they are corner-nailed, metal gusset plates in wooden trusses can warp and fail quickly when exposed to heat. Although both steel and wooden trusses may be protected with fire-retardant treatments to enhance their fire resistance, most lack this protection. Engineered wood beams are also used in lightweight construction **(Figure 4.28)**. They have fire characteristics, such as early failure, similar to wooden trusses.

More modern homes may use engineered joists that burn more quickly and fail before the fire affects the roof decking. Therefore, the plywood or OSB used for roof sheathing may not show any signs of sagging during a fire. When the trusses fail first, entire pieces of the decking may fall into the fire **(Figure 4.29)**.

> **WARNING:** When lightweight trusses fail, entire pieces of decking may fall. There may be no indications from the exterior that the trusses no longer support the roof decking.

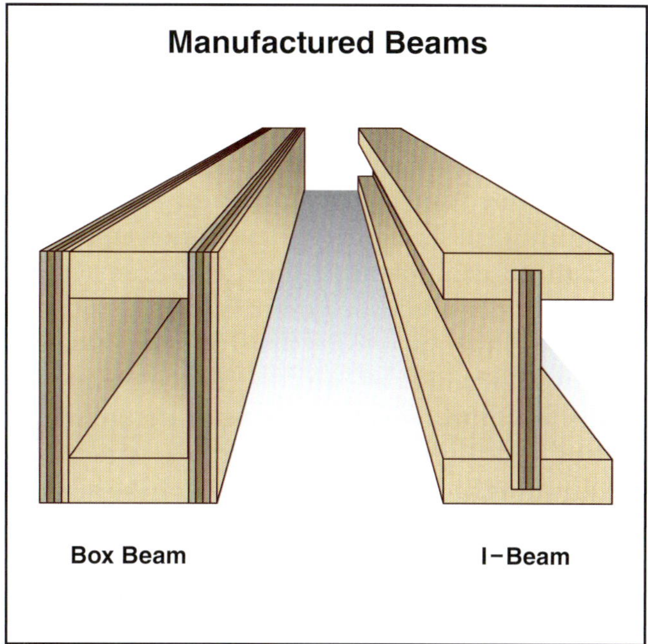

Figure 4.28 Box beams and I-beams use materials arranged strategically for a determined purpose.

Figure 4.29 Parts of the roof of this modern home failed during a fire and fell into the fire within the structure.

Fire investigators at a fire scene must have an ongoing concern for potential building collapse, or partial collapse, especially where significant structural damage has occurred. A fire can significantly weaken a structure by compromising the load bearing components of the building, or even structural components which support other structural components, such as walls and chimneys. While investigating fire scenes, investigators should also be aware of changing conditions, such as a result of weather (snow, rain, wind). Similarly, the removal of fire debris can weaken the stability of floors or walls and can also expose holes burned in floors.

Wall Assemblies

Wall assemblies are an integral part of any building design. A common method of construction uses the walls of a building to support spanning elements such as beams, trusses, and precast concrete slabs. These are appropriately known as *load-bearing walls,* which are usually the exterior walls with an interior support system consisting of columns and beams. However, it is possible to use interior walls for structural support. Bearing walls provide lateral support to the structure along the direction of the wall. Should columns and/or the connections of beams be compromised, shoring or denial of entry is advised. Fire investigators should measure the height of walls during the documentation of the scene. This information can be used to reconstruct a model of the fire conditions at a later time. Investigators can also obtain flashover calculations from this information.

Figure 4.30 A door kept closed during a fire was able to keep fire damage outside of an apartment. *Courtesy of Ed Prendergast, PE.*

Walls constructed for any purpose affect the spread and development of fire in a building **(Figure 4.30)**. Fire-rated partition walls act as barriers to the spread of fire. Walls that have a flammable finish will contribute to the spread of fire. For example, a dramatic difference can be seen between a fire involving gypsum covering versus one involving wood-panel covering.

Exterior Walls without Fire Stops (Balloon-Frame Construction)

Balloon-frame construction features exterior walls without fire stops. This feature can contribute to the spread of fire from floor to floor in a multistory building and can contribute to the early collapse of a structure during a fire. In balloon-frame construction, the exterior wall studs are continuous from the ground-floor assembly to the roof. The joists that support the second floor are supported by ribbon boards that are recessed into the vertical stud. The vertical combustible spaces between the studs in balloon-frame construction provide a channel for the rapid communication of fire from floor to floor **(Figure 4.31)**.

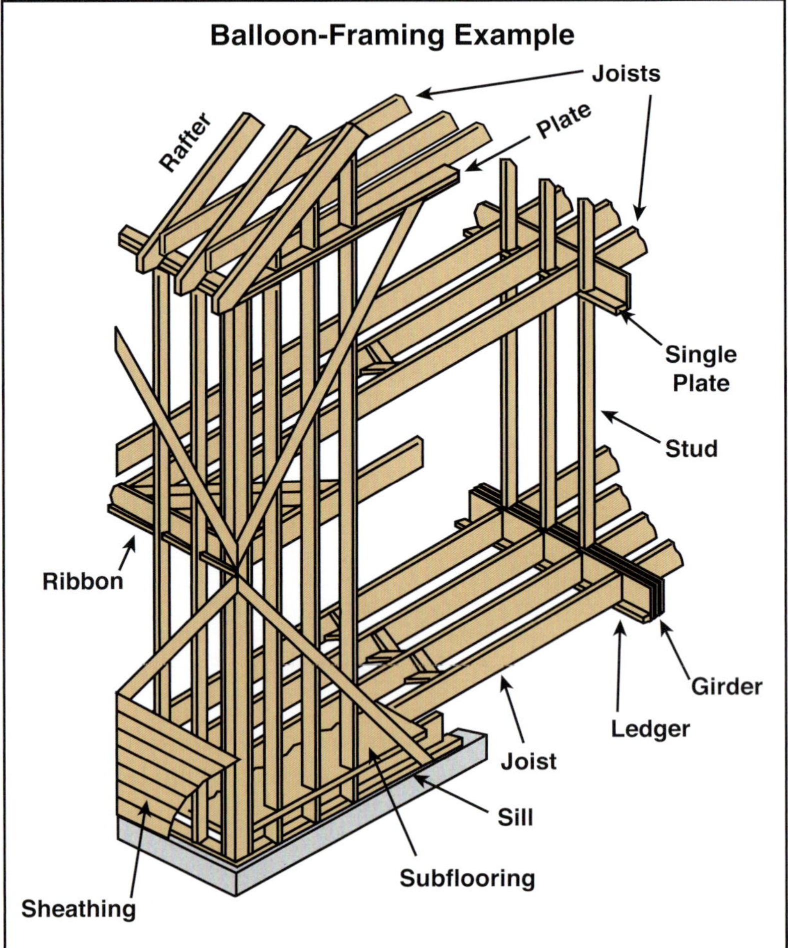

Figure 4.31 Illustration of typical balloon frame construction that connects floors with no fire stops.

Platform-Frame Construction (Western Framing)

In platform-frame construction (also sometimes known as *western framing*), the exterior wall vertical studs are not continuous to the second floor. The first floor is constructed as a platform upon which the exterior vertical studs are erected. The second floor is also constructed as a platform, and the second floor studs are erected on the second floor **(Figure 4.32)**. Because of these platforms, exterior walls in this construction include **fire stops** at each level, which prevent fire from easily spreading between floors.

The construction of a platform-frame building begins with the following steps:

Step 1: Attach a wood **sill** to the foundation, usually with bolts.

Step 2: Attach a header and floor joists or trusses to the sill.

Step 3: Attach subflooring to the floor joists to form a floor deck.

The first floor wall framing, with a top and bottom plate, is usually laid out horizontally on the floor deck and then raised into its vertical position. When the first floor walls are in position and braced, the second floor joists are erected on the top plates of the first floor walls.

Compartmentation

NFPA 1033 (2022) 4.2.2, 4.2.3

Compartmentation is defined as the series of barriers designed to keep flames, smoke, and heat from spreading from one room or floor to another. Barriers may be doors, walls or partitions, fire-stopping materials inside walls or other concealed spaces, or floors. In some occupancies, such as hotels with numerous rooms, a degree of compartmentation is inherent in the architecture of the building **(Figure 4.33, p. 120)**. In other occupancies, such as a large warehouse, there is little or no compartmentation. A lack of compartmentation can result in a rapid spread of fire horizontally and vertically through a building. Examples of this type are open vertical shafts and undivided attics.

Almost any floor and ceiling assembly, wall construction, or door acts as a barrier to some degree to products of combustion (smoke, heat, flame, or fire gases). However, not every wall, partition, or door in a building is fire rated. For example, most partitions and doors separating individual rooms within an apartment are not fire rated, but they can still be effective in preventing the passage of smoke and heat. The degree of fire resistance required by the building code of a wall or partition depends on its purpose.

The arrangement and size of compartments in a building directly affects fire development, severity, possible duration, and intensity. Building

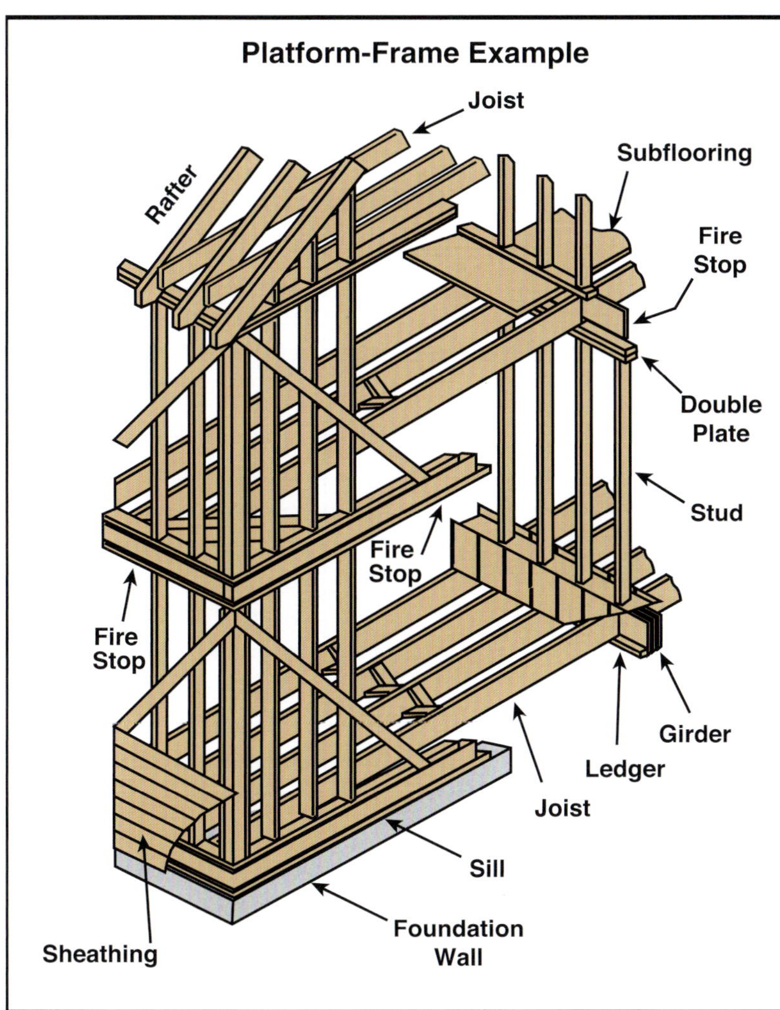

Figure 4.32 Platform framing includes fire stops between floors.

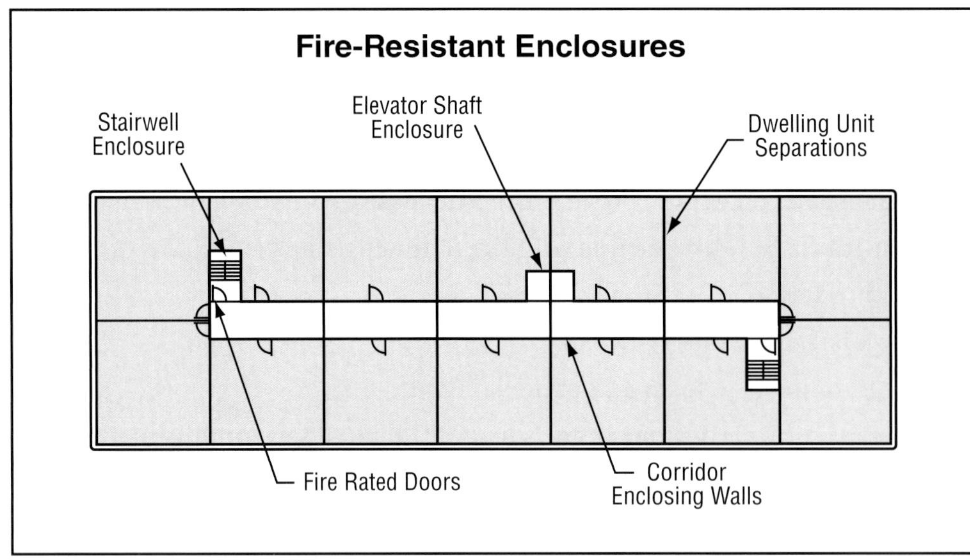

Figure 4.33 Fire-rated partitions are located in strategic areas of an occupancy.

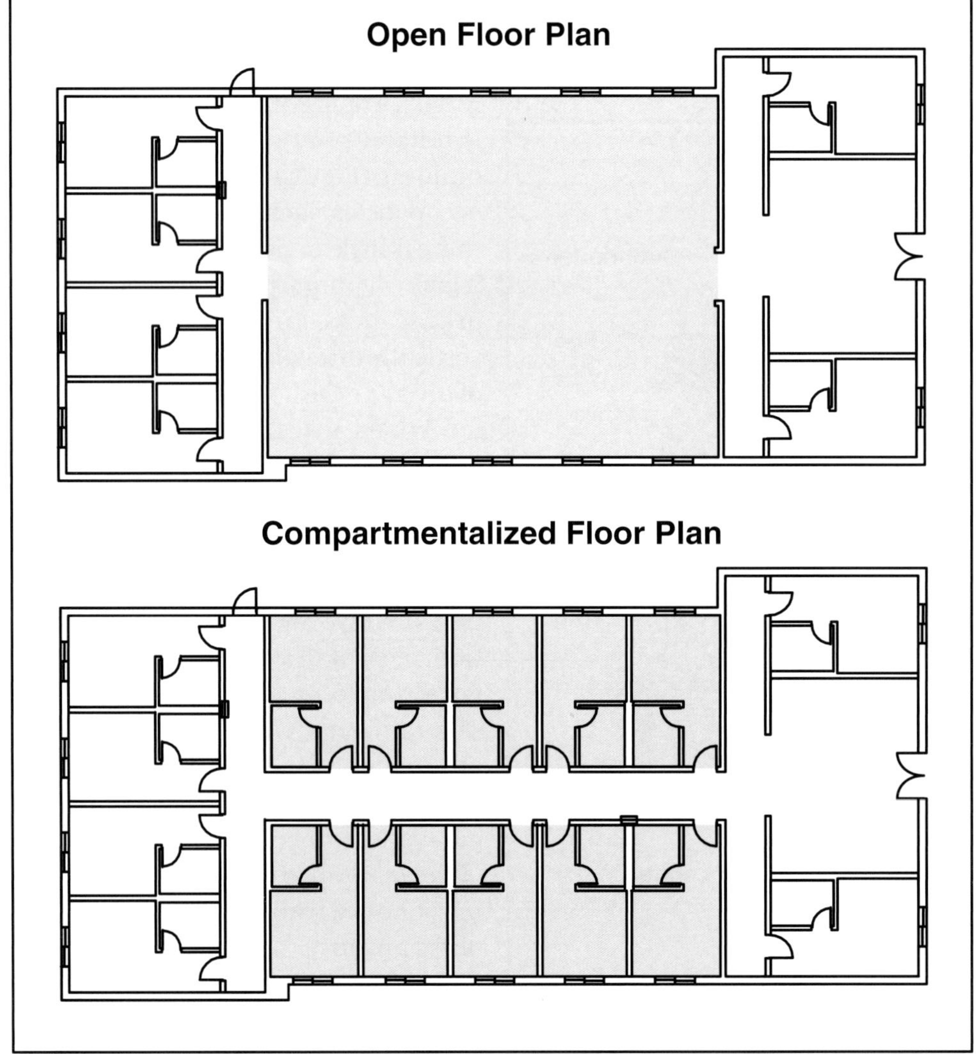

Figure 4.34 Open floor plans have non-fire benefits compared to compartmentalized floor plans.

compartmentation is the layout of the various open spaces in a structure and includes:

- Floor plan
- Barriers to fire spread
- Openings between floors
- Continuous voids or concealed spaces
- Number of stories above or below ground

Each of these elements may contribute to fire spread or containment. For example, an open floor-plan space may contain furnishings that provide fuel sources on all sides of a point of ignition. Conversely, a compartmentalized configuration may have fire-rated barriers, such as walls, ceilings, and doors, separating fuel sources and limiting fire development to an individual compartment **(Figure 4.34)**.

Any open space with no complete fire barrier dividing it is considered a *compartment*. Connected rooms are considered to be separate compartments if a barrier (door) between them is closed. When the door is open, a fire can draw in the oxygen from the adjoining room. Given enough available fuel, fire will move toward the air supply along any available flow path **(Figure 4.35)**.

A fire in a large compartment will develop toward flashover or full-room involvement more slowly than a fire in a small compartment. As discussed in Chapter 3, fire primarily spreads within a compartment along the hot gas layer. Radiant heat energy from the hot gas layer ignites secondary fuels.

Fire in rooms with high ceilings and walls far apart will develop more slowly than rooms with low ceilings and walls closer together, because of the development of the hot gas layer. Low ceilings and walls closer together allow for the ceiling layer to develop and thicken more quickly than a fire in a larger room with high ceilings and walls farther apart.

Concealed and Interstitial Spaces

Concealed and interstitial spaces in building construction, such as spaces for building systems, wiring, or plumbing, are common in most structural wall, ceiling, and floor systems. In addition, concealed and interstitial spaces may be byproducts of additional construction due to a renovation. The following sections describe these areas of building construction.

Interstitial Ceiling Spaces

Virtually all buildings include interstitial spaces (voids) in the ceiling assembly. Interstitial spaces may be commonly found in buildings to:

- Create space for HVAC air movement
- Create space when ceiling levels are lowered during remodeling
- Create space to accommodate building services above the ceiling; may be sufficient for a person to stand to access the equipment

The amount of combustibles in an interstitial space is usually low in new commercial construction because of the requirements for fire-rated cables. In fire-resistive buildings, automatic sprinklers usually are not provided in the space. Over a period of time or in the case of remodeled ceilings, combustible materials may accumulate in the space. Similarly, intermediary spaces may include intentional voids in building construction that are used for storage when the occupancy is in use **(Figure 4.36)**.

Figure 4.35 Burning behind a door can simulate a smaller compartment. This pattern may form a plume and a ceiling jet. *Courtesy of Jennifer Johnson.*

Figure 4.36 Intentional void spaces may be used as storage. *Courtesy of Nelson Mascarenhas.*

NOTE: Some older types of construction, such as balloon construction previously discussed in the wall assembly section, include interstitial spaces as a factor of their construction.

Utility Chases

Utility chase is a term generally applied to the vertical pathways in a building that contain building services. A utility chase may contain combustible materials such as plastic pipe and electrical insulation.

These pathways include:

- Data cableways
- Plumbing chases
- Electrical raceways
- Telecommunication routes
- Ductwork for HVAC systems

Chutes and Hoistways

Shaft enclosures and **chutes** are addressed together in model building codes. These features may affect fire spread through a structure. A **refuse chute** (also known as *trash chute*) provides for the removal of unwanted material from upper floors of buildings. A large vertical chute extends through the building and has openings on each floor for depositing material. The chute terminates at grade level or in a basement where the refuse is collected and possibly compacted for disposal **(Figure 4.37)**.

A linen chute is similar to a refuse chute in construction and potential fire spread. Linen chutes provide for the removal of soiled linen from upper floors of hotels, health care facilities, and similar occupancies. Chutes are often limited in height to one or two stories and often are not sprinklered.

Figure 4.37 Refuse and laundry chutes channel deposited items from a designated place to another destination for removal or processing.

Current model codes require sprinkler systems in chutes over a certain height, generally three stories. An improperly constructed chute is a potentially severe fire problem. A chute can become jammed, resulting in the possibility of a large quantity of combustible material being lodged in the chute. Further, the trash deposited in the chute is mostly combustible, and it is easy to drop an ignition source, such as a cigarette, down the chute. Smoke from a fire in a chute can be carried through all floors, and heavy smoke may be present in the upper floors. Fire that spreads from chutes can indicate that sprinkler systems are inoperative.

In addition to chutes, some occupancies may include dumbwaiters. Comparably to an elevator, a dumbwaiter moves along a hoistway and can transport small, light items between floors. Dumbwaiters may include an electric motor, especially in modern structures. Hoistways may not be sprinklered, depending on the construction authority having jurisdiction. If they are classified with elevators, the hoistway may be constructed with only the wiring, ductwork, or piping needed for the dumbwaiter itself, within a fire rated enclosure.

Ventilation

As discussed in Chapter 3, Fire Dynamics, sufficient air must be available for a fire to develop beyond the incipient stage. Building construction plays an important role in ventilation in that it can contribute to air flow or hinder it. Fire investigators should examine a scene and determine how building construction has affected the air flow to a fire and how this air flow, in turn, affected the course and development of the fire. Restricting or enhancing ventilation into a compartment can have both positive and negative consequences. For example, a fire in an enclosed room may fail to develop due to lack of adequate air.

Ventilation to a fire can be provided via mechanical means such as heating, ventilating, and air conditioning (HVAC) flows, as described in Chapter 5, Building Systems and Utilities, or through natural openings such as windows and open doors. The amount of ventilation available to a fire has an effect on fire growth and spread.

A significant factor in construction trends includes the prevalence of open floor plans. Fires in structures with minimal compartmentation are often not ventilation limited, depending on the size of the compartment and the heat release rate of the fire. Yet even with larger compartments, a fire in a large open space provided with enough fuel will show a slowed heat release rate after enough time, based on a reduction in oxygen available for combustion.

Ceiling Assemblies

Ceilings provide an appropriate interior finish and may be used to conceal building systems such as wiring, ventilation, and sprinkler piping. In multistory buildings, floors and ceilings act as the primary barrier to vertical fire spread **(Figure 4.38)**. A fire investigator should consider the combustion properties of specific ceiling/floor assemblies and how the height of the ceiling may have affected fire development and fire patterns. Penetrations in the ceiling assembly can also affect the direction of fire spread.

A ceiling is often rated as part of a floor and ceiling assembly. The materials used to construct the ceiling are rated independently to confirm their function in an assembly. Varying levels of ceiling heights can also affect time lags concerning the time fire-suppression and warning systems are activated as well as full room involvement. When a space exists between a ceiling and the floor above, a concealed space is created through which fire can spread.

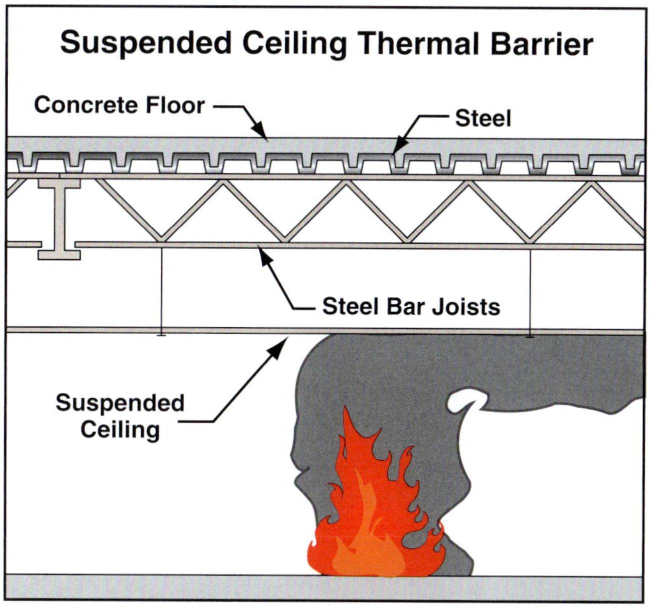

Ceilings are constructed using a variety of materials, but the most common ones are plaster, gypsum board, and mineral tile. When an exotic interior decor is desired, materials such as polished aluminum and mirrors may be used. Occasionally, the following combustible ceiling materials are used:

- Solid wood boards

- Fiberboard

- Oriented Strand Board (OSB)

- Plywood

- Styrofoam® extruded polystyrene foam tiles

Figure 4.38 A suspended ceiling protects roof framing systems.

These combustible coverings significantly increase the **fuel load** and can also increase the rate of fire spread in a structure. Combustible materials may fall during the fire onto other combustible materials within the compartment causing additional fires that may look as though there are multiple points of origin.

Roof Systems

The basic purpose of a roof is to protect the inside of a building from exposure to snow, wind, rain, etc. However, the roof also provides for a controllable interior environment, enhances the architectural style of a building, and contributes to the functional purpose of the building. For example, roofing materials prevent unplanned moisture within a building, and some types of roof supports help hold walls in place. Roofs are frequently described by their style or shape **(Figure 4.39)**.

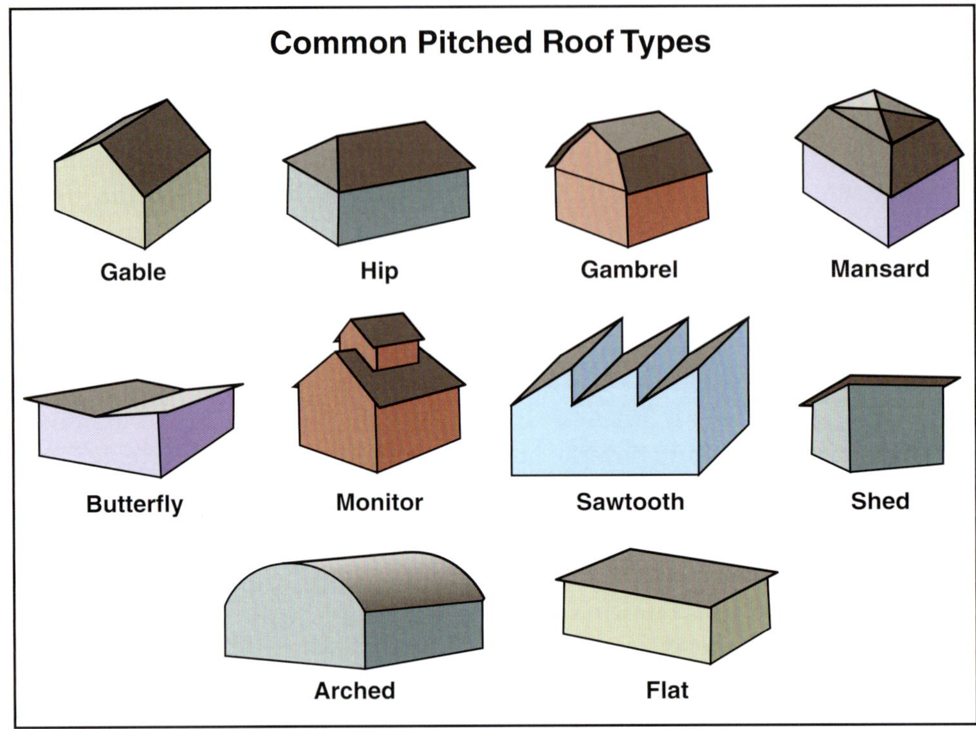

Figure 4.39 Common pitched roof types are designed for specific applications.

A fire investigator should use caution when examining a roof after a fire because the structural stability of the system may be compromised. Roof systems can collapse in large sections, making the examination of the interior of a building very difficult **(Figure 4.40)**. The use of heavy construction equipment may be the only safe method to remove the debris. If possible, an excavator with a thumb attachment may be the preferred method of removing sections of the roof. Using heavy equipment with a thumb attachment will allow the investigator to examine the section removed in larger pieces with less damage caused by the use of heavy equipment **(Figure 4.41)**.

Figure 4.40 Roofs may collapse in sections, which can impair access to those areas and adjacent sections. *Courtesy of Nicole Brewer.*

Figure 4.41 A digger with a thumb attachment on the bucket can manipulate a larger range of materials at a structural collapse site. *Courtesy of Mike Makela.*

Also of interest to fire investigators are roof systems built from products that will melt and flow when they are exposed to the heat of a fire. Many of these products are made from hydrocarbon-based materials and will burn when exposed to fire. The flowing material can result in burning that is remote from the area of origin and give indicators that are similar to those of ignitable liquids. The sections that follow address specific considerations for investigators regarding roof systems.

Arched and Domed Roofs

Arches are suitable for supporting roofs with large, clear spaces such as exhibition halls and field houses **(Figure 4.42)**. Arches were originally constructed of masonry, but modern arches can be made of wood, concrete, or steel. A dome roof can be used when the area to be enclosed is circular **(Figure 4.43)**. A dome roof produces structural forces similar to those of an arch.

Fire Behavior. Fire investigators should understand that fire behavior in arched or domed buildings will be different from rectangular roofed buildings. Heated air from a fire gathers at the highest point, which is the center of the arch. This collected heat combined with the compressive forces used to support the arch or dome can make collapse likely.

Figure 4.42 Arched roofs provide a large space underneath with minimal need for supports.

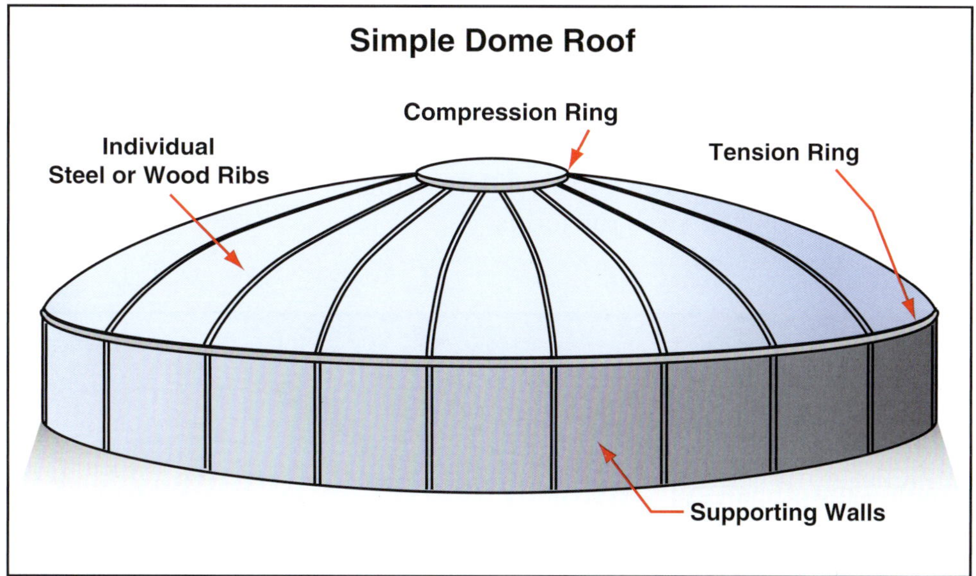

Figure 4.43 Circular roofs are often constructed as a series of half arches that connect at a central point.

Bowstring Truss Roof. An arched or curved outline often indicates a bowstring truss roof. Before 1960, the bowstring truss roof design was one of the most common design types for large commercial and industrial structures. The bowstring truss roof was commonly used in facilities wherever large open floor spaces with limited interior supports were needed, such as:

- Automobile dealerships and repair facilities
- Bowling alleys
- Grocery stores
- Industrial complexes

Bowstring Truss Construction. The principles of bowstring truss construction are similar to other types of truss construction. Web members form a series of triangles that transfer tension from the bottom chord and compression from the top chord of the truss onto the load-bearing walls (**Figure 4.44**). One difference with the bowstring truss is that the compressional forces within the top chord act to force the load-bearing walls outward as well as downward. Another difference is that the space between trusses is greater than the spaces in other types of trusses. Bowstring truss roof systems constructed before the late 1960s have a common code deficiency: the bottom chord members may have inadequate tensile strength to support code-prescribed roof loads.

Figure 4.44 A bowstring truss allows a roof to cover a wide span without internal supports.

Some construction features, such as parapets, obstruct the ability to easily identify bowstring truss systems. Preincident surveys should provide information on hidden construction features and should be consulted during size-up and operations. Aerial mapping or photographs can also help identify bowstring truss structures.

Roof Coverings

Fire investigators must consider roof coverings and their burning properties when conducting investigations. A wide variety of roof coverings are in use. Materials that may be used to cover a roof include (**Figure 4.45 a-e**):

- Slate
- Clay tile
- Wood shingles
- Plastic coatings
- Felt and asphalt in shingle and roll forms
- Sheet metal (galvanized steel, copper, aluminum, and lead)

Figure 4.45a-e Five examples of typical roof coverings: a) shake shingles, b) asphalt shingles, c) roof tiles, d) metal roofs, and e) flat membrane and asphalt.

In dry climates, wood shingles are often applied to the structural members of the roof using no intermediate sheathing. Wood shingles are a significant hazard if flying embers or other burning materials fall on them and cause them to ignite.

In some cases, one type of roof assembly may be installed on top of an existing but different roof material. In other cases, the original roof membrane may be covered with metal sheathing. When roofs are added over existing roofs, hazards may increase such as **(Figure 4.46)**:

- Lack of normal venting

- Trapped heat can lead to rekindle

- Added fuel load, such as with foam insulation

Membrane Roofing

Many membrane roofs contain **bituminous materials** that are hydrocarbon based and can liquefy and spread a fire rapidly across a roof. The burning, liquefied material may create fire patterns that resemble those associated with incendiary fires but should not be mistaken as such.

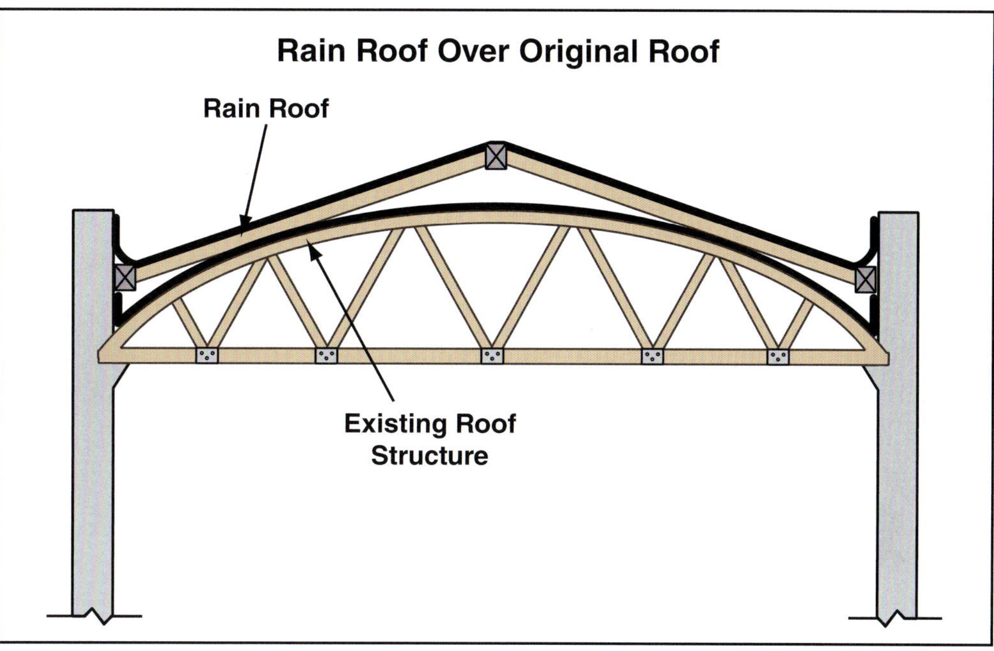

Figure 4.46 Unpredictable voids may be created when a rain roof covers an older roof.

Some common properties of a fire involving roof membranes include:

- Spreads across the entire roof with great speed

- Occurs during the application phase due to the use of propane torches to attach the materials

- Smolders and remains concealed and undetected for some time after the material has been applied

- Liquefies some materials that can drop down through holes in the roof decking onto the contents below

- Occurs at seams with an exterior wall where wooden shingles, sheathing or other combustible materials may be present. Fires in these instances typically travel within the wall cavity and may extend to an upper roof before they are noticed.

Roofing Materials and Hydrocarbons

Samples are taken for many reasons at a fire investigation scene. Samples of roofing materials taken for laboratory analysis are likely to show hydrocarbon components based on the construction type and materials used. These results may not indicate whether a fire was intentionally set. Depending on SOPs, a second sample may be collected for the lab to make a comparison. Investigators must work under the guidance of their testing facilities and legal authorities. Laboratory testing is discussed in greater detail in Chapter 13, Cause Determination.

NOTE: Fire investigators should determine what activity happened on the roof during the installation.

The following general categories of membranes are used:

- **Built-up membranes** — Use several overlapping layers of roofing felt applied to a roof deck with intervening layers of roofing cement. The layers are then saturated with a bituminous material that may be either tar or asphalt **(Figure 4.47)**.

- **Single-ply membranes** — Consist of a single membrane (made from several materials) laid in sheets on a roof deck. The most common is a synthetic rubber material, ethylene propylene diene monomer (EPDM). Other materials include polyvinyl chloride (PVC) and chlorinated polyethylene (CPE). The single-ply materials can be stretched and consequently will accommodate shifting in a building. They can be applied over decks in new buildings and over existing roofs. Application:

  — Single-ply membranes are rolled out and attached to roofs by means of adhesives, gravel ballast, or mechanical fasteners.

  — Adhesion to the roof can also be accomplished by using a propane torch to heat the underside of the membrane as it is unrolled.

  — Fluid-applied membranes — Consist of a material applied as a liquid and allowed to cure; usually several coatings are applied. The materials used include neoprene, silicone, polyurethane, and butyl rubber. Fluid-applied membranes are useful for buildings with curved roof surfaces, such as domes, that would be difficult to cover with other materials.

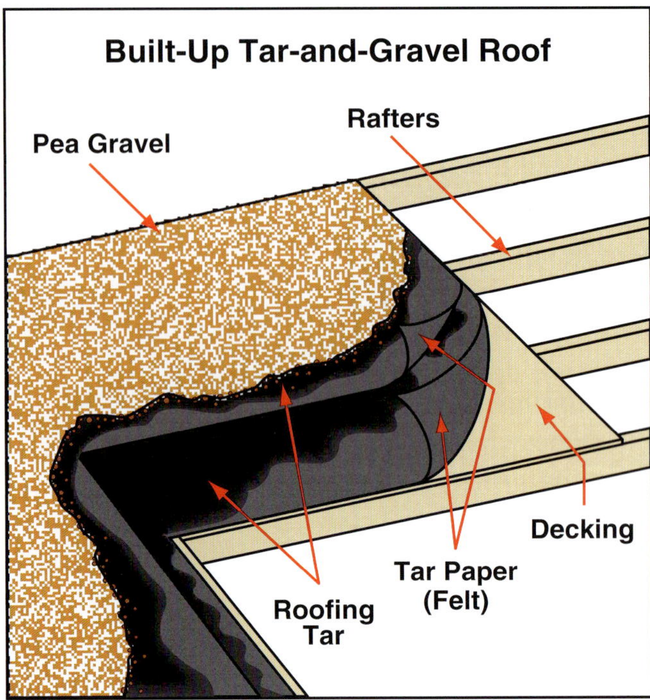

Figure 4.47 Built-up membrane roofs use materials intended to insulate, waterproof, and provide long-term durability.

Reaction of Building Construction to Fire

NFPA 1033 (2022) 4.2.2, 4.2.3, 4.2.4, 4.2.5

Resources may be available for an investigator to preemptively research information about a specific building. For example, building inspectors may be able to confirm building plans and uses that were registered with their agencies or departments.

In addition to use and design, investigators may discover information that drives future fire response and development of passive fire mitigation. Factors that may be relevant include determining:

- Whether the structure was legally able to be occupied

- The type of resources that should have been available within the structure, such as mechanical door closures

NOTE: More information about items that may be discovered during an inspection is included in the IFSTA manual, **Fire Inspection and Code Enforcement**. An investigator can outsource this level of evaluation at the structure owner's expense.

The following sections refer to additional resources that investigators can use to gain further perspective on the conditions they found, or can expect to find, at an investigation scene. This information can be used in building models, and explaining fire effects.

Construction Type and Elapsed Time of Structural Integrity

Most building codes rate the various construction types according to how long each construction type maintains its structural integrity over a certain period of time. **Table 4.2** shows some examples of the expected fire resistance of the five types of construction. How long a building will maintain structural integrity cannot always be determined by the type of construction or the fire resistance rating of various assemblies.

Table 4.2
Fire Resistance Ratings for Type I through Type V Construction (In Hours)

	Type I		Type II			Type III		Type IV	Type V	
	442	332	222	111	000	211	200	2HH	111	000
Exterior Bearing Walls										
Supporting more than one floor, columns, or other bearing walls	4	3	2	1	0	2	2	2	1	0
Supporting one floor only	4	3	2	1	0	2	2	2	1	0
Supporting a roof only	4	3	1	1	0	2	2	2	1	0
Interior Bearing Walls										
Supporting more than one floor, columns, or other bearing walls	4	3	2	1	0	1	0	2	1	0
Supporting one floor only	3	2	2	1	0	1	0	1	1	0
Supporting roofs only	3	2	1	1	0	1	0	1	1	
Columns										
Supporting more than one floor, columns, or other bearing walls	4	3	2	1	0	1	0	H	1	0
Supporting one floor only	3	2	2	1	0	1	0	H	1	0
Supporting roofs only	3	2	1	1	0	1	0	H	1	0
Beams, Girders, Trusses, and Arches										
Supporting more than one floor, columns, or other bearing walls	4	3	2	1	0	1	0	H	1	0
Supporting one floor only	2	2	2	1	0	1	0	H	1	0
Supporting roofs only	2	2	1	1	0	1	0	H	1	0
Floor-Ceiling Assemblies	2	2	2	1	0	1	0	H	1	0
Roof-Ceiling Assemblies	2	1½	1	1	0	1	0	H	1	0
Interior Nonbearing Walls	0	0	0	0	0	0	0	0	0	0
Exterior Nonbearing Walls	0	0	0	0	0	0	0	0	0	0

H: Heavy timber members
Adapted from Table 4.1.1 of NFPA 220.

Building and Fire Code Analysis

An investigator may need to conduct a building and code fire analysis to help document conditions and explain the damage at the scene. A building and code fire analysis is separate from an origin and cause analysis.

Information gathered at the scene provides indicators of structural integrity. The investigator can then compare this immediate knowledge with any known information about the structure, such as its construction type, to form opinions about safety at the scene.

Fuel Load of Structural Members and Contents

NFPA 1033 (2022) 4.2.2, 4.2.4

The total quantity of combustible contents of a building, space, or fire area is referred to as the fuel load (some documents may use the term fire load). All combustible materials in the building's construction comprise the fuel load, such as:

- Floors
- Ceilings
- Furnishings
- Wood framing

- Interior finishes
- Decorations including seasonal items
- Combustible materials within the building

The more materials that are combustible, the more fuel is available to pyrolize and burn, provided there is an adequate supply of oxygen for combustion. Knowledge of building construction and type of occupancies can assist in developing hypotheses regarding potential fuel loads (i.e. structure and contents), fire development, and fire spread in a room or structure.

The orientation of the fuels as well as their chemical composition and surface-to-mass ratio will also influence the rate and intensity of fire spread. The contents of a structure are often the most readily available fuel source, significantly influencing fire development in a compartment fire. When contents release a large amount of heat rapidly, both the intensity of the fire and speed of development will increase. For example, synthetic furnishings, such as polyurethane foam, begin to pyrolize rapidly under fire conditions even when the contents are located some distance from the fire's origin. The chemical makeup of the foam and its high surface-to-mass ratio speed the process of fire development **(Figure 4.48)**.

Figure 4.48 These images demonstrate how the polyurethane foam in the chair pyrolizes rapidly under fire conditions. *Courtesy of Dan Madrzykowski, NIST.*

The proximity and continuity of contents and structural fuels also influence fire development. Fuels located in the upper level of adjacent compartments will pyrolize more quickly because of heat radiating from the hot gas layer. Continuous fuels, such as combustible interior finishes, will rapidly spread the fire through compartments.

Similarly, the fire's location within the building will influence fire development. Fires originating on upper levels generally extend downward much more slowly following the fuel path or as a result of structural collapse. When the fire originates in a low level of the building, such as in the basement or on the first floor, convected heat currents will cause vertical extension through:

- Atriums
- Stairways
- Vertical shafts
- Concealed spaces

Additionally, if the structural elements of the building become involved in the fire, not only does the structure itself provide a new source of fuel, but the fire may be burning in hidden cavities at several places within the building. These hidden spaces make finding and extinguishing the fire more difficult and increase the potential risk of building collapse.

In commercial, industrial, and storage facilities with large fuel loads, the fire may overwhelm the capabilities of a fire suppression system and make it difficult for firefighters to gain access during fire suppression operations **(Figure 4.49)**. Performing and updating preincident surveys is the most effective means of establishing awareness of these hazards.

Figure 4.49 These rolls of printing paper are an example of a heavy fuel load.

Assuming that there is available oxygen, the higher the fuel load is, the more likely the fire will behave in the following ways:

- The fire may have a higher heat-release rate.
- The longer the fire burns, the more fire spread accelerates.
- If structural members are part of the fuel load, structural integrity of the building will deteriorate faster.
- The structure may self-ventilate, introducing even more oxygen to the ventilation-limited fire, and accelerating fire development and involvement of combustible structural members.

If fires are ventilation-limited, higher fuel loads indicate a greater amount of unburned fuel that could reignite with the introduction of a new oxygen source. There may also be a greater amount of unburned fuel gases in the air if fuel packages pyrolized but did not begin combustion before the building became ventilation-limited. Such buildings may have undergone backdrafts and flashovers if firefighters did not coordinate ventilation.

Interior Furnishings and Finishes

An important consideration for a fire investigator is the combustibility of the materials used for interior finishes. Combustible interior-finish materials have contributed to the fire spread in many fires involving multiple deaths and injuries. Interior finishes include window drapes or coverings, wallpaper and paneling, and carpet. Furnishings may include items found in occupancies such as:

- Beds
- Sofas
- Desks
- Tables

The speed of flame spread over an interior finish material is influenced by such factors as the chemical composition of the material used as a backing, whether the material is in the horizontal or vertical orientation, and the geometry of the space in which it is installed.

Modern building codes for many types of occupancies list the flame-spread index on interior-finish materials. These listings are used to prioritize the use of materials that will prevent rapid fire spread. Some construction may not meet modern standards, regardless of their date of construction.

Hazards presented by interior-finish materials include:

- Can speed the onset of flashover
- Can increase fire growth and spread
- Can increase the intensity of the fire
- Can increase the flame spread index of a surface
- Can pose a life safety hazard because of the additional smoke and toxic gases released when they burn

Combustible Exterior Wall and Roof Coverings

Flammable materials that contribute to the structure's fuel load often cover exterior walls. Exterior wall coverings may add carbon fuels (wooden siding) or petroleum-based fuels (combustible sheathing and foam-based insulation installed behind vinyl siding) to the fuel load. The wall coverings may be installed atop exterior insulation which, in turn, is another fuel source.

Exterior Cladding

As indicated in Chapter 3, Fire Dynamics, flame spread moves vertically significantly faster than it moves laterally. In cases where fire stopping is not in place, flames can propagate based on fuel and oxygen access.

On June 14, 2017, a fire in Grenfell Tower (in London, England) escaped the origin apartment and travelled vertically on the exterior façade. The exterior facade was clad with aluminum-covered combustible insulation. The flame temperatures were greater than the melting temperatures of aluminum, and flames propagated through the interior cavities of the facade system.

When exterior coverings become exposed to heat and catch fire, they can spread the fire to other areas of the structure or to adjacent exposures such as vegetation or neighboring buildings **(Figure 4.50)**. Wood shakes, even when treated with fire retardant, can significantly contribute to fire spread. This is a problem in wildland/urban interface fires where wood shake roofs have contributed to large fires. The following exterior coverings will limit air movement, retain heat in the structure, and add to the fuel load.

- Plastic wrap
- Cellulosic insulation
- Foam plastic insulation

Fire-resistant metal roof decking may be covered with combustible layers of foam insulation and felt paper covered with asphalt waterproofing. A fire below the metal deck can melt and ignite combustible materials, causing a second fire above the roof.

Figure 4.50 The exterior finish, insulation, and sheathing on this building enabled rapid exterior fire spread. *Courtesy of Ron Jeffers, Union City, N.J.*

Effects of Building Construction Features

Building features such as lightweight materials or open floor plans have direct effects on how fire will spread in the structure. Remember, a structure fire is the place where fire dynamics and building construction interact.

Modern vs. Legacy Construction

Over the past 50 years, building construction in North America has changed drastically. In single-family residential structures, the square footage of houses increased over 150 percent between 1973 and 2008. At the same time, lot sizes have shrunk, reducing access and increasing potential exposure risks (Kerber, "Analysis ..." 2012).

Residential interior layouts and construction materials have also changed. Older structures (prior to approximately 1970) had the following features:

- Smaller compartments
- Windows that could be opened for ventilation
- Air pockets used as insulation in empty wall cavities
- More compartments within the same square footage as modern homes

Modern single-family structures may feature:

- Atriums
- High ceilings
- Open floor plans
- Sealed windows
- Synthetic insulation
- Lightweight manufactured structural components
- Wall cavities often used to hide wiring, ductwork, and piping

Because of energy-efficient designs, the structures also tend to contain fires for a longer period of time, thus creating fuel-rich, ventilation-limited environments. These problems are magnified in large-area residential structures.

Commercial, institutional, educational, and multifamily residential structures also rely on energy conservation measures that increase the intensity of a fire and make the use of tactical ventilation difficult. Open plan commercial structures, such as warehouse-style stores, have high fuel loads in the contents and no physical barriers to prevent the spread of fire and smoke in the space **(Figure 4.51)**.

The use of plastics and other synthetic materials has also dramatically increased the fuel load in all types of occupancies. These synthetic materials produce large quantities of toxic and combustible gases. Fires in these fuel types can escalate rapidly, reach high temperatures, and consume the structure's available oxygen quickly.

Figure 4.51 A commercial structure with an open floor plan and a high fuel load should have a fire suppression system suited to the conditions.

Thermal Properties of the Building

The thermal properties of the building can contribute to rapid fire development. The thermal properties can also make extinguishment more difficult and reignition possible. Thermal properties of a building include:

- **Insulation** — Contains heat within the building which causes a localized increase in the temperature and fire growth and may introduce an additional fuel source.

- **Heat reflectivity** — Increases fire spread through the transfer of radiant heat from wall surfaces to adjacent fuel sources.

- **Retention** — Maintains temperature by slowly absorbing and releasing large amounts of heat.

Construction, Renovation, and Demolition Hazards

The risk of fire rises sharply when a structure is under construction, being renovated, or awaiting demolition. Contributing factors include the additional fuel loads and ignition sources such as open flames from cutting torches and sparks from grinding or welding operations.

Occupancy Change and Fire Load

Some expectations of fuel load can be developed based on the types of activities that take place in a building. Should the use or occupancy change, the hazards may also change. The investigator should be aware of the possibility that a structure's protective systems may not have been upgraded to protect against an increased hazard.

For example, a change in occupancy may require an upgrade to a sprinkler and alarm system. If those improvements are not made, the structure may be at risk from higher hazards. Changes in use or occupancy may also affect whether other fire safety measures may have been required such as the construction of fire walls or fire separations.

Buildings under construction are subject to rapid fire spread when they are partially completed because many of the protective features, such as gypsum wallboard and automatic fire suppression systems, are not yet in place. Buildings under construction with exposed wooden framing are often thought of as the equivalent of a vertical lumberyard **(Figure 4.52)**. The lack of doors or other barriers that would normally slow fire spread also contribute to rapid fire growth.

Abandoned buildings or structures undergoing renovation or demolition are also subject to faster-than-normal fire growth. Abandoned structures may have missing or altered components, may be in a deteriorated state, and may be more susceptible to structural collapse.

Figure 4.52 The exposed wood in this building would contribute to rapid fire spread.

Other potential contributors to fast fire growth include:

- Missing doors
- Breached walls
- Open stairwells
- Deactivated fire suppression systems

Hazardous situations may arise during renovation because occupants and their belongings may remain in one part of the building while work continues in another **(Figure 4.53)**. Fire detection or alarm systems may be taken out of service or damaged during renovation. If good housekeeping is not maintained, accumulations of debris and construction materials can block exits. This debris may impede occupants trying to escape from the building in an emergency. The contractors or owner/occupants performing the renovations do not always follow local building codes.

Figure 4.53 A number of hazardous situations that might contribute to fire spread could arise during the renovation of this hospital. *Courtesy of Ron Moore, McKinney (TX) FD.*

Some local fire codes mandate that standpipe systems must remain in operation during the demolition of multistory buildings. Unfortunately, contractors do not always adhere to these requirements. The result is that inoperative standpipes and sprinkler systems have become a contributing factor in fires in buildings under demolition.

Review Questions

1. How does fire resistance of building materials affect fire development?

2. How does a knowledge of different types of building materials help the fire investigator?

3. Which types of building materials are most affected by exposure to fire?

4. Which types of building materials are likely to retain their structural stability when exposed to fire?

5. How do the five basic construction types differ from each other?

6. How do manufactured, prefabricated, and other unique construction types affect fire behavior?

7. How do the various structural features and assemblies impact fire growth and development?

8. How can understanding construction types and occupancy classifications help a fire investigator?

9. How does fuel load influence fire development?

10. How do building construction features impact fire spread?

11. What types of hazards exist for buildings under construction or undergoing renovation?

Discussion Questions

1. What types of building materials are most commonly used for residential structures in your area?

2. Which construction types are most commonly used in your area?

3. Are there any buildings in your area undergoing construction or renovation that might be especially hazardous if a fire were to start there?

Chapter References

Kerber, Stephen. 2012. "Analysis of Changing Residential Fire Dynamics and its Implications on Firefighter Operational Timeframes," Underwriters Laboratories, Inc.

Kerber, S., B. Backstrom, J. Dalton, and D. Madrzykowski. January 2012. "Full-Scale Floor System Field and Laboratory Fire Experiments." Underwriters Laboratories, Northbrook, Illinois.

Kerber, Stephen; Daniel Madrzykowski; James Dalton; Bob Backstrom. 2012. "Improving Fire Safety by Understanding the Fire Performance of Engineered Floor Systems and Providing the Fire Service with Information for Tactical Decision Making." Underwriters Laboratories, Inc.

Mowrer, F. P. June 1998. *Window Breakage Induced By Exterior Fires.* NISTGCR-98-751, Washington, DC: U.S. Dep L of Commerce.

Sanderson, J. L. 1995."Tests Add Further Doubt to Concrete Spalling Theories," FireFindings, Fall. volume 3.

SFPE Handbook of Fire Protection Engineering. 2016. Hurley, M.J. et al. editors, Springer.

Chapter Key Terms

Balloon-Frame Construction — Type of structural framing used in some single-story and multistory wood frame buildings; studs are continuous from the foundation to the roof, and there may be no fire stops between the studs.

Bituminous Material — Materials that contain tar or asphalt-like materials. Bituminous materials used in roofing include tar and tar-impregnated papers.

Chase — Vertical or horizontal space in a building used to route pipes, wires, ducts, or other utility or mechanical systems. Usually surrounded with a fire-rated enclosure.

Chute — Salvage cover arrangement that channels excess water from a building. A modified version can be made with larger sizes of fire hose.

Compartmentation — Series of barriers designed to keep flames, smoke, and heat from spreading from one room or floor to another.

Compression — Vertical and/or horizontal forces that tend to push the mass of a material together; for example, the force exerted on the top chord of a truss.

Dumbwaiter — Small freight elevators that carry items, not people, and generally have a small weight and size capacity.

Fire Cut — Angled cut made at the end of a wood joist or wood beam that rests in a masonry wall to allow the beam to fall away freely from the wall in case of failure of the beam. This helps prevent the beam from acting as a lever to push against the masonry.

Fire Resistance — The ability of a structural assembly or material to maintain its load-bearing ability under fire conditions.

Fire Stop — Solid materials, such as wood blocks, used to prevent or limit the vertical and horizontal spread of fire and the products of combustion; installed in hollow walls or floors, above false ceilings, in penetrations for plumbing or electrical installations, in penetrations of a fire-rated assembly, or in cocklofts and crawl spaces.

Fuel Load — The total quantity of combustible contents of a building, space, or fire area, including interior finish and trim, expressed in heat units of the equivalent weight in wood.

Gusset Plate — Metal or wooden plates used to connect and strengthen the joints of two or more separate components (such as metal or wooden truss components or roof or floor components) into a load-bearing unit.

Noncombustible — Incapable of supporting combustion under normal circumstances.

Parapet — (1) Portion of the exterior walls of a building that extends above the roof. A low wall at the edge of a roof. (2) Any required fire walls surrounding or dividing a roof or surrounding roof openings such as light/ventilation shafts.

Platform Frame Construction — Type of framing in which each floor is built as a separate platform, and the studs are not continuous beyond each floor; also called *western-frame construction*.

Polyvinyl Chloride (PVC) — Synthetic chemical used in the manufacture of plastics and single-ply membrane roofs.

Refuse Chute — Vertical shaft with a self-closing access door on every floor; usually extending from the basement or ground floor to the top floor of multistory buildings.

Sill — (1) Bottom rough structural member that rests on the foundation. (2) Bottom exterior member of a window or door or the masonry below.

Spalling — Expansion of excess moisture within masonry materials due to exposure to the heat of a fire, resulting in tensile forces within the material, and causing it to break apart. The expansion causes sections of the material's surface to violently disintegrate, resulting in explosive pitting or chipping of the material's surface.

Thermoplastic — Plastic that softens with an increase of temperature and hardens with a decrease of temperature but does not undergo any chemical change. Synthetic material made from the polymerization of organic compounds that become soft when heated and hard when cooled.

Thermoset Plastics — Plastics that are hardened into a permanent shape in the manufacturing process and are not subject to softening when heated again.

Photo courtesy of Donny Howard, Agent, Oklahoma State Fire Marshal's Office.

Chapter Contents

NFPA 1033 JPRs addressed in this chapter

This chapter provides information that addresses the following job performance requirements (JPRs) of NFPA 1033, *Standard for Professional Qualifications for Fire Investigator* (2022):

4.2.2	4.2.8	4.5.1
4.2.3	4.4.2	4.6.1

Learning Objectives

1. Explain how building systems relate to fire investigation. [NFPA 1033, 4.2.8]

2. Describe stairs and their importance to fire investigation. [NFPA 1033, 4.2.8]

3. Describe how transportation and conveyance systems can play a role in fire growth and development. [NFPA 1033, 4.2.8]

4. Describe how environmental systems can play a role in fire growth and development. [NFPA 1033, 4.2.2, 4.2.3, 4.2.8]

5. Describe how electrical systems can play a role in fire growth and development. [NFPA 1033, 4.2.8]

6. Describe the importance of fire protection systems to a fire investigation. [NFPA 1033, 4.2.2, 4.2.3, 4.2.8, 4.4.2, 4.6.1]

7. Describe the importance of recording and transmitting systems to a fire investigation. [NFPA 1033, 4.2.8, 4.4.2, 4.5.1, 4.6.1]

Chapter 5
Building Systems and Utilities

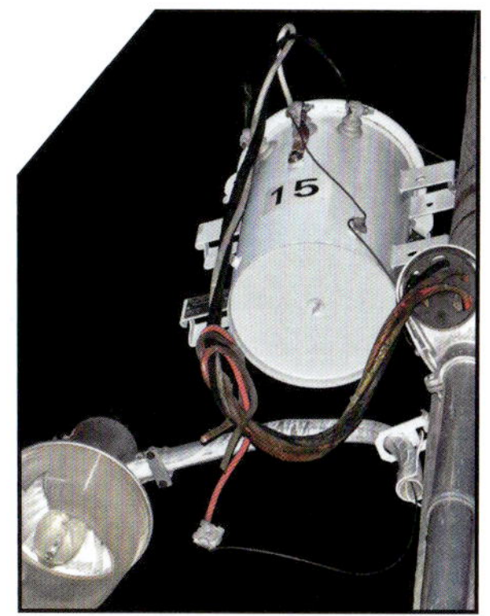

Courtesy of Donny Howard, Agent, Oklahoma State Fire Marshal's Office.

This chapter provides an overview of systems and utilities that serve functions within a building, and may contribute to fire spread and intensity. These systems and utilities are discussed in terms of their intended functions, the way they can affect fire conditions, and ways that they are interconnected with other building systems. Investigators should note the following as relevant to the incident site:

- Building systems related to fire investigation
- Stairs
- Transportation and conveyance systems
- Environmental systems
- Electrical systems
- Fire protection systems
- Recording and transmitting systems

Building Systems Related to Fire Investigation

NFPA 1033 (2022): 4.2.8

Buildings, like complex machines, often contain many basic systems to address the convenience, access, comfort, efficiency, and, most important, life safety for occupants. All building systems may significantly influence any fire event. Any defects in a system can impair the building's overall functionality. The building design team is responsible to ensure that the necessary building systems initially provide the intended level of fire and life safety. Subsequently, the building owner and/or management must maintain these systems over time.

Fire investigators should remember that all of a building's systems and utilities compose the total system. Fire investigators will have to determine how the systems in the building being examined may have been interconnected and what, if any, effect that interconnection had on the fire. A system that is improperly connected, designed, modified, or installed may contribute to the ignition and/or spread of the fire. This chapter references topics that may be explained in more detail in other IFSTA manuals:

- **Plans Examiner**
- **Fire Inspection and Code Enforcement**
- **Building Construction Related to the Fire Service**
- **Fire Protection, Detection, and Suppression Systems**

Building systems and the methods used for their installations may have a direct involvement in the ignition of fires in buildings. The installation and location of these systems can also affect how a fire and the products of combustion are transported in a building. An investigator should examine all building systems to determine if there was a failure or if the system was defeated.

Generally speaking, building systems fail or are defeated through the following:

- Human error, such as blocked venting in heat-producing appliances

- Deliberate tampering, such as pennies inserted in Edison Base fuse holders **(Figure 5.1)**

- Improper installation and maintenance, such as inadequate fireplace or chimney cleaning

NOTE: Investigators should avoid the operation or manipulation or alteration of any component of a building system. Doing so may change the evidentiary value of that component.

A fire investigator should address the following regarding building systems in fires involving residential properties:

- **Housekeeping** — Was the general housekeeping around heat-producing devices visibly maintained?

- **Electrical wiring** — Are there any indicators that electrical wiring could have been involved in the ignition of combustibles?

- **Combustion clearances** — Were the clearances from the furnace, fireplace, or wood stove according to applicable code?

- **Exhaust clearances** — Were the clearances provided for the furnace exhaust stack and the chimney for the wood stove proper to prevent ignition of structural components?

- **Piping or wiring** — Did penetrations for piping or wiring allow fire extension or smoke travel?

Figure 5.1 Pennies used as a conductor in place of actual fuses can cause an electrical system to fail.

One- and Two-Family Occupancies

The systems found in most one- and two-family occupancies may not be as complex as those found in larger buildings, but they can still affect a fire and the subsequent investigation **(Figure 5.2)**. The fire investigator must evaluate the effect the system or service had on a fire in a dwelling.

Questions a fire investigator should address regarding building services in fires involving residential properties include the following:

- Did penetrations for piping or wiring allow fire extension or smoke travel?
- Was the electrical wiring involved in the ignition of combustibles?
- Were the clearances from the furnace, fireplace, or wood stove according to applicable code?
- What was the housekeeping like in the area of the heat-producing devices?
- Were the clearances provided for the furnace exhaust stack and the chimney for the wood stove proper to prevent ignition of structural components?

Because heating devices and their related chimneys are such a common issue faced by fire investigators in residential properties, they should have a very good understanding of the codes that regulate them. Investigators should make themselves familiar with the codes and standards applicable in their jurisdiction. Investigators should understand that not all documents will be applicable to their jurisdiction and circumstances.

Stairs

NFPA 1033 (2022): 4.2.8

Most stairs provide a dual role as a building system. First, they enable access to multiple levels of the structure. Second, they serve as a basic component of building egress during an emergency.

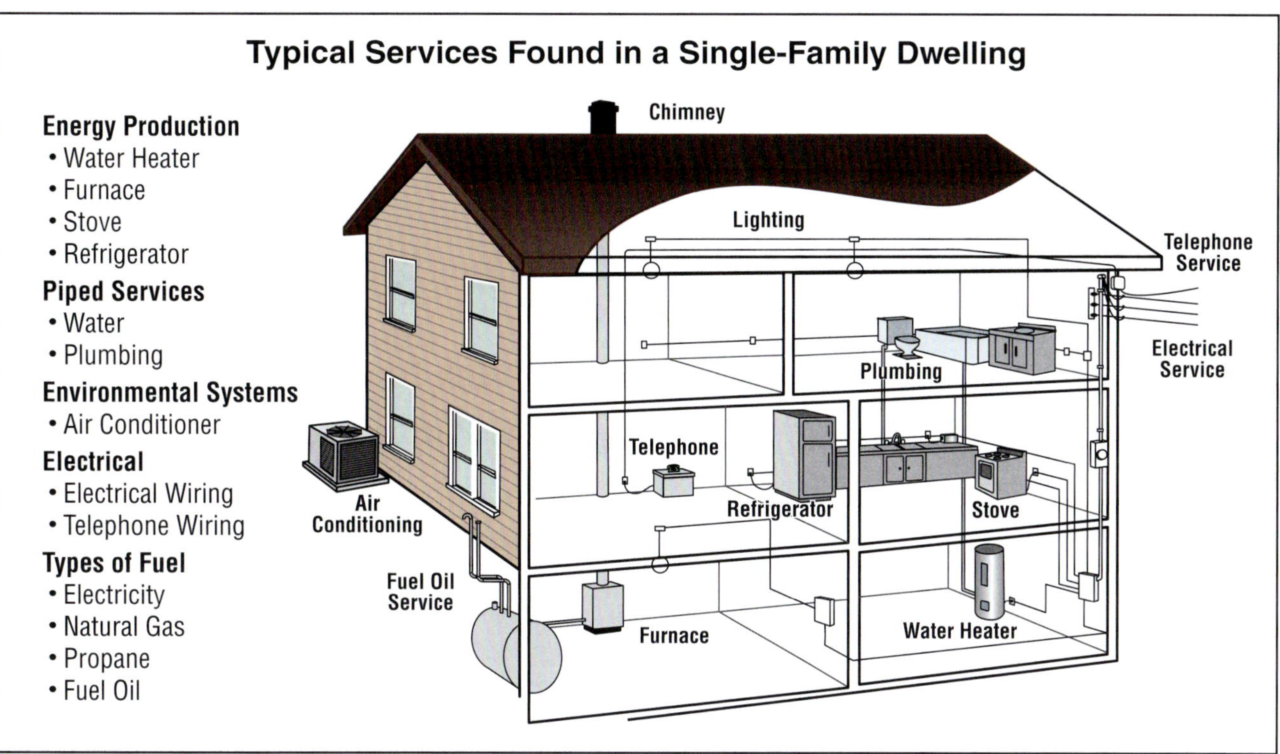

Figure 5.2 Services within a residence are typically found in similar places.

Fire investigators should consider whether stairs, their enclosures, or any associated systems, contributed to fire spread.

Stairs that are not part of the means of egress typically only connect two levels, per code limitations. Codes may require buildings with four or more stories in height to have a stairway, identified by signage, that extends to the roof for access to mechanical equipment.

Building systems and utilities that are interconnected with stairs can include:

- Electrical systems for lighting

- Pipe chases and utility runs for ease of installation

- **Heating, ventilation, and air conditioning (HVAC) systems** to pressurize stairwells

- Fire protection, detection, and suppression systems for convenient access and to minimize fire spread **(Figure 5.3)**

Stairwells may be designed to pressurize when the fire alarm system activates **(Figure 5.4, p.144)**. Pressurized stairwells minimize the use of space while still protecting stairways from the products of combustion. Blowers or fans provide a slightly greater pressure in the stairwell than the corridor to prevent the infiltration of smoke from the corridor into the stairwell.

Figure 5.3 Positive pressure stairwells keep smoke out of spaces where it could otherwise easily spread between floors.

In contrast, another style of **smoke control** for stairways is the use of a **smoke tower** separated from a building corridor by a vestibule that is open to the atmosphere **(Figure 5.5, p.144)**. Smoke entering the vestibule from the corridor is exhausted to the atmosphere and the stairwell remains free of smoke.

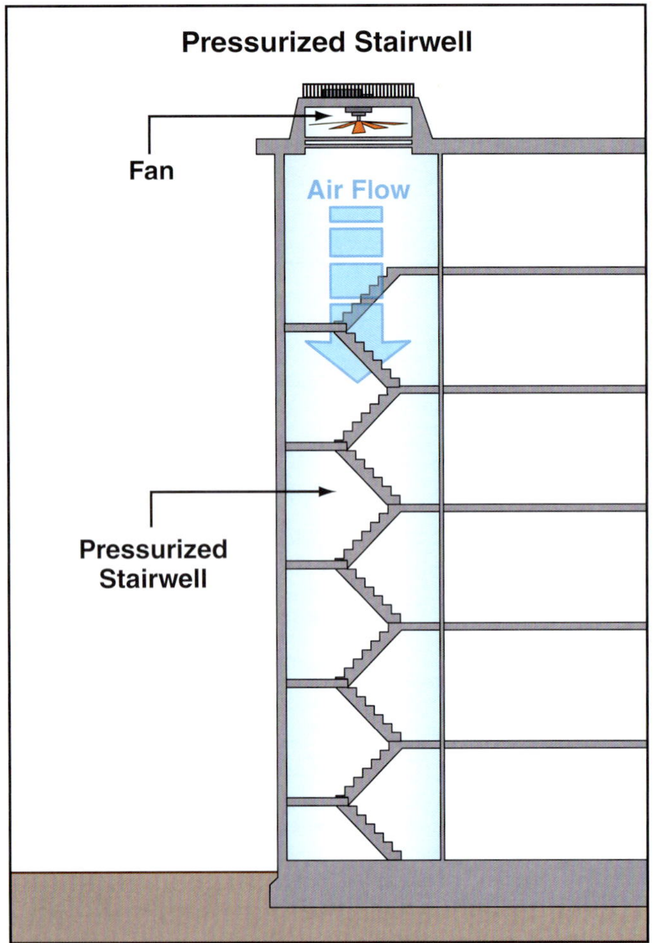

Figure 5.4 A pressurized stairwell uses a ventilation system to maintain a higher air pressure in the stairwell than on the individual floors to prevent smoke from spreading via that space.

Transportation and Conveyance Systems

NFPA 1033 (2022): 4.2.8

Transportation and conveyance systems within a building can include the following, whether separately or in combination:

- Elevators
- Escalators
- Conveyors
- Trash and laundry chutes
- Other materials handling systems

These systems move occupants, materials, and components of other systems vertically and horizontally within buildings. Shafts for ductwork, piping, and wiring are also included in this category.

Building systems and utilities that are interconnected with transportation and conveyance systems can include:

- Electrical systems for lighting and powering equipment throughout the structure
- Piping systems for heating, ventilation, and air conditioning throughout the structure
- Fire protection, detection, and suppression systems as required by codes
- Stairwells for ease of installation

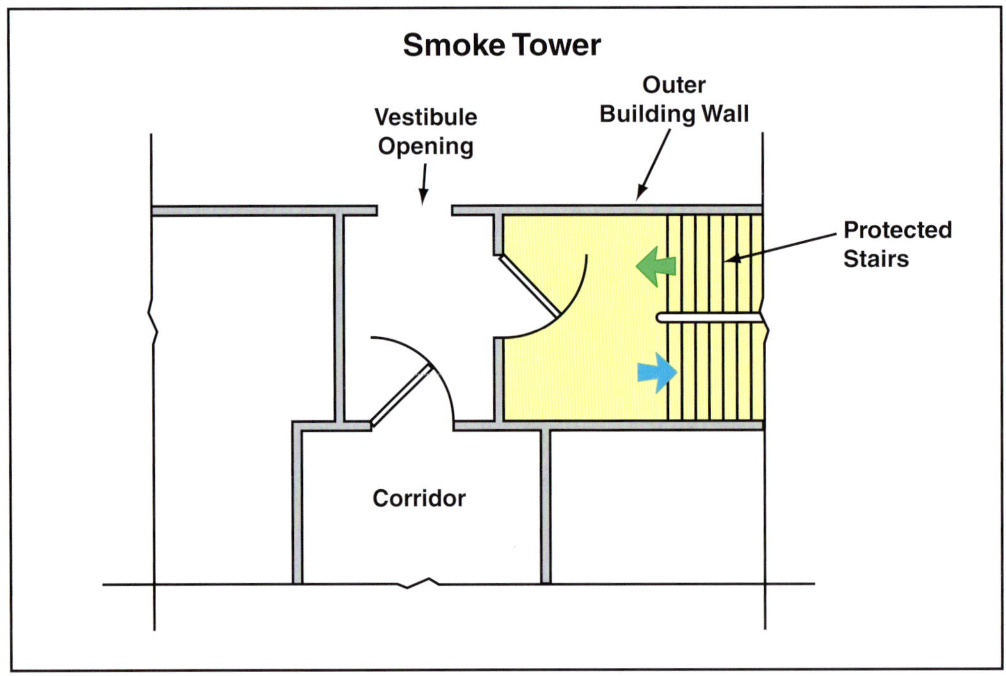

Figure 5.5 Similar to stairs and vestibules, smoke towers located at the periphery of buildings keep smoke out of a means of egress.

The transportation and conveyance systems provided in a building almost always require the penetration of floors and walls. These system accommodations can provide a path through which fire and smoke can travel if not properly protected with automatic **sprinklers** or **fire dampers** controlled by **fusible links** or smoke detectors. After installation, these penetrations must continue to be maintained to ensure that they do not provide a path of travel for smoke and fire.

Shafts, Chutes, and Chases

Building and fire codes contain provisions for the fire resistance of shafts that span multiple floors. These requirements establish a level of protection to the remainder of the building if smoke or fire enters the shafts.

Common fire-related problems with wiring and piping systems include:

- Penetrations in walls and floors can allow fire spread around them if the openings are not properly fire-stopped or otherwise protected.

- Heat can be transferred by conduction and cause the ignition of adjacent combustible materials some distance from the initial heat source.

- The increased use of plastic in pipes and wire insulation increases the potential that the fire can consume the wiring or piping and pass through a barrier.

For horizontal penetrations of building compartmentation or penetrations of floors with no shafts, the codes may require that the openings be fire-stopped or otherwise protected. Fire-stopping should prevent smoke and fire from traveling through the openings. Other methods of protection include the use of water spray sprinklers on both sides of a horizontal opening and the use of doors or shutters that close on exposure to fire products.

Wiring

In concept, all electrical systems have the potential to ignite combustibles. This potential is managed a number of ways to prevent unintended fires. For example, wiring itself is combustible and must be protected from physical damage. When wire is run in spaces that could present a life safety risk should it burn, a range of applicable codes may require the use of materials with a low smoke-production rating.

Telecommunications cable and equipment installed in buildings present a number of fire-related problems, including:

- Installation is sometimes completed by unlicensed individuals with little knowledge of fire safety or other building systems.

- The volume of wire necessary in a building may represent a significant fuel load.

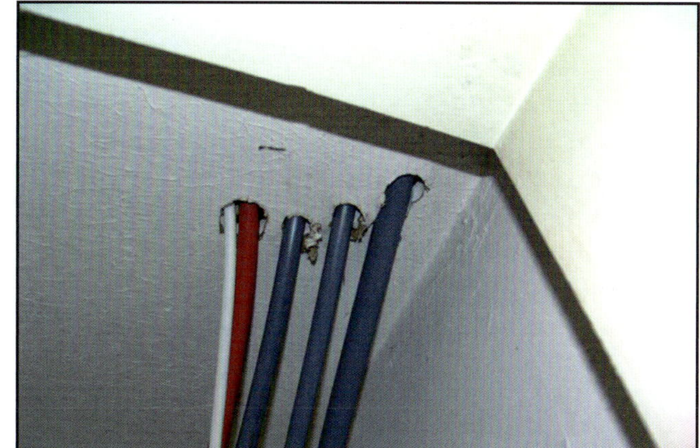

Figure 5.6 Anywhere a fire or smoke barrier is breached, structural accommodations and protections should be placed.

- Cables are often installed unprotected, tied together, or not listed for use in concealed spaces.

- Cable installation paths frequently breach fire and smoke barriers **(Figure 5.6)**.

- Associated equipment and workstations can be significant fuel packages.

Piping Systems

Piping systems transport a variety of gases and liquids within a building. Common building systems that use pipes include:

- Water utility
- Medical gases
- Fire protection **(Figure 5.7)**
- Compressed air
- Fuel gas or liquid fuels
- Wastewater/sanitary sewer
- Heating systems (hot water or steam)

Figure 5.7 Fire protection systems use pipes to move water as needed.

Residential and light industrial fuel-gas systems typically consist of natural gas (methane) or liquefied petroleum gas (butane or propane). The fuel-gas system provides a combustible mixture to the intended energy release point (heat-producing appliance). The fuel is contained in piping, and the total system consists of all the equipment and piping from the distribution point (tank or meter set) through the appliance. The installation and design of these systems are regulated by the codes such as *National Fuel Gas Code* (NFPA 54) and *Liquefied Petroleum Gas Code* (NFPA 58).

NOTE: The current edition of a code or standard may not be the one that applies to a building system. Investigators must always determine which editions of the codes are applicable in their jurisdictions or to the particular structure or occupancy being investigated.

In fuel-gas systems, the operating pressure (expressed in pounds per square inch [psi] or kilopascals [kPa] of water column) is regulated before entering a structure. Piping may be metal or plastic (or a combination of both). Failures within the piping typically occur at joints or appliance connections. Piping system failures may likewise occur within the appliance or pressure regulator, resulting in an accidental release of fuel and a potential of ignition at any heat- or energy-producing device.

Ducting and Void Spaces

Environmental systems often use ductwork and void spaces to move air around a building. The ductwork, as with other systems that service multiple spaces in a building, is installed in shafts and other concealed spaces in the structure. Air duct void spaces may penetrate fire-rated assemblies.

Building, mechanical, and fire codes contain many requirements for ducts, including allowable materials and the requirement for smoke and fire dampers to maintain the integrity of the fire-rated assemblies **(Figure 5.8)**. In order to prevent smoke or fire from

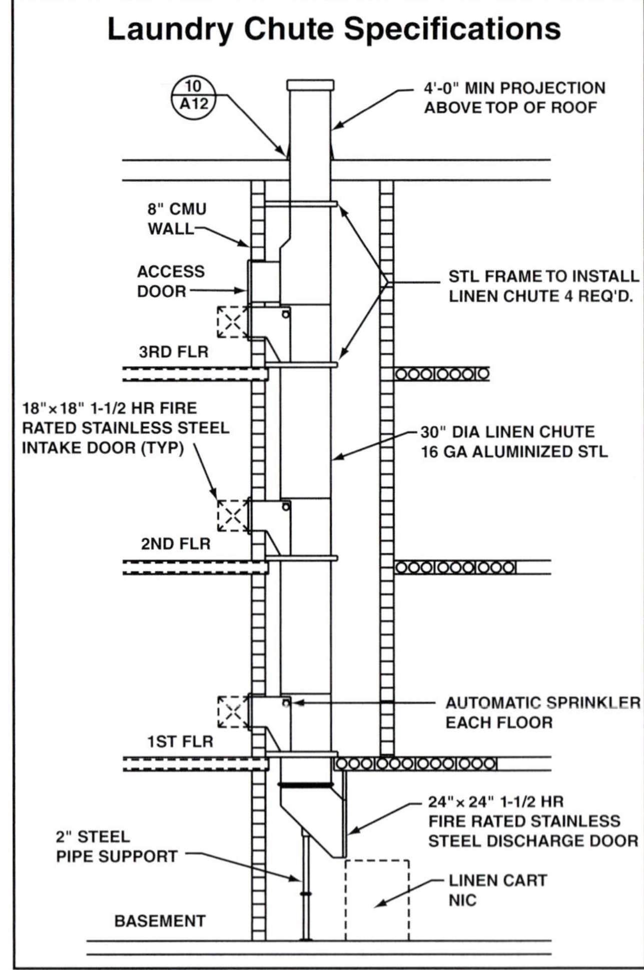

Figure 5.8 Construction of refuse and laundry chutes includes requirements for the enclosure, intake points, and fire resistance.

traveling through a building via the ductwork, building and fire codes may require the installation of smoke and/or fire dampers within the ducts. Some air-handling systems are also interconnected to the fire alarm system and turn off automatically when products of combustion are detected.

While the transport of air through ductwork presents little potential for developing sufficient heat to cause ignition, environmental systems may generate heat due to the fuels used for heating. The belts used to operate large fan units may produce mechanical heating. In residential occupancies, return-air plenum may be used as the cold air return for the system, providing avenues for fire spread.

Elevators, Escalators, and Conveyors

Elevators are an essential system in many multilevel buildings, providing a means of transportation for both people and materials **(Figure 5.9)**. Applicable building and life safety codes provide requirements for elevators that are used as part of the egress system for a building. These requirements include provisions for a smokeproof shaft for the equipment and automatic sprinkler protection throughout the building.

Hazards presented by elevators, escalators, and conveyors include:

- The equipment or its associated fuel may be combustible.

- Combustible materials can accumulate at the base of system shafts.

- Quantities of lubricant may be used for elevators and escalators.

- Motors may be at various locations and in mechanical spaces that support the systems.

- Electrical components of the system may produce heat sufficient to ignite combustibles.

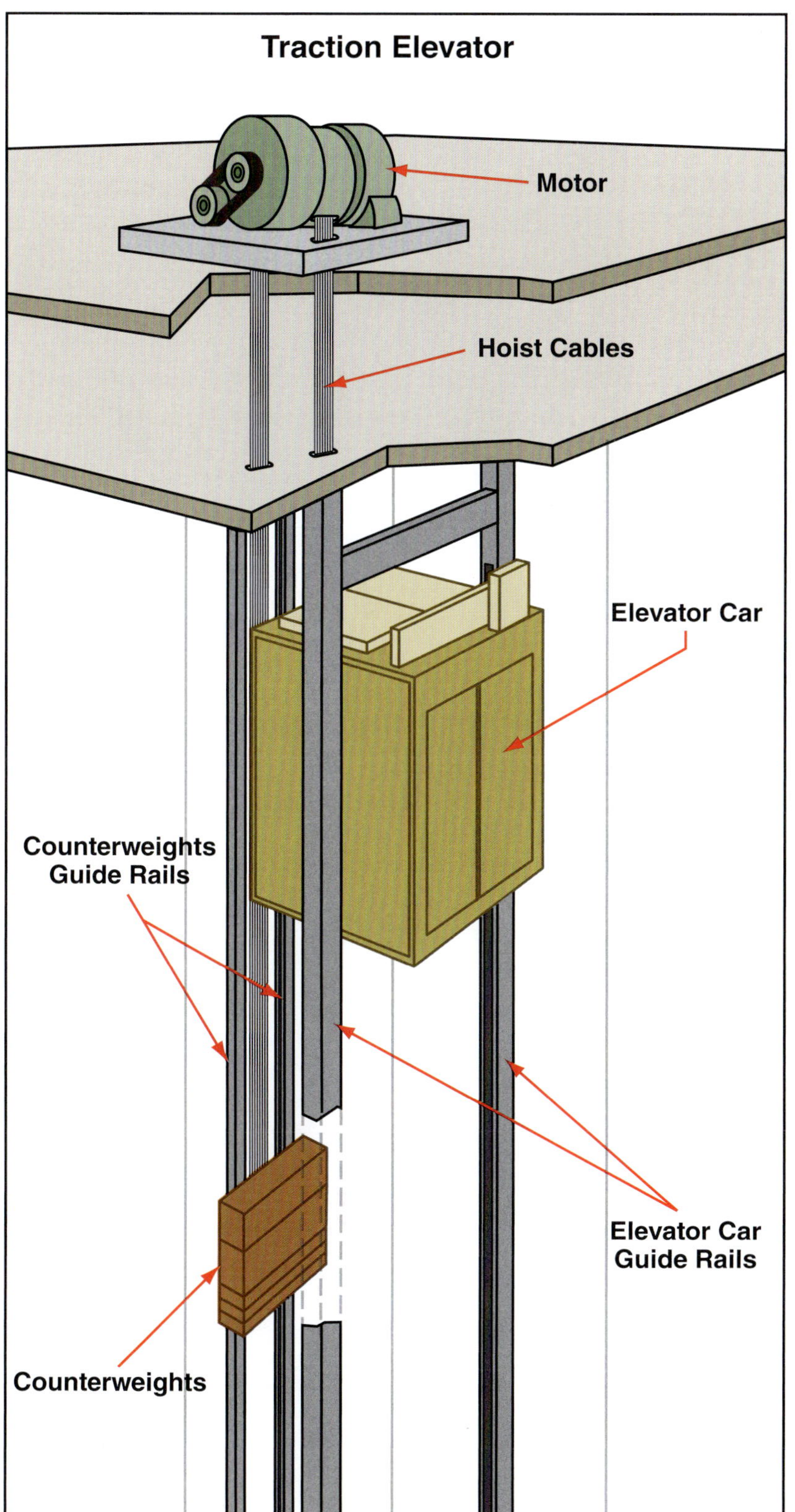

Figure 5.9 Traction elevators use a series of components to safely raise and lower an elevator car in normal and emergency modes.

Elevators present serious issues in the event of a fire due to the large shafts in which they operate. There is also the potential for occupants to be trapped as a result of fire or a loss of power.

NOTE: Because of legislation related to access to public buildings by individuals with physical disabilities, research is currently underway to make elevators available for evacuating individuals in the event of a fire.

In horizontal systems, such as conveyors, the penetration of fire walls can provide a means for fire and smoke to travel to other portions of a building **(Figure 5.10)**. This potential can increase when there are flammable or combustible materials transported on or stored near the conveyor. Similarly, escalators can provide a means for fire and smoke to travel horizontally and vertically.

Environmental Systems

NFPA 1033 (2022): 4.2.2, 4.2.3, 4.2.8

The primary purpose of environmental systems in a building is to provide conditioned air to the building spaces for the comfort and safety of its occupants. Environmental systems may use *make-up air* from the outside or filter and circulate the air within the building. Environmental systems may range from small air handling units that move relatively small volumes of air to large fans designed to move thousands of cubic feet of air per minute (m³/min) **(Figure 5.11)**. Environmental systems may also be designed to remove odors and particulate materials from the air.

Building systems and utilities that are interconnected with environmental systems can include:

- Conveyance systems to organize and protect piping, ducting, and wiring.

- Fire protection, detection, and suppression systems as installed for fire response

- Electrical systems for lighting and powering equipment throughout the structure

Combustible materials within environmental systems can include:

- Tape
- Filters
- Ducting
- Insulation
- Duct coverings
- Dust and other debris collected from the environment

Figure 5.10 Escalators and moving walkways transport passengers between portions of a building that may otherwise be protected with fire walls.

Figure 5.11 Industrial fans can move large volumes of air through a structure in a short amount of time. *Courtesy of Yates & Associates.*

While the flame spread and smoke production of component materials are regulated by codes, there is a potential for combustible materials to become involved in fire. Should these materials ignite in sufficient amount and continuity, there is a potential for an explosion or fire spread over a wide area of a building.

Investigators should evaluate whether environmental systems were designed, installed, and maintained properly to minimize the transport of products of combustion throughout the building. Environmental systems may include the following systems:

- Air handling systems
- Smoke control systems
- Smoke and heat vents

Air Handling Systems

The primary purpose of air handling systems is to create and maintain a comfortable environment for occupants. These systems can significantly affect fire conditions and the resulting fire patterns for reasons including:

- Wide accessibility of heating, ventilating, and air conditioning (HVAC) systems means that many buildings have windows that do not open, or no windows at all. Modern buildings may be nearly airtight aside from these systems.

- An outside air intake may draw in contaminated air **(Figure 5.12)**.

- Fans used to draw supply air and to expel exhaust air will also move contaminants that are introduced into the system.

- Air filters may use liquid adhesives that can be combustible. Electrostatic filtration systems may present a significant electrical equipment hazard.

- Heating and cooling equipment may be fueled with natural gas or oil. These fuels may be stored within the structure.

Figure 5.12 Air handling systems may draw in contaminants from outside of a structure.

Smoke Control Systems

The purpose of a smoke control system is to ventilate and exhaust combustion products from a building. Removing the smoke decreases the loss of life due to smoke inhalation and creates a more workable environment for firefighters who must enter the structure to fight fires.

Design of Smoke Control Systems

The design of many modern buildings can make traditional ventilation methods difficult or impossible, especially in high-rise construction. Some types of windows may not open at all. In addition, newer glazing types, such as **hurricane glazing**, may be very difficult to remove or ventilate. In these conditions, some HVAC systems may be used in a smoke control capacity. They can automatically or manually be transferred from normal operating mode to **smoke control mode**. This transfer can be done to prevent the recirculation of smoke through the system and to facilitate removal of smoke. For more information on smoke control systems, refer to the IFSTA manual, **Fire Protection, Detection, and Suppression Systems**.

The design of a smoke control system requires extensive engineering analysis as well as provision for a situation in which smoke detectors operate in more than one zone. Any equipment, such as fans and ducts, used to exhaust the products of combustion must be capable of withstanding the anticipated temperatures. Other factors the designer must anticipate include:

- Fire size

- Volume of a fire zone

- Outside weather conditions

- Maximum pressure differences across barriers, such as stairwell doors

Operation of Smoke Control Systems

Smoke control systems can be simple or complex. A smoke exhaust fan located at the top of an atrium is an example of a relatively simple smoke control system. A smoke control system may also use the entire HVAC system of a building to control the movement of smoke. Some systems may also manage additional products of combustion.

Smoke control systems may be automatic or manual or have activators for both. Automatic systems may be activated in a building with fire alarms when the presence of fire is detected. Manual systems can be activated in case of emergencies by personnel in the building. Smoke control systems can create ventilation, which may have a profound effect on fire and burn patterns. Investigators should carefully analyze the fire effects and fire patterns to ensure they correctly interpret the fire and smoke spread throughout the structure when these systems were in use during the fire **(Figure 5.13)**.

An HVAC system operating in smoke control mode discharges smoke through the exhaust fan from the fire floor to the outside without returning air to the supply fan **(Figure 5.14)**. The continuance of air supply to the non-fire floors creates a "pressure sandwich" of higher air pressure on floors above and below the fire floor(s), reducing the movement of smoke into those areas. Another example of smoke control is the use of fans in various configurations to pressurize stairwells to keep them free of smoke.

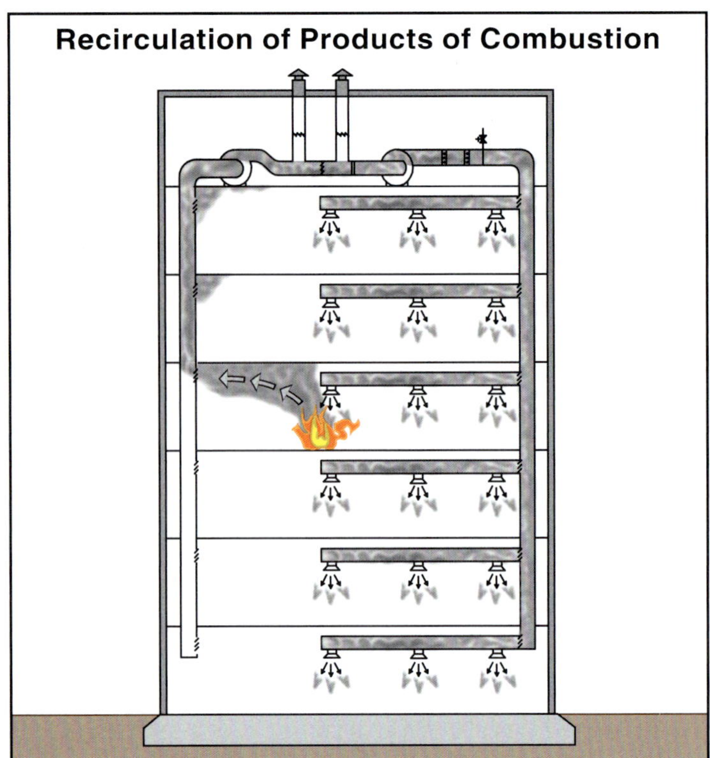

Figure 5.13 An HVAC system operating in normal mode may draw products of combustion into the ducts and transport them throughout the building.

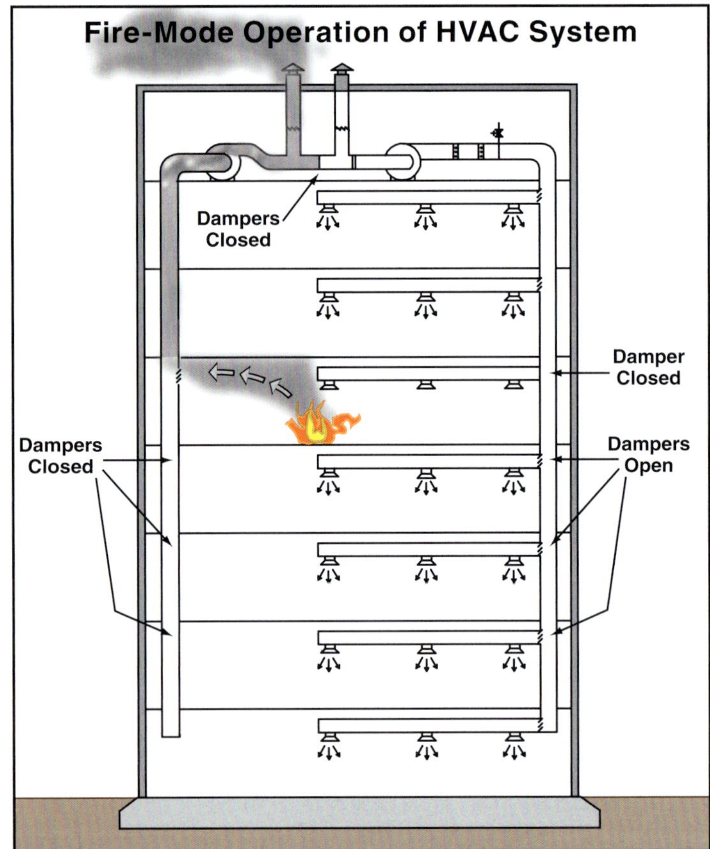

Figure 5.14 An HVAC system operating in the fire mode will exhaust smoke from the fire zone and supply fresh air to adjacent zones.

Through the use of a damper, air supply is restricted to the fire floor and continued to adjacent floors. Dampers can be opened or closed, depending on the location of the fire, to redirect the flow of air and to exhaust the smoke. The system detectors must be carefully designed and placed. This eliminates the possibility of a detector outside the fire area being activated first, resulting in the misdirection of the system.

Smoke and Heat Vents

Fire control and occupant safety are enhanced through venting the products of combustion to the outside via the shortest route possible. Vent openings may be placed on a roof to help remove air or contaminants from a building's HVAC system. Fire investigators should note whether smoke and heat vents were utilized.

Even in buildings without an HVAC system, roof ventilation openings are a simple and rapid means of ejecting products of combustion from inside the building. Built-in smoke and heat vents enable firefighters to make a faster and safer interior attack and dissipate some of the thermal energy of a fire. Without a means of venting the heat and smoke, firefighters could be forced to withdraw to an exterior attack while the fire continued to burn in the interior. Rooftop smoke and heat vents are typically required on the roofs of large-area buildings and in buildings with few windows including industrial, storage, and mercantile buildings **(Figure 5.15)**.

Figure 5.15 Smoke vents may operate manually or automatically.

Electrical Systems

NFPA 1033 (2022): 4.2.8

The electrical (utility) service provides electric current to power many of the systems within the building, including lights and machinery. The investigator should be able to recognize whether the electrical service is a single-phase or three-phase system. The primary difference between the two systems is the amount of voltage supplied. A single-phase system will have at least three wires — two of which are insulated and one may be bare **(Figure 5.16)**. A three-phase system will have at least four wires — one of which is insulated and one may be bare **(Figure 5.17)**. Electrical systems are further discussed in Chapter 6.

Figure 5.16 Single-phase electric service uses three lines to serve a residence. *Courtesy of Donny Howard, Agent, Oklahoma Fire Marshal's Office.*

Figure 5.17 A three-phase electric service supplies a higher voltage than a single-phase system. This higher level of service is commonly used by business or industrial occupancies. *Courtesy of Mary Maurath.*

Building systems and utilities that are interconnected with electrical systems can include:

- Heating, ventilation, and air conditioning systems throughout the building

- Fire protection, detection, and suppression systems as installed for fire response

- Mechanical systems including transportation and conveyance systems

The electrical (function) system consists of all equipment and wiring necessary to distribute power from the distribution point (part of the energy production system) to the remaining spaces. Standards such as NFPA 70®, *National Electrical Code*®, regulate the design and installation of electrical systems.

Suitability to the Purpose

A fire investigator should research whether the electrical service to a building was designed to handle the power demands of the current use. For example, a yoga studio converted to a commercial cooking operation will demand increased power draw from appliances, which may necessitate an upgrade in the utility service. A failure for the demand to be accommodated by the supply may cause adverse resistance heating.

Energy Production Systems

In contrast to electrical service that is provided via a grid system, some structures are equipped with energy production and storage systems. Energy-production systems convert utilities that enter a building, such as electricity, gas, or oil, into a usable form or distribute them for use in the remainder of the building. These systems produce heat and cooling and provide electricity to buildings.

Energy production systems can be located almost anywhere within a building or within the wider incident site. Examples include:

- Boilers
- Chillers
- Air heaters
- Heat pumps
- Water heaters
- HVAC systems
- Geothermal systems
- Electrical transformers
- Steam reducing stations
- Electricity producing generators
  - Gasoline, propane, natural gas, or diesel powered backup generators
  - Solar energy systems
  - Wind turbines
  - Hydroelectric turbine generator
- Uninterrupted power supplies (UPS)
- Equipment used for refrigeration and cooling

NOTE: Nuclear power generators are beyond this manual's scope.

Energy-production systems can provide either the heat energy or fuel necessary for a fire, especially when:

- Equipment is not properly operated or maintained.

- Any combustible surfaces or materials are in close proximity.

- Good housekeeping practices are not followed near the devices.

- The fuel leaks from the device or associated piping that delivers it to the system.

In addition to the heating devices themselves, the chimney or *stack* is another element of many energy-production systems used to transport products of combustion through the building to the outside atmosphere. Chimneys are often installed in chases that penetrate building floors and/or adjacent to wall assemblies. They must be properly insulated or isolated to prevent the buildup of excessive heat **(Figure 5.18)**. Both the chimney and the chase must be constructed and maintained to minimize the leakage of hot and potentially toxic gases produced during the combustion process. They also must be designed and installed so that any heat within them does not transfer to combustible components of the building.

Investigators should have a thorough understanding of the codes that regulate these systems. Investigators should make themselves familiar with the codes and standards applicable in their jurisdiction while understanding that not all documents will be applicable to their jurisdiction and circumstances.

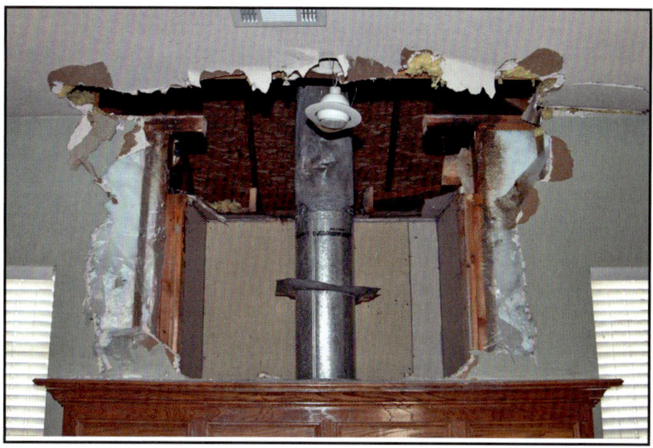

Figure 5.18 Chimneys can radiate heat to surrounding structural members. *Courtesy of Yates & Associates.*

Photovoltaic Systems

A **photovoltaic (PV) system** is a reliable source for solar energy used in a wide range of applications, including within a building's systems. The collection of photovoltaic energy is a rapidly developing field. Fire investigators should note the presence of such systems.

NOTE: Several jurisdictions in North America have seen significant increases in the prevalence of **clean energy** technology. Each jurisdiction must research the specific systems relevant to an area, develop appropriate response tactics, and train response personnel.

Solar Cell Terminology

A single unit of a solar energy collector is referred to as a "cell." A series of solar cells arranged in a self-contained unit are referred to as a "panel" or "module." Individual cells may also be connected in series, as is common with solar shingles. A system of solar panels or shingles is called an "array" **(Figure 5.19)**.

Figure 5.19 Photovoltaic arrays may not be readily visible in low-visibility conditions. *Courtesy of McKinney (TX) FD.*

Typically, photovoltaic cells are shaped as panels and shingles. They can be installed on top of a roof or embedded in the roof covering. Recent findings and updates on photovoltaic systems include:

- **Solar energy applications** — Have increased the overall prevalence and cost-efficiency of photovoltaic systems in residential applications.

- **Price-matching** — Photovoltaic shingles can cost-match with more traditional types of shingles.

- **Standalone roofing system** — Some solar shingles are designed as a standalone roofing system that can be walked on.

- **New technology** — Includes the development of photovoltaic paints and other systems.

WARNING: Photovoltaic systems are producing power if they are uncovered and collecting any light source, including response vehicles' lights.

PV technology is continuing to develop, and the utilities are becoming more commonly available, also meaning that incidents involving these materials are becoming more prevalent. For example, one type of solar panel is designed to be rolled out as a sheet on a flat membrane roof. In low light, it may look like a maintenance walkway. Where panels cover the entire roof, space for rooftop access, travel, and ventilation may be unavailable. Raised panels may create a significant traction, tripping, and falling hazard. Raised panels may also include void spaces. It can be difficult to identify photovoltaic cells in the following conditions:

- In low light
- On flat roofs
- During adverse weather
- During fireground operations

Model codes and other authorities continue to propose and add parameters regarding photovoltaic applications, including requirements for:

- Clear space along the edges and ridge of a roof
- Clear aisle space between arrays with certain dimensions
- Capability for rapid shutdown of PV systems on buildings
- Minimum visibility requirements of photovoltaic components
- Uniform labeling of structures with photovoltaic related materials
- External shut-off switches to reduce or eliminate electrocution danger

CAUTION: Rooftop photovoltaic applications may be found on many structures, regardless of the occupancy, roof type, or pitch.

The energy feed from photovoltaic cells cannot be isolated or shut off. Even when electric power to the building is shut off and the cells are isolated from the building's electrical system, the cells continue to store and generate electric power **(Figure 5.20)**. These systems should be handled with the same care as any other electrical system: all wiring and conduits must be assumed to be live. Although not common, PV systems can transfer an electrical charge to metal roofing materials if the supports and other components are compromised. Failures of these systems will be similar to other electrical sources.

Figure 5.20 Rooftop PV arrays may blend in with roof features and should be assumed to be energized at all times. *Courtesy of McKinney (TX) FD.*

Fire Protection Systems

NFPA 1033 (2022): 4.2.2, 4.2.3, 4.2.8, 4.4.2, 4.6.1

An investigator will need to evaluate all present fire protection systems for proper functioning. For more information on fire protection, detection, and suppression systems, refer to the IFSTA manual, **Fire Protection, Detection, and Suppression Systems**.

Fire protection systems are broadly categorized as passive and active. Passive systems include:

- Firewalls
- Fire doors **(Figure 5.21)**
- Built-in or sprayed-on structural fire protection

Figure 5.21 Fire doors undergoing fire resistance testing.

Passive systems perform with no outside intervention or mechanical support. Passive fire protection cannot warn occupants of the dangers of an unwanted fire or suppress a growing fire. Passive systems are described in the IFSTA manual, **Building Construction Related to the Fire Service**.

In contrast, active fire protection systems require some type of outside intervention or mechanical support, such as electricity or a water supply. Active fire protection systems include detection and suppression systems that activate and operate automatically or manually, and alert occupants upon the presence of smoke or fire. These systems, when activated, are designed to change the course and outcome of a fire in a building **(Figure 5.22)**. These systems work to the benefit of the building owner, its occupants, and the responding fire department personnel.

Figure 5.22 A sprinkler is an active fire protection device.

Building systems and utilities that are interconnected with fire protection systems can include:

- Mechanical systems including transportation and conveyance systems
- Heating, ventilation, and air conditioning systems throughout the building
- Electrical systems to power components of the protection and detection systems

An investigator should request installation, maintenance, and inspection records for active fire protection components to show whether they were functioning correctly at the time the fire occurred. The investigator should also get schematics for initial design diagrams and schematics from utility company/ installation. Utility records can also show whether services were connected at the time of the incident.

When fire protection systems are present, investigators should evaluate:

- Overall system operation
- Use by occupants who did not have proper training
- Possible tampering, obstructions, or incorrect design
- System deficiencies including insufficient water supply
- System installation, inspection history, and maintenance
- A change in occupancy or use that could affect the system

Active fire protection systems can be divided into two broad categories: fire detection and fire suppression. Fire detection systems alert occupants to hazardous conditions, smoke, or changing fire conditions by automatic or manual fire alarm equipment. The early detection of a fire and the signaling of an appropriate alarm remains the most significant factor in preventing large losses due to fire.

Overall, fire protection, detection, and suppression systems can be divided into four basic categories:

- **Automatic fire protection systems** are activated by smoke, high temperatures, radiant heat, or flame signatures, and alert the occupants in the building and/or the fire department to the fire. These systems do not slow the growth of the fire or reduce the amount of smoke produced.

- **Automatic fire suppression systems** work to suppress the fire hazard by applying a fire-suppressing agent to the fire. Human intervention is not needed for their effectiveness. This process reduces the hazard to the occupants, responding firefighters, and damage to the building and its contents.

- **Manual fire alarm systems** require activation by individuals upon discovery of a fire. Devices such as manual pull stations are common components of manual systems.

- **Manual fire suppression systems** include standpipes (both vertical and horizontal), fire hoses, and fire extinguishers **(Figure 5.23)**. These mechanisms require an individual to operate them in order to perform their function; they are not automatic.

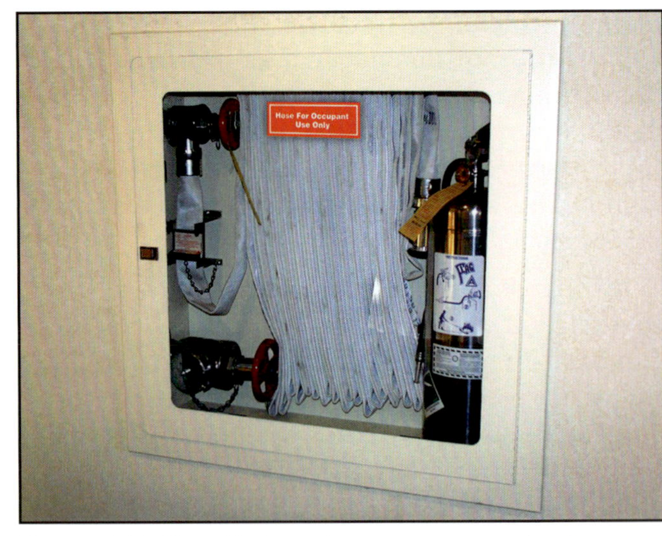

Figure 5.23 Class III standpipes are an example of a manual fire suppression system.

Fire Detection and Alarm Systems

A fire detection and alarm system may consist of manual and/or automatic **alarm-initiating devices** that are activated by the presence of the following, whether separately or in combination:

- Fire
- Heat
- Flame
- Smoke
- Human intervention **(Figure 5.24)**

Functions of Fire Detection and Alarm Systems

When activated, an alarm-initiating device activates and sends a signal to a **fire alarm control panel (FACP) (Figure 5.25)**. Alarm-initiating devices (both automatic and manual) include the following types:

- Heat detectors
- Flame detectors
- Intrusion alarms
- Smoke detectors
- Tamper switches
- Water-flow devices
- Manual pull stations
- Combination detectors
- Fire gas detectors (carbon monoxide, carbon dioxide, etc.)

Figure 5.24 This manual pull station requires activation by an individual upon discovery of fire.

Figure 5.25 FACPs vary in appearance but all serve the function of routing alarms to central stations or fire departments.

Figure 5.26 Magnetic door closure units are designed to deenergize and automatically shut doors when a fire alarm sounds.

An FACP processes the signal and takes appropriate action by activating audible and visual local alarm devices. Simultaneously, the FACP transmits an emergency signal to an alarm-signaling system, such as:

- A proprietary system
- Central station service
- Fire department telecommunications center

An investigator may be able to use the timing, sequence, and location of these alarms to assist in determining the origin of the fire and its spread. Fire investigators should request all records from the alarm monitoring service. These records are typically very helpful in determining any problems with the system prior to the fire. The records will also typically record timelines of device activation and notification.

Fire investigators are cautioned that not all fire alarm systems are required to be monitored and an on-site person may have a key to the alarm panel. It is common for such persons to silence these systems while they investigate the cause of the alarm. This improper action may result in additional fire spread or a delay in alerting the fire department.

Building codes have special requirements for some types of occupancies in the event of a possible fire condition. These additional functions may affect building ventilation, fire patterns, fire spread, and fire growth. Fire detection and alarm systems can be designed to perform the following special functions:

- Operate heat and smoke vents
- Close **smoke dampers** and/or fire doors **(Figure 5.26)**
- Pressurize stairwell(s) for evacuation purposes
- Deactivate the HVAC system for smoke control
- Return the elevator to a designated evacuation floor
- Override control of elevators and prevent them from opening on the fire floor
- Deactivate or stop the fuel supply to specific pieces of equipment such as a deep fat fryer
- Activate special fire-extinguishing systems, such as preaction and deluge sprinkler systems, or a variety of special-agent fire-extinguishing systems

Fire Detection and Alarm System Failures

Fire detection and alarm systems are only effective when they function properly. After a system failure, investigators should be familiar enough with these systems to determine which components of the system failed, especially in the case of injuries or fatalities. Questions that investigators may need to answer after a fire in a building with a fire/smoke detection or alarm system are included in **Appendix D**. Fire detection and alarm system activations, times, or sequences may provide useful information for determining:

- Cause

- Area of origin

- Initiation of the fire

- Other circumstances related to the incident

Water-Based Suppression Systems

Fire suppression systems control and contain hazardous conditions until manual suppression can be achieved. Many of these systems are water-based. Working together, fire detection and suppression systems save lives and decrease property loss by notification and early mitigation and control of the fire.

Building codes in a particular jurisdiction may mandate fire protection systems for various types of construction and occupancies. Local ordinances may also require these systems based on a building's height, area, configuration, and building contents.

Functions of Water-Based Suppression Systems

Water-based fire-suppression systems successfully reduce loss of life and property from fires. Automatic sprinkler systems are considered effective fire protection devices. Fire-loss data reveals that in buildings equipped with automatic sprinklers, the system controls or extinguishes about 96 percent of all fires in which they operated (Ahrens, 2017).

An automatic sprinkler system consists of a series of discharge devices (sprinklers) systematically arranged to distribute sufficient quantities of water to extinguish a fire or prevent its spread until firefighters arrive. A network of pipes supplies water to the sprinklers. Most sprinklers (except deluge system sprinklers) are kept closed by fusible links or other thermally sensitive devices **(Figure 5.27)**. When thermally sensitive devices are used, heat from a fire causes affected sprinklers to activate automatically. **Deluge**- or preaction-type **sprinkler systems** operate when an electronic detector or manual control device activates.

Figure 5.27 When fusible links weaken due to heat, water is released from the sprinkler system.

Automatic Sprinklers in Homes and other Residences

Recent code modifications may require automatic fire sprinkler system installation in local residential construction. Most jurisdictions require sprinkler systems for residential structures over a specified size. The National Fire Sprinkler Association (NFSA) has estimated that fire deaths in one- and two-family dwellings can be reduced by up to 82 percent with the installation of residential fire sprinklers. The NFSA, NFPA, American Fire Sprinkler Association (AFSA), and Home Safety Council (HSC) provide local jurisdictions with assistance in adopting or developing model building codes that require the installation of sprinklers in all new residential construction. These installations will be based on NFPA 13D, *Standard for the Installation of Sprinkler Systems in One- and Two-Family Dwellings*

Failures and Testing of Water-Based Suppression Systems

Although sprinkler systems have an impressive success rating, investigators should be able to identify system failures such as when the systems fail to suppress or control the fire, operate improperly, or show evidence of tampering. A fire that exceeds the operational capacity of the sprinkler system may continue to burn even though the system activated.

In addition, the installation, inspection, testing, and maintenance of water-based fire protection systems and equipment must be in compliance with local, state/provincial and/or federal codes, ordinances, and standards as well as the manufacturer's instructions. The owner must perform acceptance and periodic inspections, testing, and maintenance of the system. An investigator should seek to determine whether the system was compliant with the manufacturer's instructions and all applicable codes as part of an investigation. Investigators may need to contact sprinkler-system experts to assist in investigations. Investigators may also review the inspection and testing records as part of their investigation.

Sprinklers can minimize the damage in the area of origin while spreading the fire to unprotected areas, which in turn may make areas of origin more difficult to identify. In addition, sprinklers may affect fire patterns. Also, investigators should remember that any sprinkler-system component may be evidence and therefore subject to contamination and spoliation just like any other piece of evidence. There are times when the building owner or restoration company may arrive at the scene and remove/replace the affected sprinkler activation devices prior to the arrival of the fire investigator. The investigator should seek out and retrieve these devices as well as gather important information as to their prefire location and information as to who removed them and what condition they were in prior to removal.

Standpipe and Hose Systems

Standpipe and hose systems are designed to provide a means for rapidly deploying and operating fire hose on all levels of multistory structures and at remote points in large area structures and facilities **(Figure 5.28)**. Investigators should be able to locate and identify these systems.

If these systems fail to function properly, firefighters may lack sufficient water or pressure for fire fighting operations. Failure of standpipe and hose systems allows fire to spread further than if the standpipe system worked properly.

Investigators should carefully examine the hose for evidence of wear and tampering. Investigators should also review the inspection records for this system.

Figure 5.28 Standpipe systems provide a water supply for firefighters in multistory structures or remote locations at a facility.

Fire Pumps

The main function of a water pump is to increase the pressure of the water that flows through it **(Figure 5.29)**. In addition, fire pumps:

- Increase the water supply pressure available from public/private water systems and gravity tanks

- Transfer water from reservoirs, lakes, ponds, swimming pools, or other such nonpotable static water sources

- Provide adequate water volume and pressure to private fire hydrants and/or water-based fire suppression systems

A fire pump may be needed to supply a sprinkler or standpipe system. If available, the fire pump should be installed within a rated enclosure so it can function as necessary in the event of the fire. Should the fire pump fail, the sprinkler system may function inadequately. Such failures could be key pieces of information in an investigation.

Non-Water Suppression Systems

Non-water based suppression systems are used in locations where water-based fire suppression systems may not be appropriate for the hazard to be protected. These suppression systems can also be an alternative where preventing water damage is necessary. Locations may have materials and activities that would react adversely to water-based suppression such as:

Figure 5.29 Pumps can increase the pressure to sprinkler systems and standpipes in large structures.

- Combustible metals

- Water reactive materials

- Food preparation, especially near grease

- Highly sensitive computer or electronic equipment

Commercial kitchen hood suppression systems may be the type of system that an investigator will encounter the most frequently. These systems use a **wet chemical system** in areas where rapid fire knockdown is required. Wet chemical agents extinguish grease and oil fires by fuel removal, cooling, smothering, and flame inhibition. All wet chemical systems should comply with NFPA 17A, *Standard for Wet Chemical Extinguishing Systems*.

Specifically, model codes require that kitchen hood extinguishing systems are installed in commercial kitchens that produce grease laden vapors. The extinguishing system works in conjunction with an exhaust hood. The investigator should evaluate the entire system.

If a fire occurs in a commercial kitchen, the investigator needs to look at installation records, maintenance records, and cleaning records. The investigator then needs to document that the appliances underneath the extinguishing system were the appliances that the system was designed to protect.

Recording and Transmitting Systems

NFPA 1033 (2022): 4.2.8, 4.4.2, 4.5.1, 4.6.1

Many structures will have specialized systems that investigators should evaluate in the course of an investigation. These specialized systems may record and/or transmit data.

Building systems and utilities that are interconnected with recording and transmitting systems can include:

- Fire protection, detection, and suppression systems
- Mechanical systems including transportation and conveyance systems
- Heating, ventilation, and air conditioning systems throughout the building
- Electrical systems to power components of the protection and detection systems

Time-Sensitive Information

The availability of information from any security, card access, alarm, or surveillance equipment may be time-sensitive. For example, captured video from security cameras may only be available for 24 hours or may *overwrite* previous data on a weekly or monthly basis. In addition, data may be lost when the system loses power. Investigators must react quickly to obtain this important evidence.

Security and Monitoring Systems

Security and monitoring systems can be found in many different types of occupancies. A security system alerts occupants or authorities to unlawful entry to a residence or facility. Such systems may be designed to alert security personnel and occupants on the secured site. Security systems may also be monitored by a central station that contacts local law enforcement in the jurisdiction. Monitoring agencies also may keep records of entry and exits at a facility.

The investigator needs to access information about an alarm system and its capabilities whenever one is present at the scene. Investigators may be able to determine whether there was unlawful entry to a structure by analyzing data from security and monitoring systems, especially if the system is monitored at an off-site location. In addition, witnesses may have heard security alarms sounding if alarms are audible. Monitoring companies and local law enforcement personnel should have access to alarm activation information including time and date; an investigator can obtain this information as part of an investigation.

The sequence of alarm activation may provide valuable information to the investigator in determining the origin of the fire or initial incident that caused the fire. For example, a glass-break alarm functioning before an infrared-motion alarm may indicate an overpressurization took place before the actual flaming combustion. All of this information can add to and enhance evidence collected at the scene.

Alarm Panels

Alarm panels within the structure may record more information than the owner/occupant has paid a company to monitor. This information is stored in the panel itself, and it can be lost when a system is running on battery power after primary power has been removed. Investigators should be aware that this additional information may be available.

An expert on the system will have to be contacted to retrieve the information after the panel is removed and sent for testing/data retrieval. Alarm panels that have been fire/water damaged may retain information stored on the hard drive and internal components.

Video Surveillance

Video surveillance systems have become commonplace in many business, office, manufacturing, residential, and industrial structures **(Figure 5.30)**. In these systems, cameras mounted at strategic locations inside and outside a building record video in a target area 24 hours a day upon detecting movement or on a predetermined time schedule. Old systems may store the video captured from cameras on VHS videotapes. Newer systems store this captured video on CD-ROMs, DVD-ROMs, DVRs/PVRs or computer hard drives. The captured video can be stored off-site and/or on-site.

Cloud-based storage of the video surveillance is becoming more commonplace. In the event of a fire, this system allows easier retrieval of the video. Other types of storage systems may also have a phone/tablet app that may automatically record and send clips to the building's designated person. These clips may prove invaluable with origin and cause determination. If the storage medium for the cameras is located on-site, it may not survive a fire. However, any salvageable stored video could reveal important evidence. Investigators may wish to retrieve the hard drives from the storage device to have the drives' data-recovery potential examined.

Video surveillance offers fire investigators video documentation of the occurrences at a fire scene. Surveillance from surrounding or unaffected buildings may also capture important video evidence.

Figure 5.30 Video surveillance systems are increasingly subtle and prevalent in public places.

Process Alarms

Process alarms are highly specialized systems often found in industrial and manufacturing facilities to monitor the activity of industrial equipment. These alarms alert workers to critical failures in machinery and may be designed to deactivate or de-energize equipment to prevent a fire. Some process alarm systems are not designed to monitor fire but are focused on other critical functions. These alarms still provide valuable information about the occurrences at a fire scene.

Process alarms may be connected to on- or off-site computer systems and may generate logs that include the date, time, and perhaps reason for their activation. To fully understand the data reported from a process alarm, investigators may need the assistance of facility personnel. Witnesses may report that process alarms sounded, prompting an evacuation of a facility before the fire occurred. This information could be a key piece of evidence in an investigator's attempts to determine the origin and cause of the fire.

Emergency Power

Some facilities may be equipped with backup or emergency power. When primary power is de-energized at such a facility, it does not necessarily mean that the facility is without power. Emergency power systems may have been activated and may still be active when investigators enter the scene. In addition, emergency power may activate after primary power has been removed, perhaps while an investigator is conducting an examination. As a result, investigators should make sure that they know whether a site has backup power. They should also attempt to determine the time of emergency activation and location of the emergency power system.

Specialized-Access Entry Systems

Secure facilities often have specialized-access entry systems such as keycard access, passcode access, or biometric access (thumbprint scanners) **(Figure 5.31)**. Only authorized personnel receive access capability. Security personnel, however, should have access to any entrance at the facility.

These systems may generate logs that track when and where the system granted access and who entered or exited. Such logs can provide an investigator with evidence instrumental in identifying witnesses, victims found at the fire scene, the last person in the building, or in the case of an arson investigation, possible suspects. This information also indicates which individuals an investigator may need to interview.

Figure 5.31 Keycard access is one of three common methods used to provide building security.

Review Questions

1. How can building systems affect fire development?

2. How do stairs act as a building system in the event of a fire?

3. What are three common hazards or problems related to shafts, chutes, and chases?

4. How can the presence of elevators in a building affect fire development?

5. How can environmental systems affect fire growth and development?

6. How can energy production systems add heat or fuel to a fire?

7. What hazards do photovoltaic systems present?

8. What should a fire investigator evaluate when fire protection systems are present?

9. Why is it beneficial for investigators to have an understanding of fire detection and alarm systems?

10. What types of information should investigators seek out about fire suppression systems?

11. How can information gathered about transmitting and recording systems be beneficial to an investigator when determining the cause of the fire?

Chapter Reference

Ahrens, Marty. July 2017. "U.S. Experience with Sprinklers." National Fire Protection Association (NFPA) Fire Statistics and Reports. Accessed online.

Chapter Key Terms

Alarm-Initiating Device — Alarm-system component that transmits a signal when a change occurs; change may be the result of an action such as the activation of a manual fire alarm box, the presence of products of combustion in the atmosphere, or the automatic activation of a supervisory switch.

Clean Energy — Energy sources that meet the needs of current consumers without compromising future resources. *Also known as* Sustainable Energy.

Deluge Sprinkler System — Fire-suppression system that consists of piping and open sprinklers. A fire detection system is used to activate the water or foam control valve. When the system activates, the extinguishing agent expels from all sprinkler heads in the designated area.

Fire Alarm Control Panel (FACP) — System component that receives input from automatic and manual fire alarm devices and may provide power to detection devices or communication devices. The fire alarm control unit can be a local control unit or a master control unit. *Also known as* Fire Alarm Control Unit.

Fire Damper — Device that automatically interrupts airflow through all or part of an air-handling system, thereby restricting the passage of heat and the spread of fire.

Fusible Link — Connecting link device that fuses or melts when exposed to fire temperatures; used to activate individual elements in active and passive fire suppression systems. Benefits include: inexpensive, rugged, easy to maintain. Disadvantages include: slower to activate than automated systems.

Heating, Ventilating, and Air Conditioning (HVAC) — Mechanical system used to provide environmental control within a structure, and the equipment necessary to make it function; usually a single, integrated unit with a complex system of ducts throughout the building.

Hurricane Glazing — Protective treatment for exterior windows designed to withstand hurricane conditions including high wind and impact.

Photovoltaic (PV) System — An arrangement of components that convey electrical power to an energy system by converting solar energy into direct current (DC) electricity.

Smoke Control — Strategic use of passive and active devices and systems to direct or stop the movement of smoke and other products of combustion.

Smoke Control Mode — Setting on an HVAC system or Fire Alarm Control Unit system that can be activated automatically or manually to initiate a programmed smoke control procedure.

Smoke Damper — Device that restricts the flow of smoke through an air-handling system; usually activated by the building's fire alarm signaling system.

Smoke Tower — Fully enclosed escape stairway that exits directly onto a public way; these enclosures are either mechanically pressurized or they require the user to exit the building onto an outside balcony before entering the stairway. Also known as *Smokeproof Enclosure* or *Smokeproof Stairway*.

Sprinkler — Water flow discharge device in a sprinkler system; consists of a threaded intake nipple, a discharge orifice, a heat-actuated plug, and a deflector that creates an effective fire stream pattern that is suitable for fire control.

Wet Chemical System — Extinguishing system that uses a wet chemical solution as the primary extinguishing agent; usually installed in range hoods and associated ducting where grease may accumulate. *Also known as* Wet Chemical Fire-Extinguishing System.

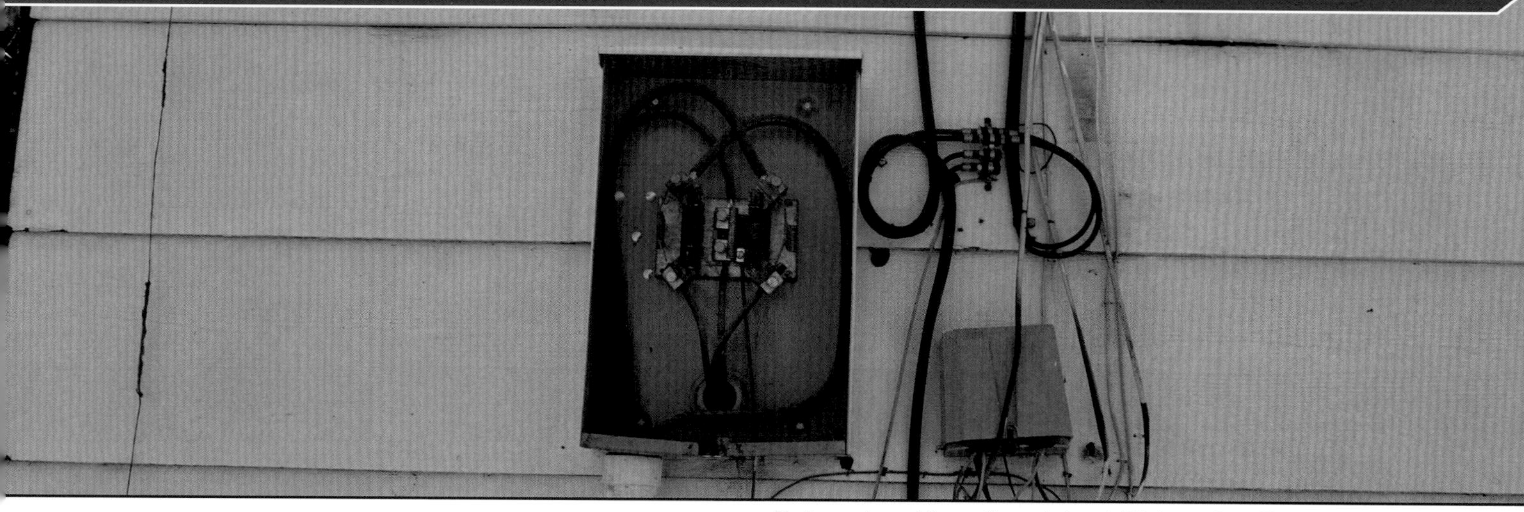

Photo courtesy of Donny Howard, Agent, Oklahoma State Fire Marshal's Office.

Chapter Contents

NFPA 1033 JPRs addressed in this chapter

This chapter provides information that addresses the following job performance requirements (JPRs) of NFPA 1033, *Standard for Professional Qualifications for Fire Investigator* (2022):

4.1.7   4.2.8   4.7.3

FESHE Outcomes

This chapter provides information that addresses the following course outcomes for the FESHE courses:

Fire Investigation II (C0284)

4. Analyze and determine the causes of fires and contributing factors.

Fire Investigation and Analysis (C0285)

2. Document the fire scene in accordance with best practice and legal requirements.

Learning Objectives

1. Explain basic electrical concepts. [NFPA 1033, 4.1.7, 4.2.8]

2. Describe electrical systems that are commonly found in buildings. [NFPA 1033, 4.1.7, 4.2.8]

3. Describe types of electrical heat sources and how they are prone to fail. [NFPA 1033, 4.1.7, 4.2.8]

4. Describe the types of damage that investigators commonly see on electrical conductors. [NFPA 1033, 4.1.7, 4.2.8]

5. Explain the purpose of grounding and the function of circuit interrupters. [NFPA 1033, 4.1.7, 4.2.8]

6. Describe investigative clues related to electrical systems. [NFPA 1033, 4.1.7, 4.2.8, 4.7.3]

Chapter 6
Basic Electricity and Electrical Analysis for Fire Investigators

Courtesy of Wayne Chapdelaine, Metro-Rural Fire Forensics.

The purpose of this chapter is to give fire investigators an understanding of terminology, functions, and applications regarding electricity. This basic, scientific information will assist investigators in evaluating potential electrical ignition sources. This chapter addresses:

- Basic concepts in electricity
- Common electrical systems in buildings
- Electrical heat sources and failures
- Damage to conductors
- Grounding and circuit interrupters
- Investigative clues

Basic Concepts in Electricity

NFPA 1033 (2022): 4.1.7, 4.2.8

Electricity is discussed in detail in NFPA 921, Chapter 9. This section is intended to present an overview of the concepts.

In its most basic form, *electricity* can be defined as the flow of **electrons** through a conductor. Electrons are small, negatively charged particles that revolve around the nucleus of atoms. Atoms compose all matter. Electrons are loosely attached to an atom, and under the right conditions, they can be pulled away from one atom and moved to another atom. This transfer results in a change in the charge of the atoms involved **(Figure 6.1)**. Materials having electrons that are free to move from one atom to another are called **conductors**. This movement of charge in a conductor is an electric current.

While copper and aluminum are the most common conductors used in wiring, most metals will also conduct electricity. Materials with atomic structures that do not allow the easy movement of electrons are not good conductors and are called **insulators**. Glass, plastics, rubber, and stone are examples of insulators.

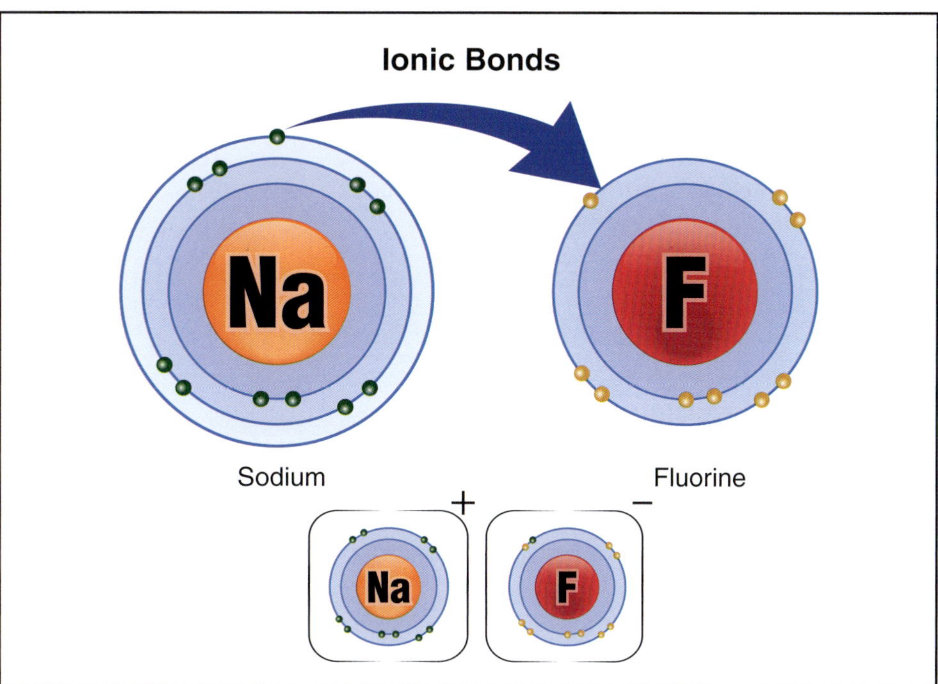

Figure 6.1 Ionic bonds form charged elements.

While the actual particles (electrons) move very short distances, their movement causes a chain reaction of very rapid movements within the conductor. When a charge enters a conductor at one end, it moves through it from atom to atom at the speed of light, virtually instantaneously. For a charge to move in a conductor, it must have an energy source that causes the charge to move initially, and a path for it to travel through. In discussing electricity, the path of travel for the charge is commonly called a **circuit**: a closed-loop system where the electrons start and end at the same power source **(Figure 6.2)**. The source of energy in a circuit can be a battery (chemical energy) or a **generator** (mechanical energy). The electrical force that causes the flow is called **voltage**.

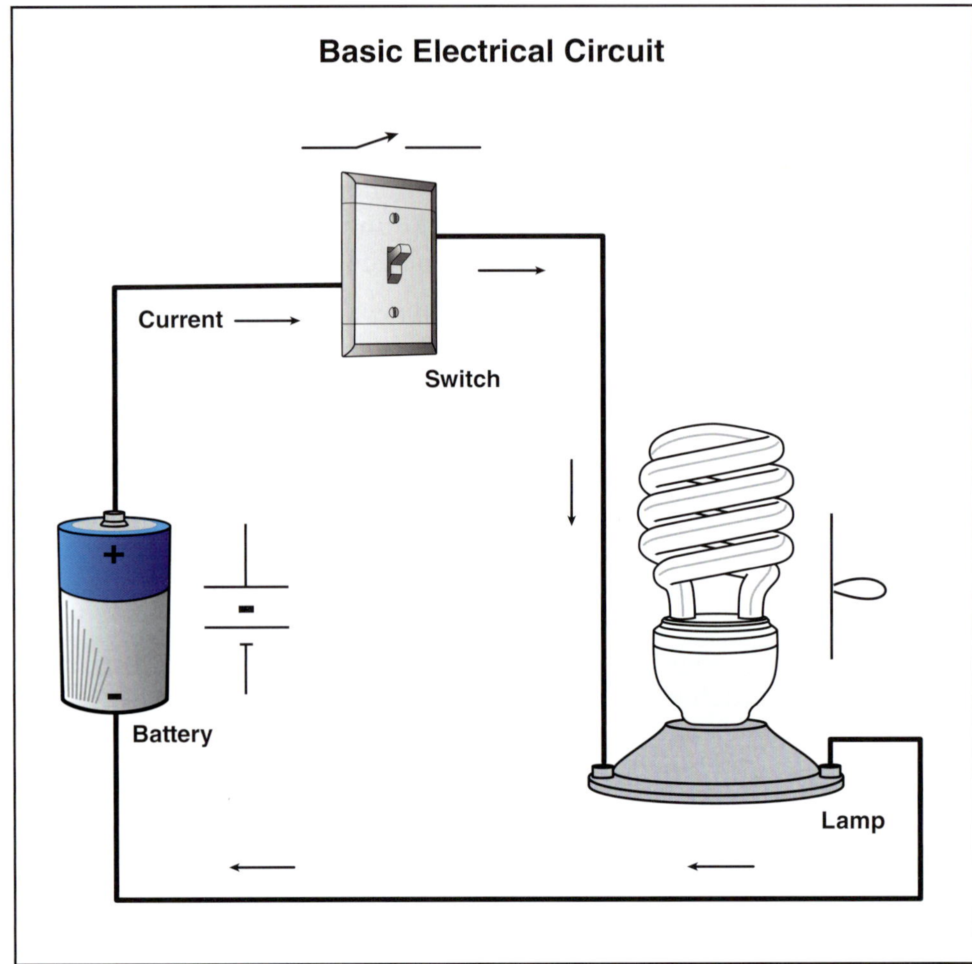

Figure 6.2 A basic electrical circuit includes a power source, current travel path, and a switch.

NFPA 70®, *National Electrical Code*®

Fire investigators are likely to find electrical wiring, appliances, or machinery involved in most of the fires they investigate. This chapter references NFPA 70®, *National Electrical Code*®, *(NEC)*®, which is widely adopted in the United States and more recently in Mexico and Central America. Part 1 of this code provides safety standards for installation, and Part 2 establishes the product standards such as those developed by Underwriters Laboratories Inc. (UL) and referenced in the *NEC*®.

NOTE: In Canada, fire investigators should refer to the *Canadian Electrical Code (CEC*®*)*, published by the Canadian Standards Association.

Water Pressure Compared to Electrical Current

Electrical circuits function similarly to water circulation systems under pressure **(Figure 6.3)**. Power can be lost in transmission the same way that water pressure is lost due to friction in a piping system **(Table 6.1)**. In this analogy, the pipe (water system) and the wire conductors (electrical circuit) both provide a flow pathway. In electricity, this flow is known as **current (I)** and expressed in **amperes (A)**. In a hydraulic system, the flow may be regulated using a valve; in an electrical circuit, current is controlled or regulated with a switch.

In a hydraulic system, a pump forces the liquid through the pipes. In an electrical system, a generator or battery provides the energy source to move current.

Hydraulic pressure is measured in pounds per square inch (psi) (or kilopascals [kPa]). Electrical pressure is called *voltage* or *electromotive force* and is expressed in **volts (V)**.

In both systems, the flow itself can provide the energy to do work. In a hydraulic system, a pipe can be arranged to pass through a turbine and turn a shaft that, in turn, can operate machinery. In the electrical system, current can be passed through a device or component, such as a bulb that converts the electrical energy to light and heat energy, or a motor that converts the electrical energy to mechanical and heat energy that can be used to operate machinery.

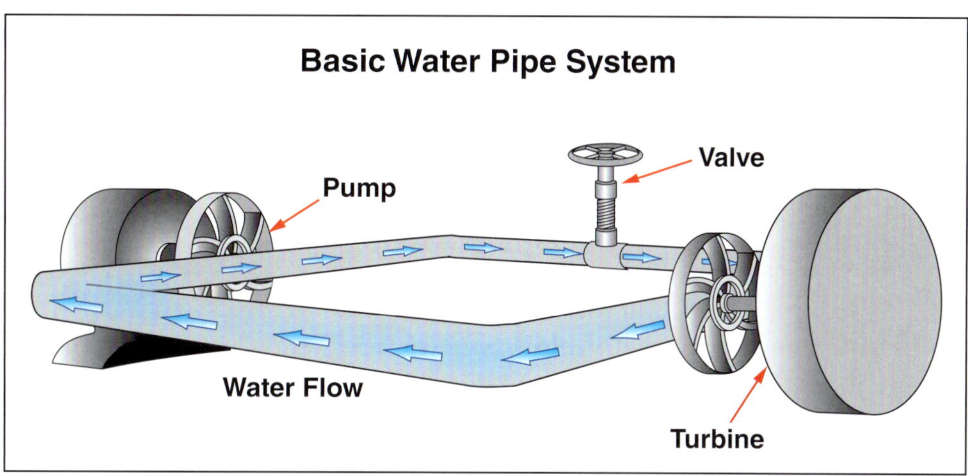

Figure 6.3 Similar to an electrical system, a water pipe system includes an energy source, travel path, and a mechanism to start or stop flow.

Table 6.1 Water and Electrical System Comparison	
Elements of the Water System	**Elements of the Electrical System**
Pump	Generator
Pressure	Voltage
Pressure Gauge	Voltmeter
Flow	Current
Flowmeter	Ammeter
Valve	Switch
Friction	Resistance (Ohms)
Friction Loss	Voltage Drop

Note also that in these examples, the flow of liquid or electricity moves in one direction. This direction of flow is analogous to a **direct current (DC) circuit**. In an **alternating current (AC) circuit**, the flow can constantly reverse direction **(Figure 6.4, p. 170)**.

And again in both systems, the pressure or voltage is reduced after performing the work. In the hydraulic system, this drop is called *friction loss*, and is measured with a pressure gauge. In an electrical system, this is called **resistance**, and is measured with a **voltmeter**. Both friction loss and resistance produce heat within their relative systems.

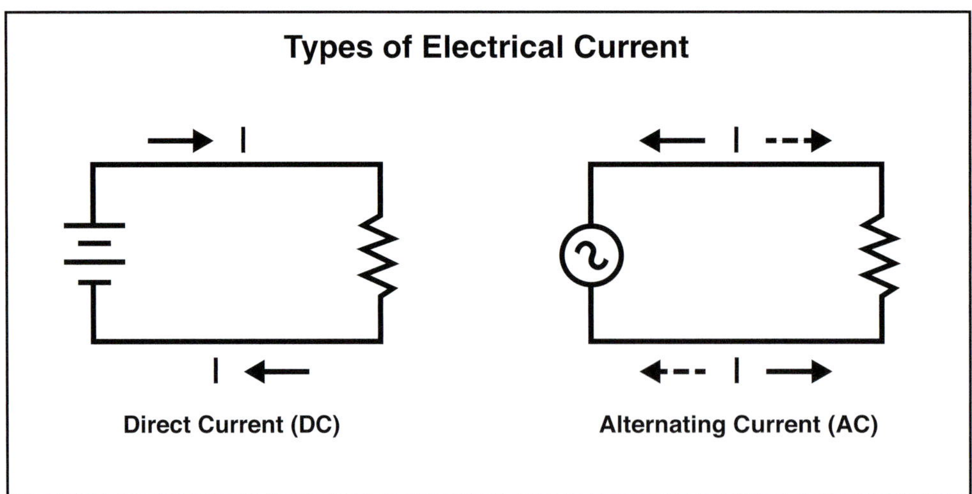

Figure 6.4 Electrical circuits where the current moves in one direction are direct current circuits. Circuits with current constantly reversing direction are alternating current circuits.

Finally, in both systems, the amount of friction loss or resistance in the system is directly related to the size (circumference) of the pipe or wire, the length (run) of the pipe or wire, and the material from which it is made. These factors will be explored throughout this chapter.

Basic Circuits

A basic circuit consists of a source path and load. There are two basic types of circuits: **series circuits** and **parallel circuits**.

Series Circuits

A series circuit flows current through all components. A series circuit is the simplest type of circuit.

In practical terms, consider vehicle headlights **(Figure 6.5)**. For the lights to operate, current must pass from the positive, through the switch, through both lights, and back to the negative terminal of the battery. When the switch is opened (off), the lights are not illuminated because there is no current in the circuit.

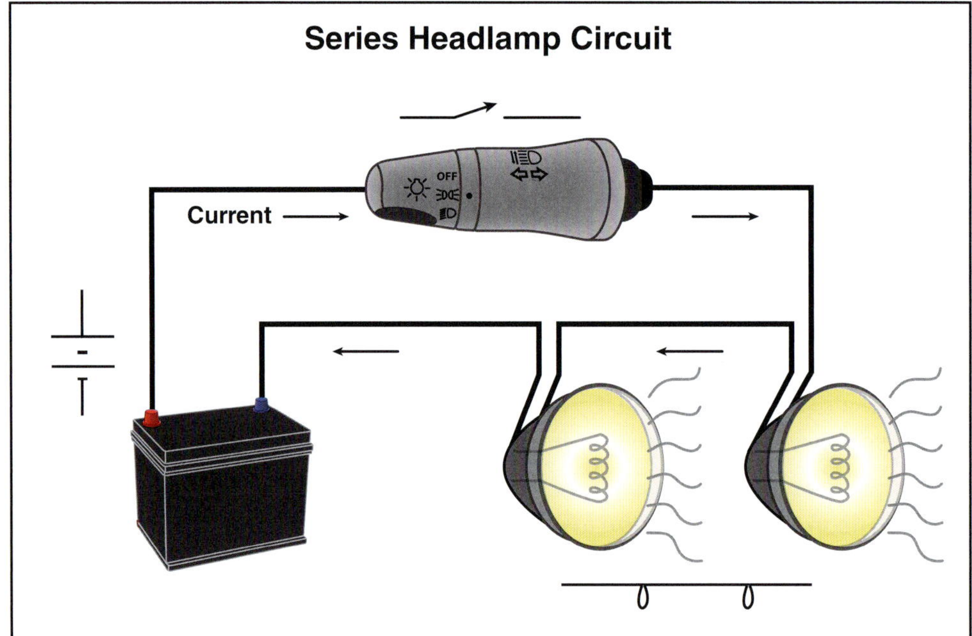

Figure 6.5 A series circuit has one path for electricity through all of the electrical components.

As wired, all of the current passes through the switch and both lights. If one light fails, it would have the same effect as opening the switch — removing the current from the circuit. Thus, if one light fails, the other will not illuminate. For this important circuit, a configuration that would allow one light to illuminate if the other one fails would be better.

Parallel Circuits

A parallel circuit provides a unique path for the current from the switch to each light **(Figure 6.6)**. If one light fails in this system, the other light will still illuminate because the current has a complete circuit or path through which to travel. Parallel circuits are the most common circuits that investigators will encounter.

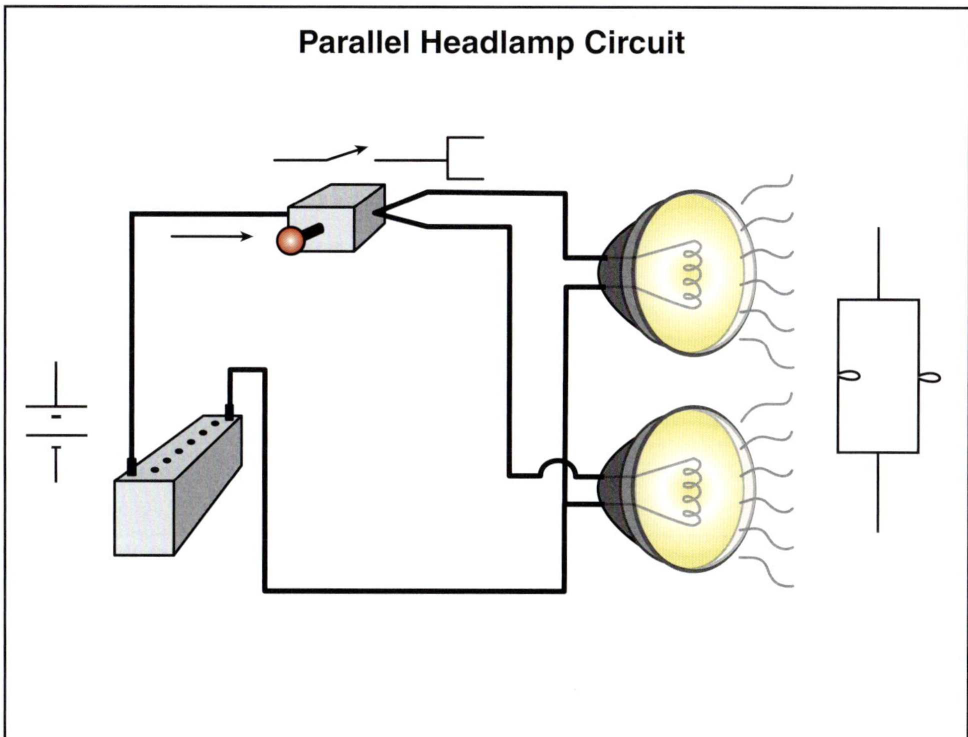

Figure 6.6 Components can be added or removed from a parallel circuit without interfering with the operation of other components.

All components connected to a parallel circuit operate at the same voltage. Also, components may be added or removed from the circuit without interfering with the operation of other components. These added components are referred to as **branch circuits**, because they connect the main electricity supply system to individual outlets **(Figure 6.7, p.172)**. Branch circuits are discussed later in this chapter.

In a simple parallel circuit, the switch controls the circuit. When the switch is open, the circuit is not complete, and the lights are off. When the switch is closed, the circuit is completed. Current flows from the battery to the lights and back to the battery. The current through the lights causes them to illuminate, thus releasing light and heat energy.

Electrical circuits often include some type of **grounding** as a safety measure. Grounding provides a path for excess electrical current to leave the system in a controlled way. Grounding is discussed later in this chapter.

Circuits also often include some kind of **fuse**, which opens/breaks a circuit during **overcurrent** (overloading) conditions. Fuses are discussed later in this chapter.

In this example of a circuit that includes grounding and a fuse, the voltage comes from the positive terminal of the vehicle's 12-volt battery to a fuse block (**Figure 6.8**). The negative terminal of the battery is connected to the vehicle frame, creating a grounded conductor. In this arrangement, the vehicle frame becomes one of the conductors in the headlight circuit.

NOTE: Circuits in homes and other buildings do not use conducting materials, such as structural steel or metal water pipes, as a return path for current.

With the addition of grounding and switches, headlights may be wired in this way:

- Wire connects the fuse block to the positive side of the battery, and then to the headlights' control switch.

- From the switch, a separate power wire is connected to each light.

- The opposite side of the light is connected to the vehicle frame or ground to complete the circuit back to the battery.

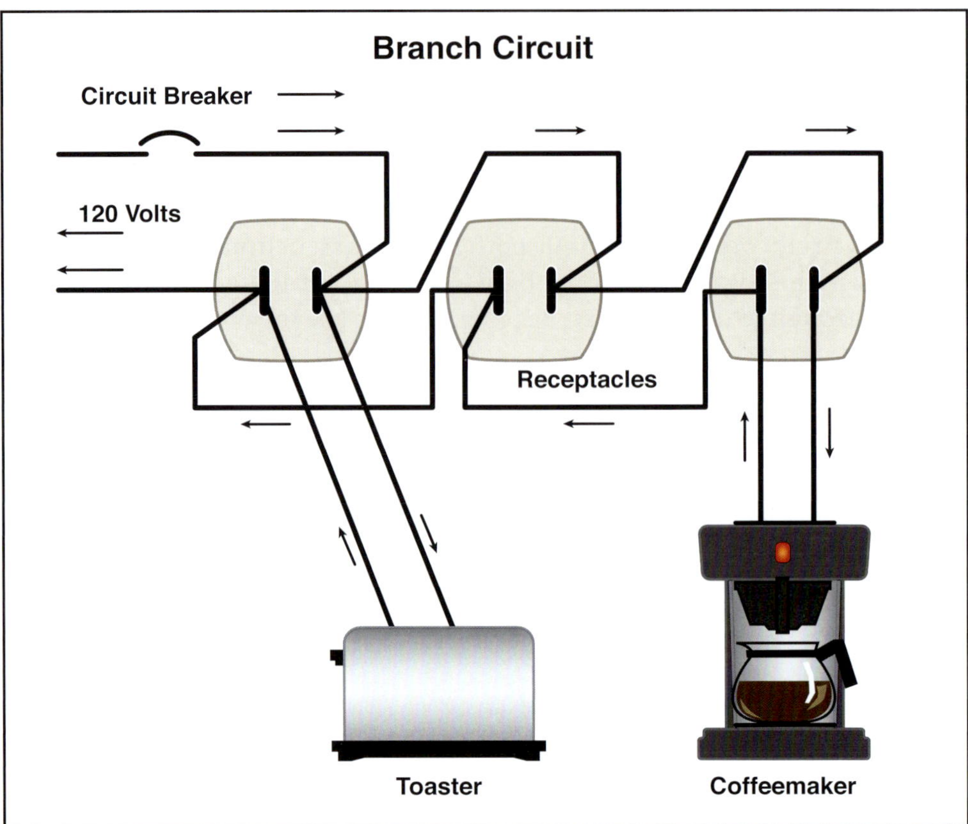

Figure 6.7 Appliances attached to the circuit use parallel wiring. Circuit conductors closer to the source potentially carry higher current.

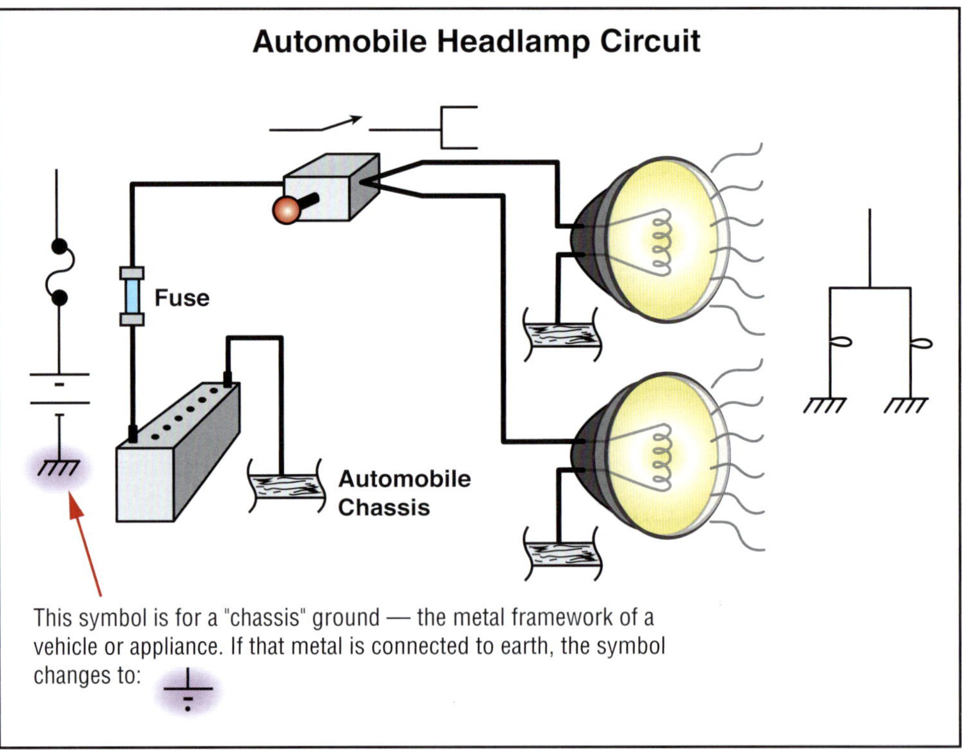

Figure 6.8 This circuit includes a ground connector at the battery and also at each lamp.

In many non-structure applications, the energy source for a circuit is a battery. To rephrase, the battery is the source of the energy that results in the electrons being pushed through the circuit, resulting in current.

To measure voltage in the circuit, a voltmeter is placed across (parallel to) the circuit (**Figure 6.9**). To measure current in the circuit, an **ammeter** is placed in the line (series) so that the current flows through it. However, fire investigators rarely use these meters. Refer to the manufacturers' instructions, if needed, for the use of ammeters and voltmeters.

Resistance

As mentioned earlier, as current passes through the circuit, it encounters some resistance from the wire and other components in the circuit. Resistance is measured in **ohms** (Ω).

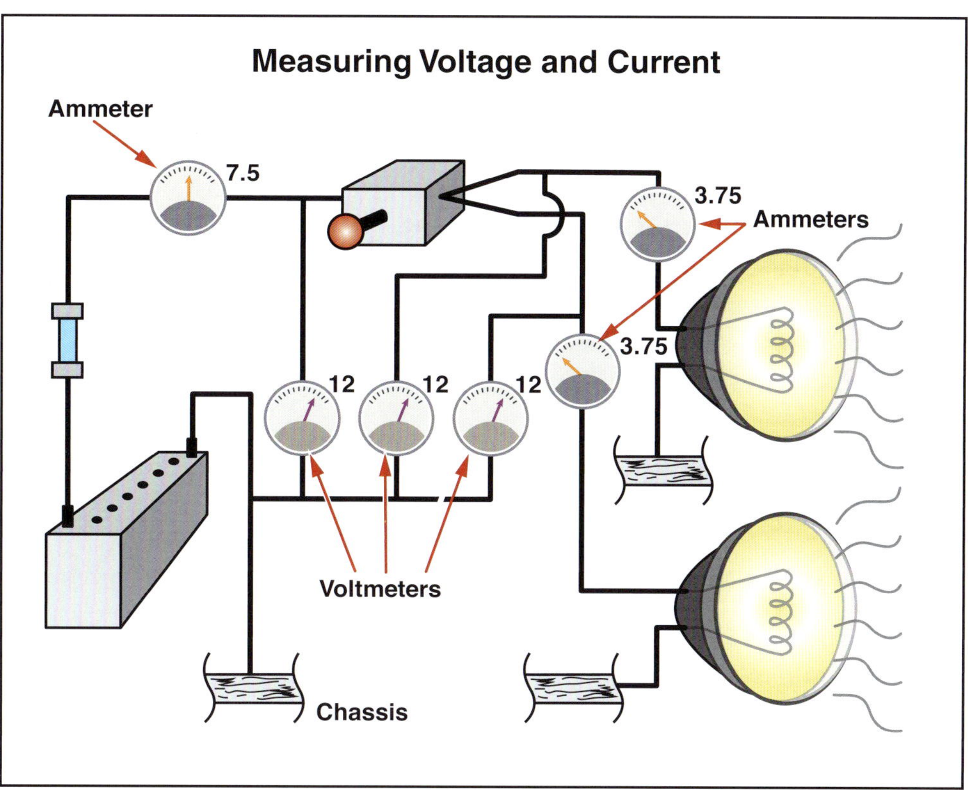

Figure 6.9 Voltmeters and ammeters can be added to a circuit to measure the voltage and current in different parts of the circuit.

The gauge, material, and length of the wire used in a circuit have some effect on the total resistance found in the circuit. **Table 6.2** provides the resistance per 1,000 feet (300 m) of wire for three common conductor sizes of both copper and aluminum.

Gauge. Wire size is given in American Wire Gauge (AWG) increments and is determined by the diameter of the wire in question. The numbers increase as the diameter of the wire decreases. For example, No. 14 AWG wire is smaller in

Table 6.2 Resistance of Copper and Aluminum Wire for Common Gauge Sizes		
Size of Wire: AWG	Resistance in Ohms: Copper	Resistance in Ohms: Aluminum
14	3.1	5.2
12	2.0	3.2
10	1.2	2.0
Resistance values per 1,000 feet (300m) of wire at 167°F (75°C)		

diameter than No. 8 AWG wire. A large-diameter wire has less resistance than a small-diameter wire and therefore can carry a greater amount of current before overheating. Conductors must be of the appropriate gauge to carry the load required. For example, a larger conductor would be required for a stove or dryer than for a residential lighting circuit.

Material. As indicated earlier, both copper and aluminum are considered good conductors because they offer relatively low resistance to the flow of current in a circuit. Between the two, copper is the better conductor because it has a lower resistance. In addition, copper conductors require less maintenance than aluminum conductors. Aluminum conductors, such as **duplex receptacles**, switches, and light fixtures, may require occasional tightening at the connections. Expansion and contraction over time can loosen connections, which can overheat and cause a fire.

Poor conductors, such as Nichrome®, nickel-chromium, offer very high resistance to current flow. Because of its resistance, Nichrome® is used as the heating element in many heat-producing appliances, such as space heaters and kitchen toasters.

Calculating Resistance. The total resistance from the wire in the circuit can be calculated separately from other components, given the gauge, material, and length. In the example of the vehicle headlight circuit, it is reasonable to estimate a total length of 20 feet (6 m) of No. 12 AWG copper wire as 0.04 ohms, per the calculation in **Table 6.3**.

Table 6.3 Resistance in Circuit Wiring		
Given	**Solving for**	**Calculation**
1,000 feet (300 m) x No.12 AWG copper = 2 Ω	Resistance in 20 feet (6 m)	2 Ω ÷ 1,000 ft = 0.002 Ω 20 feet (6 m) x 0.002 Ω = 0.04 Ω

The resistance from the wire in this circuit is negligible, considering the resistance of a 45-watt headlight bulb is 3.2 ohms. Exceptions where wiring has more than negligible resistance include:

- Long runs of wire
- Very small gauge wire
- Poor connections

Ohm's Law

Voltage and current are directly proportional, that is, when one increases, the other also increases (**Figure 6.10**). **Ohm's Law** is used to calculate the voltage, current, or resistance in a circuit, using the formula shown in **Table 6.4**.

If two of the three values are known, the third can be calculated using variations on that formula. In the example of the vehicle headlight circuit, the current on one circuit branch can be calculated, given voltage and resistance as shown in **Table 6.5**.

Table 6.4 Symbols for Units in Ohm's Law		
Name	**Symbol**	**Unit**
Voltage	V	Volt (V)
Current	I	Ampere (A)
Resistance	R	Ohm (Ω)
Power	P	Watt (W)

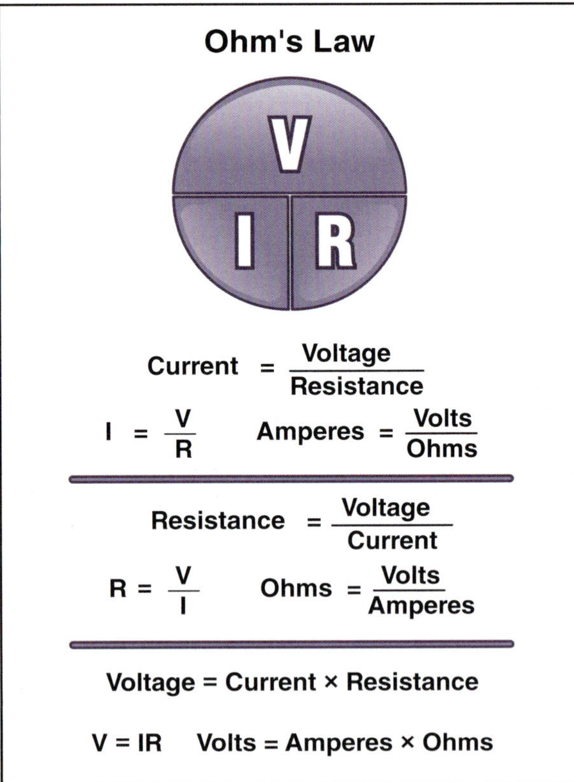

Figure 6.10 Ohm's Law can be shown in a pie chart.

Table 6.5		
Headlight Circuit Current		
Given	Solving for	Calculation
12-volt battery: 12 V Headlight resistance: 3.2 Ω	Current (I) = amperes (A)	Current (I) = Voltage (V) ÷ Resistance (R) I = 12 V ÷ 3.2 Ω I = 3.75 amperes (A)

Power

As previously discussed, energy is spent when current moves through a circuit. *Power* is defined as the amount of energy delivered over a given period of time. In electrical circuits, the unit used for power is the watt (W).

This energy expenditure may appear in any number of ways, including light coming from headlights, a vehicle horn sounding, or the starter motor in a vehicle spinning, or toast being made in a toaster. These units should be familiar because most electrical devices provide a wattage rating. By using Ohm's Law, the relationships between these values can be developed and applied, depending on which of the quantities is known by the investigator **(Table 6.6)**.

Table 6.6
Power Equation and Relationships
Power (P) = Voltage (V) × Current (I)
Current (I) = Power (P) ÷ Voltage (V)
Voltage (V) = Power (P) ÷ Current (I)
Power (P) = Current^2 (I^2) × Resistance (R)
Power (P) = Voltage^2 (V^2) ÷ Resistance (R)

Continuing the headlight circuit example, the power consumption of one of the lights can be calculated using the given information **(Table 6.7)**. The calculation for amperes in this circuit will ignore the 0.04 ohms of resistance in the wire because it is so small in comparison to the resistance of the headlight.

Table 6.7		
Power Consumption in One Headlight		
Given	Solving for	Calculation
Voltage (V) = 12 volts (V) Current (I) = 3.75 amperes (A)	Power (P) = watts (W)	P = 12 V × 3.75 A P = 45 watts (W)

A fire investigator can determine the load on a branch circuit by simply adding the wattages of the components connected to the circuit **(Table 6.8, p. 176)**. With this information, Ohm's Law can be used to calculate the other variables. In the headlight circuit example, the calculation is simplified to only include the two lights, which consume 90 watts total when operating **(Table 6.9, p. 176)**.

Common branch circuits found in most homes have either a 15 or 20 ampere-rated capacity. Most modern installations use copper in branch circuits. The load capacity for these circuits can be calculated as shown in **Table 6.10, p. 176**.

Table 6.8	
Power Consumed by Typical Appliances	
Appliance	**Watts**
Air conditioner (room type)	800-1500
Electric blanket	175
Coffeemaker	600
Dishwasher	1800
Hair dryer	1500
Clothes dryer	4500
Portable heater	1200
Household iron	1000
Electric motor (per hp)	1000
Built-in oven	4000
Electric stove (all burners and oven on)	8000-16000
Refrigerator	250
Television	250
Electric hot water heater (80 gallon)	4500

Table 6.9	
Circuit Load with Two Headlights	
Given	**Calculation**
Power (P) = 90 watts (W) Voltage (V) = 12 volts (V) Current (I) = amperes (A)	I = 90 W ÷ 12 V I = 7.5 A

Table 6.10		
Circuit Load Capacities		
Given	**Solving for**	**Calculation**
Current (I) = 15 and 20 ampere (A) capacities Voltage (V) = 120 volts (V)	Power (P) = watts (W)	15 A × 120 V = 1,800 W 20 A × 120 V = 2,400 W

Investigators should remember that the wire sizes for electrical wiring used in branch circuits in buildings and other locations are determined by the requirements of NFPA 70®, *National Electrical Code*®. Minimum wire sizes in the code are very conservative and have a significant safety factor built into them. Wiring requirements are based on:

- Wire gauge
- Type of insulation on the conductor
- Wire material
- Temperature at which it is intended to operate

Under some circumstances, electricity and the wiring used to carry it can pose an ignition hazard. However, investigators should understand that the presence of electrical equipment or wiring in or near the area of origin does not indicate that the ignition source was electrical. Before the source of ignition is established, all factors must be evaluated. Simply finding a circuit that appears to be overloaded may not cause sufficient heat production to result in an ignition.

Alternating Current/ Direct Current

The concept of direct current (DC) and alternating current (AC) power was introduced earlier in this chapter. In DC, electrons flow in only one direction, from negative to positive. The automobile headlight circuit example is a DC circuit with energy supplied by a battery.

AC is a common type of current found in the electrical systems in most buildings. Thus, it is the type that fire investigators will see most frequently when investigating building fires. One of the few exceptions is the source of energy for elevators, which are often operated using DC power. The primary difference between alternating and direct current is the method used to produce the energy.

Alternating current is produced using a generator; basically, a coil of conductors is rotated within a magnetic field:

- As the coil is turned in the field, a voltage is produced.

- As the coil turns, the voltage and its charge "polarity" is constantly changing.

- The direction of the flow of electrons is constantly changing (**Figure 6.11**).

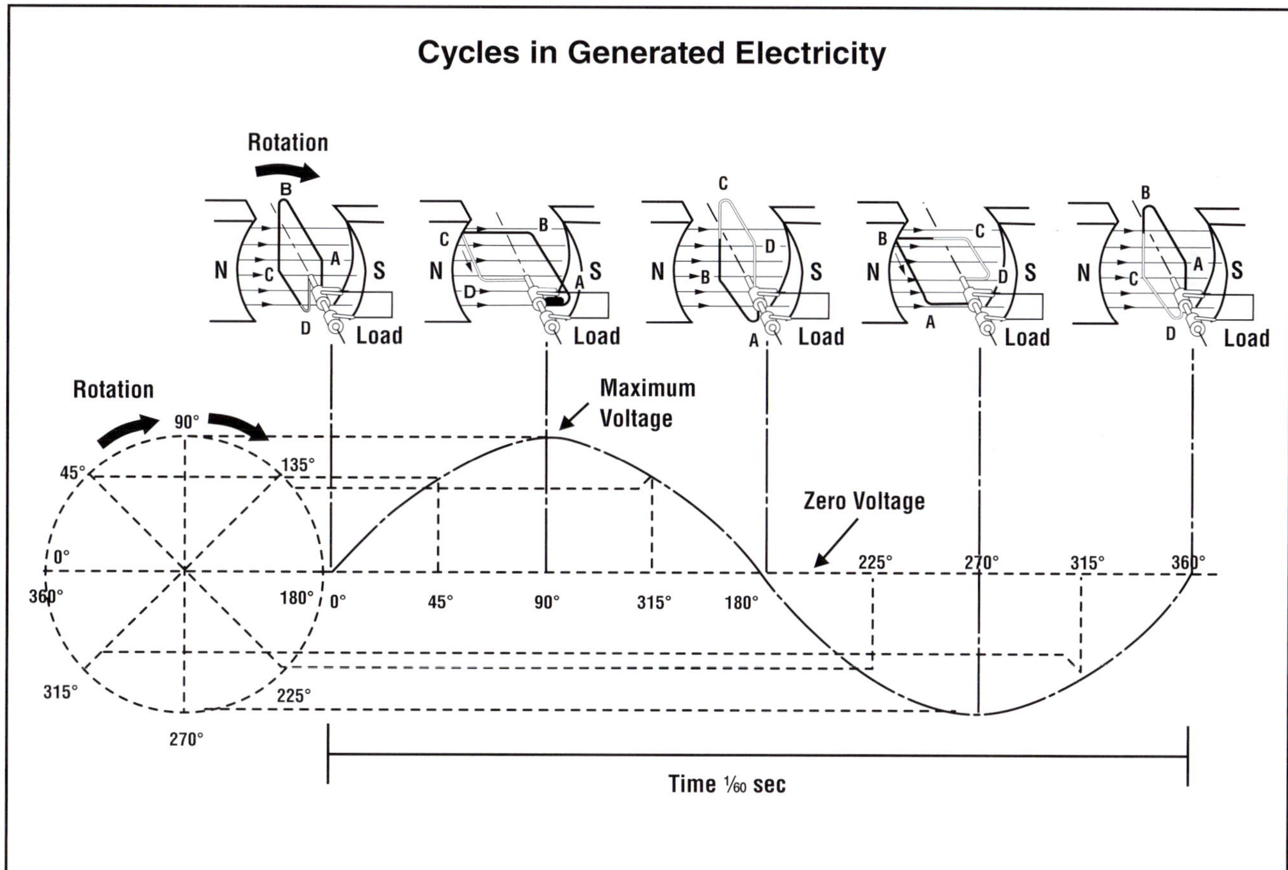

Figure 6.11 As the coil in a generator turns, the voltage rises and falls.

Kitchen Toaster Example

The previous calculations worked with a circuit supplied with direct current power. The basic calculations using a circuit supplied with alternating current power are the same. Given manufacturer's data for a domestic toaster appliance, **Table 6.11** and **Table 6.12** show calculations for current and power consumption.

Table 6.11
Kitchen Toaster Current

Given	Solving for	Calculation
Resistance (R): 10 ohms (Ω) Voltage (V): 120 volts	Current (I) = amperes (A)	120 V = I × 10 Ω I = 120 V ÷ 10 Ω I = 12 A

Table 6.12
Power Consumption in a Kitchen Toaster

Given	Solving for	Calculation
Voltage (V) = 120 volts (V) Current (I) = 12 amperes (A)	Power (P) = watts (W)	P = 120 V × 12 A P = 1,440 watts (W)

Distribution Lines

Electric companies supply AC service to most customers via distribution lines. The large electric generators at electric companies incorporate turbines that are turned by steam or water to generate large amounts of energy efficiently. When the electrical energy is produced, it can be transmitted at high voltages. Higher voltages help in the distribution of electricity. When the transmission voltage is higher, the current in the lines will be lower **(Figure 6.12)**.

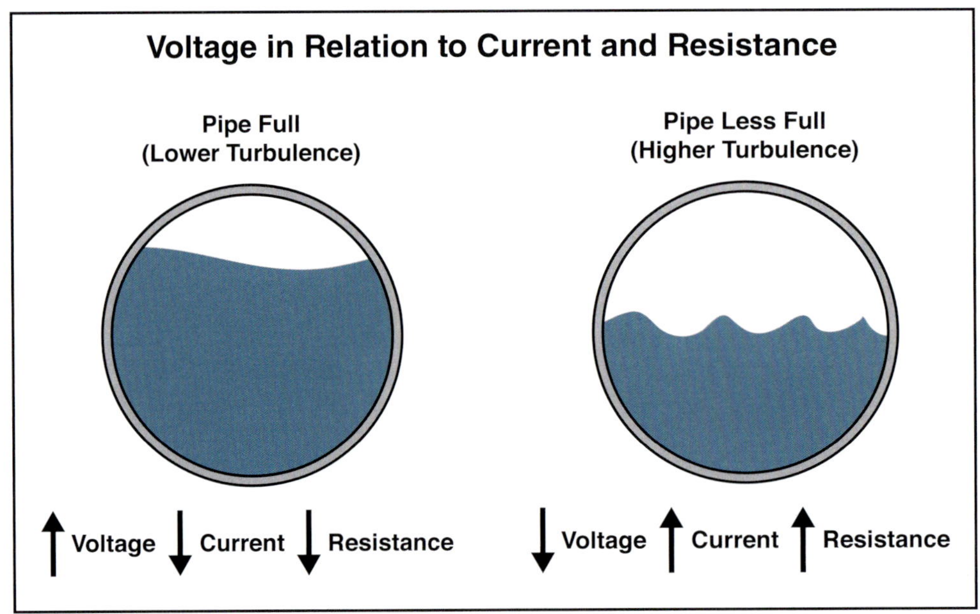

Figure 6.12 Comparison between the three factors in Ohm's Law using a pipe as a visual example.

The resistance of long transmission lines results in a loss of energy, primarily due to the electrical energy being converted to heat via resistance. As the current in a transmission line is decreased, the resistance also decreases. Compensating for this power loss is accomplished by using high voltages. Resistance increases as voltage decreases.

Although power is distributed at high voltages, the voltage must be reduced before residential and commercial customers can use it. This is accomplished using step-down **transformers** (**Figure 6.13**).

Basic Functions of Transformers

Fire investigators should have a basic understanding of the function of a transformer. Transformers reduce commercial/industrial voltages of up to 600 volts to typical 120/240 volt levels. These units will be found in electrical substations and throughout the power distribution system in every community (**Figure 6.14**). Transformers may also be located as part of the facilities for occupancies that require a large amount of power to operate such as:

- Health care institutions

- Commercial complexes

- Industrial complexes

The investigation of fires in such equipment will more than likely require that a fire investigator seek expert assistance such as an electrician or utility provider. Fire investigators may also encounter step-down transformers used to reduce 120-volt AC electricity to various lower AC or DC voltages used to power appliances or devices such as cell phone chargers and laptop computers.

Step-up and step-down transformers operate as follows:

- Primary and secondary coils are wrapped around an iron core.

- As the AC passes through the primary coil, a magnetic field is created.

- The rapid variation from positive to negative and back creates a voltage in the secondary coil through the magnetic properties of the iron core. Depending on the ratio of windings in the primary and secondary coils, the voltage created in the secondary coil will be higher (step-up) or lower (step-down) than the voltage in the primary coil.

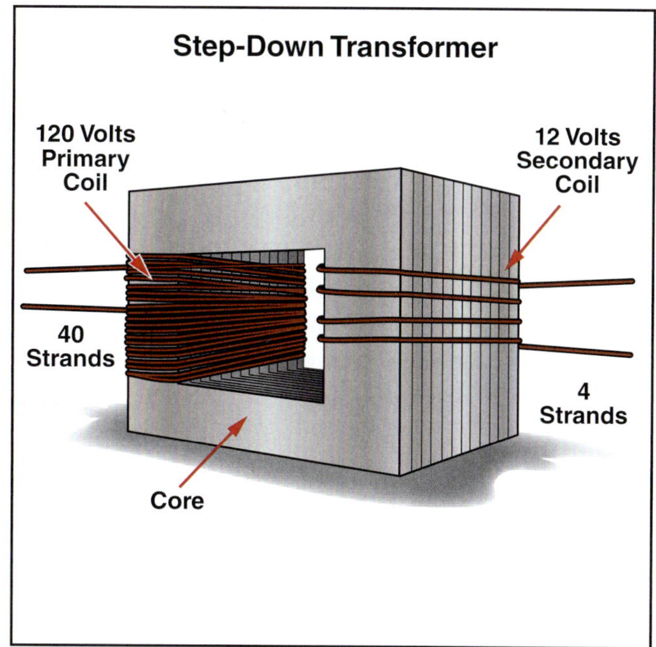

Figure 6.13 A step-down transformer converts a high voltage current flow to a lower-voltage current flow.

Figure 6.14 Electrical transformers are commonly found in rural and urban areas. *Courtesy of Wayne Chapdelaine, Metro-Rural Fire Forensics.*

Electrical Conversions

Fire investigators should be aware of the following common electrical conversions:

1 kilowatt (kW) = 1,000 watts (W)

1 kilovolt (kV) = 1,000 volts (V)

Power Loss in Transmission

When learning about transformers, fire investigators should keep in mind that the power into and out of a transformer is the same because of the law of **conservation of energy**. That is, energy cannot be created nor destroyed, but can be transferred or transformed.

For example, consider a step-down transformer that has a primary winding of two loops with a one loop secondary. If 10 volts were applied to the primary, the resulting voltage in the secondary would be five volts **(Table 6.13)**. Assuming perfect efficiency, 1 ampere entering the primary coil would yield 2 amperes in the secondary. Because the energy is conserved, there must also be 10 watts in the secondary winding.

Table 6.13 Energy Conservation in a Transformer	
Primary Winding (Power)	**Secondary Winding (Current)**
10 V × 1 A = 10 W	*10 W ÷ 5 V = 2 A*

The current required for 10,000 watts at 120 volts is 83 amperes. The current required for the same number of watts at 10,000 volts is 1 ampere. This comparison is calculated in **Table 6.14**.

Table 6.14 Calculating Current at Two Voltages			
Given	**Solving for**	**Calculation**	
Power (P) = 10,000 watts (W) Voltage (V) = 120 volts (V) and 10,000 volts	Current (I) = amperes (A)	I = 10,000 W ÷ 120 V I = 83 A	I = 10,000 W ÷ 10,000 V I = 1 A

Electrical systems are never perfectly efficient. The only reduction in power in a transformer is converted to another form of energy, such as heat due to resistance. The *NEC®* states that 1 mile (1.6 km) of No. 1 AWG wire has a resistance of 0.665 Ω. The power loss due to resistance can be calculated at the two voltages **(Table 6.15)**.

Table 6.15 Line Loss in Transmission			
Given	**Solving for**	**Calculation**	
Resistance = No. 1 AWG cable for 1 mile (1.6 km) at 0.665 Ω Current (I) = 83 amperes (A) and 1 ampere (A)	Power (P) = watts (W)	P = (83 A)2 × 0.665 Ω P = 4,580 W	P = (1 A)2 × 0.665 Ω P = 0.665 W

Common Electrical Systems in Buildings

NFPA 1033 (2022): 4.1.7, 4.2.8

Most jurisdictions in the United States have adopted NFPA 70®, *National Electrical Code*®, to regulate the installation of electrical systems. The electrical code is an essential reference for any investigator examining a fire scene where there is a possibility that the electrical system was involved in the ignition of the fire.

NOTE: The *Canadian Electrical Code*® *(CEC*®*)* is a federal document. Most Canadian provinces/territories adopt their own electrical codes based on the CEC®.

Fire investigators can find additional information regarding electrical systems and their components in these codes. The following sections provide investigators with an overview of a code-compliant electrical system in a residential property.

Service

As introduced in Chapter 5, Building Systems and Utilities, electricity enters the building via a service **(Figure 6.15)**. The service may be aboveground (service drop), but services are increasingly installed underground (service lateral). The service wires will include at least three conductors that lead from the power company's lines to the building. In rare instances, investigators may encounter two-wire service in old structures. *NEC*® provides complete information on service wires.

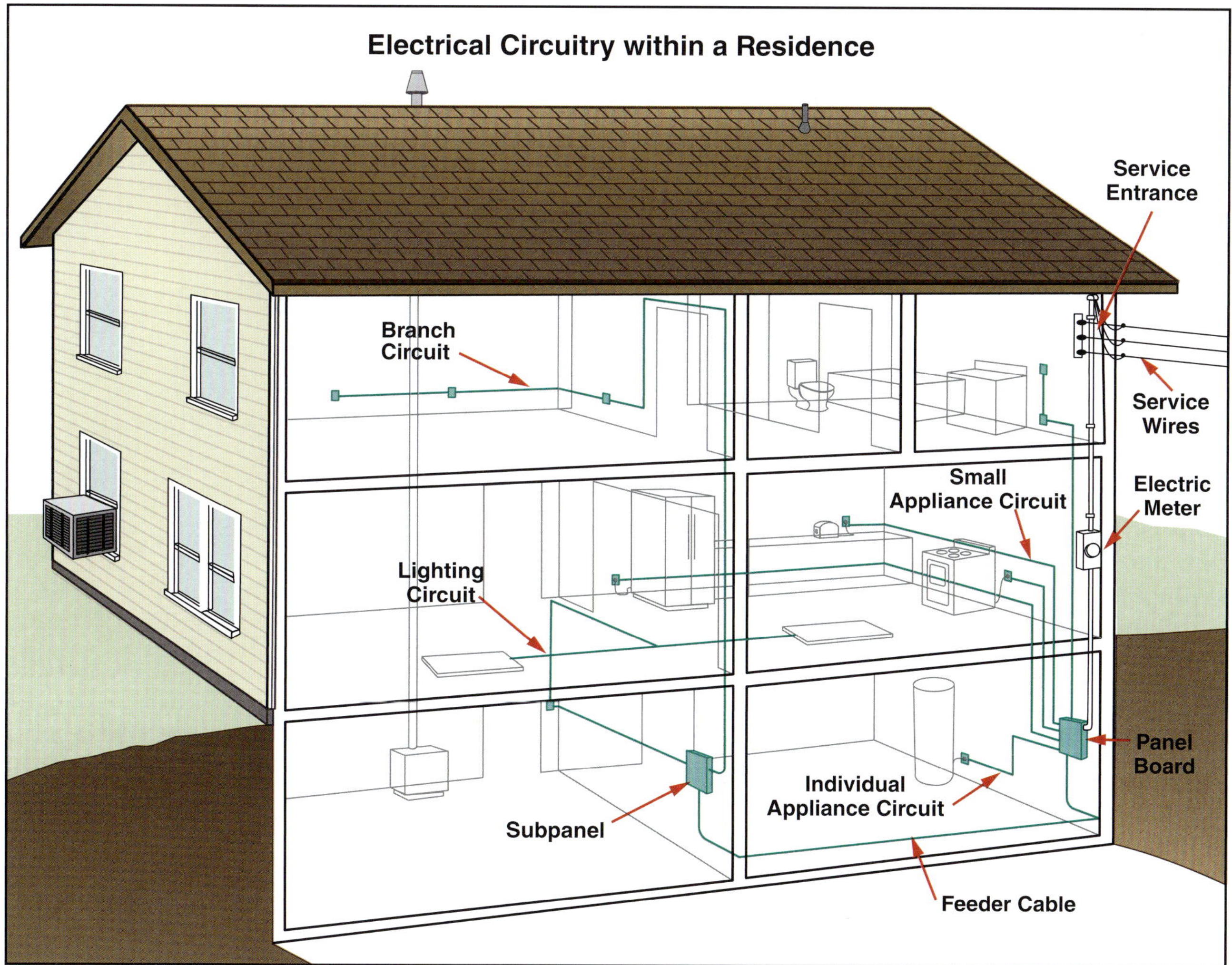

Figure 6.15 Depiction of electrical service entering a structure and the circuits that distribute electricity throughout the residence.

Service Conductor

The size of the service conductor is determined by the **ampacity** of the service. Most modern service connections to residential properties are made with three No. 2 AWG aluminum conductors **(Table 6.16)**. Two of the cables will be insulated with the third uninsulated conductor serving as a **grounded neutral conductor**. The neutral conductor will share a voltage of 120 volts with either of the insulated conductors. The insulated conductors will share 240 volts.

Table 6.16 Branch Circuit Requirements					
Circuit Rating	15 Amp	20 Amp	30 Amp	40 Amp	50 Amp
Min. conductor size	14 AWG	12 AWG	10 AWG	8 AWG	6 AWG
Outlet devices	Any type	Any type	Heavy duty	Heavy duty	Heavy duty
Receptacle rating	15 amp	15 or 20 amp	30 amp	40 or 50 amp	50 amp
Type of load permitted	Lighting or utilization equip.	Lighting or utilization equip.	Utilization equip.	Fixed cooking equipment	Fixed cooking equipment
Overcurrent protection	15 amperes	20 amperes	30 amperes	40 amperes	50 amperes

Note: 30-ampere and greater branch circuits may be used for lighting circuits provided with heavy duty lamp holders in occupancies other than dwelling units.

Note: Utilization equipment is defined in NFPA 70® as "equipment that utilizes electric energy for electronic, electromechanical, chemical, heating, lighting, or similar purposes."

Service Lines

The service lines are installed to the building in such a way as to prevent damage or unintentional contact with the conductors. Aboveground cables are securely attached to the structure at the service entrance. From the service entrance, the line connects to a meter that measures the consumption of energy by the customer **(Figure 6.16)**. The meter quantifies the total energy flow into the building in kilowatt-hours (kW-h) and is in place for billing purposes.

In most cases, current flows through the meter as part of the electrical system. If the meter is disconnected or "pulled," power is also disconnected. In some services, the current does not flow through the meter but the meter only reads the flow. In those cases, pulling the meter does not disconnect the power to the building and does not stop the electrical flow.

Figure 6.16 Electric meters for most applications have a number of similarities. *Courtesy of Mike Makela.*

Disconnecting Service

From the meter, the service conductors connect to an electrical distribution panel where the power is distributed to the electrical system (branch circuits) in the building. At some point between the meter and the branch circuit protective devices, there is a disconnect switch or switches for the service. The means for disconnecting the service may be located in an electrical distribution panel (circuit breaker/fuse panel) or a separate box. Article 230 of the *NEC*®allows the service disconnect panel or box to have up to six separate switches or circuit breakers **(Figure 6.17)**. More information about electrical distribution panels and circuit breakers is presented later in this chapter.

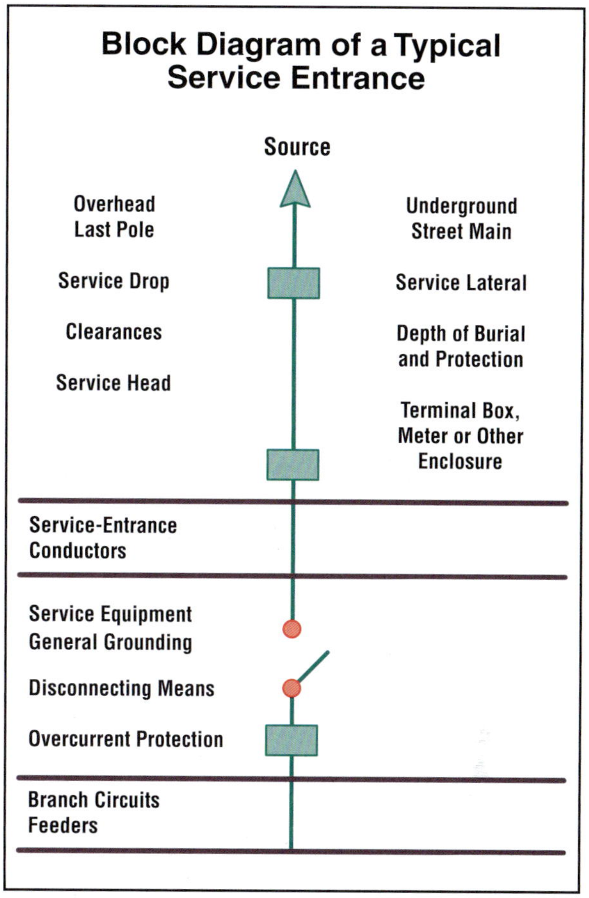

Power for Mobile Homes

Power for older mobile homes is provided using a feeder assembly that is composed of a four-wire cable that attaches to the mobile home with an attachment plug cap. The capacity of the plug cap is normally 50 amperes. In these installations, grounding and grounded (neutral) conductors are separated past the metering point. New mobile homes have the same electrical services as ordinary residential structures.

Branch Circuits

Branch circuits lead from an overcurrent protection device in the circuit breaker panel to service the rooms and equipment in a building. Branch circuits in a building are used for:

- Lighting

- Operation of nonfixed electrical equipment such as appliances that are plugged into an electrical receptacle

- Devices, such as fixed cooking equipment in a home, that are attached to the circuit that feeds them **(Figure 6.18, p. 184)**

Residential electrical systems will often include several 15- or 20-ampere general-purpose branch circuits. These circuits are provided in most rooms for lighting and to supply wall receptacles. Additional 20-ampere branch circuits are provided as appliance circuits in kitchens, bathrooms, and laundry areas.

Individual 30-ampere branch circuits (these may be 240 volt circuits) may supply power directly to a specific device or appliance such as:

- Electric water heater

- Clothes dryer

- Commercial garbage disposal

- Water pump for a well

Circuits rated at 40 or 50 amperes are permitted in dwellings to supply fixed cooking appliances. These circuits have only one heavy-duty receptacle located near the rear of the stove unit. For workshops or other locations where there may be a heavy power demand, subpanels fed by feeder lines from the electrical distribution panel are permitted.

Circuit Wiring

The *NEC*®defines the acceptable materials and methods for the wiring of branch circuits in a building. The most common cable found in residential properties is **nonmetallic (NM) shielded cable**. One

Figure 6.17 Service disconnect panels or boxes often include a typical service entrance.

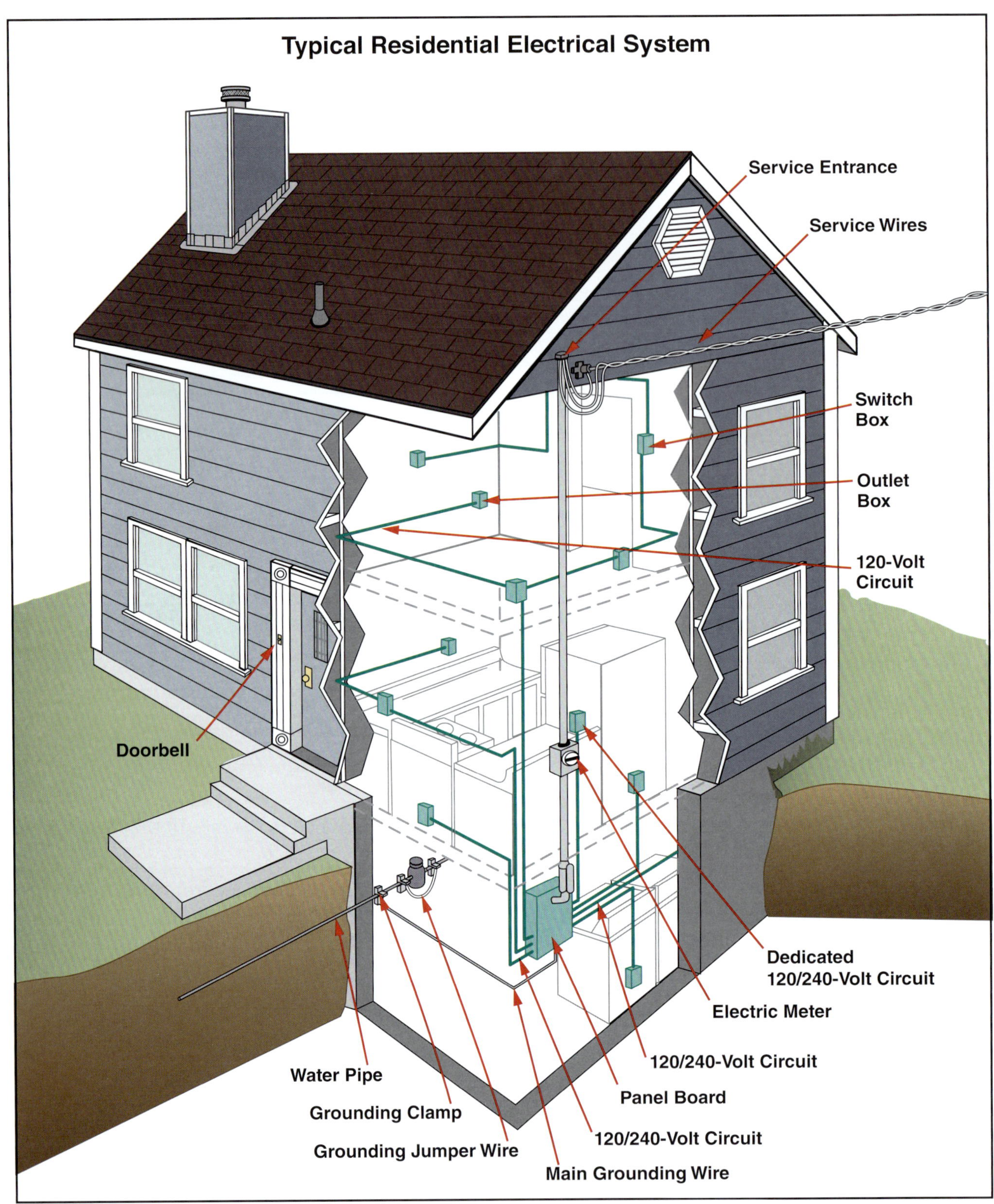

Figure 6.18 Residential electrical systems often include common features.

popular brand is Romex® plastic-sheathed wiring. These cables are factory assemblies with two or more wires, and they have a nonmetallic outer sheath that is moisture resistant and fire retardant. This type of cable may be installed directly or threaded into partitions and other spaces.

The *NEC*® Article 334 lists locations where nonmetallic sheathed cable is not permitted, including:

- Dwelling or structure that exceeds three stories above grade
- Exposed within a dropped or suspended ceiling cavity
- As service-entrance cable
- Commercial garages having hazardous locations (as defined in *NEC*® Article 511)
- Theaters or similar occupancies
- Motion picture studios
- Storage battery rooms
- Escalators or elevator hoistways
- Inside (embedded in) poured cement, concrete, or aggregate
- Hazardous locations (as defined in Article 500 of the *NEC*®)

Specific wiring methods and requirements are found in Article 300 of the *NEC*®. All conductors serving branch circuits must be protected from physical damage. The *NEC*® allows cable to be installed on or through:

- A surface
- Bored holes and notches in wood framing
- Holes in metal framing that are protected with bushings or grommets

The *NEC*® requires that metal plates be provided for protection when cable access holes are located where there is a potential for nails to touch the cable. Other than in one- and two-family dwellings, branch circuits are protected with conduit, cable raceways, or other approved means.

NEC® Section 300.15 requires that all splices and connections be made in approved electrical boxes. Thus, any splice in a wire or connection to a device, such as a switch or receptacle, must be made in an electrical box. Electrical boxes may be either metal or plastic depending upon the jurisdiction. The size of the box required is determined by the size and number of conductors that enter or leave the box. The conductors entering or leaving a box must be secured with a clamp or other approved means at the point of entry or exit.

Within the electrical box, the actual connections are made by using terminal screws on a device. Examples of the terminal screws include:

- A switch
- A receptacle
- An approved connector, such as a solderless connector, commonly called a **wire nut** (**Figure 6.19**)

Connections are a common point of failure. Poor connections can result in increased resistance. Under the right conditions, increased heat can cause the ignition of nearby combustibles. Proper connections are tight, and they are made using approved devices and methods.

Figure 6.19 Wire nut connectors must be approved for the wires they connect and be nonconductive. *Courtesy of Wayne Chapdelaine, Metro-Rural Fire Forensics.*

A point of special interest is a connection made with aluminum wire. For this application, devices specifically approved for aluminum must be used. Current devices are marked with *AL/CU*; older devices are marked with *CO/ALR*. The joining of copper and aluminum wire directly together is prohibited due to the different **thermal expansion** coefficients of the two metals, which could cause resistance heating. Devices approved for splicing copper to aluminum provide for the separation of the two metals within the device.

Energized Systems

An electrical system is normally deenergized during fire-suppression activities. Often, the local electric utility will remove the meter on a single-family dwelling, cut the overhead electrical service, or open the disconnects at larger facilities. However, building wiring systems can remain energized in the following ways:

- Generators are automatic or manually activated.

- Customers bypassed metering equipment (utility theft).

- Customers ran extension cords from adjacent electrical sources.

- Computer or communication equipment has standby or uninterruptible power supplies.

In addition, if investigation activities take place a few hours (or longer) after a fire response is terminated, temporary service may have been restored to some circuits. Appliances and other utilities may remain powered well after they are disconnected if they have an alternative power source such as a battery or capacitor. Investigator interviews with first responders are addressed later in this manual and will include confirming the actions that were taken during the initial response.

Before examining any part of an electrical system at a scene, fire investigators should confirm that the system has been deenergized. Contact and noncontact devices can be used to test whether a system is energized.

A contact device may consist of two leads and a neon light that visually indicates the presence of any voltage from 120 to 600 volts. The advantage of this device is that no batteries are required for it to operate. The disadvantage is that it requires a direct connection from the line being tested to the neutral conductor or grounded object. In addition, the light may be difficult to see in direct sunlight **(Figure 6.20)**.

A noncontact device uses a battery operated sensor to pick up current from a live wire, which then triggers an indicator. The advantage of this device is its safety. The disadvantage is that batteries require maintenance or replacement.

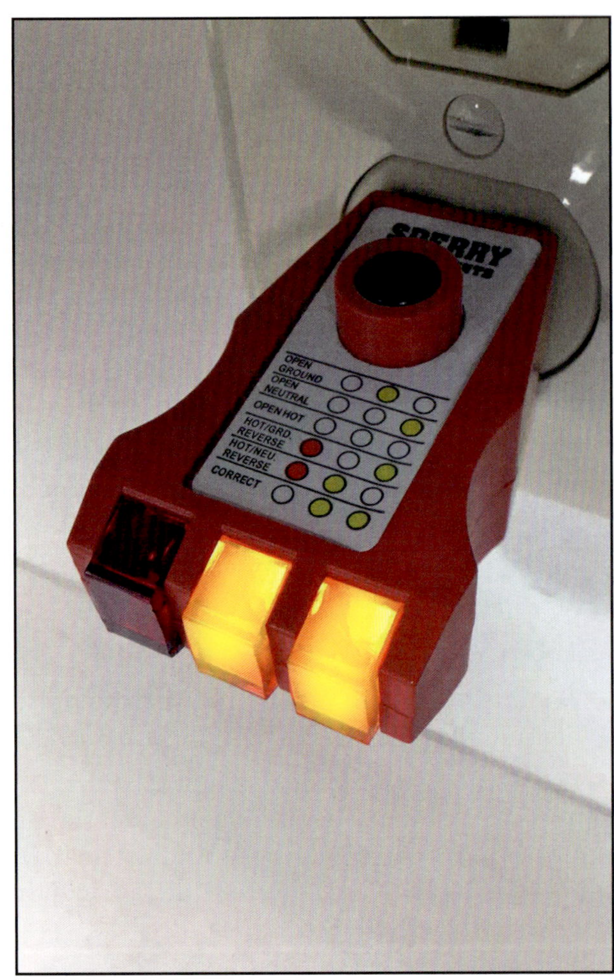

Figure 6.20 An outlet tester can show whether an intact outlet is energized.

Electrical circuit testers have a range of capabilities. Investigators should know the capabilities and limitations of their instruments. Manufacturer's recommendations should always be followed.

One type of circuit tester, a multimeter, can measure voltage, current, and resistance. A multimeter may be used to test whether a circuit is energized. The device normally applies a low voltage to the circuit or component being tested. If there is a path for current to flow, the meter on the device gives an indication.

A multimeter set to resistance/ohms (**ohmmeter**) can be used to:

- Trace branch circuits.

- Test fuses and circuit breakers.

- Check thermostats and other contacts in a circuit.

When using a multimeter, consider important facts including:

- A multimeter set to resistance/ohms is NEVER used on any circuit or device that might be energized. For example, if a multimeter's leads are inserted into a receptacle to check for a short in the wiring and the branch circuit is energized at its normal 120 volts, the ohmmeter will be damaged and the user could be injured.

- In a fire, wiring is coated with various forms of oxidation and combustion products. These products must be removed from the conductor in order to obtain an accurate reading.

- Contacts on items, such as switches, thermostats, and other electrical equipment, may also be coated as a result of a fire and may measure as open even though they are actually closed.

Circuits may also measure as closed even though they were open before the fire, because the contact material may melt and fuse the surfaces. It is preferable to preserve the device for later laboratory analysis than to chance irreparably altering fragile components of the involved electrical system. Spoliation and evidence preservation are further explained in Chapter 9, Evidence Collection and Preservation.

Electrical Heat Sources and Failure

NFPA 1033 (2022): 4.1.7, 4.2.8

Common electrical failures may be found on electrical conductors or at electrical connections. Electrical failures that can cause fires will be detailed more thoroughly in Chapter 13, Cause Determination. Examples of common electrical failures include:

- **Damaged insulation on conductors** — Damage from pinching, stretching, or an overdriven nail or screw. This damage to the insulation may allow the conductors to touch, resulting in a **short circuit** or electrical **arc**.

- **Loose connections** — Loose connections at terminal screws inside duplex receptacles and switches may result in resistance heating as an oxide interface is formed on the terminal screw, conductor, and metal plate. This resistance heating may cause the terminal screw and associated components to eventually glow red hot and fail.

- **Damaged electrical appliance cords** — Damage to electrical appliance cords may occur due to wear or mechanical fracture. Investigators should fully examine the appliance cord for signs of damage including the male plug blades for signs of micro arcing, damage to the conductors, and damage where the cord attaches to the appliance.

Electrical systems and their components may become competent sources of ignition for a fire when they fail as a result of one of the following common electrical incidents:

- Electrical arcs
- Sparks
- Resistance heating

Electrical Arcs

An electrical arc occurs when electricity flows across a gap or through a conductive path in a manner that was not intended, resulting in the production of high-temperature heat energy **(Figure 6.21)**. Common occurrences of this type of electrical event include:

- Short circuits

- **Parting arcs**

- Static electrical discharge

- Arcing through char, such as conductive matter, including charred insulation/ carbonized path

- Arcing across materials that have been contaminated with a conductive medium (arc tracking)

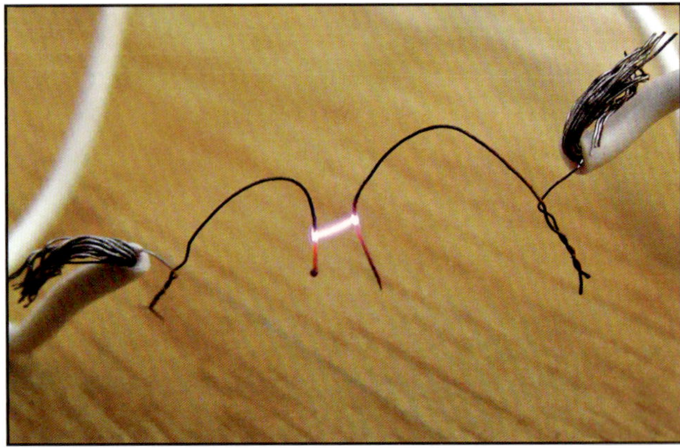

Figure 6.21 Electrical arcs can occur in gaps between charged conductors.

CAUTION: Arcing occurs normally in various electrical appliances and equipment. These arcs may be capable of igniting vapors, dust, and gases.

Electrical systems and their components are often damaged by a fire, and may result in electrical arcing **(Figure 6.22)**. This damage may result in melting that can be difficult to distinguish from electrical arcing, especially in the field. In these cases, it often is necessary to evaluate such damage under magnification. The field use of portable USB microscopes connected to a laptop or tablet can produce very good quality images with magnification up to 1000X. Experts may also be consulted.

NOTE: Evidence of electrical arcing within the origin area of a particular fire does not mean that arcing is the source of ignition for the fire.

Figure 6.22 As insulation around wiring heats and chars, the charred insulation can act as a conductor rather than an insulator, resulting in electrical arcing through char. *Courtesy of Yates & Associates.*

Sparks

Sparks are small, solid particles that have been heated to luminescence and are often formed when a short circuit occurs between metal objects that were normally insulated from one another. Electrical arcing may be very brief. While arcing persists, droplets of molten material may be thrown or projected. Sparks begin cooling the moment an arc occurs and may be an ignition source for only a brief period of time.

Investigators should recognize that sparks from different sources may have different properties (Lewis, 2019). Properties include:

- Sparks from an aluminum conductor may remain hotter for a longer period of time than sparks from a copper conductor.

- Sparks produced during an electrical arc event tend to be very small.

- Sparks resulting from welding or cutting may be larger and therefore are more prone to causing ignition of secondary fuels that they contact.

- Sparks from mechanical friction, such as failed bearings in an electric motor, can become an ignition source, but these sparks would not indicate an electrical failure.

Resistance Heating

Electrical heat energy occurs whenever electricity flows through an electrical system or one of its components. Most often, if properly designed and/or installed, this heat energy is safely dissipated or controlled so that it does not result in a fire.

The following conditions may generate higher resistance than intended:

- Loose connections **(Figure 6.23)**

- Undersized wiring

- Overloaded circuits

- Misapplications of electrical equipment

Resistance heating may also accumulate when an electrical source is overly insulated and heat cannot safely dissipate from the electrical source. For example, branch circuit wiring in a confined space that carries a continuously high load could cause heat to become trapped at sufficient rates to ignite combustible materials.

Figure 6.23 Loose connections can lead to resistance heating. *Courtesy of Yates & Associates.*

Short Circuits

A short circuit is an abnormal low-resistance path between conductors that allow a high current flow. The reduction in resistance resulting from the short circuit results in a very significant increase in current flow in the circuit; much more than the conductors can safely carry. Short circuits can result from several conditions including:

- Improper wiring of branch circuits or components

- Malfunctions in appliances or electrical components

- Physical damage to conductors that result in conductors firmly contacting each other (bolted fault)

To continue the earlier example of applying Ohm's Law to a kitchen toaster, it was found that with 10 ohms of resistance, the device would draw 12 amperes. In a short-circuit situation, the resistance would be greatly reduced. To measure the current in the shorted circuit, Ohm's Law must be reapplied using a much smaller resistance that could occur as a result of damage to the insulation of one of the conductors, allowing the energized conductor to come into contact with the neutral conductor **(Table 6.17)**.

Table 6.17 Current in a Circuit		
Given	**Solving for**	**Calculation**
Voltage (V): 120 volts Resistance (R): 0.5 ohms (Ω)	Current (I) = amperes (A)	Current (I) = Voltage (V) ÷ Resistance (R) I = 120 V ÷ 0.5 Ω I = 240 A

Overloads

Overloads occur when components requiring more power than a circuit can safely carry are connected to a circuit. While circuits meeting the requirements of the *NEC*® have a significant safety margin built into them, overloading a circuit stresses all components in the circuit. Before tripping a circuit breaker or blowing a fuse, an overload condition could cause poor connections in a circuit to overheat and potentially ignite nearby combustibles. When the overload condition reaches the rating of the device, it should trip a circuit breaker or blow a fuse, which opens the circuit, and removes the current **(Figure 6.24)**.

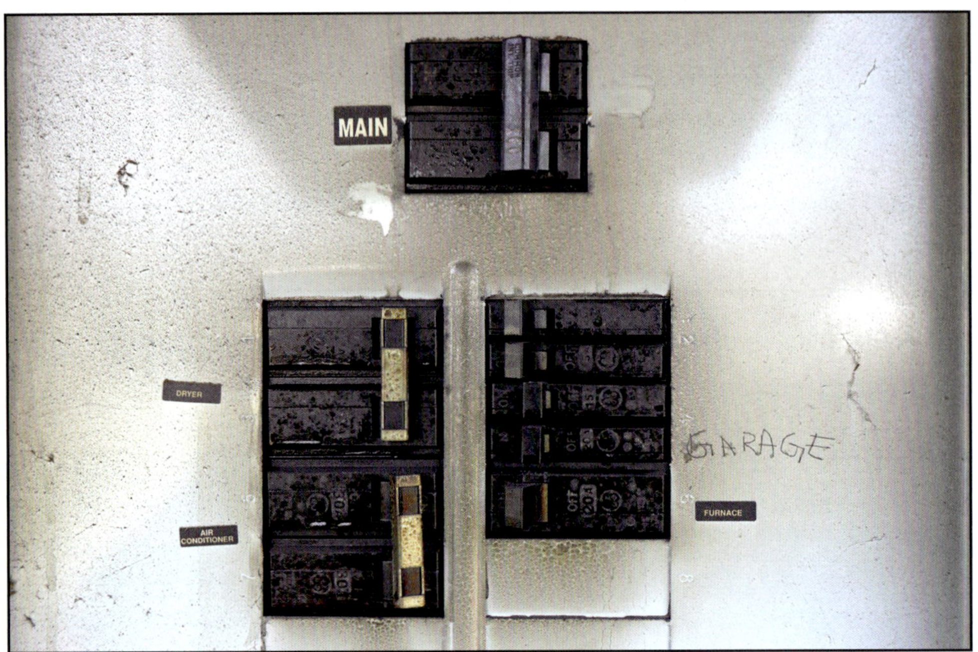

Figure 6.24 An electrical distribution panel showing tripped breakers gives an investigator an idea of the area in which the electrical system failed. *Courtesy of Donny Howard, Agent, Oklahoma State Fire Marshal's Office.*

The time required for a circuit breaker to trip depends on the amount of overcurrent. A typical 20-ampere circuit breaker might take 15 minutes to trip with a current of 30 amperes, 30 seconds for 40 amperes, and less than 0.1 second for 200 amperes. This trip time versus amount of overcurrent is called a **trip curve** and is available from manufacturers of the devices.

Damage to Conductors

NFPA 1033 (2022): 4.1.7, 4.2.8

Careful examination of the surface of the conductors is necessary to determine the cause of damage observed on a conductor. Investigators need to be aware that visual analysis may not be conclusive. Other methods of analysis may need to be utilized to make an accurate determination as to the cause of the damage. Methods may include low or high power magnification, and the use of a scanning electron microscope or CT scanner. Further discussion on arc mapping and cause determination is in Chapter 13, Fire Cause.

Arcing and Melting Damage

In order to categorize the damage to conductors such as arc damage or melt damage, the investigator should be familiar with the characteristics of an arc site versus damage to a conductor from melting. Consider commonly observed damage found on a conductor. For example, copper conductors may melt when exposed to fire.

Evidence of melting and arcing may both be present on a conductor. If a conductor is close to an arc site, *spatter* may be present on its surface. This spatter is due to the conductor melting during the arc event. The molten metal conductor particles (spatter) are ejected back onto the surface of the conductor and resolidify.

On a severed conductor, the ends may have visible damage **(Figure 6.25)**:

- *Beads* are a result of localized heating from electrical activity. Beads will have a characteristic sharp line of demarcation between the damaged and undamaged portion of the conductor.

- *Globules* are caused by melting as a result of non-localized heating such as from being exposed to a fire or an overloaded circuit. Globules caused by melting will not have a sharp line of demarcation and the damage may be extended across the conductor, also referred to as *necking*.

Arcing Damage

Arcing will show adjacent or corresponding damage between two surfaces such as:

- The hot wire and the grounded neutral

- The hot wire and the grounding wire

- The hot wire and a grounding metal such as metal conduit, a metal junction box, or metal case of an appliance.

Conductors that are not severed may have localized damage caused by arcing. This may present as corresponding damage where two or more energized conductors have come in contact with each other resulting in a notch in one conductor and the other conductor having the notched section welded to it. Non-severed conductors may also have a localized point of contact where the conductor causes an arc with another metal.

Figure 6.25 A bead or globe at the end of a melted wire indicates certain conditions at the fire scene. *Courtesy of Donny Howard, Agent, Oklahoma State Fire Marshal's Office.*

Melting Damage

Damage associated with melting may show the following characteristics **(Figure 6.26)**:

- Damage is over a wider area on the conductor with no defined line of demarcation. The conductor may become thinner where this damage is present (necking).

- Surface of the conductor may have blistering.

- The striations on the surface of the conductor created during the manufacturing process may disappear.

- The conductor may melt, causing the metal to flow, resulting in the appearance of gravitational drops on the surface of the conductor. This may eventually cause the conductor to become thinner in the area where it has melted.

Figure 6.26 Wires melted to each other provide an indication of fire scene conditions. *Courtesy of Donny Howard, Agent, Oklahoma State Fire Marshal's Office.*

- The conductor may also have low internal porosity when viewed in a cross section with a scanning electron microscope.

- The conductor may also have resolidification waves present within it when viewed with a scanning electron microscope.

- Stranded conductors that have melted will often become stiffer and less flexible. They may also exhibit a change in color on the surface with a reddish hue. Other colors may also be present on the conductors such as blue/green. The presence of a blue/green hue on the conductor is the result of the PVC insulation burning during the fire and producing copper chloride ($CuCl_2$). Copper chloride is an acid and absorbs moisture as it forms, changing the color from brown to a blue/green. This acid may further damage the conductor over time.

Alloying

Alloying occurs on conductors where dissimilar metals have come in contact with each other as a result of melting, forming a new material. Fire investigators commonly note a copper-aluminum alloy. This new alloy will have a lower melting temperature than either of the pure metals which formed the alloy. A shiny or grey area that is not consistent with colors observed on the remainder of the conductors may be on the newly formed alloy. The alloyed area may also be more brittle than the other areas of the conductors. Determining whether an alloy has been formed on the conductors may require chemical analysis.

Other Damage

Other damage may be present on conductors as a result of the fire that is not related to arcing or melting **(Figure 6.27)**. Examples of these types of damage include:

- **Mechanical damage** — Conductors may have become frayed and separated as a result of being pulled apart. This damage is typically observed on stranded conductors. Mechanical damage may also be observed on solid conductors where the conductors were broken either during the fire or during fire suppression efforts.

- **Tool marks** — During overhaul operations firefighters will sometimes use tools, such as side cutters, to cut conductors that are low hanging and posing a hazard. The investigator should examine the ends for evidence of tool marks, and identify the opposite end if possible.

- **Damaged insulation** — Insulation on conductors is rated to withstand a certain voltage before it breaks down. Voltage past the rated threshold of the insulation (including the safety margin) may damage the insulation, causing it to char. This charring may become conductive, allowing current to flow to an unintended path. Insulation may also become damaged as a result of a misplaced nail or staple that pierces the insulation. This damage may also allow current to flow to an unintended path or cause localized heating at the nail or staple.

Figure 6.27 Non-melting and non-arcing damage also provides an indication of fire scene conditions. *Courtesy of Donny Howard, Agent, Oklahoma State Fire Marshal's Office.*

Grounding and Circuit Interrupters

NFPA 1033 (2022): 4.1.7, 4.2.8

A fundamental principle of electricity is its tendency to go to ground, unless guided through a circuit or electrical system. The previous section addressed ways that unguided electricity can create damage. This section presents systems and devices that prevent damage.

Electrical systems are artificially grounded by connecting part of the circuit to the earth through a ground wire. The grounding of an electrical system is required to reduce the hazard of electrical shock. Grounding is provided for the protection of those using an electrical device.

> **CAUTION:** Electricity finds the shortest path to the earth through any available conductor.

In modern construction, the entire electrical system is grounded, from the service entrance through all of the branch circuits **(Table 6.18)**. In older homes, fire investigators may find two-wire, ungrounded branch circuits. The system itself is still grounded, even in two-conductor circuits.

Table 6.18 Wire Colors and Purposes	
Wire Color	**Purpose**
Black, Red	Live electrical load from power source
Blue, Yellow	Switch connections
Green, Bare Copper	Ground
White, Gray	Neutral

Note: All wires may carry an electrical current and should be treated as live.
Source: *https://www.pacmanelectric.com/blog/electrical-wires-color-meaning#*

A *ground fault* is an unintended path of current flow between a conductor and the earth. A **ground fault circuit interrupter (GFCI)** is a device used to protect against ground faults. The following sections discuss grounding features and protection devices.

NOTE: All wires may carry an electrical current and should be treated as energized.

Grounding

System grounding is used to protect against:

- Line surges
- Damage from lightning
- Any contact between an energized conductor and grounded objects
- Accidental contact of high-voltage conductors with lower-voltage conductors

In newer systems, one of the three connectors that provide the service is a grounding wire (normally bare wire). The grounding wire provides a low-resistance path for the current to follow in case of accidental contact. This flow is normally sufficient to cause the circuit protection to operate and stop the current flow in the branch containing the fault. More information on circuit protection is included later in this section.

The electrical distribution panel is then grounded by connecting the panel to the electric utility grounded neutral conductor. As a backup, the panel can be connected to the following:

- Metal frame of a building.

- A grounding plate near the sill plate or under the foundation

- Metal underground water pipe on the utility side of the meter

- A series of grounding rods driven into the ground near the service entrance **(Figure 6.28)**

From the circuit breaker in the electrical distribution panel, a modern branch circuit is grounded throughout. The grounding (bare or green-colored) conductors leave the panel from the **ground bus** and go to the circuit boxes. At each box, the grounding conductor is attached to the box, if it is metal, and then attached to the grounding wires of the cables leading into and out of the box **(Figure 6.29)**. If plastic boxes are used, the ground wires are connected within the appropriate boxes and to the appropriate connector on the receptacle, device, or switch.

The case or housing of the appliance is connected to a grounding conductor in the cord. The ground conductor is then connected to a three-prong receptacle on a branch circuit. The grounding conductor from the appliance is in contact with the system ground at this point. If the case becomes energized, a short circuit (low-resistance path allowing high current flow) develops between the case and the ground, causing the overcurrent protection device on the circuit to operate.

Figure 6.28 Example of a grounding rod. *Courtesy of Donny Howard, Agent, Oklahoma State Fire Marshal's Office.*

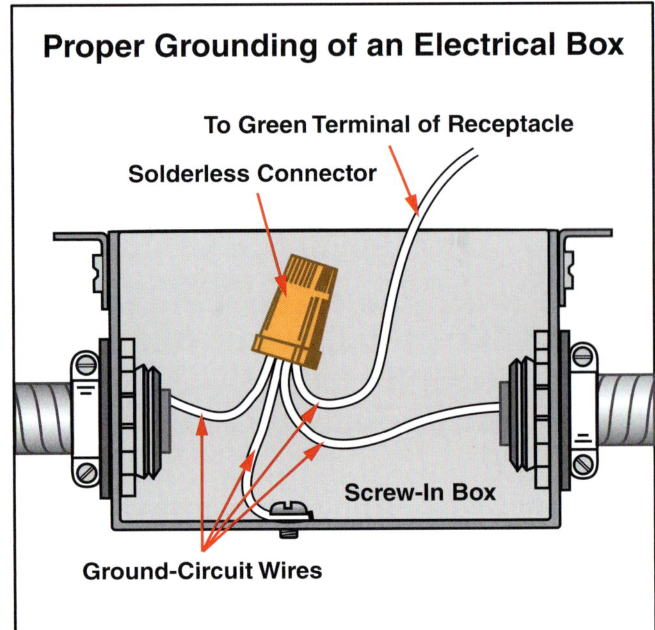

Figure 6.29 Each electrical box must be grounded properly.

Grounded Versus Grounding Terminology

Any confusion between "grounded conductor" and "grounding conductor" can lead to potentially lethal mistakes. In addition, a misunderstanding in the terminology may lead to incorrect labels during documentation and decreased credibility in reporting.

> **WARNING:** The bare (or green) wire should ONLY be used as the grounding conductor, and the white wire should ONLY be used as the grounded conductor.

Simply stated, a circuit is made up of a hot wire and a grounded neutral wire. The hot wire and grounded neutral wires both carry electricity under typical conditions.

According to the terminology in the *CEC*® and *NEC*®, the grounding conductor refers to the safety ground. The *NEC*® requires the grounding wires to be bare, or insulated green, yellow, or green with a yellow stripe.

According to the *CEC*®, a grounding conductor is the main electrode driven into the ground. The grounding conductor protects the electrical equipment from overloading when a power surge occurs or lightning strikes the electrical wires. The *NEC*® refers to this as a "grounding electrode conductor." Its function is to serve as an alternate source for carrying potentially dangerous electricity from the appliance or outlet back to the source.

More technically, current may flow through the grounding conductor in the event of a short circuit where the hot wire or the neutral wire come in contact with a metal component. The grounding wire is a safety component to take this unwanted current to ground. The grounding wire will be connected to the grounding bus in the electric panel. The grounding wire is a safety wire that is attached or bonded directly to earth and does not normally carry current.

The grounded neutral wire is required to be white or gray on the customer side. The grounded neutral conductor is often bare on the utility side of the meter (upstream). This is often also called the "ground wire" or "neutral wire." The word "neutral" is reserved for the white wire when you have a circuit with more than one "hot" wire. Since the white wire is connected to neutral and the grounding conductor inside the panel, the proper term is "grounded conductor." In common usage, the word "neutral" is used for "grounded conductor."

According to the *NEC*®, a grounded conductor is a wire which runs through the electrical system, commonly referred to as a neutral wire and serves as a current return path for electrical services. The *CEC*® refers to this as an "identified conductor."

More technically, the purpose of the grounded neutral conductor is to complete the 120V circuit by providing a path back to the electrical distribution panel where the neutral conductor is connected and bonded to the earth ground. This is the return path for the current provided by the hot (normally black) conductor. The neutral wire has an insulation coating because it is part of the circuit which flows electrical current. The grounded neutral conductor will be attached to the neutral bus in the electric panel.

Insufficient Neutral Connections

Not all municipalities test the residential grounding system as part of the electrical inspection of new services. In the event of a broken or disconnected neutral on the utility side, the ground (safety system) may not be able to compensate.

Although the neutral (white or gray) conductor is not energized, it does carry current, and remains isolated from the grounding conductor, except at the electrical distribution panel. A resulting imbalance of the electrical load within the electrical distribution panel may result in the electrical service becoming imbalanced, which results in abnormal voltages within the panel and on some circuits. Homeowners may report that some lights were very bright while others were dim, and appliances and lighting may turn on and off. When an imbalance of the electrical load occurs, many appliances will be damaged due to insufficient or excessive voltages.

Damaged Insulation

If the insulation on the internal wiring were to fail or be damaged and come into contact with a metal case or housing, a severe shock could occur. Damaged insulation can lead to energy flow to unexpected materials and the potential for a lethal shock.

Improper or Poor Grounding

An investigator should note the presence and quality of grounding. Improper grounding may lead to shock dangers and potentially an increase in fire danger. Poor grounding may indicate a poor quality wiring installation in general. If an appliance or extension cord is supposed to have had a grounding conductor, care should be taken to see if it was still in place at the time of the fire.

Ground-Fault Protection Devices

NFPA 921, *Guide for Fire and Explosion Investigations*, defines a ground fault as a current that flows outside the normal circuit path such as:

- Through the equipment grounding connector

- Through conductive material in contact with lower potential (such as earth), other than the electrical system grounded conductor (metal water or plumbing pipes, etc.)

- Through a combination of these ground return paths

The ground fault circuit interrupter (GFCI) operates very rapidly in response to small amounts of current leakage **(Figure 6.30)**. For example, a Class A GFCI responds to approximately 5 milliamps (0.005 amps). The primary purpose of the device is to provide protection against electric shocks. GFCIs use electronic-monitoring circuits to monitor the current flow in both the energized (hot) and neutral conductors supplying the load. If the circuit detects less current in the neutral conductor, this indicates that current is being diverted (possibly through a person). As a result, the interrupter trips and opens the circuit, stopping the current.

NOTE: GFCIs protect the people using devices connected to the circuit; fuses and circuit breakers protect the conductors in a circuit.

Figure 6.30 GFCI-equipped electrical outlets include a reset switch.

> **WARNING:** Because the resistance of the human body is so high, the current with a body in a circuit is usually well below the rating of the fuse or circuit breaker in the circuit. As a result, the normal overcurrent protection provided will not operate in these life-threatening conditions.

GFCI circuitry may be incorporated into receptacles for localized protection, into breakers that protect an entire branch circuit, or into power cords for appliances such as hair dryers. The presence of a GFCI will not protect against all short circuits. Contact between the energized conductor and the neutral conductor can still lead to arcing/sparking. The GFCI protects against contact between the energized conductor and any grounded object.

Circuit Protection Devices

As described earlier in this chapter, power for a building is brought into an electrical distribution panel where it is distributed to branch circuits. The *NEC®* requires overcurrent protection for both equipment and wiring in an electrical system. Excess current in a circuit can result from a short circuit or the overloading of a circuit with a load that exceeds its rated capacity.

Fire investigators will normally see circuit breakers in modern residential installations. Some circuit breakers can also be used to control a specific circuit, such as an electrical line, that serves a specific compartment within a building. Investigators may find circuits protected with fuses in commercial/industrial and older residential systems. Circuit protection devices and disconnect switch(es) may be at the same location as the meter base, or in a separate disconnect panel within the structure.

Some type of overcurrent protection device is provided at the origination point of a branch circuit. The protective devices in the electrical circuit prevent the overheating of conductors in the circuit. Overcurrent protection that is properly sized and operating helps to prevent the electrical system from overheating. Should the protective device be sized incorrectly or not operate properly, resistance heating may damage the conductor's insulation and create an ignition source.

Fuses

A fuse is a device that provides protection to the circuit using a fusible element that melts from the heat generated by excessive current flow **(Figure 6.31)**. Cartridge-type and plug-type fuses are common in most structures. Cartridge fuses are normally used in applications where protection in excess of 30 amperes is required **(Figure 6.32)**.

Plug-type fuses screw into a socket in the electrical distribution panel:

- **Edison Base (Type T, brass threads)** —Edison base fuses may be interchanged. For example, a 30-ampere fuse may be incorrectly installed on a circuit that requires a 15-ampere fuse. This "overfusing" may result in a condition in which the larger size fuse does not adequately protect the circuit when needed.

- **Rejection Base (Type S, porcelain threads)** — Rejection base fuses have a tamper resistant base that screws into the electrical distribution panel. Each size fuse has a different diameter base so that different size fuses cannot be interchanged. These devices are still commonly found in older installations.

Circuit Breakers

Circuit breakers provide overcurrent protection and, unlike fuses, can be reset after they have "tripped" **(Figure 6.33)**. The tripped position of the handle in many circuit breakers is in the center or middle position. The handles of circuit breakers cannot be manually placed into the center or tripped position. However, some circuit breakers fail to the OFF position.

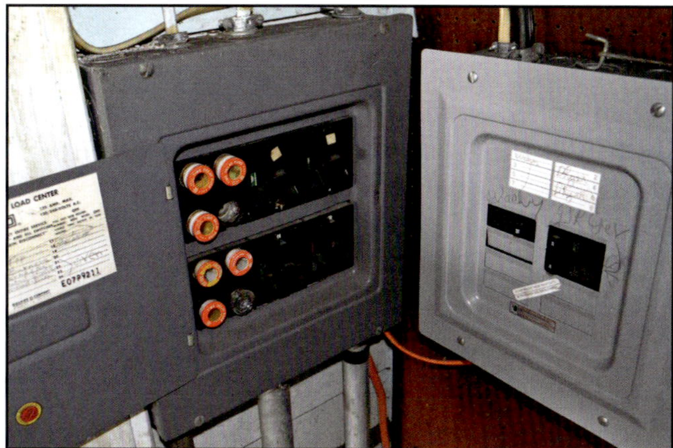

Figure 6.31 Plug-type fuses cannot be reset after they are tripped. *Courtesy of Donny Howard, Agent, Oklahoma State Fire Marshal's Office.*

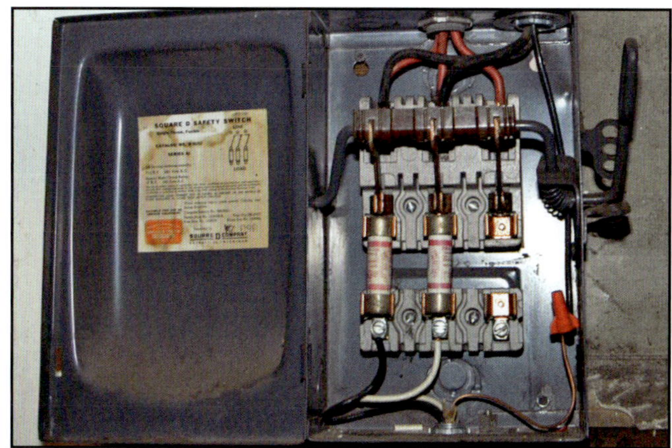

Figure 6.32 Cartridge fuses are found where higher than normal voltage is required. *Courtesy of Yates & Associates.*

Figure 6.33 Circuit breakers provide overcurrent protection, and can be reset after they have tripped. *Courtesy of Mike Makela.*

Most residential circuit breakers operate by both a thermal and magnetic means, depending on the type of fault to which they are exposed:

- Two dissimilar metals are bonded together into a bimetallic strip.

- The strip is part of the circuit and the total current in the branch circuit flows through it.

- As it is exposed to heat, the strip bends due to the different thermal expansion rates of the two metals, and interrupts the circuit.

- The device is calibrated to open the circuit when the amount of heat generated by the current flow exceeds the trip curve of the breaker.

- A magnet activates the trip bar in response to the magnetic field created by the large overcurrent condition.

The circuit breaker handle is independent of the internal mechanisms and is designed to trip even if the breaker handle is wedged into the ON position (trip-free operation). The thermal trip mechanism will operate in response to high ambient temperatures in the vicinity of the breaker; in some cases, even if the breaker is in the OFF position. Thus, if a fire is in close proximity to a panel, the fire investigator might find all breakers in the tripped position after the fire. In addition, circuit breakers that are subjected to a blow, such as from firefighter operations or debris falling and striking the electrical distribution panel, may also move into the tripped position.

In most installations, the main disconnecting means and the overcurrent protection for the branch circuits are located in the same electrical distribution panel **(Figure 6.34)**. This panel is commonly called the **circuit breaker panel**. Some older types of circuit breakers may not show whether a circuit has tripped. When these breakers do not trip under certain fault conditions, they can cause fires. In marine or corrosive environments, circuit breakers may fail more quickly than in other environments.

Arc Fault Circuit Interrupter (AFCI)

Arc-fault circuit interrupters (AFCIs) are electronic devices, generally a part of a circuit breaker, that detect arcing conditions caused by the energized conductor contacting either the neutral conductor or a grounded object. They serve as protection against fires when the arcing condition is of too low a current or for too little time to trip an ordinary circuit breaker.

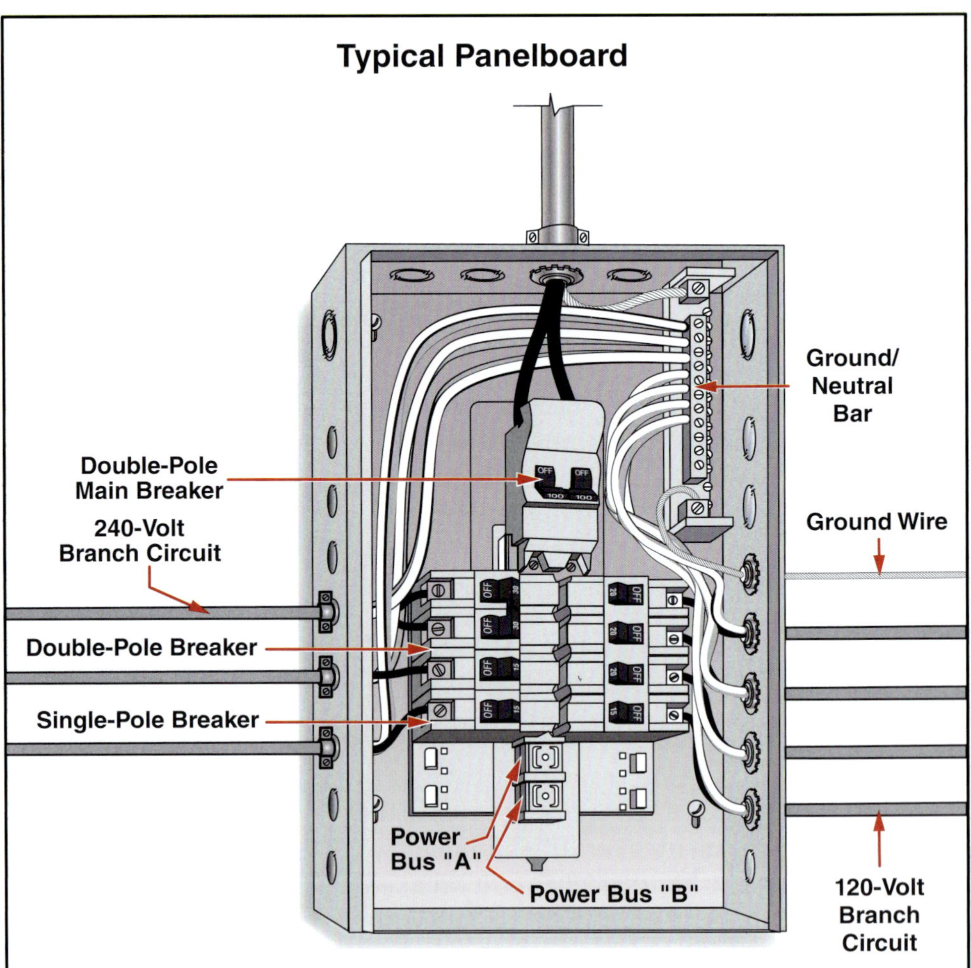

Figure 6.34 Panel boards show the status of included equipment and circuits.

A circuit board within the AFCI monitors current flow on the circuit. This circuitry is able to differentiate between normal arcing on the circuit, such as a parting arc from opening or closing a switch, and abnormal arcing, such as a failure of a conductor.

Once the internal circuitry of the AFCI recognizes an abnormal arcing event, the circuitry opens the contacts within the AFCI, de-energizing the circuit. This occurs exceptionally quickly which reduces the likelihood for a fire to occur. AFCI devices may also be capable of performing other functions such as overcurrent protection, ground-fault circuit interruption, and surge suppression.

Arc fault circuit interrupters are used as part of residential/dwelling (single phase) electrical systems, and are used to prevent electrical fires caused by arc faults in the residential wiring. AFCIs may be found in locations such as bedrooms, living rooms, or other areas of the residence. The *NEC*® requires that all branch circuits supplying 120-volt outlets installed in most rooms in dwelling units must be protected by an AFCI. Additional locations may be added to the *NEC*® over time.

Investigative Clues

NFPA 1033 (2022): 4.1.7, 4.2.8, 4.7.3

Fire investigators should look for physical evidence that an adverse electrical event has occurred. The operation of electrical safety devices, such as fuses, circuit breakers, arc-fault interrupters, and ground fault interrupters, may be clues that an electrical failure has occurred or safety devices have been circumvented **(Figure 6.35)**. Damage to electrical equipment attached to an electrical system or its components may also indicate that an electrical event had occurred. Resistive heating may be the result of incomplete electrical connections, partial shorting from various sources, or failed equipment or appliances. These signs do not suggest that the fuse or breaker or other device was the cause of the electrical event, but they do indicate that an electrical event may have occurred.

Figure 6.35 Damage on one side of this panel is severe, but the damage does not extend to the other side because the tripped circuit breakers removed the circuit of travel for arcing electricity. *Courtesy of Wayne Chapdelaine, Metro-Rural Fire Forensics.*

Indicators of resistive heating shall also be investigated and assessed. Previously reported problems with or malfunctions of the electrical system or its components may indicate that an electrical failure has occurred. Indicators specified in NFPA 921 are:

- Fire patterns
- Arc mapping
- Fire dynamics
- Witness statements

Review Questions

1. What is the difference between a conductor and an insulator?

2. What is the difference between direct current and alternating current?

3. How does electrical current flow through a series circuit?

4. How does electrical current flow through a parallel circuit?

5. How are electrical voltage and resistance related?

6. What is the purpose of a transformer?

7. What are three major components of an electrical system that delivers power to a residential utility customer?

8. What types of materials will investigators commonly encounter when inspecting building electrical systems?

9. Why are connections considered a common point of failure?

10. What can an investigator do to reduces hazards posed by an energized electrical system?

11. What are three common types of electrical failures?

12. What are some common types of damage to electrical conductors?

13. What is the purpose of grounding a circuit?

14. How do circuit protection devices function?

15. What types of clues may indicate that an electrical malfunction has occurred?

Chapter Reference

Lewis, Kevin, PE, CFI, CFEI. July 2, 2019. "Are Hot Particles Common Culprits Of Wildland Fires?" Jensen Hughes Insights. Accessed online.

Chapter Key Terms

Alternating Current (AC) Circuit — Electrical circuit in which the current can move through the circuit in both directions and the flow can be constantly reversing.

Ammeter — Instrument for measuring electric current in amperes (amps or A).

Ampacity — Current-carrying capacity of conductors or equipment, expressed in amperes.

Ampere (A) — Basic unit of electrical current; amount of current sent by one volt through one ohm of resistance. May be abbreviated either as A or I.

Arc — High-temperature luminous electric discharge across a gap or through a medium such as charred insulation.

Arc-Fault Circuit Interrupter (AFCI) — Electronic device, generally part of a circuit breaker, that detects arcing conditions caused by an energized conductor contacting either a neutral conductor or a grounded object.

Branch Circuit — Wiring between the point of application (outlets) and the final overcurrent device protecting the circuit.

Circuit — Complete path of an electrical current.

Circuit Breaker Panel — Device (basically an on/off switch) designed to allow a circuit to be opened or closed manually, and to automatically interrupt the flow of electricity in a circuit when it becomes overloaded.

Conductor — Substance or material that transmits electrical or thermal energy.

Conservation of Energy — Law of physics that states that the total amount of energy in an isolated system remains constant. As a result, energy cannot be created or destroyed.

Current (I) — Rate of electrical flow in a conductor; measured in amperes.

Direct Current (DC) Circuit — Electrical circuit in which the current moves through the circuit in only one direction.

Duplex Receptacle — An electrical outlet providing space for two electrical receptacles. One of the most common outlet designs in occupancies.

Electron — Subatomic particle that possesses a negative electric charge.

Fuse — Single-acting protective device designed to open a circuit on a predetermined overcurrent; types include Edison-based fuses and cartridge fuses.

Generator — Any coil of conductors that is rotated within a magnetic field. As the coil is turned in the field, a voltage is produced.

Ground Bus — Part of an electrical service panel where the neutral service wire is connected to the earth by a ground wire.

Ground Fault Circuit Interrupter (GFCI) — Device designed to protect against electrical shock; when grounding occurs, the device opens a circuit to shut off the flow of electricity. *Also known as* Ground Fault Indicator (GFI) Receptacle.

Grounded Neutral Conductor — Conductor in a branch circuit that carries the return current but that is not energized. The covering will be white or natural gray in color, and will be connected to the service neutral. *Also known as* Common Conductor *or* Neutral Conductor.

Grounding — Reducing the difference in electrical potential between an object and the ground by the use of conductors.

Insulator — Material with atomic structure that does not allow the easy movement of electrons. *Opposite of* conductor.

Nonmetallic (NM) Shielded Cable — Factory assemblies with two or more wires that have a nonmetallic outer sheath that is moisture resistant and fire retardant.

Ohm (Ω) — Basic unit of measurement of electrical resistance, symbolized either by Ω or R. One ohm is the resistance between two points in a conductor when one volt produces one ampere of current.

Ohm's Law — Mathematical relationship between a circuit's voltage (V), current (I), and resistance (R): $V = IR$.

Ohmmeter — Device designed to measure electrical resistance in a circuit.

Overcurrent — Current in excess of the rated current of equipment or the ampacity of a conductor; may be caused by an overload, short circuit, or ground fault.

Parallel Circuit — Circuit configuration in which different components in the circuit receive current from different pathways, which allows individual components in the circuit to continue to function even if another component fails.

Parting Arc — A brief discharge that occurs as an energized electrical path is opened while current is flowing. Ordinary parting arcs in electrical systems are usually so brief and of low enough energy that only combustible gases, vapors, and dusts can be ignited. (Reproduced with permission from NFPA 921-2017, *Guide for Fire and Explosion Investigations*, Copyright 2018, National Fire Protection Association.)

Resistance — Opposition to the flow of an electric current in a conductor or component; measured in ohms (Ω).

Series Circuit — Circuit configuration in which current flows in one path through all the components.

Short Circuit —An abnormal connection of low resistance between normal circuit conductors where the resistance is normally much greater; this is an overcurrent situation but it is not an overload.

Thermal Expansion — Elongation or expansion of materials, such as steel, when exposed to heat.

Transformer — Device that uses coils and magnetic fields to increase (step-up) or decrease (step-down) incoming voltages.

Trip Curve — Relationship between current and time that determines when circuit protective devices operate.

Volt (V) — Basic unit of electrical potential; difference in potential (electromotive force) needed to create a current of one ampere through the resistance of one ohm. Most commonly abbreviated V, but may also be abbreviated E.

Voltage — Electrical force that causes a charge (electrons) to move through a conductor. Measured in volts (V). *Also known as* Electromotive Force (EMF).

Voltmeter — Device used for measuring existing voltage in an electrical system.

Wire Nut — Approved, solderless, nonconductive connector used to connect one wire to another.

Photo courtesy of Donny Howard, Agent, Oklahoma State Fire Marshal's Office.

Chapter Contents

NFPA 1033 JPRs addressed in this chapter

This chapter provides information that addresses the following job performance requirements (JPRs) of NFPA 1033, *Standard for Professional Qualifications for Fire Investigator* (2022):

4.1.2	4.2.2	4.4.1	4.5.1	4.5.3
4.2.1	4.2.3	4.4.2	4.5.2	4.6.5

This chapter provides information that addresses the following course outcomes for the FESHE courses:

Fire Investigation I (C0283)

1. Demonstrate the importance of documentation, evidence collection, and scene security process needed for successful resolution.
3. Identify the responsibilities of a firefighter when responding to the scene of a fire, including scene security and evidence preservation.

Fire Investigation II (C0284)

4. Analyze and determine the causes of fires and contributing factors.
6. List the procedures for fire scene documentation, including sources and technology available for fire investigations.

Fire Investigation and Analysis (C0285)

2. Document the fire scene in accordance with best practice and legal requirements.
4. Analyze the legal foundation for conducting a systematic incendiary fire investigation and case preparation.

Learning Objectives

1. Explain how the scientific method is used in fire investigations. [NFPA 1033, 4.1.2, 4.6.5]

2. Differentiate among the ways that an investigator has the legal authority to enter and investigate a scene. [NFPA 1033, 4.2.1, 4.2.2, 4.2.3, 4.4.2]

3. Describe the types of information gathered during preliminary interviews. [NFPA 1033, 4.2.3 4.5.1, 4.5.2, 4.5.3, 4.6.5]

4. Explain the purpose of establishing scene security. [NFPA 1033, 4.2.1, 4.2.2, 4.2.3, 4.4.1]

5. Skill Sheet 7-1: Secure the scene. [NFPA 1033, 4.2.1]

Chapter 7
Initial Actions in the Investigative Process

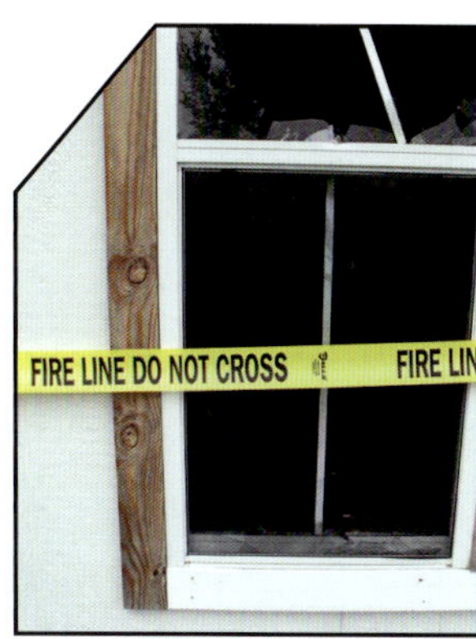

Like solving a riddle, fire investigators must acquire evidence, facts, and data about the fire, and fit these pieces together to form opinions regarding the origin, cause, and responsibility for the fire. A fire investigator's conclusions are then based on an analysis of all information collected. Initial actions in the investigative process include knowing key pieces of information before applying them at an incident scene, including:

- Scientific method
- Legal rights of entry
- Preliminary information and interviews
- Scene security

Scientific Method

NFPA 1033 (2022): 4.1.2, 4.6.5

Fire investigators must conduct investigations using a systematic approach. One such approach is the scientific method, as required by NFPA 1033 and recommended by NFPA 921. The scientific method has been recognized by courts across the United States and Canada as a valid methodology for fire investigation.

The Scientific Method

Several versions of the scientific method exist; the investigator can use the following general steps to answer a wide variety of questions **(Figure 7.1, p. 206)**:

Step 1 **Recognize the need** — A fire or explosion has occurred and both the origin of the fire and the cause of the fire must be determined.

Step 2 **Define the problem** — The mission of the investigation is to determine the origin, cause, and/or responsibility of the fire or explosion.

Step 3 **Collect data** — During the investigation, the data related to the incident are collected via observation or other direct means.

Step 4 **Analyze the data** — Collected data are then analyzed and evaluated using the fire investigator's knowledge, training, experience and expertise. Data analyses must be completed before developing a hypothesis. If the investigator does not have the expertise to properly collect or analyze data, expertise should be sought from other qualified sources.

Step 5 **Develop hypotheses** — An explanation or explanations (hypotheses) for the fire or various segments of the fire such as fire spread, fire effects and patterns, fire origin(s), and fire cause are developed based on the data analysis. Hypotheses should be based upon the data collected and analyzed from the various aspects of the investigation.

Figure 7.1 The scientific method is often depicted as a closed loop to show that all steps can restart the process of evaluating information.

Step 6 **Test the hypotheses** — The hypotheses must be able to withstand rigorous testing in order to be considered valid. The investigator must collate all of the known data and information with respect to the issue and the hypotheses. Testing of hypotheses must be designed to disprove the hypotheses rather than to prove that the hypotheses are true. An investigator can test a hypothesis several ways:

— Experimentation

— Analytically by testing the hypothesis against known scientific principles

— Through published research

Step 7 **Select a final hypothesis** — In order to select a final hypothesis, the investigator should review the entire process for completeness of data collection and analysis and the testing method(s) used for the hypothesis. The investigator must ensure that any different viable hypotheses have also been developed, tested, and eliminated (refuted). Failure to consider and eliminate other viable hypotheses may be a critical error.

Research after the fire may assist the investigator with determining the fire area of origin. Individual pieces of information collected during an investigation may include:

- Analysis of fire patterns
- Knowledge of fire effects
- The results of arc mapping

- Statements made by witnesses or involved parties
- The analysis of the fire dynamics, which occurred during the fire
- Electronic data such as from security cameras or fire alarm activations

Maintaining Objectivity

The purpose of an investigation is to find out what happened and to document it clearly for future reference. The investigator should let the investigation point to a conclusion, rather than trying to prove or disprove a conclusion.

A significant way to maintain objectivity is to conduct a systematic, thorough, consistent investigation using the scientific method. To aid in maintaining objectivity, a discussion of the perils of **bias** can be helpful.

Bias is the tendency to believe and collect data that confirms expectations and to discard, disbelieve, or disregard corresponding data that appears to conflict with those expectations. This may result in selective attention to evidence. The less scientific and more subjective the analysis, the more it is subject to bias-induced errors.

Confirmation bias occurs when investigators fail to account for additional or contradictory data. Investigators must consider that the same data may support multiple different hypotheses.

When collecting information for an investigation, different sources have distinctive strengths and weaknesses. Witness statements provide valuable, timely observations about the event, but may be incomplete or unreliable. Formal records such as building inspections and video recordings are more likely to be reliable, but still may not show a complete overview of what occurred. Witness and first responder interviews can be compared to scene conditions observed by the investigator, which can also be compared against other available resources. When facts in these sources do not align, investigators should continue to search for more information that can corroborate or refute the previously gathered facts. Using a systematic approach to evaluate the information gathered also helps to eliminate bias.

Example of Hypothesis Testing

Upon examining a scene at a residence, an investigator formed a hypothesis that an electrical appliance failed and generated sufficient heat to ignite a nearby fuel. The investigator referenced outside research confirming the following facts:

- The appliance in question had a history of generating enough heat to cause a fire and was therefore competent to ignite some materials.
- The first fuel ignited at the scene would normally ignite at the levels of heat produced by the failed appliance.

The investigator discovered, however, that the outside researchers conducted testing under circumstances and conditions that were not similar to the conditions of the fire being investigated. As a result, the investigator had to discard the data that was dissimilar to the case in question and complete additional research, experimentation, and testing to determine if the hypothesis was viable.

An investigator must ensure that the test conditions and circumstances of outside research are similar to the hypothesized conditions at the scene:

- Materials found
- Equipment used
- Distances recorded
- Temperatures recorded
- Heat energy determined
- Incident duration from the scene

Should the investigator wish to establish a test for the original hypothesis, that the appliance had ignited materials in the identified **possible** area of origin, the investigator would need to determine whether the maximum heat of the appliance was sufficient to ignite the potential first fuel. This hypothesis requires two pieces of data in order to be tested:

- The maximum temperature of the heat source

- The ignition temperature of the hypothetical first fuel

If the heat source's maximum possible temperature cannot reach the ignition temperature of the hypothetical first fuel, the investigator can eliminate the appliance as a possible ignition source. The investigator should consider all possible appliances and all possible first fuels in the hypothetical area of origin in this analysis. If the investigator cannot find a competent ignition source among potential fuels near the hypothetical origin, the investigator should reevaluate the origin hypothesis and consider alternate points of origin.

Ignition conditions are discussed in Chapter 3, Fire Dynamics. Fire origin and cause are discussed in Chapters 12 and 13, respectively. NFPA 921 addresses origin testing in Section 18.6 and 18.6.1.1.

Legal Rights of Entry

NFPA 1033 (2022): 4.2.1, 4.2.2, 4.2.3, 4.4.2

An essential component of an investigation is the proper examination of both the exteriors and interiors of structures involved in fire. Laws and codes often require that the authority having jurisdiction (AHJ) investigate a fire within a reasonable time. Because conditions at a fire scene will change as a result of postfire activities, the examination and documentation of the scene must begin as soon as possible. Considerations as to how quickly an investigation can begin include:

- Site safety

- Legal right of entry

- Environmental issues

- Status of fire suppression **(Figure 7.2)**

Investigators must obtain legal entry to do their jobs. Most commonly, public sector investigators enter under exigent (urgent) circumstances in which an immediate need is being met; therefore, a warrant is unnecessary. However, investigators should understand the point at which circumstances at a scene change and they must obtain a warrant. Private investigators often have contractual permission to enter a scene and may not need a warrant. In either case, exigent or contractual, legal entry is an issue that all investigators should understand in order to perform their duties within the scope of the law.

NOTE: Public sector investigators should obtain legal advice regarding right-of-entry issues so that guidelines and procedures can be established to address legal right of entry from their local authority having jurisdiction or agency.

Figure 7.2 Investigators should follow firefighter recommendations regarding whether it is safe to enter an incident scene. *Courtesy of Ron Jeffers, Union City, NJ.*

Fire investigators may obtain the right to enter a scene through several means or methods including the following:

- Consent
- Contractual entry
- Exigent circumstances
- Criminal search warrant
- State or provincial legislation
- Administrative search warrant (some jurisdictions do not provide the means to obtain administrative warrants)

Limited Access to the Scene

All investigators need to remember that, for a variety of reasons, they may have only one opportunity to:

- Conduct interviews
- Collect physical evidence
- View and document the scene

Public sector investigators should also be aware that they may be the only investigators to observe the scene before its alteration. As a result, they should also be aware of the importance of properly documenting the scene. Their documentation of the scene may be the only information available to all subsequent investigators and investigations.

Exigent Circumstances

In the United States, the fire department is not required to obtain a warrant to enter a property to suppress a fire. A fire is an exigent circumstance and exception to the warrant requirements of the Fourth Amendment of the U.S. Constitution.

Canadian Right-of-Entry Laws

In Canada, the Canadian Charter of Rights and Freedoms guarantees individuals protection from unlawful search and seizure. Provincial and Territorial legislation govern the circumstances in which an investigator may enter a structure. Canadian fire investigators are encouraged to review their applicable provincial and territorial legislation pertaining to administrative-type warrants, which is normally found in their Fire Safety legislation. Information and criteria required for a criminal search warrant is found in The Criminal Code of Canada. Only sworn peace officers can obtain a criminal warrant.

Entering and Remaining at a Property

Public sector investigators entering under exigent circumstances may also remain on the scene for a reasonable amount of time to determine the fire's origin and cause. Once firefighters or investigators leave the scene or the fire scene has been released, investigators may be prevented from reentering the property without first gaining the consent of the owner or obtaining a warrant.

Firefighters occasionally remain at fire scenes to prevent unauthorized entry as well as to ensure that the area is not disturbed until an investigator arrives **(Figure 7.3, p. 210)**. In certain circumstances, an investigator may not need to obtain a search warrant to enter a fire-damaged building to conduct a scene examination. The right-of-legal entry without a search warrant was reinforced by the U.S. Supreme Court decision in *Michigan vs. Tyler* (1978). The U.S. Supreme Court held in that case that "once in a building [to extinguish a fire], firefighters may seize [without a warrant] evidence of arson that is in plain view .... [and] officials need no warrant to remain in a building for a reasonable time to investigate the cause of a blaze after it has been extinguished."

Figure 7.3 Keeping other individuals from entering the hazardous area is one of the more important tasks that Awareness Level personnel can do. Using barrier tape is an effective way to prevent entrance. *Courtesy of Ron Moore/McKinney (TX) Fire Department.*

The U.S. Supreme Court agreed, with modification, with the Michigan State Supreme Court's statement that "[if] there has been a fire, the blaze extinguished, and the firefighters have left the premises, a warrant is required to reenter and search the premises, unless there is consent ..." See **Appendix C** for more discussion on the *Michigan vs. Tyler* decision. If investigators anticipate needing to maintain a presence at a fire scene for a long period (several days) or anticipate reentry after leaving the scene, they should be aware of their legal authority to do so and should seek appropriate legal advice.

NOTE: The *Michigan vs. Tyler* decision is only applicable in the United States. Investigators should become familiar with right-of-entry case law from their own countries.

Search Warrant Procedures

Another decision, *Michigan vs. Clifford* **(Appendix C)**, stressed the importance of following proper search warrant procedures to ensure the admissibility of evidence in a court of law. In the Clifford case, fire department personnel had left the premises, and the property owners had arranged for structure security. Investigators entered the structure to determine the origin and cause of the fire approximately five hours after the departure of fire department personnel. The investigators entered without consent, an administrative search warrant, or exigent circumstances.

The court deemed the search improperly conducted, making the evidence obtained inside the structure inadmissible. The court decision ruled investigators needed an administrative search warrant to reenter the structure to determine origin and cause of the fire. However, once investigators determined the cause to be arson, any additional search beyond the area of origin would require a criminal search warrant.

In the *Clifford* case, the area of origin was determined to be the basement, and the fire cause was deemed *intentional*. After investigators made the preliminary finding of the incendiary cause, they searched other parts of the building. The court held that the evidence found elsewhere in the structure was beyond the scope of determining the origin and cause of the fire. A public investigator entering a scene under exigent circumstances should take that opportunity to perform a thorough scene examination of the entire incident site before making any **probable cause** determinations.

Consent

A person in lawful control of a property can grant consent to enter the property. The person in lawful control may be an individual tenant in a building and not necessarily the building owner. In multiple occupancy buildings, the building owner can grant access to common areas of the building, while occupants may have to grant consent to their individual tenant spaces **(Figure 7.4)**.

An important consideration when conducting searches under consent: The person granting consent can withdraw that consent at any time. When consent is withdrawn, the search *cannot* continue.

However, when investigators have reasonable cause to believe that evidence will be destroyed, the following should occur:

- Obtain a warrant.

- Request that the scene be secured.

- Seek protection for evidence that could be destroyed.

Figure 7.4 Rooftop projections show the location of a fire wall in a multiple occupancy building. *Courtesy of Dave Coombs.*

Contractual Entry

Private investigators representing insurance companies or other private interests must operate under their employer's jurisdiction. Most insurance policies contain language that allows the insurer to investigate insured losses. Despite this fact, the insured has the right to not allow an investigator on the scene. Therefore, fire investigators should ensure that permission to be on the premises has been obtained through court order or by verbal or written consent prior to conducting an investigation or collecting evidence. In the case of an insurance claim investigation, the insured has an obligation to cooperate with the insurer in the gathering of facts related to the loss.

Private investigators may be requested to conduct origin-and-cause investigations at loss sites for entities other than the owner or first-party insurance carrier. In these cases, the investigator must ensure that appropriate permission or consent has been obtained to enter the investigation scene. If any removal of evidence or major alteration of the scene is anticipated, further permission is often obtained.

When a private investigator's actions exceed legal authority, potential consequences may include:

- Evidence collected may be excluded from use in court.

- The investigator and client may face potential civil liability.

- Punitive damages or an increased award of cost could be claimed against the investigator's client.

- Investigator and client are exposed to a civil trespass action, or they could be personally named in a bad-faith action.

- Collusion (civil conspiracy) or malicious prosecution could be alleged when the private investigator is acting in cooperation with law enforcement or a public fire investigator and enters a fire scene without permission or a court order.

Arson Immunity Laws

Most states have *arson immunity laws* that require insurance companies to release all information and documentation when requested, to a public entity investigating whether a fire was intentionally set. Some states may not require a determination of arson (or suspected arson) for the exchange of information to take place. The released information is considered confidential until it is used in court proceedings. The insurance company receives legal immunity from breaking any confidentiality requirements it may have to its clients. Some jurisdictions also allow insurers to request information regarding the authorities' investigation and findings. As most of these statutes are specific to each state/province or territory, investigators should verify their jurisdictions' legislation.

NOTE: Canada does not have arson immunity laws. Law enforcement officers must apply to a court for a production order.

Administrative Search Warrants

An administrative warrant is required in most jurisdictions in order to obtain entry after the fire department has left the scene. An administrative warrant can be obtained in some jurisdictions with only a showing of statutory authority to conduct an investigation regarding the origin and cause of a recent fire. However, some jurisdictions do not have a legal mechanism to obtain an administrative warrant.

Michigan vs. Tyler: Administrative Search Warrants

The following text is taken from the *Michigan vs. Tyler* judicial opinion concerning administrative warrants: "To secure a warrant to investigate the fire, an official must show more than the bare fact that the fire has occurred. The magistrate's duty is to ensure that the proposed search will be reasonable, a determination that requires inquiry into the need for the intrusion on the one hand, and the threat of disruption to the occupant on the other. The number of prior entries, the scope of the search, the time of day when it is proposed to be made, the lapse of time since the fire, the continued use of the building, and the owner's efforts to secure it against intruders might all be relevant factors. Even though a fire victim's privacy must normally yield to the vital social objective of ascertaining the cause of the fire, the magistrate can perform the important function of preventing harassment by keeping that invasion to a minimum."

Criminal Search Warrants

A criminal search warrant or consent must be obtained once an investigator has probable cause that a crime has been committed and additional entries into the scene are necessary. A criminal warrant may be needed at any time during an investigation.

Any evidence found during entries under exigent circumstances can usually be collected and preserved. However, an investigator entering a scene with an administrative warrant may have to temporarily halt an investigation to obtain a criminal search warrant if evidence of a crime is found. The evidence found during the *administrative search* can be presented as evidence of probable cause when requesting the issuance of a criminal search warrant.

Michigan vs. Tyler

In summary of search warrants, the *Michigan vs. Tyler* opinion states: "We hold that an entry to fight a fire requires no warrant, and that once in the building, officials may remain there for a reasonable amount of time to investigate the cause of the blaze. Thereafter, additional entries to investigate the cause of the fire must be made pursuant to the warrant procedures governing administrative searches. Evidence of arson discovered in the course of such investigations is admissible at trial, but if the investigating officials find probable cause to believe that arson has occurred and require further access to gather evidence for a possible prosecution, they may obtain a warrant only upon a traditional showing of probable cause applicable to searches for evidence of crime."

Use the following guidelines regarding warrants:

- Entry by the fire department to suppress fire does not require a search warrant.

- In most cases, if investigators arrive after the scene has been released, they must obtain consent or a warrant to reenter.

- Any evidence found in plain view during fire suppression or origin and cause investigations is usually admissible **(Figure 7.5)**.

- Investigators entering subsequent to a fire can stay on the fire scene for a reasonable period of time to determine the origin and cause of fire.

- During the course of an investigation, indicators of an intentional fire may arise. Until probable cause has been reached that a fire is incendiary, the investigator may continue investigating the scene.

- When it is determined that a fire was intentionally set and a search for further evidence is needed, consent or a criminal search warrant should be obtained for each reentry unless an exigent circumstance exists.

- If during a scene examination (conducted with an administrative warrant) the investigator determines that the fire was intentionally set, the investigator should stop the investigation and obtain a criminal search warrant or the consent of the owner.

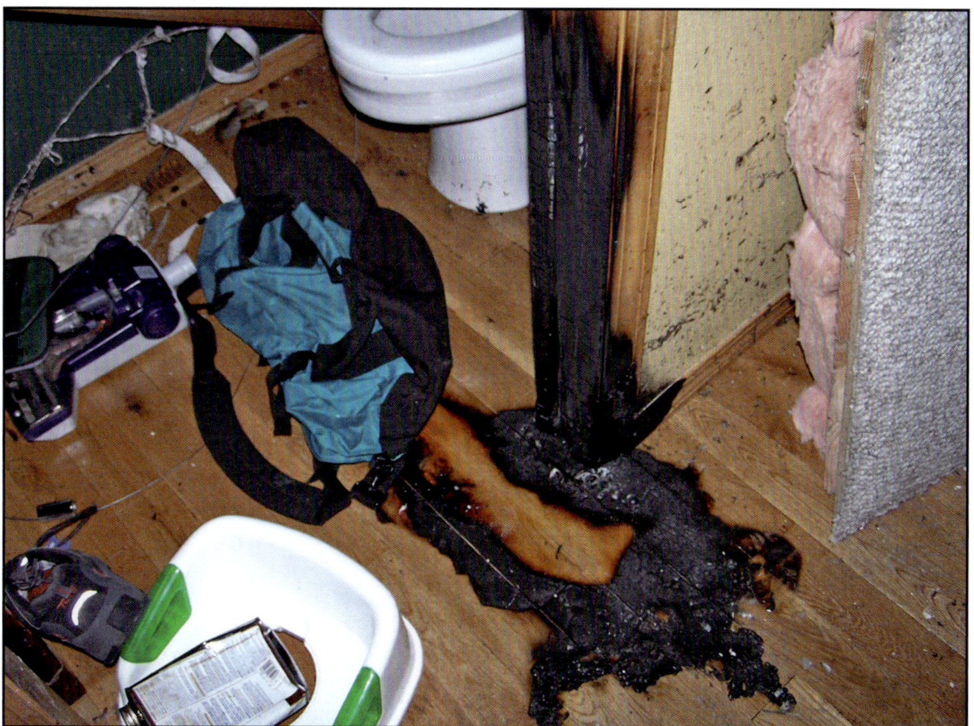

Figure 7.5 This irregularly shaped fire pattern was caused by the ignition and burning of ignitable fluids. Similar patterns may have different causes and first fuels. *Courtesy of Donny Howard, Agent, Oklahoma State Fire Marshal's Office.*

Preliminary Information and Interviews

NFPA 1033 (2022): 4.2.3, 4.5.1, 4.5.2, 4.5.3, 4.6.5

Each investigator assigned to a scene will most likely have received preliminary information regarding the incident. The investigator uses this information to begin to establish a timeline for the incident.

Interview questions an investigator may ask, based on the type of incident, are included in **Appendix D**. Types of information requested include:

- Dispatch information — Received by the first investigator assigned to the fire

- Case information — Provided to an insurance investigator assigned later in the investigation

- Location of the incident

- Time and date of the fire

- Type of occupancy of the fire-involved building

- Who first discovered the fire — Occupant vs passerby vs alarm system. This information may help the investigator determine a timeline for fire growth.

- Other information to help establish a timeline include:
  — History of the building
  — Witness statements
  — Fire fighting response
  — Data from alarm or security systems
  — 9-1-1 calls
  — Cell phone video
  — Weather events

Multiple Investigators

Investigators should not work alone if possible. Multiple investigators at a scene enhance overall investigator safety.

If additional investigators are available, it may be beneficial to include them in the investigation. Investigators should share information while conducting their duties to both verify witness observations at the scene and provide a basis for asking additional questions. For example, one investigator can start initial interviews while a second begins to examine the fire scene.

Nonverbal Communication

Investigators must practice good listening skills and understand how nonverbal communications add to a witness' statement. Listening is an active part of the communication process.

Nonverbal clues can be as important as a witness' verbal message, and the nonverbal message may overpower the verbal message. Understanding the importance of each of the elements of nonverbal communication will assist the investigator in recognizing and interpreting nonverbal signals, therefore improving nonverbal communication **(Figure 7.6)**. Nonverbal clues include:

- **Kinesics** — Use of body motion and position

- **Paralanguage or vocalics** — Vowel sounds or tones used to create the verbal message

- **Self-presentation** — Clothing, touch, use of time, and control of the speaker's environment

Witness Interviews

Fire investigators should conduct initial interviews of any available eyewitnesses, such as neighbors, occupants, and suspects, to the fire. During the initial interview, the investigator should identify all witnesses, particularly in the event that further

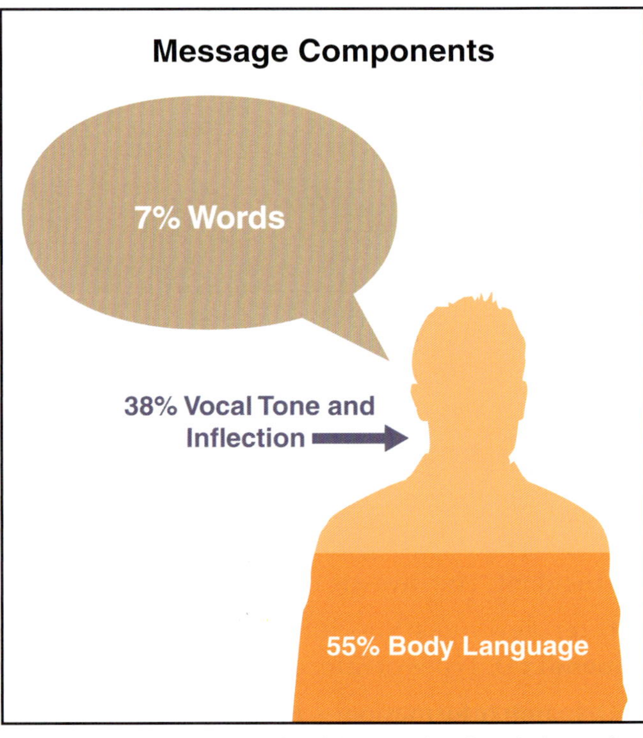

Figure 7.6 Investigators should recognize that their words only account for a small percentage of interpersonal communication.

interviews are necessary. Investigators should interview witnesses separately to ensure individual observations are not influenced by other witnesses' perspectives. The investigator should document the witnesses' observations including vantage points, if possible. Documentation of these observances can take the form of written notes, audio or video recordings, or signed statements. These interviews are conducted to obtain the following types of information:

- Building security, alarms, or surveillance systems

- Environmental factors such as weather, earthquakes, floods, etc.

- Names and number of people in the structure at the time of the event

- Fire size and growth and overall amount of structure involved in the fire

- Time of witnesses' observations verified by phone calls, text, photos, etc.

- Recent construction, occupancy changes, equipment changes, or services

- Witness name, address, date of birth, contact phone numbers, and email address

- Personal activities of the witness or witnesses before the fire up until discovery of the fire

- Observations regarding status of windows and doors, such as open, closed, locked, broken

- Any issues with utilities, including electrical, propane, natural gas, or water in or around the building

- Initial observed location of the fire or smoke and location and vantage point of the witness or witnesses

- Homeowner's insurance carrier and insurance agent policy number (if homeowner/occupant)

- Building construction, structural layout, condition, and type and locations of contents and building systems in the structure

  Interviewers should also ask about circumstances or activities before the discovery of the fire, such as:

- Domestic disputes

- Behavior of occupants

- Unusual sounds or odors

- Lack or change of normal activity

- Persons in the area (how many, who)

- Vehicles or persons leaving the scene

- Processes (controlled burns, welding operations)

The investigator determines the timing of these interviews. Considerations should include whether witnesses will remain on the scene, and the condition of the scene itself. In addition, investigators may prefer to examine the scene before conducting interviews, as physical findings can assist in verifying witness observations. Prior examination can provide a basis for asking additional questions and allows investigators to note when witness answers support or conflict with physical findings.

Developing a process flowchart can aid investigators in a logical order of investigation **(Figure 7.7, p. 216)**. Investigators should keep in mind that almost every incident is different and that an investigation may have to be accomplished in a different order than presented in this chart. This chapter explains each of the components of the process related to processing the fire scene.

Firefighter and First Responder Interviews
NFPA 1033 (2022): 4.2.3

First responders should prioritize their responsibilities over interviews. Investigators should schedule interviews with firefighters and other first responders only when they are available. Regardless of when interviews are conducted, first responders should follow established organizational procedures to maintain scene safety. Similar to other interviews, investigators should interview first responders separately to ensure individual observations are not influenced by other responders' assertions.

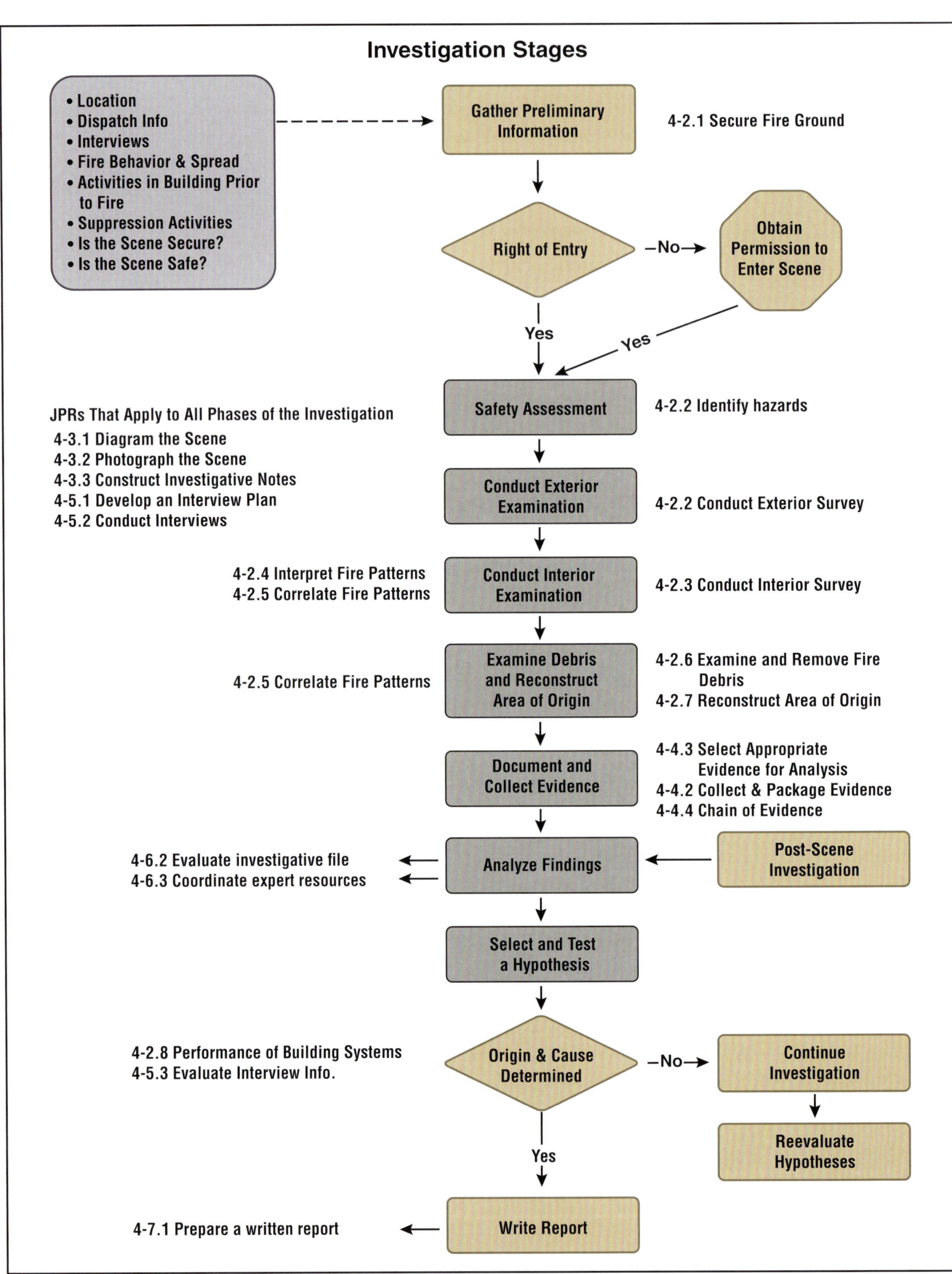

Figure 7.7 This flowchart represents the typical stages of a fire investigation.

During the interviews, investigators will document witness observations regarding the fire **(Figure 7.8)**. Investigators must follow their organizations' SOPs for interviews, if available. Interview questions for a variety of scenarios are included in **Appendix D** of this manual.

Questions for first responders may refer to any of the following observations:

- Observations conducted during 360-degree size up

- Times associated with the fire including:
  - Time of call
  - Time of dispatch
  - Time of arrival

- Observations of crowds and traffic

- Observations of human activity on the scene

- Location of the fire on arrival

- Security of the building

- Smoke conditions (color, volume, etc)

- Size of the fire

- Spread of the fire

- Fire fighting activities such as:
  - Breaking windows
  - Forcing doors
  - Conducting positive-pressure ventilation
  - Determining initial attack location

- Other factors that may have altered the ventilation patterns

- Weather and environmental conditions

- Method of reporting the fire

- Condition of the interior of the structure

- Location, type, and amount of contents

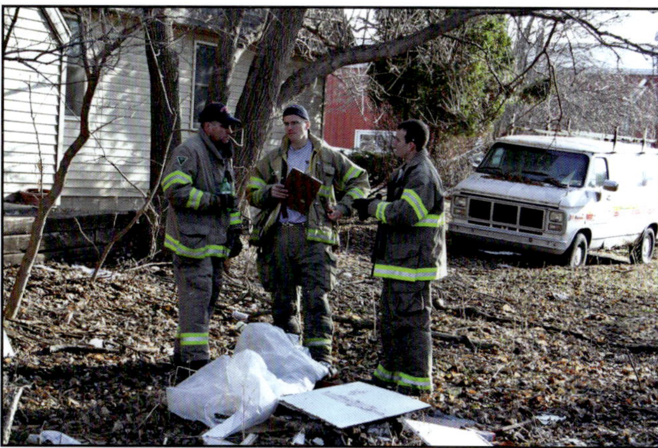

Figure 7.8 Interviewing fire suppression personnel after a fire can provide important evidence about an incident. *Courtesy of Iowa State Fire Training Bureau.*

Fire Condition Notes

Firefighters should be able to offer specific information that other witnesses, including first responders, may not be able to provide. First-arriving firefighters should be able to provide information regarding fire conditions and other observations (size-up) upon their arrival and prior to fire suppression operations began. In addition to the effects or effectiveness of fire suppression, firefighters may also be able to provide information regarding:

- Fire development in a room or compartment

- Fire progression or fire spread between rooms, areas, or floors

Video Recordings at the Scene

Many agencies have equipped the pump operator with a camera to photograph bystanders and the fire scene from their point of view. This may be helpful later in showing fire progression and identification of suspects or witnesses.

Firefighters and other witnesses may know of a video showing the scene conditions and/or development. This video can include the timeframe before suppression operations begin, during incident response, and/or after suppression activities. Sources of video can include security footage, firefighter or law enforcement officer cameras, and bystander video.

Other Indicators from Firefighters

Firefighters may be able to provide specific information about the following indicators:

- **Building indicators** — Interior configuration, fuel load, thermal properties, and ventilation
- **Smoke indicators** — Rapidly increasing volume, turbulence, darkening color, optical density, and lowering of the hot gas layer and/or neutral plane **(Figure 7.9)**

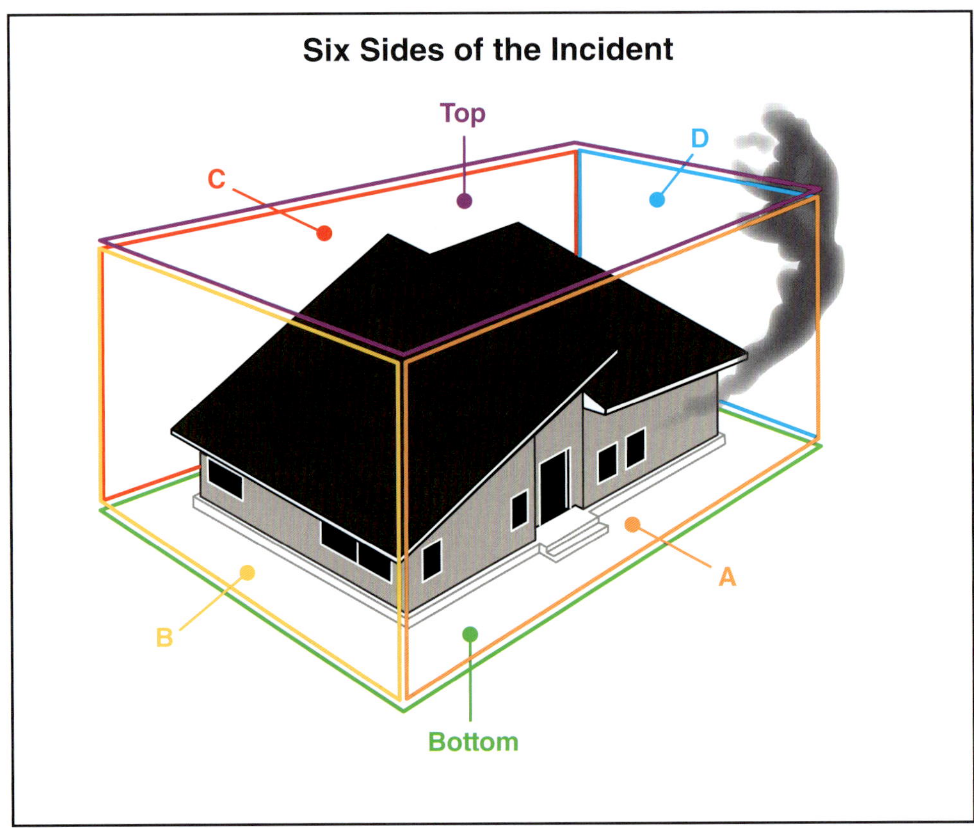

Figure 7.9 Size-up must consider all six sides of an incident, in context..

- **Heat indicators** — Post extinguishment indicators of rapidly increasing temperature in the compartment, such as pyrolysis of contents or fuel packages located some distance away from the fire, or hot surfaces
- **Flame indicators** — Isolated flames or rollover in the hot gas layers or near the ceiling

Firefighters and other witnesses may describe an event or conditions consistent with the onset of flashover (e.g. a rapid increase in fire intensity) in language such as:

- A sudden, rapid transition from smoke to flames
- Windows suddenly broke ("exploded"), then "fire everywhere"
- Rapidly developing conditions, from a "small amount of fire" to a "large volume of fire"

Scene Security

NFPA 1033 (2022): 4.2.1, 4.2.2, 4.2.3, 4.4.1

A secure investigation scene has a recognizable perimeter and a means to maintain that perimeter. The security of the scene is initially the responsibility of the fire-suppression personnel who respond to extinguish the fire. Early security measures should include the following:

- Restricting access to the scene

- Protecting any potential evidence located in the area

- Documenting evidence prior to moving or altering, when circumstances require it

- Minimizing fire-suppression and overhaul activities that could destroy important information regarding the origin and cause of the fire

The first-arriving fire investigator may have to adjust the security measures already in place or implement additional measures to protect the scene. This decision will be based on the assessment of the scene, the circumstances surrounding the fire, and the measures in place. At incidents involving the assistance of other law enforcement agencies, such as when injuries and/or fatalities are involved, an investigator would be prudent to request an officer to monitor each entrance and exit to the scene to document every individual who has entered the secured area and the times of entry and exit.

Establishing Effective Perimeters

NFPA 1033 (2022): 4.2.1

Part of an investigator's preliminary assessment is to establish the boundaries or the perimeter of the fire scene. For various models of how to establish effective perimeters, refer to the IFSTA manual, **Fire and Emergency Services Company Officer**. **Skill Sheet 7-1** shows the steps for securing the scene.

The size of the perimeter may change during an incident based on the situation and progression of the investigation. Two guidelines for establishing a perimeter of the proper size are:

- **Explosions** — Establish the perimeter for explosions at 1.5 times the distance from the farthest piece of debris found. As the investigation continues, this perimeter may expand as additional debris is located **(Figure 7.10)**.

- **Structure fires** — Firefighters may begin the initial fire-scene perimeter to allow their access to the fire and provide safety to bystanders. Investigators may expand the perimeter to encompass the area surrounding the building and any potential evidence that is located outside the building. The perimeter should extend beyond the farthest piece of evidence located during the exterior examination of the structure. If no evidence is found outside the building, it may be sufficient to restrict access into the building or an area of the building. The restricted area should provide investigators with room to work and protect all known evidence without being hampered by unnecessary personnel or bystanders.

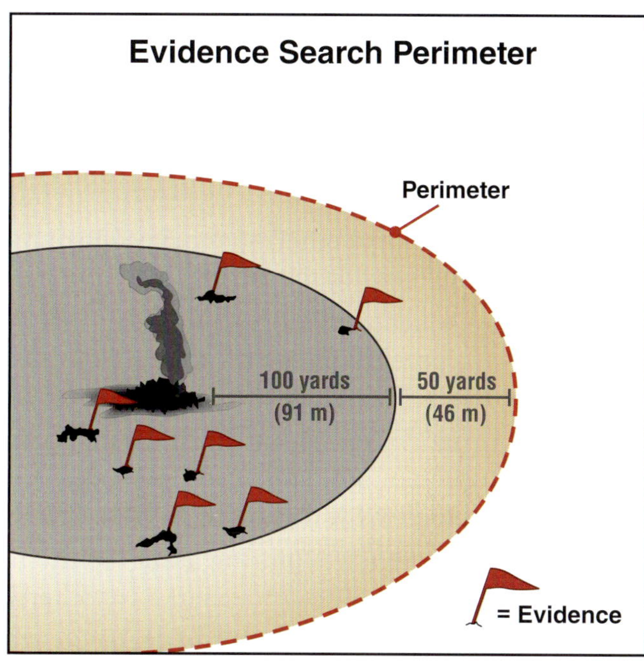

Figure 7.10 Explosion scenes require an expanded perimeter because evidence can be scattered over a much larger area than most fire scenes.

To be effective, fire-scene perimeters must be both recognizable and enforceable. Common ways to accomplish these ends are as follows:

- Ensure that the initial perimeter is larger than necessary for investigations. If needed, the investigator can more easily reduce the perimeter than increase the size of a perimeter.

- Make certain that the perimeter is visible and recognizable to everyone on the scene. Many public safety organizations use rope, traffic cones, or marked barrier tape for this purpose **(Figure 7.11)**.

- Task uniformed law enforcement officers or firefighters to control access into the established perimeter. For long-term operations, consider employing private security guards and/or installing construction barriers or fences **(Figure 7.12)**.

Figure 7.11 Fire line tape is a visible and commonly recognized barrier for establishing a perimeter. *Courtesy of Donny Howard, Agent Oklahoma State Fire Marshal's Office.*

Figure 7.12 Construction fencing provides a more long-term barrier around an active investigation scene. *Courtesy of Hugh W. Graham, Yates & Associates.*

Treat all fire scenes as though any outcome is possible. Perform the following procedures to protect the scene integrity:

- Ensure that others brought into the area are always escorted.

- Permit access only to those individuals who are authorized to be in the area.

- Keep a log of all persons who enter and leave the incident perimeter. This log can be used to show the number of personnel in the scene at any given time.

- When firefighters and other emergency personnel have completed their tasks in the area, move them to a staging area outside the perimeter to wait for additional assignments or release from the incident.

- Station a firefighter or investigator with evidence until it can be processed if personnel are still operating and there is a potential for damage or contamination due to foot traffic or from the operations being conducted.

- Mark potential evidence located within the perimeter so that it will not be disturbed before detailed examination, documentation, and collection. Use available materials, such as rope, traffic cones, or barrier tape, to provide this protection.

A primary objective for limiting access to the fire scene is safety. As the perimeter is established, known hazardous areas should be marked or otherwise secured to prevent injury to personnel operating in the area and bystanders **(Figure 7.13)**.

Another purpose of scene security is to control access to the scene to protect against potential alteration, damage, or loss of evidence. Necessary emergency personnel should document their entry via a scene access log. They should be trained in techniques to minimize the negative effects of their actions **(Figure 7.14)**.

Figure 7.13 It may be necessary to cordon off certain areas of a scene that present unique hazards. *Courtesy of Jocelyn Augustino, FEMA News.*

Figure 7.14 Establishing perimeters restricts access to a scene. *Courtesy of Wayne Chapdelaine, Metro-Rural Fire Forensics.*

Investigators should recognize alterations to a scene caused by responders in the performance of duties including:

- Investigation
- Emergency medical service
- Fire suppression, salvage, and overhaul

Canine (K9) Units and Setting Perimeters

The use of scent-tracking canines (K9s) may be useful in determining a suspect's direction of travel away from a scene **(Figure 7.15)**. Consider establishing a perimeter that would protect this scent line from all involved at the scene. Usually this involves a perimeter that completely surrounds the scene. Enforcement of this perimeter is essential toward protecting this evidence.

Figure 7.15 Canine (K9) Units can determine where ignitable liquids are located to track individuals who have been at the scene. *Courtesy of New South Wales Fire Brigades.*

Protecting Evidence

Secure the scene to protect evidence from being contaminated, damaged, or destroyed before it can be documented, preserved, or collected. Scene security should also prevent the alteration of the evidence at the scene such as the changing of circuit breaker or appliance controls and/ or switch positions. Evidence preservation is further described in Chapter 11, Explosion Dynamics and Investigation.

An investigator should recognize the need to preserve evidence. The Incident Commander may be able to help with the following actions:

- Limit the use of fire hoselines during the overhaul stage.

- Identify potential evidence with the use of markers or tags **(Figure 7.16)**.

- Minimize destruction during overhaul, reinforcing, or shoring operations.

- Minimize and limit overhaul to essential areas until an investigator has observed and documented the conditions.

- Limit access to fire-suppression and other related emergency personnel to the fire area and especially any suspected area of origin.

- Spread tarps over portions of the scene that are exposed to the weather and that otherwise would be adversely affected from exposure to the elements **(Figure 7.17)**. Where fatalities are part of the incident scene, confirm appropriate handling of remains BEFORE laying out tarps or other coverings. Fatalities are addressed in Chapter 9.

Figure 7.16 One way to preserving evidence is to photograph an item in its context. The use of markers or tags can allow an investigator to call attention to small items in a larger space. *Courtesy of Donny Howard, Agent Oklahoma State Fire Marshal's Office.*

Figure 7.17 Tarps can protect areas from further damage during a scene investigation. *Courtesy of Hugh W. Graham, Yates & Associates.*

Contamination and Spoliation

Even if a scene is marked and access is restricted to emergency responders or other authorized personnel, contamination or spoliation could occur that would threaten the evidentiary value of evidence **(Figure 7.18)**. Sometimes these contaminations arise as unforeseen consequences of routine activities such as overhaul.

Contamination is a broad concept, encompassing anything that can taint physical evidence. Examples of potential sources of contamination include:

- Smoking materials, such as cigarette butts or matchsticks, that have been dropped by firefighters, spectators, or investigators

- Ignitable liquid traces introduced into the scene by items or equipment that have been used or stored at the scene including:

  — Boots/gloves

  — Power cords

  — Tools

  — Power equipment

Spoliation refers to evidence that is destroyed, damaged, altered, or otherwise not preserved by someone who has the responsibility to preserve it. Spoliation occurs when the movement, change, or alteration of the evidence prevents another investigator or interested party from obtaining the same evidentiary, interpretive, or analytical value from the evidence as the initial investigator. The legal definition and the consequences of spoliation vary among jurisdictions. It is each fire investigator's responsibility to understand what constitutes spoliation in their jurisdiction.

Outside Assistance

Certain investigative activities may exceed an investigator's training, knowledge, and experience. In such cases and especially if that activity is intrusive or destructive in nature, fire investigators should consider other parties who may have legal interests in that activity. Such consideration may result in the need to consult with other experts.

Handling Evidence

Evidence at the incident site may assist the investigator in determining the origin, spread of fire, and the cause and responsibility for the fire. Proper evidence identification and handling are extremely important technical skills for an investigator. Collected physical evidence may serve a number of purposes, including:

Figure 7.18 Any operations at an incident scene have the potential to damage evidence.

- To test for the presence of an ignitable liquid

- To determine if a failure of a product had occurred

The investigator should be aware of the legal requirements and standards on how to handle any physical item considered evidence. ASTM has published several standards on evidence collection, handling, storage, and examination.

Marking Evidence

During the examination of the fire scene, investigators will often discover items of potential evidentiary value. Identification of the evidence can ensure that the items can be properly protected and documented as part of the scene examination.

Depending on the type of evidence and the scene conditions, different methods may be used in order to identify the evidence and provide adequate protection. Such methods may include:

- Placing a flag next to the object.

- Covering the evidence with a box or similar object.

- Taping off the area where the evidence was located.

- Placing brightly colored tape on evidence such as electrical wires or gas lines. This will allow photographs of the routing of the conductors or gas lines and also show areas of interest including damage.

Review Questions

1. What are the steps in the scientific method?

2. In what type of situation may an investigator enter a property under exigent circumstances?

3. Who can provide consent for an investigator to enter a property?

4. What are potential consequences if a private investigator's actions at an investigation exceed his or her legal authority?

5. When must a criminal search warrant be obtained?

6. What types of information can an investigator gather to help determine a timeline for the incident?

7. How can understanding nonverbal clues be beneficial to investigators?

8. What information should be gathered from witnesses?

9. What types of information can firefighters provide that other types of witnesses may not be able to provide?

10. What are the two primary purposes for ensuring scene security?

Chapter References

Michigan v. Clifford. 1984. 464 U.S. 287.

Michigan v. Tyler. 1978. 436 U.S. 499.

Chapter Key Terms

Bias — Highly personal or unreasoned distortion of judgment; prejudice.

Confirmation Bias — The tendency to prefer and prioritize information that matches existing expectations.

Possible — Legal classification for an occurrence that is less likely than 50 percent.

Probable Cause — Sufficient information or facts to believe that it is probable (more likely than not) that a certain party is responsible for committing a felony (indictable offense).

NOTE: This skill provides basic steps for securing a fire scene for investigative purposes. Certain scenes may require other tasks due to complexity or time required for investigation. Ensure that the scene is continuously protected or monitored, if possible. This skill assumes that the fire investigator will arrive on scene after other first responders who have established initial scene security.

Step 1: Report to the Incident Commander and initiate the investigation process.

Step 2: Assess the scene.

> **a.** Check to see that the initial perimeter has been established and maintained effectively **(Figure 7.19)**.
>
> **b.** Ensure that the perimeter is visible and recognizable.
>
> **c.** Determine what additional scene control measures may be necessary.

Step 3: Enlarge, change, or reinforce the perimeter as necessary.

Step 4: Control access to the perimeter.

> **a.** Permit only authorized individuals.
>
> **b.** Enlist law enforcement personnel to assist with perimeter control, if necessary.
>
> **c.** Keep a log of all persons who enter the scene, if necessary.

Step 5: Use appropriate marking devices to mark potential evidence within the perimeter so that it will not be disturbed before the detailed scene examination begins.

Figure 7.19

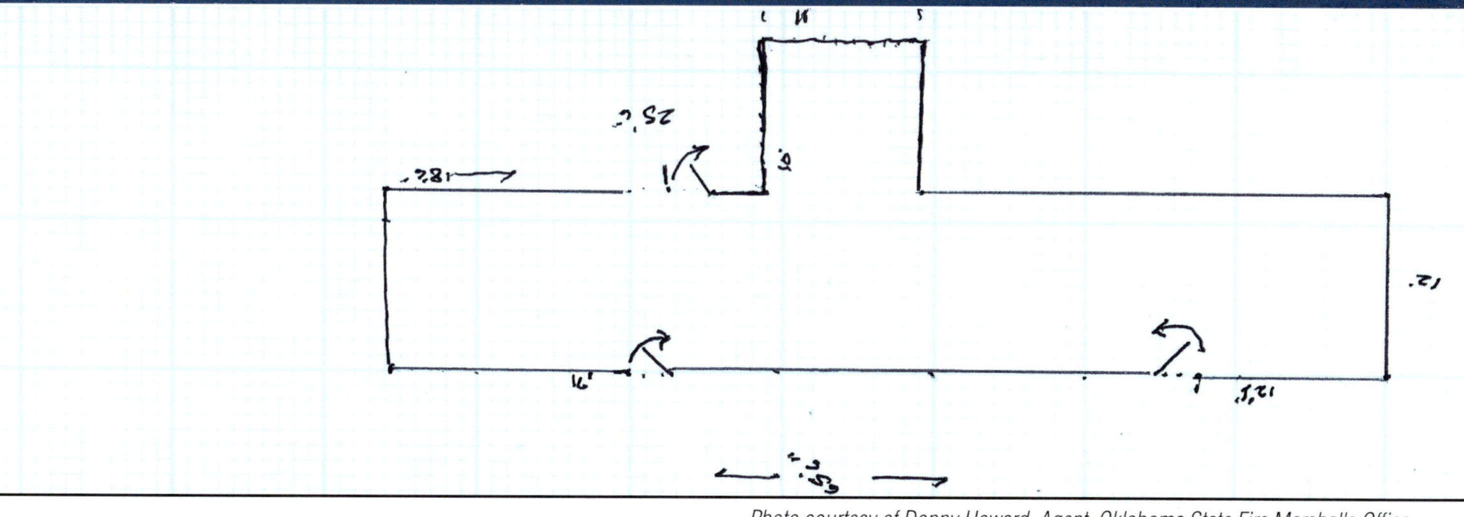

Photo courtesy of Donny Howard, Agent, Oklahoma State Fire Marshal's Office.

Chapter Contents

NFPA 1033 JPRs addressed in this chapter

This chapter provides information that addresses the following job performance requirements (JPRs) of NFPA 1033, *Standard for Professional Qualifications for Fire Investigator* (2022):

4.1.7   4.2.6   4.3.1   4.3.2   4.3.3   4.5.2

This chapter provides information that addresses the following course outcomes for the FESHE courses:

Fire Investigation I (C0283)

1. Demonstrate the importance of documentation, evidence collection, and scene security process needed for successful resolution.

Fire Investigation II (C0284)

6. List the procedures for fire scene documentation, including sources and technology available for fire investigations.

Fire Investigation and Analysis (C0285)

2. Document the fire scene in accordance with best practice and legal requirements.
4. Analyze the legal foundation for conducting a systematic incendiary fire investigation and case preparation.

Learning Objectives

1. Explain the basic principles of scene documentation. [NFPA 1033, 4.1.7, 4.2.6, 4.3.1]

2. Explain how investigators compile observations into investigative field notes. [NFPA 1033, 4.3.3]

3. Describe the types of information that should be included in a scene sketch. [NFPA 1033, 4.3.1]

4. Explain how investigators use drawings, diagrams, and maps to document scene information. [NFPA 1033, 4.3.1, 4.3.3]

5. Describe the types of equipment used for photographic scene documentation. [NFPA 1033, 4.3.2]

6. Explain the photographic processes that investigators must understand to correctly photographically document the scene. [NFPA 1033, 4.2.6, 4.3.2, 4.3.3]

7. Explain how video and audio recording devices are used to document the scene. [NFPA 1033, 4.5.2]

8. Skill Sheet 8-1: Document the scene. [NFPA 1033, 4.3.1, 4.3.2, 4.3.3]

Chapter 8
Scene Documentation

Documenting a scene is one of the most important tasks a fire investigator will perform. Using a variety of methods, an investigator records information regarding the incident to recall the conditions observed, activities undertaken, and any changes made to the scene that might occur during the investigation. This chapter addresses several factors in documentation, including:

- Basics of scene documentation
- Investigative field notes
- Sketches

- Drawings, diagrams, and maps
- Photographs
- Recording devices

Basics of Scene Documentation

NFPA 1033 (2022): 4.1.7, 4.2.6, 4.3.1

Once the scene examination is complete, fire investigators may not receive another opportunity to observe the scene, obtain information, or collect evidence. In addition, the investigator who completed the examination may be the only person to do so.

As part of thorough documentation, investigators should document the overall scene before making alterations. Chapter 14, Presentation of Investigative Findings. **Skill Sheet 8-1** shows the steps for documenting the scene.

When the overall scene has been documented, any potential evidence should be documented prior to collection. Proper documentation will ensure that the investigator has recorded the location, condition, and times associated with the identification and collection of the evidence. Evidence identification and collection are briefly discussed in this chapter, and in more detail in in Chapter 9, Evidence Collection and Preservation.

In fire and explosion investigations, the most common methods used to document the scene and support investigative findings include:

- Sketches
- 2D and 3D photographs
- Investigative field notes

- Audio and visual recordings
- 3D models and laser scanning
- Drawings, diagrams, and maps

These methods should be used in tandem, using the strengths of each to offset the weaknesses of others. For example, in tandem with the photography work, investigators should maintain detailed notes at the scene with respect to the identification, preservation, and collection of the physical evidence. Formats for those notes may vary, depending on the investigator's resources, preferences, and AHJ. Examples of formats include a lined notebook with numbered pages or an electronic device such as a tablet or smartphone. Where conditions are prohibitive for writing, the investigator may consider narrating an audio or video recording. Examples of poor scene conditions can include:

- Extreme cold temperatures
- Moisture dripping from ceilings
- A confined space or other hazardous environment

Scene Documentation

During the investigation process, observations will help determine:

- Scene inspection for safety
- Condition of building utilities
- Type of building construction
- Interior ceiling, wall, and floor covering
- Preparation for conducting scene reconstruction

  The overall documentation should include the following details:
- Explosion damage
- Evidence of collapse
- Specific location of items at the scene
- Forcible entry damage and/or broken windows
- Date and time that the item was located and/or collected
- Evidence that will be collected including unique markings or labeling
- Identity of the individual who located the evidence and removed it from the scene
- Photograph or sketch showing the item as it was found and its relationship to other items
- The fire damage, or lack thereof, on the exterior and interior of the structure, vehicle, or equipment and surrounding areas

Evidence Documentation

When completed, the overall evidence documentation notes should provide a clear overview of the findings, with details such as:

- Time the evidence was discovered
- Who collected the evidence, and when
- How and where the evidence was secured after collection
- How the evidence was collected (i.e, tweezers, gauze pad)
- Where the evidence was discovered (i.e., living room, exterior, floor)
- How the evidence was documented (i.e., notes, sketches, photographs)
- Who discovered the evidence (i.e., name of the firefighter, police, fire investigator)
- Type of container used for collection (i.e., common plastic bag, paper bag, metal can) **(Figure 8.1)**

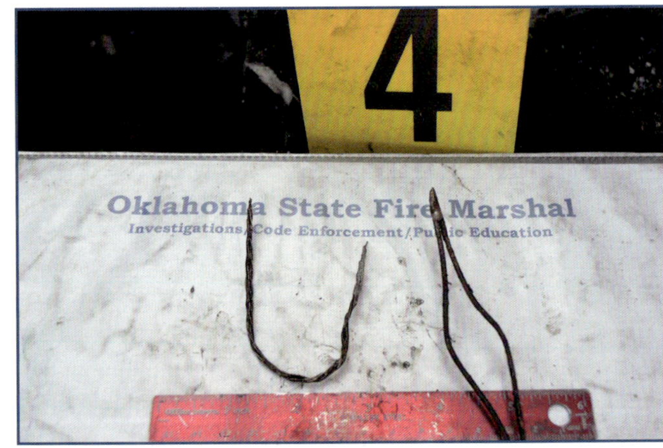

Figure 8.1 The photograph of some materials may be aided with an item, such as a small ruler, to show scale. *Courtesy of Donny Howard, Agent, Oklahoma State Fire Marshal's Office.*

The investigator should also give the evidence a unique identifying number, letter, or combination of these as part of the documentation process. Some investigators use sequential numbers for physical evidence and letters for other objects.

Evidence tags or labels should be affixed to each piece of evidence. Items that should be included on each piece of evidence include:

- Date of collection

- Time of collection

- Evidence description

- Unique identification number

- Original location of the evidence, or where it was collected

- Initials or other identifying mark of the individual who collected the evidence

- Tape used to seal the container (evidence packaging), per the lab's specifications **(Figure 8.2)**

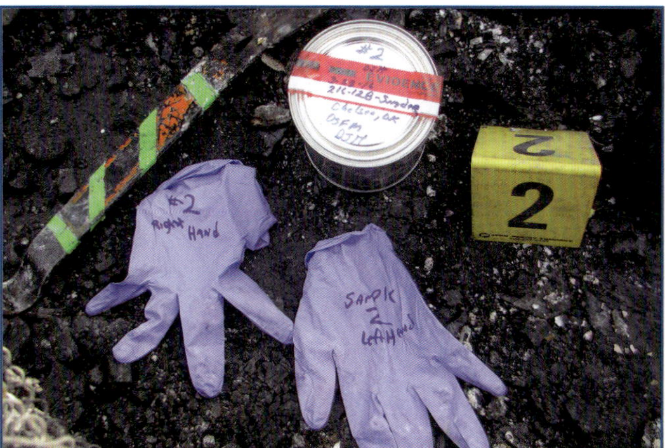

Figure 8.2 An investigator should photograph sealed and labeled evidence and the gloves used to collect that evidence. *Courtesy of Donny Howard, Agent, Oklahoma State Fire Marshal's Office.*

Investigative Field Notes

NFPA 1033 (2022): 4.3.3

Field notes are the most common type of documentation that investigators use. They provide a written record of an investigator's observations and findings. Professional field notes are accurate, complete, concise, and written in the format required by the authority having jurisdiction. These notes assist the investigator in recalling scene observations used to develop the final incident report and to provide background information when called to testify in a criminal or civil trial. In most cases, the investigator handwrites the field notes while at the scene or while continuing the investigation beyond the scene.

Organization is critical for accurate report development and compilation of a complete and organized case file. Investigative field notes can also be used to provide an overview of the incident for fire department statistical records. Whatever format is used, fire investigators must develop enough information so that a report can be prepared and findings can be accurately and effectively supported.

While developing field notes, investigators should adhere to the following guidelines:

- Compile a separate set of notes for each incident.

- Record only facts and actual observations related to the incident.

- Remember that field notes may be read by others, including a jury.

- Be complete — An investigator may not get the chance to obtain the information again.

- Include information pertinent to the investigation, such as observations, comments, and hypotheses.

- Use a systematic, consistent method that enables thorough recollection and straightforward interpretation **(Figure 8.3)**.

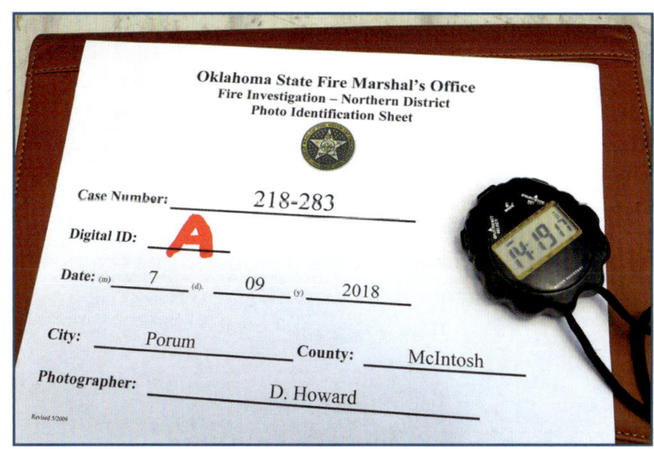

Figure 8.3 A photographer may capture a photo start page as part of a systematic record of an element of an investigation. *Courtesy of Donny Howard, Agent, Oklahoma State Fire Marshal's Office.*

Investigators may find it helpful to use a notepad with numbered pages. This ensures that any information was not altered by pages being torn out. If the investigator takes notes on an electronic device such as a tablet, the app or software should be set to record the date, time, and history of each entry.

Investigation Guidance Forms

NFPA 921, *Guide for Fire and Explosion Investigations*, contains forms suitable for photocopying to assist in the collection of investigative information. While every investigation may not require the level of detail found in these forms, the forms provide an excellent overview of the information the investigator should collect. A system such as the one presented in NFPA 921 also assists in the organization of other incident documentation such as photographs, sketches, collected evidence, and interview data. Investigators should not be restricted by the forms' design. The NFPA 921 forms are provided in Annex A to assist investigators in developing their own field note format.

Sketches

NFPA 1033 (2022): 4.3.1

Investigators should construct scene sketches identifying the location and types of evidence. These sketches should be done at the scene and can be used to create more detailed diagrams later if required **(Figure 8.4)**.

According to NFPA 921, a simple sketch is the minimum drawing fire investigators should develop as part of every investigation. A more detailed sketch may be necessary, depending on the complexities of the incident. The sketch provides a graphical representation of the scene that is proportional but not necessarily to scale. Fire investigators should use common symbols in the sketch and provide a legend explaining the symbols used. Whatever symbols an investigator chooses, the markings should be used consistently in all sketches and drawings related to the investigation **(Table 8.1)**.

Grid paper (graph paper) is an excellent aid to a fire investigator while sketching. The paper's grid can be assigned a scale and the drawing can easily be completed to an approximate scale **(Figure 8.5)**. Investigators may also use items such as straightedges or measuring devices and clipboards to assist in the sketching. When using tablets or other electronic devices to create sketches, a stylus may be used for drawing and making any annotations such as symbols.

Figure 8.4 The investigator's vehicle may supply sufficient resources, such as climate control and lighting, for an investigator to complete sketches and other scene paperwork. *Courtesy of Ron Jeffers, Union City, NJ.*

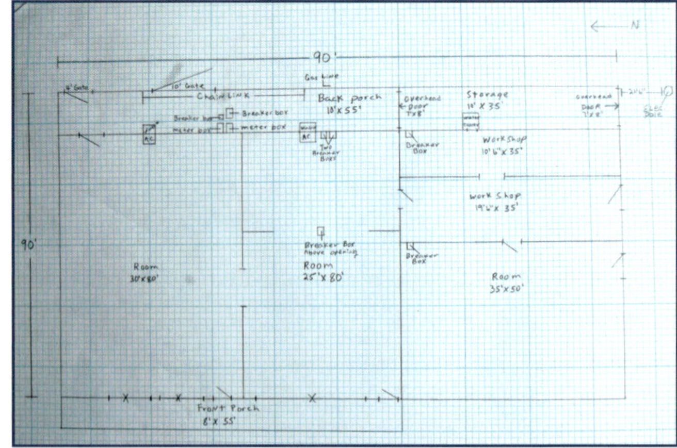

Figure 8.5 Producing sketches on grid or graph paper allows the investigator to accurately gauge scale and distance. *Courtesy of Hugh W. Graham, Yates & Associates.*

Access Features, Assessment Features, Ventilation

△ Features, and Utility Shutoffs

△FD Fire Department Access Point

△K Fire Department Key Box

△RA Roof Access

△AP Fire Alarm Annunciator Panel

△RP Fire Alarm Reset Panel

△CP Fire Alarm Voice Communication Panel

△WB Sprinkler System Water Flow Bell

△SV Smoke Vent

Utility Shutoffs

△E Electric Shutoff

△W Domestic Water Shutoff

△G Gas Shutoff

△LPG LP-Gas Shutoff

△NG Natural Gas Shutoff

△CNG Compressed Natural Gas Shutoff

△SP Smoke Control and Pressurization Panel

△SL Skylight

◇ Detection/Extinguishing Equipment

◇HD Heat Detector

◇FS Flow Switch (Water)

◇TS Tamper Switch

◇DC Dry Chemical System

◇WC Wet Chemical System

◇CA Clean Agent System

◇DD Duct Detector

◇SD Smoke Detector

◇PS Manual Pull Station

◇HL Halon System

◇CO₂ CO_2 System

◇FO Foam System

◇BSD Beam Smoke Detector

(PIV) Post-Indicator Valve

(ZV) Sprinkler Zone Valve

(WH) Wall Hydrant

(TC) Inspector's Test Connection

(FDC) Fire Department Connection

(WT) Water Tank

(RV) Riser Valve

(HC) Hose Cabinet or Connection

(TH) Test Header (Fire Pump)

(FH) Fire Hydrant

(DS) Drafting Site

Equipment Rooms

[AC] Air-Conditioning Equipment Room

[EG] Emergency Generator Room

[EE] Elevator Equipment Room

[FP] Fire Pump Room

The investigator's basic sketch should provide an overall representation of the scene and may include the following items:

- Fire patterns

- Ceiling height

- Layout of the scene

- Area/point of origin

- Nearby paths or roads

- Direction of fire travel

- Smoke patterns and travel

- Entry points into the scene

- Locations of furniture and other contents

- Locations and positions of any fire victims

- Locations and dimensions of windows or openings

- Locations of witnesses when observations were made

- Direction, methods, and effects of fire-suppression activities

- The scale used to make the sketch (i.e., 1 inch equals 10 feet)

- Overall measurements of the structure and its individual rooms

- Locations and types of fire detection or suppression systems and devices

- Locations of important information or evidence related to the investigation

- Locations of anomalies or other electrical activity found during arc mapping

Sketches should accurately capture critical incident information in a format that is easy to understand and can be translated into three dimensions. A key consideration when developing sketches is to keep them uncluttered. It is better to use multiple sketches to show various pieces of information than to include too much on one drawing.

Whatever method is used to produce field sketches of the scene, the investigator will also have to use a reliable means of locating items on the sketch. The collection of this information is essential in the reconstruction of the scene, documentation of evidence collection points, and the development of more detailed diagrams after the scene examination is completed (**Figure 8.6**).

Drawings, Diagrams, and Maps

NFPA 1033 (2022): 4.3.1, 4.3.3

For many investigations, the simple sketch completed on the scene may not be adequate. It may be necessary to develop a formal set of drawings, diagrams, or maps that present detailed information

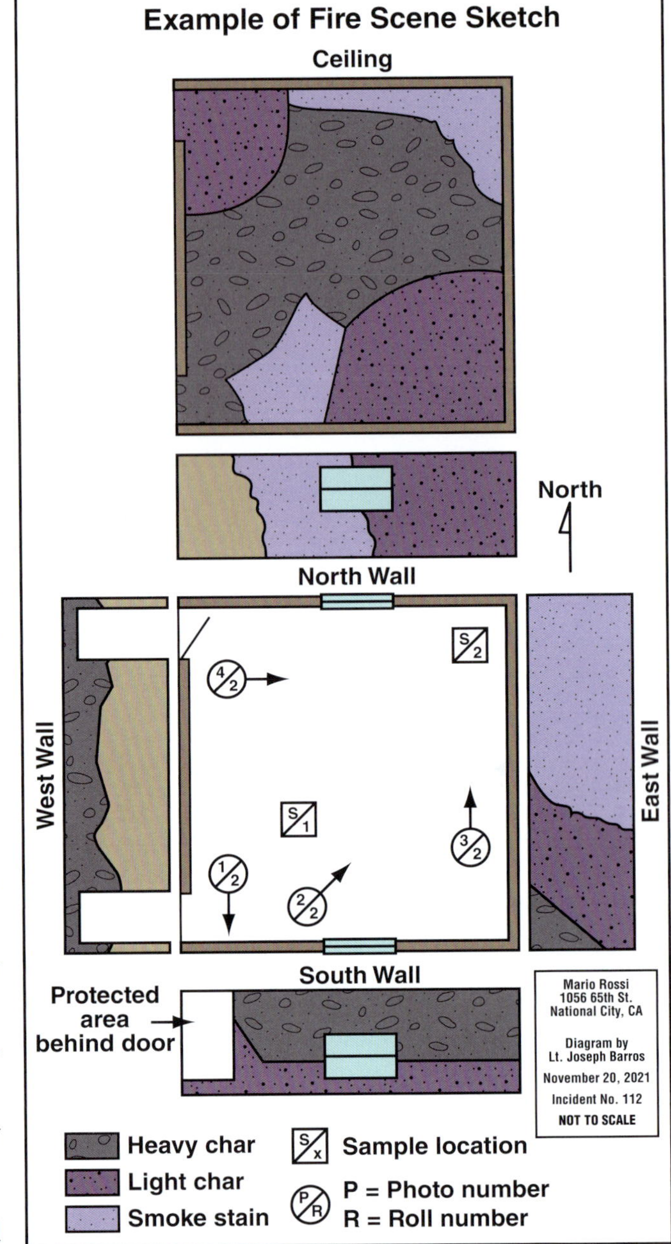

Figure 8.6 A fire scene diagram depicts where photos were taken and the presence of notable fire effects in specific areas.

related to the fire. Drawings may depict the floor plan of a fire building and have various levels of detail regarding the building, its construction, and contents. Information regarding the fire can then be placed on the drawing and several versions can be generated showing different types of data.

Using high-quality equipment or computers to generate drawings, diagrams, or maps for the investigative report will only be successful if a fire investigator collects details and accurate information during the scene examination. Almost all of the information included in specialized artwork related to an incident will be based on the basic sketches and field notes produced by an investigator while on the scene.

Drawings

In addition to developing their own materials, fire investigators may find external resources that prove helpful in documenting the incident. These resources can include:

- Drawings of a building used in the plans review, permit, and construction processes
- Street maps of the incident area from the jurisdiction's street or engineering department
- Sketches and building details from a county or municipal auditor/assessment office or website

Although these drawings may not be to scale or include all necessary details, they normally provide a great deal of information about a building and its systems. A fire investigator using building plans for the basis of a drawing should verify the accuracy of the plans and include any structural changes made during construction or renovation **(Table 8.2)**.

Table 8.2	
Design and Construction Drawings and Their Sources	
Type of plan, drawing, or diagram	**Source**
Topographical Plans	Surveyor
Site Plans	Civil Engineer
Floor Plans	Architect
Plumbing Diagrams	Mechanical Engineer
Electrical Diagrams	Electrical Engineer
Mechanical System Diagrams	Mechanical Engineer
Sprinkler/Fire Alarm Diagrams	Fire Protection Engineer
Structural Plans	Structural Engineer
Elevations	Architect
Cross Sections	Architect
Detailed Close-ups	Multiple Sources

Diagrams

Diagrams of specific areas or systems involved in a fire may be needed to provide details for the investigation. Many computer-based drawing and **computer-aided design (CAD)** programs or applications for tablets allow investigators to develop high-quality materials with little special training **(Figure 8.7, p. 236)**. Additionally, scene drawing and diagramming programs can quickly import measurements taken from measuring devices using lasers and convert drawings and diagrams into two-dimensional (2D) or three-dimensional (3D) animations. More advanced programs can also import scene photographs and apply them to animated 3D computer models.

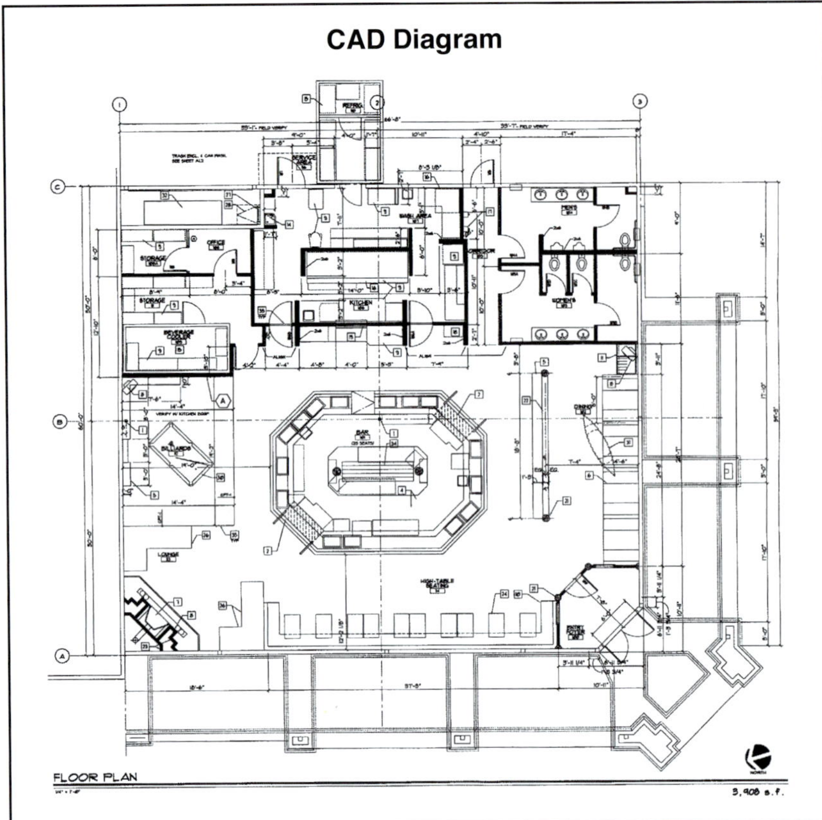

CAD Diagram

Figure 8.7 Computer-assisted design drawings can yield detailed schematics of buildings under investigation.

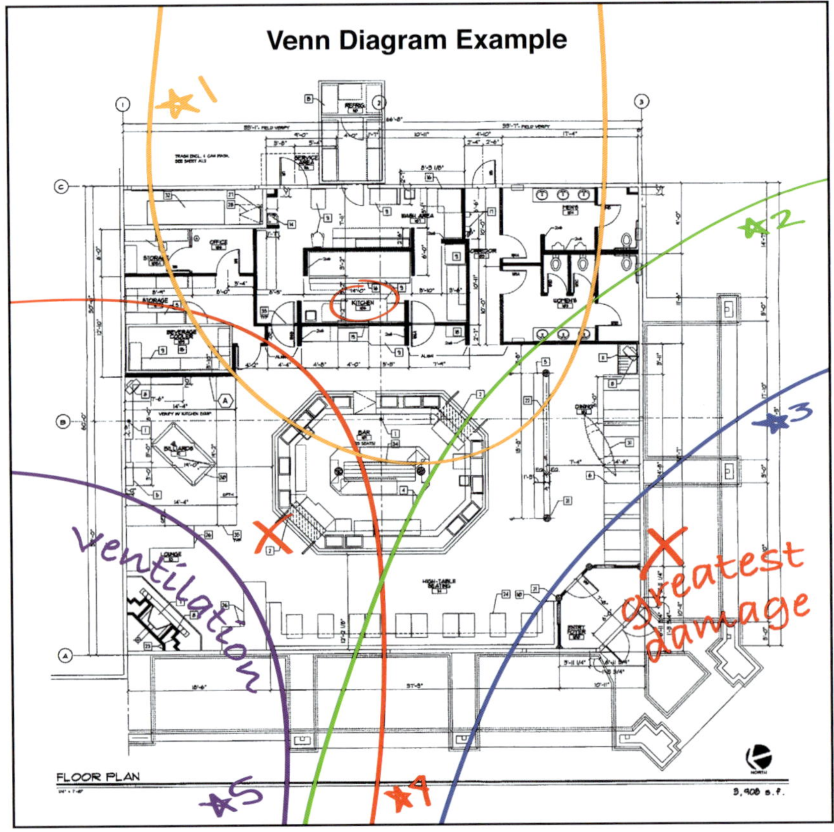

Figure 8.8 A Venn diagram may be overlaid on an existing diagram, sketch, or photo to show overlapping areas of damage and other points of interest.

Fire investigators may create a **Venn diagram** to show how they determined the hypothetical area of origin. The Venn diagram can be used to show the fire damage, witness information, and any arc sites within the compartment of fire origin **(Figure 8.8)**.

Maps

Street maps, site plans, or aerial (including drone) photographs may be used to show the location of the incident relative to the surrounding area, including nearby streets, locations of other facilities, water supply, and other similar details. Remote image maps such as **Geographic Information Systems (GIS)** or Internet-based images may also be useful **(Figure 8.9)**. Maps can be used to show the locations of fire-suppression units and specific fire-suppression activities.

For incidents that affect large areas such as explosions or hazardous materials releases, maps can be used to depict the damaged or affected area, weather conditions such as wind direction, or any other specialized information. Investigative information may be overlaid on existing drawings, maps, and aerial photographs or created as wholly original documents.

Photographs

NFPA 1033 (2022): 4.2.6, 4.3.2, 4.3.3

While field notes and sketches of the scene are important components of the documentation process, photographs are the key components. Good photographic documentation of a scene assists a fire investigator throughout the investigation. Photographs allow investigators to show the scene as it was found and depict the investigative process at the

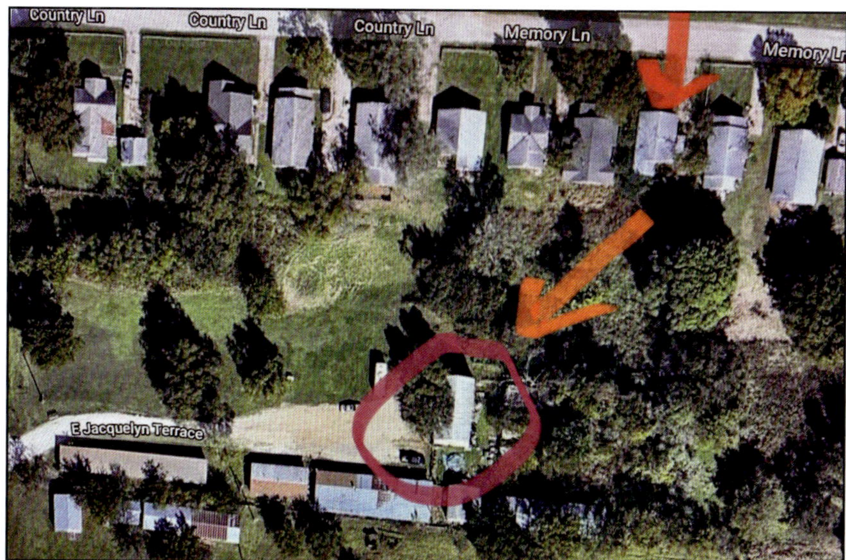

Figure 8.9 As a simple measure, an online topographic map can be edited to include indications of notable features at or near a fire scene.

scene through debris removal, reconstruction of the area of origin, and identification and collection of evidence. When the investigation is completed and the scene is cleaned, reconstructed, or demolished, the photographs taken during the scene examination may be all that remains.

In large organizations, or cases involving several investigative teams, scene photography may be assigned to a designated photographer. However, most fire investigators are responsible for taking their own photos while on the scene.

Photographic Logs

NFPA 921 provides a photographic log that may assist with organizing fire investigation photos. This form or one similar aids the investigator if the photos are used in court. Most digital cameras provide information (metadata) embedded in each photograph that includes the date, time, and location of each photograph. Additional information is stored within each photograph's metadata. Examples include the type of camera, type of lens, and the camera setting when the image was created.

The metadata feature provides a detailed timeline regarding the conditions and activities at the time a photograph was taken. The investigator should make sure the date and time set for the camera are accurate.

Photographing the Scene

Fire investigators must remember that they may have only one chance to obtain photographs that document the scene. They should take as many photographs while on the scene as they need to fully document the fire damage, evidence, and fire spread as well as the scene in general.

In order to accurately depict and support their findings using photographs, fire investigators must have a basic understanding of the equipment and techniques used in fire-investigation photography. The sections that follow discuss the operation of the camera and associated equipment. They also discuss using this equipment to get quality photographs in the difficult circumstances present at a scene.

Some jurisdictions may still use a 35 millimeter (mm) **single-lens reflex (SLR)** camera. ISO settings for SLR cameras are discussed later in this chapter. In most applications, the most significant advantages of **digital single-lens reflex (DSLR)** cameras over SLR cameras are the lack of film and the ability to view photos immediately after their capture.

Photographing the Exterior

To document a scene, investigators should begin photographing on the outside of the fire structure. These photographs should establish the scene location, any landmarks near the building, the structure's specific location on the property, and any visible utilities.

During this process, the investigator should photograph:

- Any damage to openings
- Apparent routes of fire travel
- All fire- or explosion-related damage
- Any glass fragments and their locations
- All openings such as windows and doors
- All potential ignition sources within the area of fire origin
- Any appliances in or near the area of origin, including the position of controls
- Any exterior utilities including overhead electrical supply lines, meters, and transformers
- Items found on the building's exterior that might have a factor in the origin and cause of the fire
- Broken glass beneath windows to show if the window has been broken due to mechanical force or from thermal effects
- All corners and sides of the structure to provide a record of the building's configuration and related structural elements

Photographing the Interior

Interior photographs should begin at the point the investigator makes entry and continue in a systematic manner **(Figure 8.10)**. This methodic approach will show how the investigator moved through the scene. If the investigator notices that an area or object was not completely photographed when reviewing photographs prior to leaving the scene, it is acceptable to re-photograph these areas.

Attempt to take photographs from a number of angles in the interior of the structure:

- Windows, doors, ceiling, floor, and contents
- The entrance to a room and all walls enclosing it
- Corners of the rooms to show two sides in each photograph

Other subjects for photographs at the scene include:

- Any victims and their locations
- All potential ignition sources within the area of fire origin
- Photographs that illustrate the witness's view of the scene
- Utility entrances to the structure, including meters and controls
- Any appliances in or near the area of origin, including the position of controls
- Property access points, surrounding area, location of vehicles, and surrounding property
- Electric service equipment including fuse or circuit breaker panels, as well as breaker position or fuse conditions

Figure 8.10 When moving to a scene's interior, an investigator must remember to update photography settings suitable to the new environment.

Photographing Activities

Following initial photo documentation at the scene, investigators should take additional photographs of their process. While sifting through debris and reconstructing the area of origin, they should photograph each step.

When arriving on the scene with fire or suppression activities in progress, the investigator should begin photographing those activities. These photographs provide a timeline of the fire's progress and location, as well as show the activities and locations of fire-suppression personnel, bystanders, and witnesses. Photography of these activities may be important during the investigation and analysis of the scene and fire patterns.

Photographing Evidence

Prior to moving or disturbing the evidence, the investigator should use a systematic manner to photograph the evidence in place. An example of photographing item locations, orientation, and perspective within a fire scene:

Step 1 Photograph the evidence from an overall perspective showing its location within the room or scene.

Step 2 Photograph a closer or mid-range perspective.

Step 3 Take a series of photographs close to the evidence showing its physical characteristics prior to removal.

Step 4 Photograph evidence after collection.

During the photography process, investigators should also use measuring devices to show the exact location of the object within the scene. Some investigators find it helpful to take two separate series of photographs – one showing the scene with measuring devices such as pocket rods, measuring tapes, or surveying sticks, the other showing the same scene without the devices **(Figure 8.11)**.

Figure 8.11 A pocket rod can be used in photographs to show scale, especially for fairly large items. *Courtesy of Dennis Smith.*

DSLR Cameras

Digital single-lens reflex (DSLR) cameras are typically used for investigative photography. DSLR and other digital cameras are relatively inexpensive. They allow the photographer to take a large number of photos and review the quality of images immediately after shooting. The DSLR camera can be automatically or manually focused, allowing the user to adjust various settings to obtain a properly focused and exposed photograph of the subject. DSLR cameras also allow users to interchange lenses and use an external flash unit **(Figure 8.12, p. 240)**. Any digital camera used by investigators should have external flash unit capability and sufficient resolution to produce a high quality, 8 × 10-inch (200 mm by 250 mm) print.

DSLR cameras include built-in light metering that may provide the settings needed for a correctly exposed photo. The camera may also include a number of different automatic light-metering options. Investigators should be familiar with these settings and choose the one that produces the best image. Most camera models provide automatic focusing and exposure control features that make the cameras easy to use under normal conditions. The difficult conditions found on a scene, including low light and lack of contrast between surfaces, may require investigators to choose an alternative automated setting or use the camera's manual mode instead of automatic settings.

Investigators should practice using manual settings in order to adapt to the various conditions found at scenes. Investigators should read any documentation that comes with the camera they will use on the scene.

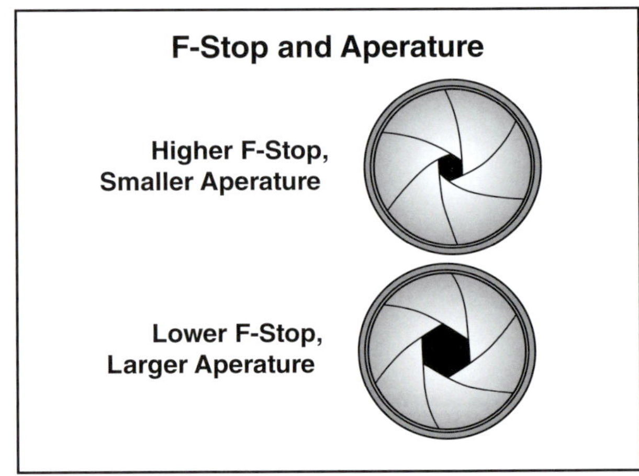

Figure 8.12 A typical DSLR camera, front and back, includes a number of features that can affect a particular photo's quality.

Most manual settings for DSLR cameras can be found by navigating the on-screen menus. These settings are arranged differently depending on camera type. Common settings include:

- Manual zoom
- Shutter speeds
- **Aperture** settings **(Figure 8.13)**
- Auto mode on or off
- Flash or no-flash options
- Picture review and deletion
- Autofocus turned on or off

Figure 8.13 F-stop settings are inversely related to the aperture size.

- Preset options for additional photo effects

- Red-eye settings for reducing eye reflections

- Manual preset options for indoor, outdoor, day, night, or panoramic photos

- Picture output size usually described in terms of **megapixel** setting or picture-resolution size such as 1028 × 768 or 1240 × 1028

Digital Imaging and Photos

In digital imaging, a **pixel** is the smallest piece of information in an image. Pixels are normally arranged in a regular two-dimensional (2-D) grid and often represented as dots, squares, or rectangles. Each pixel is a sample of an original image, where more samples typically provide a more accurate representation of the original. When the original image from the camera is saved, compression occurs and information is lost. A popular image-compression method is Joint Photographic Experts Group (JPEG or .jpg) format for both Macintosh/Windows (MAC/WIN) operating systems, a common file format compresses the original, raw image file for easier storage. However, this photograph loses some quality every time it is compressed, saved, or altered. Loss of image quality may result in information loss beyond the point of forensic analysis to determine the image's accuracy. One method of addressing this issue is to immediately store the image on a server or other electronic storage device as the first copy of the original image.

Table 8.3 shows a comparison between the megapixel capacity of a camera and the maximum print size associated with the megapixel capacity. The numbers in the chart are derived from a standard formula for calculating print size and are readily available on the Internet. As mentioned earlier, investigators should use a camera that creates high quality 8-× 10-inch (200 mm by 250 mm) prints. According to the chart, megapixel capacities of eight or higher will produce these photos. The print-size settings are listed in **pixels per inch (ppi)**. The more ppi, the higher the resolution of the printed photos. Investigators should print at 300 ppi or higher for best photo quality. However, if this level of ppi is not available, higher megapixel size can compensate for lower ppi when printing.

	Table 8.3 Megapixels vs. Maximum Print Size Chart				
Megapixels	Pixel Resolution	Print Size at 300 ppi	Print Size at 200 ppi	Print Size at 150 ppi
3	2048 x 1536	6.82" x 5.12"	10.24" x 7.68"	13.65" x 10.24"
4	2464 x 1632	8.21" x 5.44"	12.32" x 8.16"	16.42" x 10.88"
6	3008 x 2000	10.02" x 6.67"	15.04" x 10.00"	20.05" x 13.34"
8	3264 x 2448	10.88" x 8.16"	16.32" x 12.24"	21.76" x 16.32"
10	3872 x 2592	12.91" x 8.64"	19.36" x 12.96"	25.81" x 17.28"
12	4290 x 2800	14.30" x 9.34"	21.45" x 14.00"	28.60" x 18.67"
16	4920 x 3264	16.40" x 10.88"	24.60" x 16.32"	32.80" x 21.76"
35mm film, Scanned	5380 x 3620	17.93" x 12.06"	26.90" x 18.10"	35.87" x 24.13"

ppi = pixels per inch
mm = millimeter

Image Creation and Exposure

In digital camera photography, light passes through the lens and is divided by a prism before reaching a digital sensor. The prism displays the image to the viewfinder and also onto the sensor, which in turn displays a preview image on the camera screen. When the picture is taken, the image displayed to the sensor is captured and stored on a digital memory card. For a properly exposed image, the timing of the exposure and size of the opening the light passes through must be controlled.

 **Bracketing**

In difficult lighting situations, many investigators will *bracket* their shots. **Bracketing** involves taking a photograph at the setting recommended by the camera meter and then manually adjusting the exposure setting one or two **f-stops** above and below the recommended exposure. DSLR cameras rarely require the use of this technique; however, when a camera's automatic settings are not producing the desired exposure results, bracketing can be useful. Some cameras give the user the option of automatically bracketing each shot a specific range above and below the metered setting **(Figure 8.14)**.

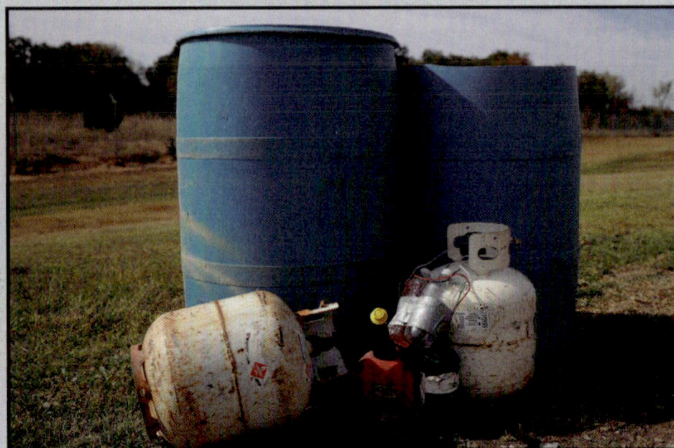

Figure 8.14 Bracketed photos can help an investigator choose a clearer photo after leaving an incident scene.

Electronic Devices as Cameras

In addition to taking notes and drawing sketches, electronic devices such as tablets and cellular phones may be used to take photographs. Apps or software built into the device allow the user to annotate the photograph with helpful information including direction of fire spread, evidence, or anything else the investigator considers important. These photographs and annotations can be transferred electronically as part of the case file.

Specialty Cameras

As technology advances, many cameras are becoming smaller and including more features such as:

- Built-in Wi-Fi
- Rugged design
- Anti-fog lenses
- Image stabilization

- Microscope modes
- Increased megapixel sensors
- Resistance to water, freezing, dust, or shock

- 4K video (video with a horizontal resolution at approximately 4,000 pixels)

Specialty cameras are increasingly available for documentation of scenes. Some cameras are capable of taking high dynamic range (HDR) images and use infrared (IR) to create a 3D model at the scene complete with a "dollhouse" type view. A tablet controls these cameras, and the data uploads into the cloud where it is processed using proprietary software. This software produces a 3D virtual walk-through model of the structure. The data also allows for automatic floor plan creation as well as accurate measurements from within the scene. The user can extract 2D images from the 3D model from any view. Links to the model can also be created and sent via email to other parties. These cameras are also capable of creating a virtual reality tour of the scene.

Another specialty camera used at scenes is a 360-degree camera. This camera can take video at over 5K (video with a horizontal resolution at approximately 5,000 pixels) and can also be used to capture

360-degree images in high resolution. This camera type is typically controlled by a cellular phone application via Wi-Fi.

Laser Scanners

Laser scanners are highly accurate and provide court-ready documentation. Scanners can include incident mapping software. Many of these scanners take HDR images and create a 3D point cloud of the structure by measuring large numbers of data points on surfaces such as contents, walls, ceilings, and other items. This point cloud is created by the use of laser or IR technology.

These types of scanners eliminate the need for personnel to reenter a scene after the initial examination. This provides additional protection against scene contamination and ensures investigators have what they need to analyze the scene and deliver compelling evidence. Another benefit of laser scanners is that investigators can collect a significant amount of information in a relatively short amount of time, which minimizes time on the scene and exposure to potentially harmful conditions.

Lenses

The function of a camera lens is to focus the light entering the camera. The **focal length** of a lens does not describe the actual length or size of the lens itself. It describes the distance between the camera's sensor and the point that light converges to form a sharply focused image when the lens is set to infinity. The DSLR camera allows the use of many different lenses of various focal lengths depending on the needs dictated by the situation. Variations in the focal length of a lens determine the size of the image focused on the sensor and the amount of the scene that is captured in the photograph (field of view).

For DSLR cameras, the 50 mm lens is considered the normal focal length because it provides an image that is close to what the human eye would see. Three lens ranges typically used with the DSLR camera include (**Figure 8.15**):

- **Normal** — Normal focal length: 35 to 55 mm

- **Wide angle** — Short focal length: 24 to 35 mm

- **Telephoto** — Long focal length: 105 mm and up

Investigators may need to take close-up photos of small objects with sharp detail. A macro lens or close-up adapter for the camera lens is recommended for

Figure 8.15 An investigator may choose to use different lenses to get a range of photos showing a point of interest in focus and with wider context.

these close-up photos. Many newer cameras will have a microscope mode to allow for extreme close-up photography. Other special lenses include the fish-eye with a short focal length and a wide field of view (17 mm and 180 degrees) and the super telephoto with a long focal length and narrow field of view (1 200 mm and 2 degrees). These lenses can be cost-prohibitive for some departments and have limited applications in normal fire-investigative photography.

Lenses may be fixed or zoom lenses. Fixed lenses do not allow the focal length to be changed. In order to zoom in or out on something with a fixed lens, the photographer must physically move closer or further away from the subject. Zoom lenses allows the photographer to remain in one place and change the focal length.

The most versatile lens a fire investigator can use is the adjustable focal length zoom lens with macro (close-up) capability. This lens allows the user to adjust the focal length of the lens to the desired detail. Zoom lenses are available in various ranges, but fire investigators may prefer a lens that is adjustable from wide-angle to normal or just above normal focal lengths such as 28 mm to 105 mm.

A special consideration when using DSLR cameras is vibration. DSLR cameras, like film cameras, are sensitive enough to detect the slightest movements, regardless of how still the photographer attempts to stand. DSLR manufacturers have reduced this problem by creating lenses with vibration control. Depending on the camera manufacturer, this feature may be referred to as vibration reduction (VR), image stabilization (IS), or optical stabilization (OS). Lenses and cameras with this feature adjust for movement as they focus on the intended image. Tripods or monopods offer the greatest stability for reduction of vibration.

Camera Settings

Three camera settings are particularly relevant to an investigator's work **(Figure 8.16)**:

- Shutter speed

- Aperture opening (**f-stop**)

- Film speed (ISO rating)

Each of the three settings can affect the camera's ability to capture a properly exposed image in many lighting and field conditions. Most DSLR cameras have automatic settings to select the proper positions for ambient lighting conditions. However, fire investigators should use a camera that includes a method for manually changing settings when necessary **(Figure 8.17)**.

Shutter Speed

The camera shutter controls the amount of time the sensor is exposed to light. The shutter speed is adjustable from several seconds in length down to thousandths of a second. This setting may be useful in conditions where stationary elements should remain the focus despite the presence of moving elements. For example, slow-shutter photography allows a photographer to manipulate the lighting in an area to highlight specific features. This technique's application for investigators is discussed later in this chapter in terms of "painting with light."

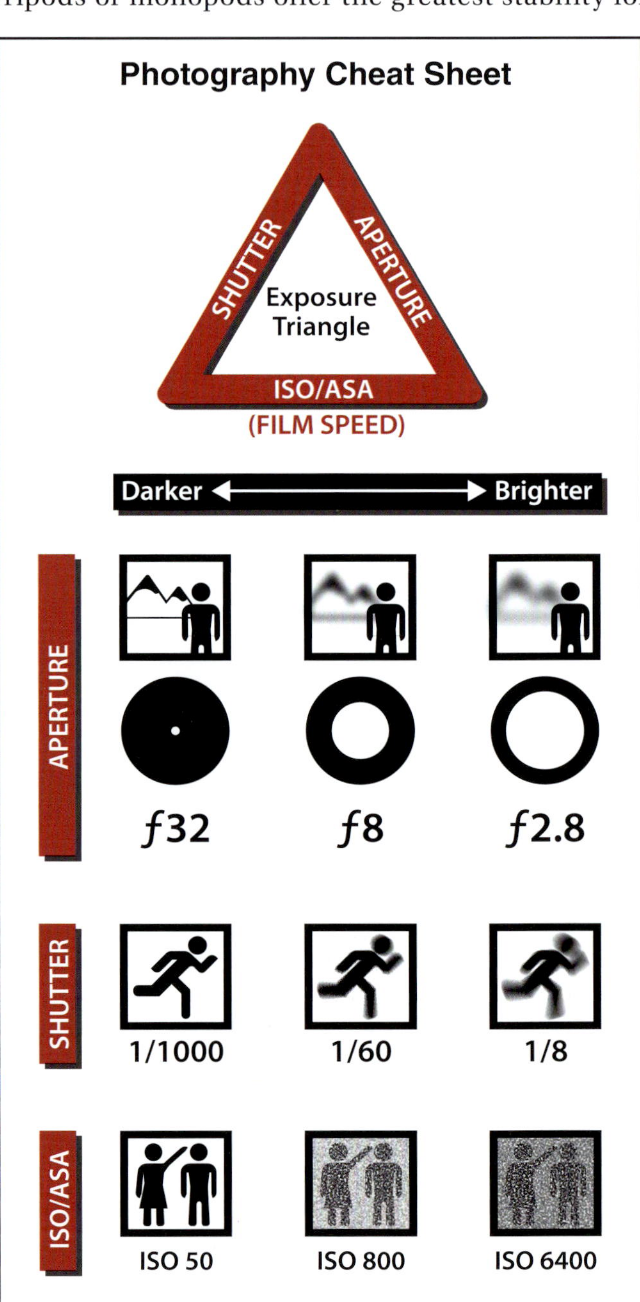

Figure 8.16 Manually changing f-stop settings affects how the camera receives light.

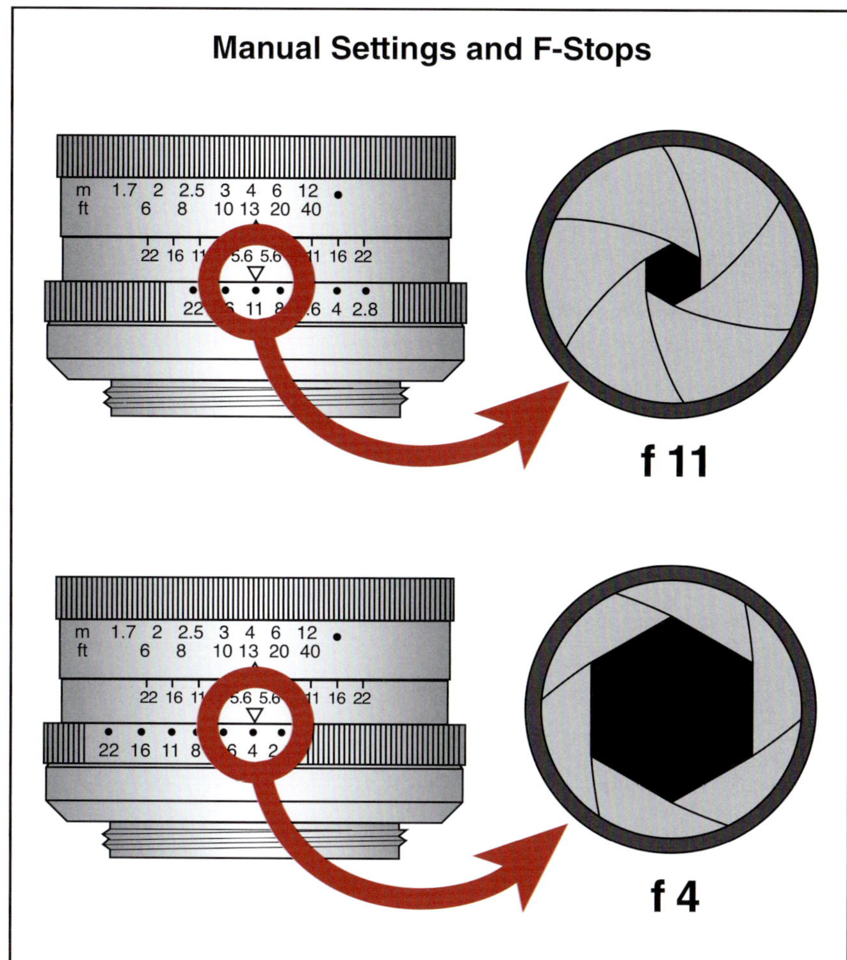

Figure 8.17 Fire investigators should use a camera that includes a method for manually changing settings when necessary.

Aperture Opening

The aperture opening is the size of the opening in the camera lens that allows the light to pass through. The aperture opening is adjustable and calibrated in adjustments known as *f-stops* (sometimes called *f-numbers*). The diameter of the aperture opening decreases as the f-stop increases. For example, an f-stop of 2.8 would allow in more light than an f-stop of 5.6. The f-stops available vary depending on the lens attached to the camera.

ISO Settings

The light sensitivity used for a 35-mm film photograph was rated according to the speed of film used to capture the photo. Film speed was set in accordance with standards established by the International Standards Organization (ISO). The higher the ISO rating of the film, the more sensitive it is to light. The more sensitive the film, the less light is required for proper exposure.

This ISO standard also applies to digital cameras. Digital cameras include sensing devices that automatically set the proper ISO setting based upon internal light meters. Many digital cameras also have settings that allow the photographer to manually set the ISO number.

ISO numbers for digital cameras use the same numbers for film stock and refer to the same light sensitivity. However, in digital cameras, changing this setting refers to changing the light exposure of the camera's sensor. ISO settings are most commonly available in the following speed ranges:

- **Slow speed** — Ranges from 25 to 100. These settings are good for use outside with natural lighting. They are not generally used for interior scene work.

- **Medium speed** — Ranges from 125 to 200. Useful for general-purpose photography where normal lighting is available. As with slower settings, the speed may not be suitable for interior fire-investigation photography.

- **High speed** — Ranges from 400 to 1,600 and higher. Provides the most versatility under varying lighting conditions; best choice for interior photographs of a scene. Be cautioned that a high ISO can affect image quality.

Depth of Field

The **depth of field** is the focus range in front of and behind the subject of the photograph **(Figure 8.18)**. Depth of field is a concept that fire investigators should understand as they photograph a scene. The factors that affect a photograph's depth of field include:

- **Aperture setting** — Depth of field increases as the aperture opening decreases (f-stop increases).

- **Focal length of lens** — Lenses with shorter focal length have longer depths of field. Wide angle lenses have deeper depth of field than telephoto lenses.

- **Distance to the subject** — The longer the distance between the camera and the subject, the greater the depth of field.

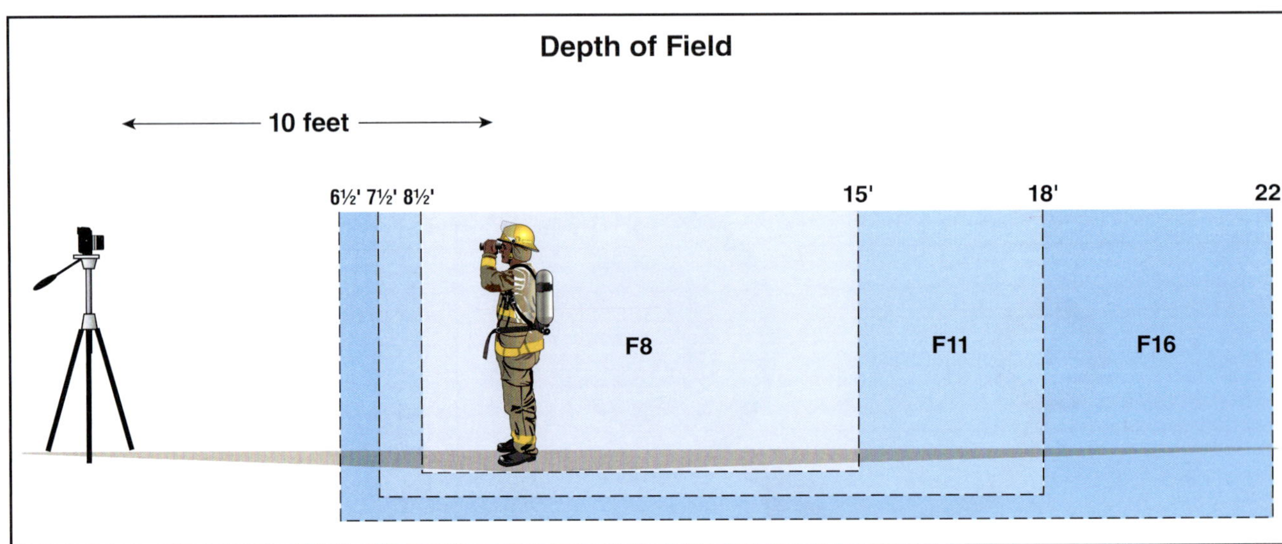

Figure 8.18 Various examples of depth of field and the f-stops associated with them.

The depth of field is greater behind the subject than in front. By setting the point of focus approximately one-third the distance into the overall image, an investigator will place the objects of interest into the area that is in focus.

The detail of a fire-scene photograph is determined by the depth of field and focus. The greater the depth of field, the more detail the photograph's viewer will see. The use of a stabilizing device, including tripods or monopods, can aid in obtaining a clear focus.

Lighting

Natural sunlight will rarely be adequate for fire investigation photography. Investigators must frequently rely on an artificial light source while documenting the scene, including an electronic flash unit compatible with the camera.

Other artificial light sources at the scene include scene lighting, the lighting provided in a building, or natural light redirected into dark areas **(Figure 8.19)**. These sources may not always be available. Investigators may carry a flashlight for tasks including focusing the camera on objects exposed to little light, and for viewing difficult-to-see camera settings. If this technique is used, the investigator should turn off the focusing light while photographing objects to avoid photo discoloration.

Figure 8.19 Large reflective surfaces can direct light into a dark area.

Using a Flash Unit

DSLR cameras feature systems in which the camera and flash unit are connected electronically and provide each other with data that supply an optimal exposure for any given photograph. Investigators should experiment with these units under conditions similar to those found at a scene to determine whether they provide acceptable results. Typically, automatic units have difficulty with bright areas or dark areas. In the bright background, the automatic flash may underexpose the photo. Large dark areas, such as burned-out rooms, may overexpose the photo (**Figure 8.20 a and b**). However, using the various DSLR camera settings can compensate for extreme conditions.

Figure 8.20a-b An investigator should recognize the difference between a photo that is likely underlit and one that plausibly has sufficient lighting.

Figure 8.21 Dark spaces at a fire scene may require manual camera adjustments to get proper exposure. Pointing the flash high and away from the subject is an example of a bounce flash.

When using a flash, investigators may want to consider removing the flash from the camera body. Holding the flash at an angle away from the camera body helps to reduce light reflection, increases the depth perception, and provides a better representation of the surface textures being photographed. This technique is useful when photographing heavy char. Another method used to reduce glare or reflection is **bounce flash**, in which the flash is directed at the ceiling above the subject and then the photo is taken. The light bounces off the ceiling and provides a softer image of the subject (**Figure 8.21**). This technique does not work well in fire-blackened areas.

Painting with Light

The technique of painting with light can be used to photograph a large area under low-light conditions (**Figure 8.22, p. 248**). The photographer places the camera on a tripod (or some other stable arrangement) and locks the shutter in the open position. The flash unit is then operated from several locations just outside the lens's field of view to illuminate the scene. Gently placing an opaque cover in front of the lens between flash operations reduces extraneous light and movement on the image.

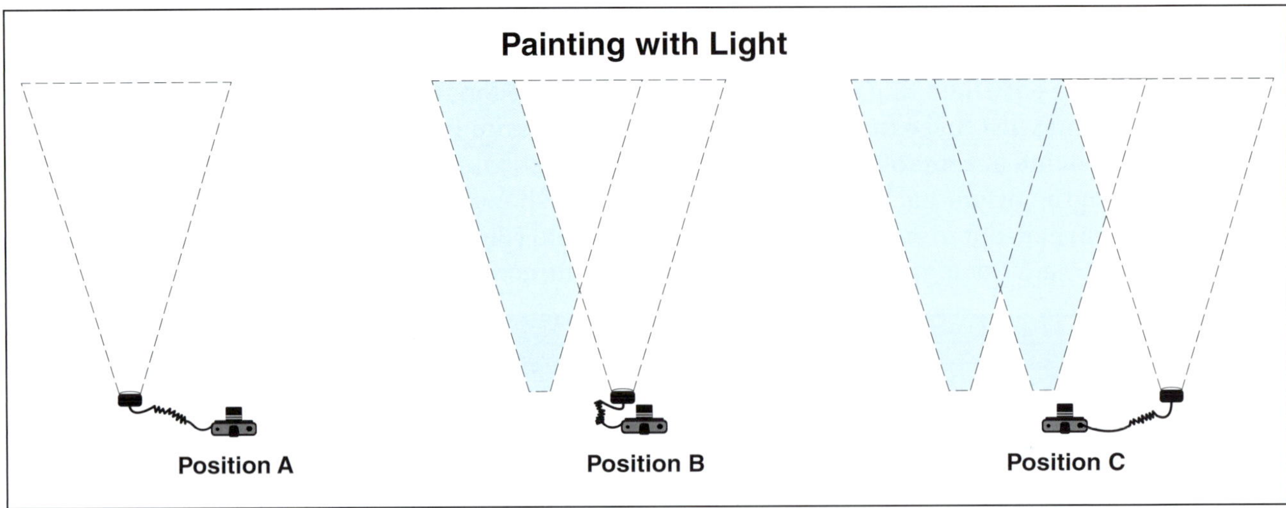

Figure 8.22 Using a long exposure, an investigator can strategically move the flash to paint with light.

Photograph Storage

Digital cameras require digital storage media to hold photos. Storage media are digital memory cards used much like a computer hard drive. CompactFlash®, SD, or micro-SD cards are the most common type of digital camera memory used with DSLR cameras because of their high transfer rates **(Figure 8.23)**. Individual DSLR cameras may use other forms of digital memory. Investigators should refer to their camera's documentation to select the correct memory format.

Figure 8.23 CompactFlash memory cards are common to many DSL cameras.

In addition to the media type, investigators should understand the memory capacity they will need for capturing photos. Memory cards commonly have capacities measured in megabytes (MB) or gigabytes (GB). In order to make an informed decision about the best digital memory to use, the investigator must understand the significance of these capacities. In computer data terms, one gigabyte is the equivalent of 1,000 megabytes. A JPEG captured at 12 megapixels (a common standard for newer DSLR cameras) is approximately 4 to 5 MB. Therefore, a 1-GB memory card can hold approximately 200 to 250 high-quality images like those needed for proper photo documentation of a scene.

Memory cards also utilize different data transfer speeds. A higher data transfer speed is recommended when using a DSLR for capturing video. Investigators should consider the needed type, size, and data transfer capabilities of the camera and its intended purpose.

Investigators using digital cameras should retain all photos, even if a particular photo is unidentifiable and a better replacement photo has been captured. Photographs taken on digital cameras are tagged with metadata that shows sequential numbers. When a photograph is deleted, gaps in the photo log show removal of documentation.

NOTE: Legal considerations are included in Chapter 9. Investigators must be familiar with their AHJ requirements regarding retention of materials collected and generated during an investigation or at the investigation scene.

After the investigator takes photos and collects them onto memory cards, the photos should be transferred to a more permanent location so that the memory card can be used again. Once photos have been removed from the camera, they should be permanently stored with the investigative file, preferably on an external storage device such as a server or external hard drive. Transferring photographs to cloud-based storage systems are also an alternative to ensure the data is not lost due to a hardware malfunction.

Photographs can also be stored on either a computer's hard drive or a central computer server, or copied to CD-ROMs or digital versatile disk-read-only memory (DVD-ROMs). Regardless of the storage device, all photos should be secure. CD-ROMs and DVD-ROMs are only as secure as where they are physically stored. Photos on hard drives and servers should be password-protected through user logins. The photo files can also be locked in an *archive-only* mode, a task that may be more time-consuming to prepare but secures the files from being edited, moved, or copied. If wishing to alter or edit photos, the investigator should work with a secondary set, and retain a set of original, unedited photos in every format taken at the scene.

Photo Editing

Numerous software applications allow computer users to edit digital photos. This software may be useful to fire investigators for enhancing photos. Enhancement features include contrast, brightness, darkness, color, or other effects such as adding circles or arrows to point out important sections of a photo. Before editing any photo, the investigator should preserve the original version. By preserving the original, the investigator or other personnel can explain and recognize any changes to the edited copy.

Recording Devices

NFPA 1033 (2022): 4.5.2

Video and audio recording devices can take a variety of forms, from standalone devices to multifunction electronic devices such as cellular phones and tablets. Regardless of the format used, an investigator who wants to use these devices must understand their operation and limitations, and when it is appropriate to use them.

Video Cameras

NFPA 1033 (2022): 4.5.2

While still photography is the primary means for documenting the scene, video cameras can also play a significant role. Where appropriate, investigators should use video to supplement the still photographs captured during the investigation. The best applications of video are for documenting the following items:

- Interviews
- Building contents
- Complex structures or floor plans
- Collection of special evidence where the collection sequence may be important to document
- Depicting events or sequences that take place over a reasonable period of time when the sequencing of events is important to the investigation

Use caution so that the images recorded provide an accurate representation of the scene. Also use the zoom feature with caution as it may provide a biased view of what the videographer wants the viewer to see. Newer video cameras include 360-degree video, which provide a unique and immersive way of videotaping the scene.

Audio Recorders

NFPA 1033 (2022): 4.5.2

Many investigators use portable audio recorders to capture field notes while in the fire building. The audio recordings can then be transcribed into written notes. Audio recorders can also be useful during interviews as an adjunct to an investigator's notes. Digital electronic devices often have a notes application preloaded; many notes apps have built-in audio recording software. An investigator may be able to record notes or interviews directly into the same app used for scene documentation. If planning to use a recorder in an interview, the investigator must comply with the laws and policies of the authority having jurisdiction (AHJ). Interview questions for a variety of incident types and scenarios are included in **Appendix D**.

As with any equipment, fire investigators should not rely only on audio recorders to capture critical investigative information. Always place backup information on paper or some other medium. Digital audio recorders or tablets with recording capabilities can upload recordings to a computer where the investigator can copy the interview or notes and keep it in the investigative file.

Review Questions

1. What are the most common methods that investigators use to document a scene?

2. What pieces of information should be included in the initial overall scene documentation?

3. What information should be included when documenting evidence?

4. What are four major guidelines that investigators should follow when constructing investigative field notes?

5. What types of information should be included in a scene sketch?

6. How are drawings, diagrams, and maps different from the initial sketch of the scene?

7. What types of resources might an investigator use to help create scene drawings?

8. How can an investigator use computer programs to enhance their scene diagramming capabilities?

9. How are maps used to provide further information during scene documentation?

10. From which vantage points and locations should an investigator photograph the scene?

11. What types of items and activities should be photographed during scene examination?

12. What guidelines should an investigator follow when photographing the scene?

13. What common camera features and settings should an investigator be familiar with prior to photographing the scene?

14. What types of specialty cameras and equipment can be used to photograph a scene?

15. What types of camera lenses are most useful to fire investigators?

16. What factors affect depth of field?

17. Which ISO settings on a camera is an investigator likely to use?

18. What types of lighting and lighting techniques can an investigator utilize to produce a properly exposed photograph of the scene?

19. How should scene photographs be secured after they are stored?

20. How can video and audio recording devices be beneficial in the documentation process?

Discussion Questions

1. How comfortable are you with operating a digital camera and taking photographs?

2. What skills do you feel you need to practice in order to take high quality scene photographs?

Chapter Key Terms

Aperture — In optics, an aperture is a hole or an opening through which light travels.

Bounce Flash — Lighting technique used to reduce glare or reflection by pointing the flash at a nearby surface rather than the subject.

Bracketing — Taking a photograph at the setting recommended by the camera meter and then manually adjusting the exposure setting one or two f-stops above and below the recommended exposure.

Computer-Aided Design (CAD) — Computer technology used to design, draw, or draft technical products, rooms, or entire buildings in two or three dimensions; it is also used to create animations from static drawings and convert existing drawings and diagrams into computer models.

Depth of Field — Range that is in focus both in front of and behind the subject of a photograph.

Digital Single-Lens Reflex (DSLR) Camera — Digital camera that uses a mechanical mirror system and prism to direct light from the lens to an optical viewfinder on the back of the camera.

Focal Length — Distance behind the lens where light from an object is sharply focused when the lens is set to infinity.

F-Stop — Setting on an SLR or DSLR camera that increases or decreases the camera's aperture size to adjust the amount of light entering the camera. *Also known as* Focal-Stop.

Geographic Information System (GIS) — Computer software application that relates physical features on the earth to a database to be used for mapping and analysis. The system captures, stores, analyzes, manages, and presents data that refers to or is linked to a location.

Megapixel — One million pixels; used as the reference for the number of pixels in a digital image. Also refers to the number of image sensor elements that a digital camera can display, which in turn describes the largest photograph that can be taken with that camera.

Pixel — Smallest item of information in a digital image; can be used to define the size of a digital image, the printing capabilities of a digital printer, or the screen resolution of high definition televisions and computer monitors.

Pixels per Inch (PPI) — Measurement of pixel density within a given space; refers to the amount of detail in or resolution of an image file.

Single-Lens Reflex (SLR) Camera — Film camera that uses a semi-automatic moving mirror system and allows the photographer to see exactly what will be captured on the film.

Venn Diagram — Graphic depiction of all possible logical relations between a finite collection of different sets.

NOTE: Documenting the scene is an ongoing process and each of these steps may be completed in a different order or may be repeated. The investigator should document the scene in a way that supports a systematic scene examination and follows AHJ policies and procedures.

Investigative Field Notes

NOTE: Field notes may be recorded using a variety of methods, including pencil and paper, audio recording, or tablet. These steps are a general guideline for things to include. Each incident is unique, and therefore may require more or less than the suggested contents. A systematic, thorough approach should be used regardless of incident type or requirements.

Step 1: Include date, time, and location of the incident.

Step 2: Systematically record observations of the scene **(Figure 8.24)**.

Step 3: Check that notes are accurate, complete, and concise.

Sketches, Drawings, Diagrams, and Maps

NOTE: These steps are a general guideline for things to include. Each incident is unique, and therefore may require more or less than the suggested contents **(Figure 8.25)**. A systematic, thorough approach should be used regardless of incident type or requirements.

Step 1: Include a key or legend to identify symbols and markings used.

Step 2: Include overall layout of the fire scene.

    a. Compartment walls and separations

    b. Entry points (doors, windows, etc.)

    c. Nearby paths or roads

    d. Water supply

Step 3: Include scene measurements and scale.

Step 4: Label location of relevant building features and contents.

    a. Furniture and other contents

    b. Evidence

    c. Witnesses

    d. Victims

    e. Fire detection or suppression systems and devices

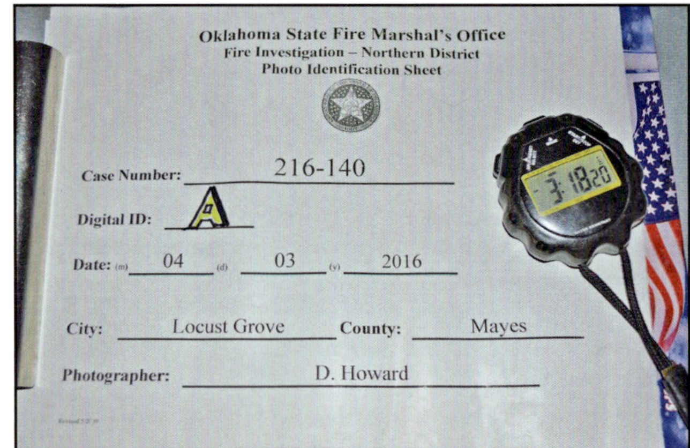

Figure 8.24 *Courtesy of Donny Howard, Agent, Oklahoma State Fire Marshal's Office.*

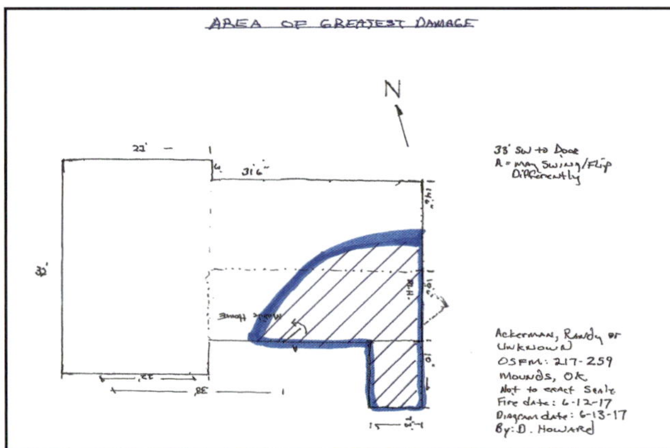

Figure 8.25 *Courtesy of Donny Howard, Agent, Oklahoma State Fire Marshal's Office.*

Step 5: Include indicators of fire behavior.

> **a.** Fire patterns
>
> **b.** Area/point of origin
>
> **c.** Direction of travel
>
> **d.** Smoke patterns
>
> **e.** Arc mapping

Step 6: Include directions, methods, and effects of fire suppression activities.

Photographs

NOTE: These steps are a general guideline for things to include. Each incident is unique, and therefore may require more or less than the suggested contents. A systematic, thorough approach should be used regardless of incident type or requirements.

Step 1: Include a log describing each photograph **(Figure 8.26)**.

Step 2: Include measurements for each item, as necessary.

Step 3: Photograph bystanders, if applicable.

Step 4: Photograph the outside of the structure.

> **a.** Location of the structure on the property
>
> **b.** Relevant landmarks
>
> **c.** Visible utilities
>
> **d.** All corners/sides of the building
>
> **e.** Position, condition, and security of exterior doors and windows
>
> **f.** Fire and explosion damage
>
> **g.** Broken windows and crazing or soot on glass fragments
>
> **h.** Exterior evidence (in position, mid-range, and close-up)
>
> **i.** Apparent routes of fire travel

Step 5: Photograph the inside of the structure.

> **a.** Begin at the point of entry and move systematically throughout
>
> **b.** Room features: Doors, walls, windows, ceilings, floors
>
> **c.** Photos taken from each corner to show multiple sides of the room
>
> **d.** Room contents

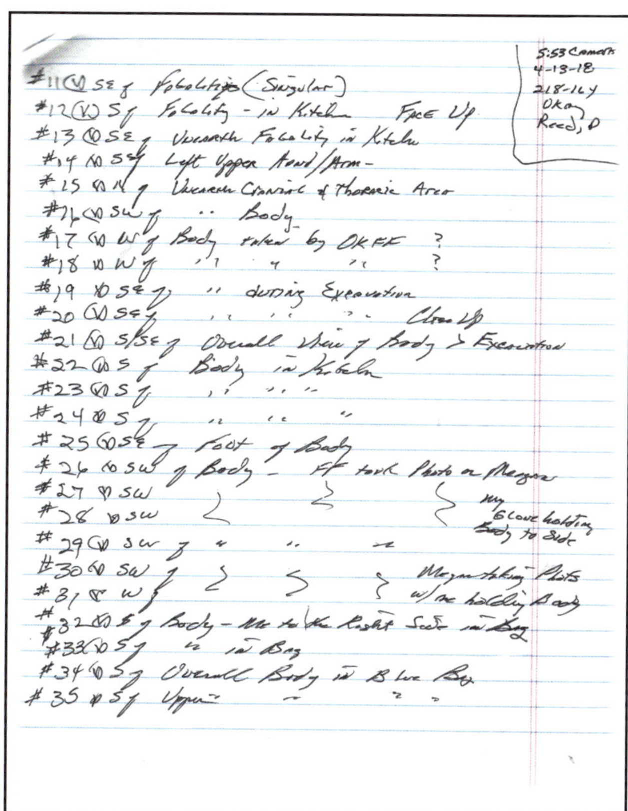

Figure 8.26 *Courtesy of Donny Howard, Agent, Oklahoma State Fire Marshal's Office.*

e. Interior evidence (in position, mid-range, and close-up)

f. Appliances near the area of origin (including position of controls)

g. Electrical breaker or fuse panels (including position of breakers and condition of fuses)

h. Ignition sources within the area of origin

i. Victims

Step 6: Photograph the investigative process as evidence is collected and fire scene reconstruction occurs.

Step 7: Rephotograph any areas as necessary.

Photo courtesy of Donny Howard, Agent, Oklahoma State Fire Marshal's Office.

Chapter Contents

NFPA 1033 JPRs addressed in this chapter

This chapter provides information that addresses the following job performance requirements (JPRs) of NFPA 1033, *Standard for Professional Qualifications for Fire Investigator* (2022):

4.1.7	4.2.2	4.2.6	4.4.2	4.4.4	4.6.3
4.2.1	4.2.3	4.4.1	4.4.3	4.4.5	

This chapter provides information that addresses the following course outcomes for the FESHE courses:

Fire Investigation I (C0283)

1. Demonstrate the importance of documentation, evidence collection, and scene security process needed for successful resolution.
3. Identify the responsibilities of a firefighter when responding to the scene of a fire, including scene security and evidence preservation.

Learning Objectives

1. Identify types of evidence commonly found at fire and explosion incidents. [NFPA 1033, 4.1.7, 4.2.1, 4.2.2, 4.2.3, 4.2.6, 4.4.1, 4.4.2, 4.4.4]

2. Explain considerations for collecting, documenting, and preserving evidence. [NFPA 1033, 4.1.7, 4.2.2, 4.2.3, 4.2.6, 4.4.1, 4.4.2, 4.4.3, 4.4.4, 4.4.5]

3. Explain the importance of preventing contamination to evidence. [NFPA 1033, 4.1.7, 4.2.1, 4.2.2, 4.2.3, 4.2.6, 4.4.1, 4.4.2]

4. Describe various types of tools and equipment used to collect evidence. [NFPA 1033, 4.1.7, 4.2.1, 4.2.2, 4.2.3, 4.2.6, 4.4.1, 4.4.2]

5. Describe methods of handling and collecting various types of evidence. [NFPA 1033, 4.1.7, 4.2.2, 4.2.3, 4.2.6, 4.4.1, 4.4.2, 4.4.3, 4.4.4]

6. Skill Sheet 9-1: Collect evidence. [NFPA 1033, 4.1.7, 4.2.6, 4.4.2, 4.4.3, 4.4.4, 4.4.5, 4.6.3]

Chapter 9
Evidence Collection and Preservation

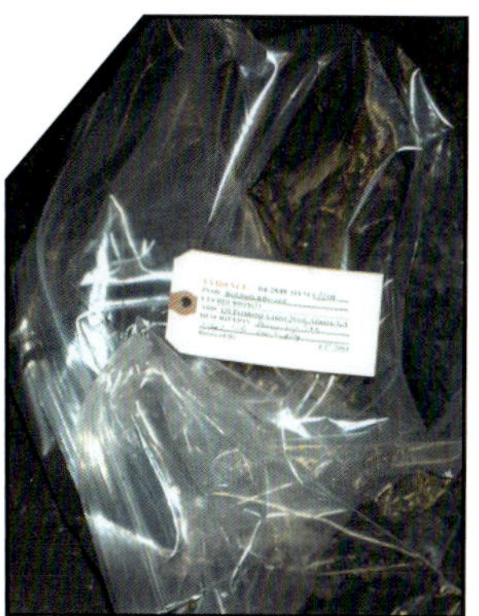

Courtesy of Yates & Associates.

The job of a fire investigator may vary, depending upon the scope of his or her assignment. Descriptions of the investigator's tasks include:

- Identifying Evidence
- Collecting and Preserving Evidence
- Preventing Contamination
- Handling Evidence

Identifying Evidence

NFPA 1033 (2022): 4.1.7, 4.2.1, 4.2.2, 4.2.3, 4.2.6, 4.4.1, 4.4.2, 4.4.4

Evidence collected by fire investigators becomes the data used to develop and test opinions or hypotheses regarding the cause of a fire or explosion. During the scene examination, as described in Chapter 7, Initial Actions in the Investigative Process, fire investigators perform the following actions in roughly this sequence:

1. Recognize items of importance to the investigation.
2. Protect those items in place.
3. Document the location of items.
4. State when items were discovered.
5. Properly collect items to analyze.
6. Transport collected items to a secure facility or lab.
7. Ensure the security and integrity of the items that have been collected.

Identifying Potential Evidence

NFPA 1033 (2022): 4.2.1, 4.2.2, 4.2.6

As the examination is conducted, fire investigators should always watch for potential evidence that may help determine the area of origin, the cause of the fire, and the identity of who may be responsible for the fire. Examples of this type of evidence include:

- Tool marks
- Foot or tire impressions
- Odor of ignitable liquids
- Indications of forced entry **(Figure 9.1, p. 258)**
- Contents in unusual locations
- Tampering with building utilities
- Absence of contents in the building
- Debris in unexplained locations
- Indications of multiple areas of origin
- Indications of multiple points of ignition
- Empty containers in the area of the scene
- Incendiary or timing devices in the building
- Trace evidence found on objects and surfaces

- Tampering, or improper, or insufficient fire protection systems

- Explosion evidence, such as clean glass shards, found some distance from the structure

- Blocked entrances that impeded fire department response or fire-suppression operations

- **Trailers** formed as a result of combustible material or ignitable liquid being used **(Figure 9.2)**

- Appliances or other devices that could have been the source of heat that ignited the fire **(Figure 9.3)**

- Fire patterns that indicate the origin and travel of the fire (see Chapter 12, Area-of-Origin Determination)

- Prefire conditions that could have increased the damage caused by the fire (open windows, holes in floors, doors open or removed, and contents piled or placed in such a way to increase the fuel load)

The scientific method used by fire investigators demands that investigators remain objective and maintain an open mind while collecting and analyzing data. Although evidence that tends to indicate a potential suspect or motive may be discovered while examining the scene, investigators are cautioned against using this evidence to the exclusion of other evidence that might indicate that the fire was not an incendiary fire.

Collecting Physical Evidence

When making the decision to collect **physical evidence**, fire investigators must determine whether the material is relevant to the investigation. For example, an investigator may collect tangible evidence such as a device and materials that show specific fire patterns to demonstrate that a person used the device to set the fire. In contrast, nonphysical evidence includes witness observations and statements, and records and documents obtained after the fire. Investigators may also collect **exclusionary evidence** to show that a particular device or scenario can be disregarded from the ignition or fire-spread scenario.

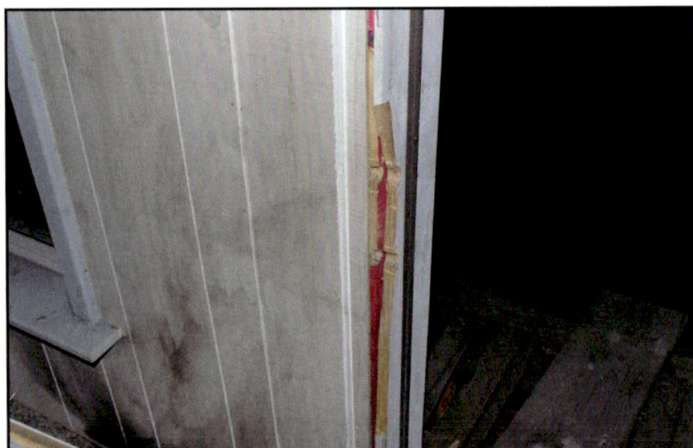

Figure 9.1 Investigators should identify evidence of forced entry at a scene. *Courtesy of Donny Howard, Agent, Oklahoma State Fire Marshal's Office.*

Figure 9.2 Combustible trailers like the burned areas of this yard where ignitable liquids were used to intentionally set a fire. *Courtesy of Donny Howard, Agent, Oklahoma State Fire Marshal's Office.*

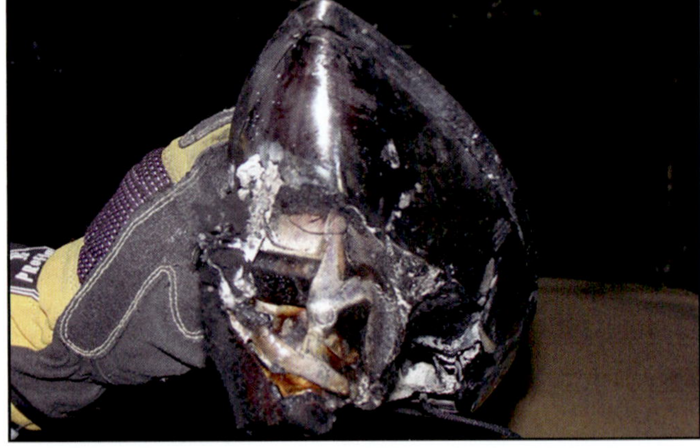

Figure 9.3 During the scene examination, fire investigators may discover small appliances that could be an ignition source. *Courtesy of Nicole Brewer, Portland Fire Investigations.*

Most evidence collected at a fire scene is in the form of **artifacts**: remains of materials involved in the fire, possibly related to the ignition, development, or spread of the fire or explosion. In relation to criminal investigations, physical evidence may include incendiary devices, ignitable liquids, and fingerprints.

The physical evidence found at a scene can take many forms including:

- Hairs/fibers
- Broken glass
- Bodily fluids
- Liquid containers
- Papers or documents
- Bedding/clothing/fabrics **(Figure 9.4)**
- Tire or shoe impressions **(Figure 9.5)**
- Tool marks on doors or windows
- Liquids thought to be accelerants
- Electrical components and systems
- Cigarette butts or other smoking materials
- Portions of incendiary or explosive devices
- Paint chips or other pieces of interior finishes
- Appliances involved in the ignition of the fire
- Pieces of charred wood, carpet, or other fuels involved in the fire
- Noncombustible items (metal objects such as door hinges or exposed nails) **(Figure 9.6)**

A fire investigator conducting the scene examination should constantly be looking for any physical evidence that will assist in the determination of the origin, cause, and responsibility for the fire, its damage, and injuries or deaths. The initial investigator on the scene has the best opportunity to recover these items in the most usable condition. As part of the investigative process, materials identified as potential evidence should be protected from further damage until they can be documented and collected for preservation and laboratory analysis.

Securing Evidence

Many accidental fires result in legal action of a civil nature. Thus, the results of investigations into noncriminal incidents may be used in court, and the materials investigators identify and collect as evidence could have a significant effect in these trials. While additional materials or items may be collected by investigators who view the scene after the initial examination, the first investigator has the best opportunity to identify and preserve critical incident-related information and materials.

Figure 9.4 Bedding and clothes are a type of evidence that investigators may need to examine, document, and preserve. *Courtesy of Nicole Brewer, Portland Fire Investigations.*

Figure 9.5 Tire impressions are a type of evidence that investigators may need to examine, document, and preserve. *Courtesy of Donny Howard, Agent, Oklahoma State Fire Marshal's Office.*

Figure 9.6 Remaining door hardware is a type of evidence that investigators may need to examine, document, and preserve. *Courtesy of Nicole Brewer, Portland Fire Investigations.*

The fire investigator who conducts the initial scene examination may be responsible for handling potential evidence intended for further analysis in the following ways:

- Storage
- Collection
- Preservation
- Identification
- Transportation
- Documentation
- Submission of items to a testing facility

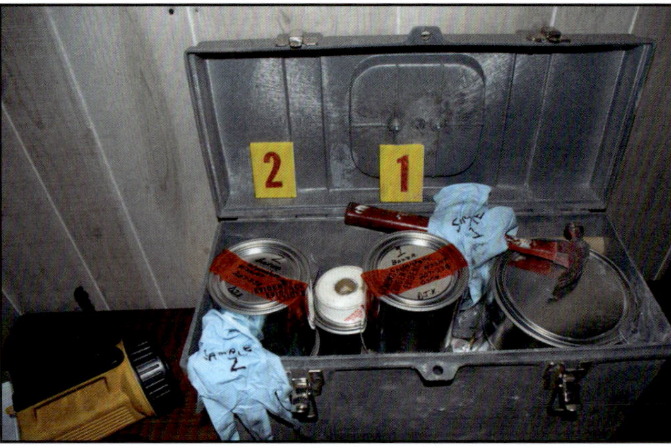

Figure 9.7 Investigators may use lockable toolboxes to secure evidence when leaving a scene. *Courtesy of Donny Howard, Agent, Oklahoma State Fire Marshal's Office.*

Investigators use a number of methods to maintain the security of all items collected. Some investigators put evidence containers in large toolboxes provided with locks; others may use lockable boxes in their vehicles or trunks **(Figure 9.7)**. Whatever method investigators use, they must be able to document that all materials collected remain secure and in their custody until securely stored or released to the custody of someone else. Using evidence tags or placing the information directly on the package in which the material is stored assists in tracking items. The evidence log from NFPA 921, Annex A, provides one example of how to track the custody of items of evidence.

Maintaining Chain of Custody

Maintaining the **chain of custody** of evidence is an essential function of a fire investigator. The chain of custody tracks an item of evidence from the time it is identified as such, until its **final disposition** **(Figure 9.8)**. Each person who has possession of an item of evidence must be able to attest to the fact that the item was not subject to tampering or contamination while it was in his or her custody.

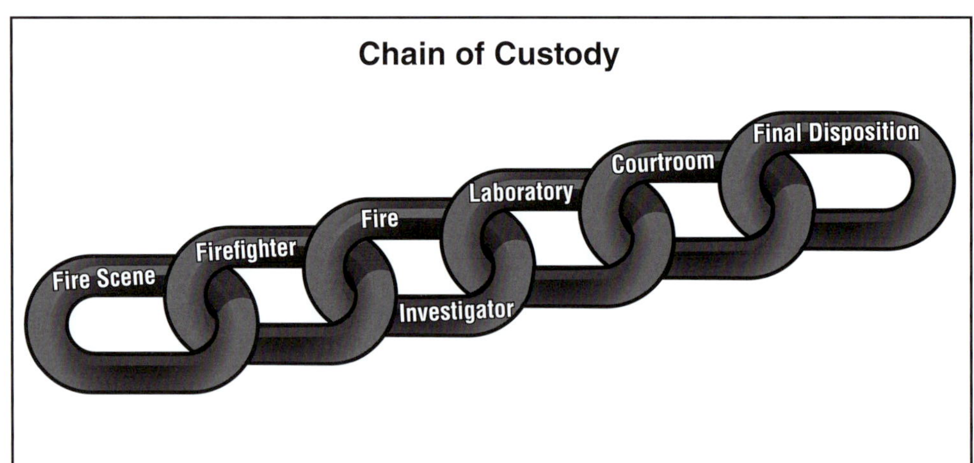

Figure 9.8 The chain of custody must remain well documented from the fire scene through the evidence's final disposition.

Anytime the custody of an item changes, the following information should be obtained:

- Date and time of item transfer and its new location
- Name and contact information of the current and prior custodian
- Condition of the item or its packaging when transferred to the new custodian
- Description of any modification, handling, testing, or other alteration that occurred while the item was in the custody of the current custodian

Chain-of-custody issues extend beyond the physical evidence collected at a fire scene. All incident-related evidence (including photographs, documents of evidentiary value, and any other items that support the findings of the investigation) should be addressed by the chain of custody. Chain of custody, however, depends on the case and policies of the jurisdiction that an investigator represents. Once an item is considered evidence, it should be properly secured, and its handling and transfer of custody should be documented from the time it comes into an investigator's possession until it reaches its final disposition.

The final disposition of a piece of evidence varies from case to case and jurisdiction to jurisdiction. If evidence is to be disposed of, the investigator must follow all appropriate policies in his or her jurisdiction. Disposal of evidence may include legal requirements including judicial review. The investigator may also identify and use any outside disposal service the AHJ recommends or requires.

Collecting and Preserving Evidence

NFPA 1033 (2022): 4.1.7, 4.2.2, 4.2.3, 4.2.6, 4.4.1, 4.4.2, 4.4.3, 4.4.4, 4.4.5

Fire investigators should follow physical evidence collection and preservation procedures established by their agencies. Prior to collecting evidence, investigators should ensure that they have clean, properly stored tools, equipment, and containers for each type of evidence. Properly contained evidence will not be further damaged, altered, destroyed, or contaminated.

If materials will be sent to a forensic laboratory for analysis, the sampling protocols of that laboratory should also be followed. Most laboratories provide guidance on the type of packaging in which they would like to receive materials and the amount of the material that they need to conduct their analyses. Investigators are encouraged to contact the laboratory to confirm these protocols or ask for guidance before collecting evidence. ASTM E 1188 2017, *Standard Practice for Collection and Preservation of Information and Physical Items by a Technical Investigator,* also provides useful information for fire investigators. **Skill Sheet 9-1** shows the steps for collecting evidence.

Choosing a Collection Method

During the evidence collection process, an investigator works to collect and preserve evidence that will be used to support a hypothesis about the origin and cause, and responsibility of the fire. Regardless of the type of material, investigators should not touch items with their bare hands. They should use tweezers or a sheet of clean white paper to retrieve an individual item from where it was found. Items that are wet should be air-dried before packaging.

NOTE: Investigators often have to improvise or use their best judgment when collecting and preserving evidence because of the variety of materials and circumstances they face in the course of conducting their investigations.

The methods used for collecting evidence and samples depend on material characteristics including:

- Location within the scene
- Size and physical characteristics
- Physical state (solid, liquid, or gas)
- Fragility, particularly if fire damaged
- Potential to evaporate or decompose after collection

Collection of Gases

The collection of gaseous samples for analysis is not a task that fire investigators usually encounter. If gaseous samples must be taken, the most effective method is to use a gas-sampling device that draws a small amount of the material over a charcoal filter.

If an item is intact, relatively undamaged, and not covered with debris, the process is simple: The item is documented, tagged, and placed into a container for transport from the scene (**Figure 9.9**). Collecting materials covered with debris and possibly damaged as a result of the fire is more complicated. The investigator should carefully expose the item, and photograph it and its surroundings at different stages of the excavation (**Figure 9.10**). When exposing the item, take care to not damage or disturb it. For items or materials that are very fragile, very large, or in some other way difficult to collect, fire investigators should consult with the forensic laboratory or a trained evidence technician for assistance.

Evidence Requiring Technical Expertise

While investigators should have a working knowledge of the various types of evidence collection described in this chapter, they should

Figure 9.9 Evidence tags on each collected piece of evidence are examples of unique identifiers. *Courtesy of Yates & Associates.*

Figure 9.10 Fire scene debris can include layers of fiberglass insulation disturbed by the fire and/or overhaul operations. *Courtesy of Donny Howard, Agent, Oklahoma State Fire Marshal's Office.*

also recognize when evidence should be collected by technical experts. For example, fire investigators may not have the expertise to properly photograph blood spatter and may require an expert to take and analyze such photographs. Technical experts are usually needed when collecting the following types of evidence:

- Hair
- Body fluids
- Fingerprints
- Fibers and textiles
- Tire or shoe impressions

Evidence Identifiers

Most investigators assign a unique identifier to each item of evidence located and collected **(Figure 9.11)**. Identifiers can include a number or letter, or combination of letters and numbers. This identifier becomes important later if the custody of the material is transferred or the material is placed in storage.

When documenting an item of evidence with an identifier, the investigator records the item in the investigation notes and on the scene diagram. Taking overall, mid-range, and close-up photographs of evidence shows its placement and size in relationship to other objects within the area.

Figure 9.11 Investigators assign an identifying number or letter to each piece of evidence for easy cross-reference. *Courtesy of Donny Howard, Agent, Oklahoma State Fire Marshal's Office.*

A fire investigator should also generate an evidence log sheet that lists each item collected. The log will become an essential portion of the documentation of the chain of custody of evidence collected during the investigation. See NFPA 921 for an example of an evidence log sheet.

Comparison Samples

During evidence collection, a **comparison sample** may be selected to show a difference between the ambient conditions at a scene and the selected evidence sample. Laboratories may need the comparison sample to rule out the presence of additional combustible or ignitable materials at a scene. The presence of hydrocarbons alone may not be definitive, since materials such as carpeting, wood, plastics, and roofing materials release volatile hydrocarbons under sufficient fire conditions.

The comparison sample must be uncontaminated and should be the same type of material that was selected as evidence. Investigators should obtain a sample from an unexposed area well away from the location of the primary sample. The locations from where the evidence and the comparison samples were removed should be documented.

For example, when selecting a section of carpet as evidence, a comparison example may be taken from the same carpet from under a piece of furniture such as a dresser. Comparison samples are packaged in the same fashion as the primary sample and marked as a comparison sample. When multiple points of origin are suspected with varying types of materials sampled, a comparison sample of each type of material should be collected.

Evidence Transportation and Storage

When transporting evidence from the fire scene to the lab for examination, the investigator should either directly deliver the evidence, or give it to a secure courier service. Evidence should be stored until it is no longer needed. The evidence should be stored in an area with secure access and environmental controls such as protection from direct sunlight or moisture. Evidence should be logged into the storage area, and chain of custody documentation should accompany each piece of evidence.

In addition to the investigator who collected and delivered the evidence, other parties that may require access to the evidence for examination and testing include:

- Lawyers
- Manufacturers
- Insurance adjusters
- Private fire investigators

Maintenance and Destruction of Notes

At the conclusion of an investigation and after completion of the incident report, notes as well as any other form of documentation should be kept with the case file **(Figure 9.12)**. Investigative notes may include information not included in the written report. Other parties may require these notes, which may be subject to subpoena should criminal or civil litigations arise.

Some agency policies mandate that investigative notes be destroyed. Investigators should know, confirm, and adhere to their agency's policy.

NOTE: Investigators should be aware that in a legal setting, the defense team may allege that an investigator who destroyed investigative notes has functionally destroyed data beneficial to the accused parties. Such allegations could have a negative effect on the perception of the investigator's opinions or testimony.

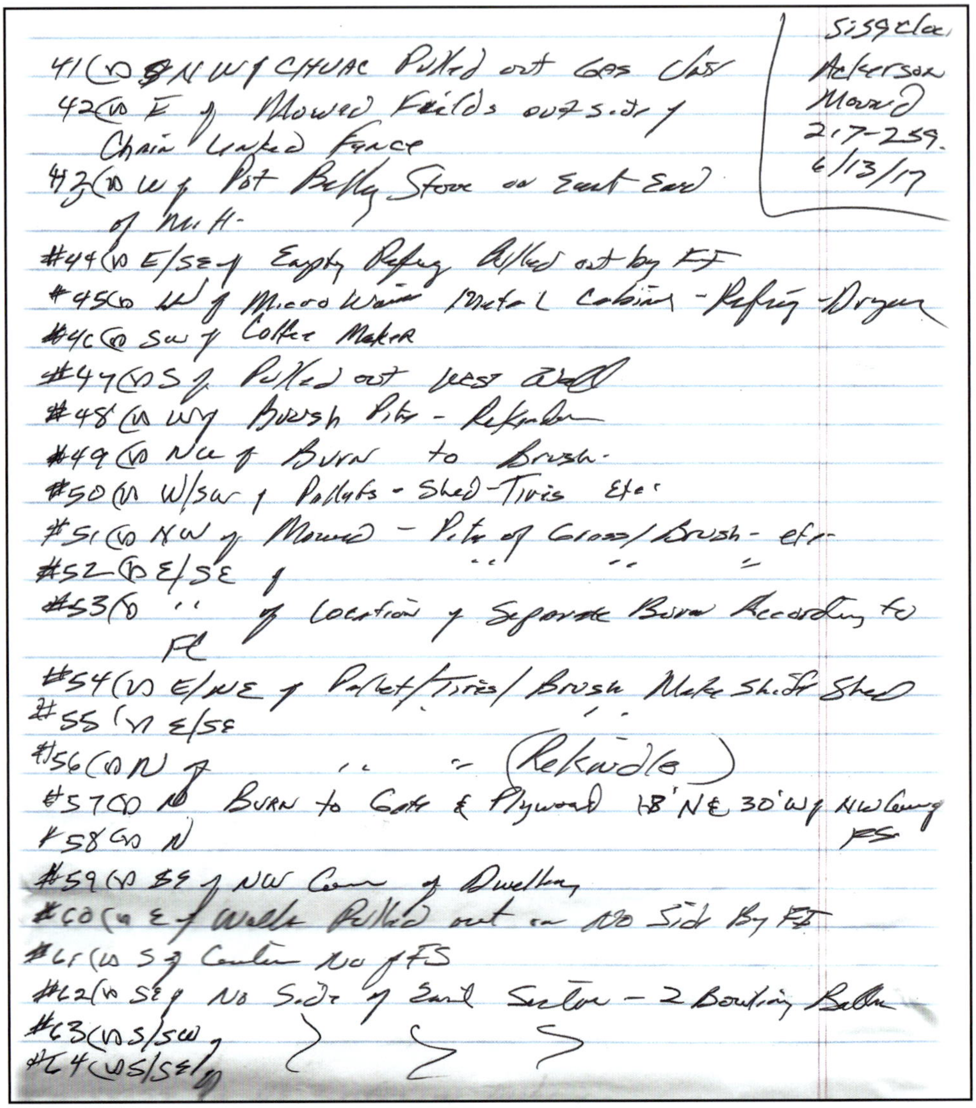

Figure 9.12 Investigation notes should be kept with the appropriate case file. *Courtesy of Donny Howard, Agent, Oklahoma State Fire Marshal's Office.*

Preventing Contamination

NFPA 1033 (2022): 4.1.7, 4.2.1, 4.2.2, 4.2.3, 4.2.6, 4.4.1, 4.4.2

The ultimate objective of collecting evidence is to have it or its significant details available in any court proceedings that might result from the investigation. Thus, once the material is collected, it must be protected from contamination or destruction before it is analyzed or presented in court. To prevent **cross-contamination**, a fire investigator must either use disposable items during the collection process or thoroughly clean all tools between each use.

Contamination of the scene or samples being collected at the scene can occur from many sources including:

- Improper collection methods

- Use of contaminated containers to store evidence

- Tools used by firefighters and investigators at the scene

- Fuel-powered equipment used at the scene during fire-suppression and investigation operations

- Cross-contamination by personnel on the scene, for example, tracking ignitable liquid residue on boots from one room to another **(Figure 9.13)**

- Protective equipment worn by firefighters and investigators working at the scene including disposable gloves that may be contaminated while in pockets

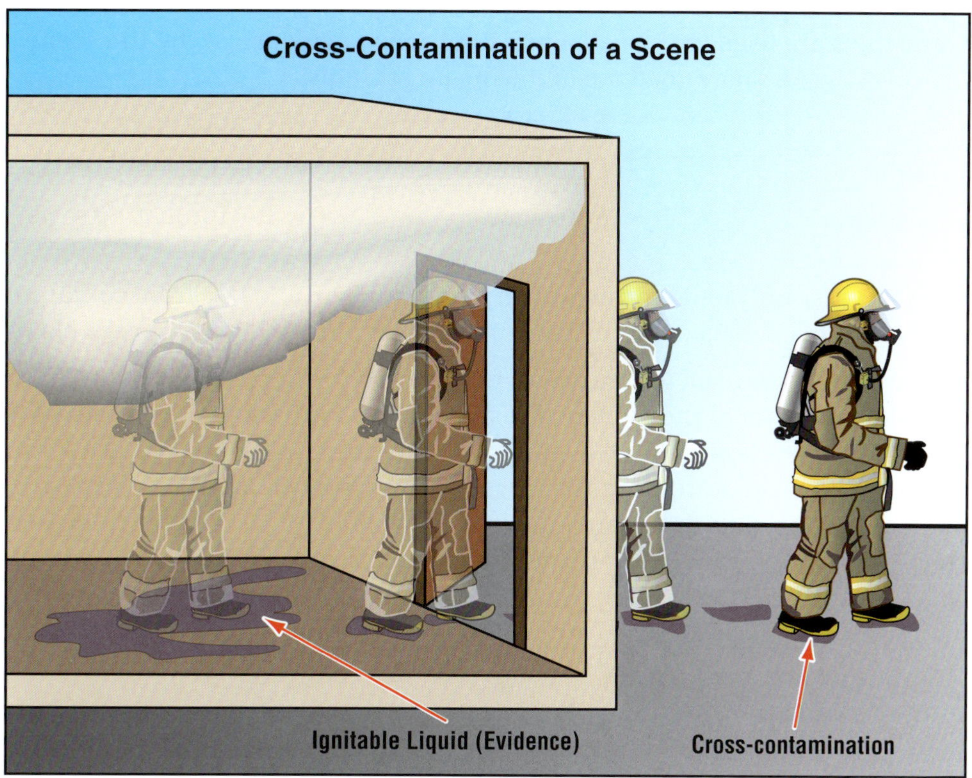

Figure 9.13 Investigators should watch for patterns that indicate fire suppression activities may have cross-contaminated a scene.

Containers

Fire investigators should have a supply of clean and unused containers and packaging materials for use in the collection and preservation of materials, such as:

- Bags **(Figure 9.14)**

- Boxes

- Glass jars

- Metal cans

Investigators also need materials to assist in packaging evidence when it is collected. These materials can include:

- Packaging tape

- Permanent markers to mark containers

Figure 9.14 The investigator should know the appropriate uses for each type of evidence collection container, including plastic bags. *Courtesy of Yates & Associates.*

- Filler material to pack glass containers and other fragile items to prevent breakage

- Plain wrapping paper to protect materials such as glass sheets, paper bags, and boxes for other items such as electrical components

- Shrink wrap to protect and secure appliances and other items where components may become dislodged during collection and transport.

Tools and Equipment

Many tools are used in the collection of physical evidence at the fire scene. In addition to hand tools, fire investigators may need the following larger tools:

- Hoes
- Drills
- Shovels
- Brooms
- Crowbars
- Squeegees
- Cutout tools
- Garden-type rakes
- Reciprocating saws

NOTE: Use only tools and containers that are clean and free from contaminants. Tools that have been used to collect evidence for the presence of an ignitable liquid must be cleaned prior to reuse.

These tools are used to clear away fire debris during the scene examination and to make collection and identification of the evidence easier and more efficient. Multiple-use tools used in the evidence collection process should be cleaned of contaminants after each use with a suitable detergent, and flushed with clean water, and thoroughly dried using clean paper towels or other suitable materials to prevent rusting. One detergent known to meet many jurisdictions' requirements is Simple Green Pro HD Heavy Duty Cleaner™. To prevent the cleaning process from contaminating evidence, investigators should clean tools away from the collection site. See Appendix E for more information on tools used during an investigation.

Fuel-Powered Tools

Fluid leaks, faulty equipment, and exhaust fumes from fuel-powered tools can potentially contaminate evidence at a scene. Once fire suppression operations conclude, investigators should document where firefighters used the tools and whether they refueled the equipment. If investigators must use fuel-powered tools during the investigation, they should take precautions to prevent contamination, such as wearing different gloves while fueling the device, and avoiding spilling fuel onto fire debris or clothing.

Multiple Use Protective Equipment and Clothing

The protective equipment worn by firefighters and investigators can contaminate evidence at the scene. Investigators should be careful to eliminate as many sources of contamination as possible.

Boots

While collecting evidence, investigators should carefully clean their boots before entering an area. As with tools, investigators can flush boots with water and a suitable detergent to remove foreign materials.

Gloves

Fire investigators should avoid using fire fighting gloves while collecting samples. Instead, a pair of unused exam gloves should be used for each sample collected. Discard the gloves after each sample is collected. Care should be taken to ensure that the unused gloves have not been contaminated before their use.

NOTE: Pockets of personal protective equipment, such as turnout gear or station wear, may contain trace amounts of petroleum products from gloves placed there after being worn while fueling vehicles.

Do not place discarded gloves in the same container as the potential evidence. To document that new gloves were used for each sample taken, some investigators number their gloves and the evidence containers sequentially. Then, they photograph the gloves with the sealed evidence container at the location the material was collected **(Figure 9.15)**. The used gloves can then be discarded.

NOTE: Some laboratories require that used gloves and disposable tools be retained in separate containers. Investigators should consult the policies of their laboratories or agencies on used gloves and disposable tools.

Handling Evidence

NFPA 1033 (2022): 4.1.7, 4.2.2, 4.2.3, 4.2.6, 4.4.1, 4.4.2, 4.4.3, 4.4.4

Depending on the investigator's jurisdiction, the types of evidence collected from a fire scene may vary. For example, some jurisdictions may have fire investigators collect fingerprints, footwear or tire track impressions, samples of a suspected ignitable liquid, and electrical evidence as part of their job requirements. In other jurisdictions, the investigator may not be tasked with collecting any physical evidence. That function may be delegated to another agency.

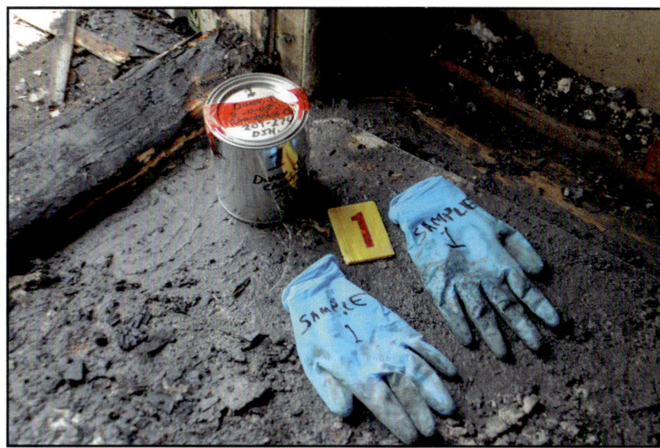

Figure 9.15 Gloves used to handle evidence should be marked in the same manner as the evidence, and photographed. The gloves can then be stored as evidence or discarded, per the AHJ. *Courtesy of Donny Howard, Agent, Oklahoma State Fire Marshal's Office.*

Objects Containing Trace Evidence

The **Locard's Exchange Principle** holds that any contact with a scene leaves something behind, and takes something away. This is an important principle in understanding trace evidence in forensic science.

Laboratory analysis may be able to work with trace amounts of material. Many solid objects may collect, absorb, or otherwise retain types of evidence including:

- Blood
- Fibers
- Saliva
- Fingerprints **(Figure 9.16)**
- Contact deoxyribonucleic acid (touch DNA)

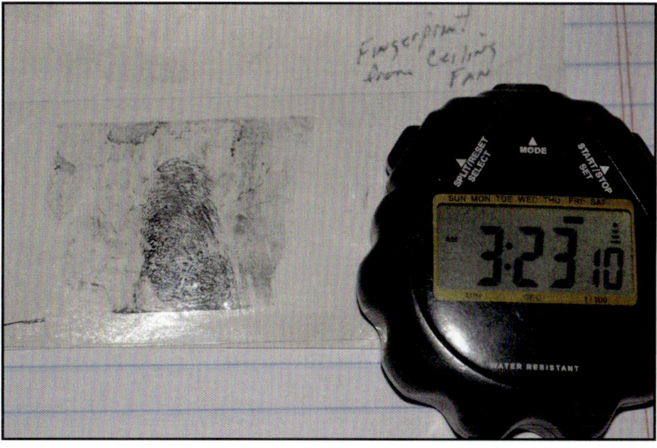

Figure 9.16 Fingerprints can be preserved for later analysis. *Courtesy of Donny Howard, Agent, Oklahoma State Fire Marshal's Office.*

Ignitable Liquids

Ignitable liquid evidence is one of the most common types of evidence a fire investigator will collect at a fire scene. Ignitable liquids include flammable and combustible liquids. Ignitable liquids found at a fire scene may be used in normal building functions and/or to intentionally accelerate a fire.

Ignitable liquids may be absorbed into porous materials such as wood, carpet and furniture, paper, concrete, and soil. Studies have shown that ignitable liquids persist after the fire when trapped in porous materials (Rankin, 2014). When investigators suspect an ignitable liquid may be present, and that the use of the ignitable liquid is foreign to the scene, they should collect it for laboratory analysis.

When considering storage materials, plastic bags, coffee cans with plastic lids, and other containers that leak vapors even when sealed should not be used for ignitable liquid samples, but may be useful for other types of evidence. Some types of plastic bags may be suitable for collecting and storing evidence **(Figure 9.17)**. Common plastic bags are NOT suitable for fire debris analysis.

Identification of Ignitable Liquid Evidence

Fire investigators must understand the physical properties of ignitable liquids to properly locate, collect, and preserve them as evidence. Properties that affect collection include:

- Materials such as alcohol and acetone mix readily with water.

- Materials such as floors, walls, or furnishings often absorb ignitable liquids.

- When trapped in a porous surface, ignitable liquids often remain intact after a fire and extinguishment efforts.

- Liquids flow to low points on surfaces and along any cracks or openings in the surface. Liquids collect in pools.

- Hydrocarbon-based ignitable liquids are less dense than water and will not mix with it; therefore, hydrocarbons tend to float on water.

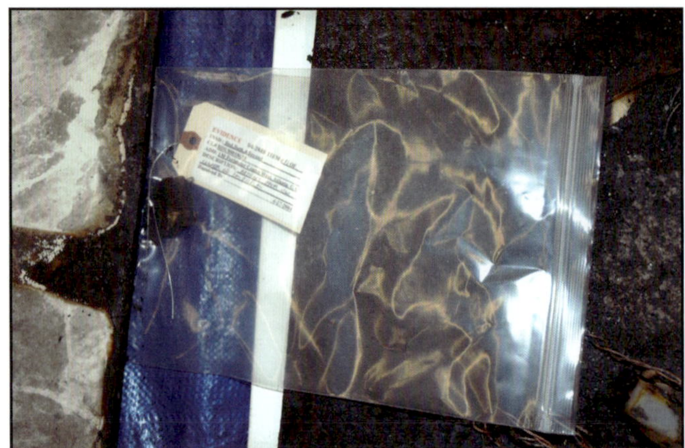

Figure 9.17 In addition to cans and larger containers, investigators should keep small bags on hand for smaller pieces of evidence. *Courtesy of Yates & Associates.*

Evidence Containers for Ignitable Liquids

To prevent evaporation, fire investigators should package ignitable liquid evidence in airtight containers. The most common container used for ignitable liquid samples is a clean, unused metal can with an airtight friction lid **(Figure 9.18)**. Metal cans are more durable than glass jars, but can eventually rust and contaminate evidence inside.

Glass mason-type jars with tight-fitting lids do not rust and retain the evidence longer without contamination. In contrast, glass jars are breakable and must be handled more carefully than metal cans.

Regardless of the type of container used, it should be approved by the forensic laboratory to which samples will be sent for analysis. All containers should be properly stored to prevent contamination before their use.

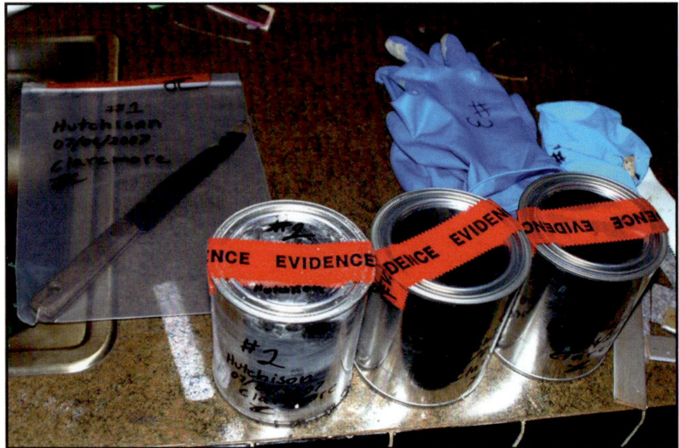

Figure 9.18 Evidence cans with friction lids can hold a range of evidentiary materials. *Courtesy of Donny Howard, Agent, Oklahoma State Fire Marshal's Office.*

Containers Found at the Scene

Upon finding a container at an incident scene, investigators must carefully document the item and the location where it was discovered. Investigators should seal openings and fill holes such as spouts with a cap or stopper to contain any liquid, residue, or vapor inside. Cork stoppers work well for this purpose because they can be easily cut and shaped to fit any size opening.

Investigators should wear gloves, and touch the container as little as possible to prevent the destruction of any fingerprint and DNA evidence on the surface. For devices such as Molotov cocktails, an investigator should package the container and any available fragments of glass in an airtight can **(Figure 9.19)**.

Collection of Ignitable Liquid Evidence

As laboratory analysis only requires a very small sample size of ignitable liquids, investigators need not collect large volumes. Due to the volatile nature of the materials, ignitable liquids should be collected as soon as possible after a fire. Ignitable liquid samples may be obtained from:

Figure 9.19 Fragments of glass and rags soaked in flammable liquids are examples of evidence left from Molotov cocktails.

- Containers found at or near the fire scene

- Fire debris (such as carpet, cloth, wood, and flooring)

- Pools on the surface of water used for extinguishment

When collecting ignitable liquid evidence, fire investigators should:

- Take precautions during the collection, packaging, and handling to prevent contamination or spoliation of the sample.

- Collect a comparison sample when necessary.

- Preserve ignitable liquid samples properly to prevent evaporation.

- Maintain the chain of custody from the time the item is discovered.

- Document all items collected before, during, and after the collection process.

- Take samples that are sufficient in size or quantity for a laboratory to conduct its analysis.

- Follow the policies and procedures of the jurisdiction and forensic laboratory that will conduct the analysis.

The investigator must change disposable gloves prior to collecting the first piece of ignitable liquid evidence and in between each subsequent piece of evidence. The investigator should not place gloves inside the evidence container after collection. The gloves may be placed beside the container after collection and photographed. This documentation shows that the gloves have been changed prior to each time evidence has been collected.

Ignitable liquid samples may be found in unburned portions of containers found at the site, or in liquid pools. After documentation, fire investigators should collect suspect liquids using the following precautions:

- Collect a small quantity of the suspect liquid.

- Collect samples of any water layer below the suspect liquid.

Free standing liquids, such as in a puddle or container, may be collected with tools such as:

- **Eye dropper** — Remove a clean eyedropper from its package and collect the evidence. The liquid may be placed inside a metal container such as a clean paint can or bag suitable for fire debris analysis.

- **Sterile gauze pads or cotton swabs** — After removing the sterile gauze pad or cotton swab from its packaging, the pad or swab may be placed inside the liquid. The pad or swab will absorb the liquid. The pad or swab may then be placed inside a metal container such as a clean paint can or plastic bag suitable for fire debris analysis.

Fire debris such as wood, dirt, or other porous material suspected of containing an ignitable liquid may be placed directly into a clean metal container such as a paint can. Fire debris suspected of containing an ignitable liquid may also be placed in a suitable plastic bag for fire debris analysis.

The investigator must ensure that the paint can or bag is filled to no more than 2/3 full before sealing the container **(Figure 9.20)**. This 1/3 air space (head space) will collect vapors from the sample. The laboratory will analyze these vapors during the testing process to determine the presence of an ignitable liquid.

Comparison samples are collected from the same material, but from areas that were protected from the fire, if possible. For example, a fire investigator may collect a section of carpet from a bedroom suspected of containing an ignitable liquid. In this case, the investigator may also wish to collect a comparison sample of carpet from under a dresser or from inside a closet. In these areas, the ignitable liquid suspected may not be present.

Figure 9.20 Evidence cans should accommodate the potential accumulation of vapors. *Courtesy of Donny Howard, Agent, Oklahoma State Fire Marshal's Office.*

Solid Debris Holding Ignitable Liquids

As with liquid samples, when collecting solid debris as ignitable liquid evidence, NFPA 921 recommends that the container be filled to no more than two-thirds of its capacity. This procedure provides the proper air or head space needed by the laboratory to conduct the analysis.

Debris samples that are relatively cohesive may be broken into smaller pieces to increase their surface area and placed on end into a container, allowing for maximum air space. Similarly, carpet samples should be cut into pieces that fit evenly into a container. The sample should be loosely rolled and placed on end into the container.

For concrete surfaces, the suspected accelerant can be absorbed from cracks and craters with a fine powder of calcium carbonate or agricultural lime, or non-clumping cat litter. A 1/8-inch (3 mm) layer is spread over the surface to be sampled and allowed to rest for approximately 30 minutes. The powder is then collected and placed into a vapor-tight container. A ½-inch (13 mm) layer of the concrete from the area should also be chipped and added to the same container.

Soil samples that are suspected of containing ignitable-liquid evidence should be collected using a small clean shovel or similar tool to obtain a sample of the material from the surface down approximately 6 to 8 inches (150 to 200 mm) **(Figure 9.21)**. As with other solid debris, soil should be collected in a container, leaving the correct amount of head space needed for laboratory analysis. Soil samples should be placed in and stored in a freezer as soon as possible after collection to inhibit any live microorganisms present in the soil from degrading any ignitable liquid residue that may be present in the sample.

Figure 9.21 Soil containing materials that appear significant to the incident should be collected as evidence. *Courtesy of Donny Howard, Agent, Oklahoma State Fire Marshal's Office.*

Electrical Evidence

As part of the scene investigation, investigators identify, examine, and collect electrical wiring and components to determine if a failure has occurred. Methods for examining electrical evidence for failures are included in Chapter 6, Basic Electricity and Electrical Analysis for Fire Investigators.

Electrical wiring suspected of a failure should be thoroughly documented prior to collection. Often this documentation includes photographs of the wiring to show several features:

- The suspected failure area

- Damage to other conductors or nearby materials

- The type of wiring including size and direction it was routed

- The type of circuit protection and its position (i.e., on/off/tripped or fuse blown)

Investigators should also document and photograph any duplex receptacles including:

- Switches and their position

- Items that may have been plugged into them

- Fixtures or other devices connected to the electrical system, especially if the circuit they were connected to is suspected of having failed

Portions of a building's or vehicle's electrical system may be collected to substantiate its relationship to the ignition of a fire. In other words, the material may be collected to prove or disprove that the electrical system was involved in the ignition sequence. Fire investigators should make certain that the system is deenergized and that all sources of power to the system are disconnected before there is any effort to analyze or collect portions of an electrical system.

Identifying Electrical System Components

Components of an electrical system that have been exposed to heat and fire may be very fragile **(Figure 9.22)**. Before collection, system items that will be collected should be thoroughly documented while still in place and undisturbed. Fire investigators should not operate controls, circuit breakers, or other components of an electrical system during the scene examination or while they are being collected and preserved.

If possible, the experts who perform the analysis of the components should view, document, and test (nondestructively) the system before it is dismantled for collection. If this is not possible, the items should be handled carefully to prevent spoliation.

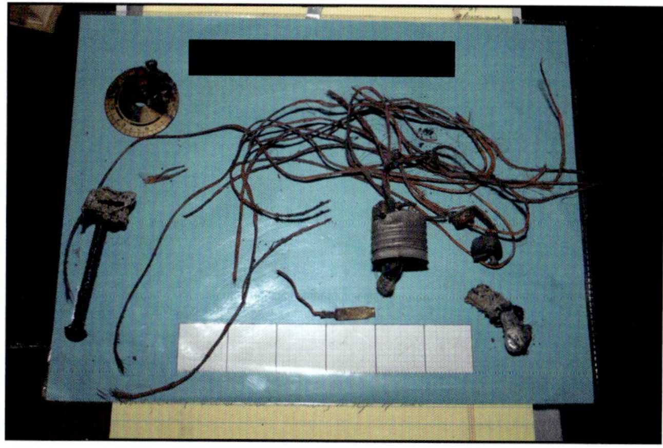

Figure 9.22 Wiring intended to be preserved as evidence may be badly damaged and fragile. *Courtesy of Yates & Associates.*

Collecting Electrical Evidence

Electrical wiring should be cut back from the suspected area of failure, preferably to where the insulation is still intact, if possible. Care should be taken to not damage the exterior surface of the conductors. The conductors themselves may be susceptible to fracture during the incident or while being handled or transported. Color changes present on the conductors may suggest that the conductor has been exposed to a higher temperature and is therefore more fragile. For example, copper conductors may turn a reddish color when subjected to increased heat **(Figure 9.23)**.

Figure 9.23 Copper wires may show a reddish color when exposed to high temperatures. *Courtesy of Donny Howard, Agent, Oklahoma State Fire Marshal's Office.*

The following precautions should be taken when collecting electrical evidence:

- Place wiring in plastic bags or boxes; wrap with paper or plastic, and seal with tape.
- Do not bend wire unnecessarily during packaging because wire may become brittle due to its exposure to heat.
- Place panel boxes and other large components in rigid containers, wrap with paper or plastic, and seal with tape.
- Package small components in boxes or bags, and pad them with cotton or paper to prevent movement in transit.
- Package any electrical system components collected as evidence in such a way so that they will not be damaged during transport and storage.

Techniques in Collecting Electrical Evidence

Fire investigators should use caution while collecting electrical items to prevent further damage. Fire investigators should always collect enough of the material, such as wiring, so that it can be properly analyzed in the laboratory or during a joint examination.

Extension Cord. When possible, items should be collected intact, the way they are connected at the incident scene. Collect the entire cord, the receptacle, and the junction box that the receptacle is in.

Branch Circuit Wiring. Collect enough wiring to show its condition. If practical, wiring from the undamaged circuit should also be collected. Document the position/condition of the fuse or circuit breaker that was protecting the circuit. Trace the circuit back to the electrical distribution panel and then to the fuse or circuit breaker using one of the following techniques:

- Expose the conductors and manually trace them back to the electrical distribution panel.

- Use an electrical circuit seeker/tracer to trace concealed conductors behind walls and through floors. These devices typically have a receiver and transmitter with both an audio and visual signal to confirm the correct circuit.

Structure or Vehicle Electrical Distribution System. Collect the component and attachment hardware, intact and in the condition they were found, if possible **(Figure 9.24)**. For example, investigators may opt to remove part of the structural elements (such as wall studs) where components or devices are attached. Again, documentation before any collection effort is critical.

Marking Electrical Evidence

Investigators may find it helpful to use colored electrical tape or colored zip ties while tracing the route of the conductors. This allows distinguishable photographs to be taken of the routing and damage on the conductors.

When marking electrical evidence, fire investigators should tag both ends of any wire collected, indicating the end connected to the device and the end connected to the electrical supply. In addition, the investigator should document the following items or conditions:

- Condition of the breaker/fuse as found

- Breaker to which the wire was connected

- Any anomalies on the conductor such as melting/arc sites

- Path of the wire between the breaker and the device to which it was connected **(Figure 9.25)**.

Examining Electrical Components

Examination techniques of electrical conductors or components may include:

Figure 9.24 Sometimes investigators may need to remove an entire electrical assembly, like this light assembly, for preservation and analysis. *Courtesy of Yates & Associates.*

Figure 9.25 Marking electrical evidence in context at a scene can help an investigator trace arcs and other electrical failures. *Courtesy of Donny Howard, Agent, Oklahoma State Fire Marshal's Office.*

- Metallurgical examinations can determine the type of metal(s) present and their microstructure.

- Examine the conductor under a microscope at low power magnification to observe and document any anomalies on the conductor such as melting or arc sites.

- Specialized high-power magnification of the conductor can determine if electrical arcing caused the observed damage. Many university or private laboratories can complete this analysis using a scanning electron microscope (SEM).

Litigation may result when a fire involves appliances and equipment, even when the fire is determined to be accidental **(Figure 9.26)**. This possibility makes the process of documentation, preservation, collection, transportation, and storage very important.

When appliances or equipment are suspected ignition sources, they should be preserved as evidence. When possible, fire investigators may seek technical assistance before moving or collecting a device. When the conditions do not permit on-site analysis of a device, fire investigators should proceed with preservation, collection, and transportation efforts.

When practical and necessary for analysis, a fire investigator should collect the entire device, electrical wiring, and fuel lines or batteries connected to the device, including the power source. For example, if the device being collected is a countertop kitchen appliance, the fire investigator would collect and preserve:

- The appliance and the electrical cord attached to it

- A section of the branch circuit wiring attached to the receptacle

- The electrical junction box or receptacle to which the cord was connected

Figure 9.26 Kitchen appliances are types of evidence that investigators may need to examine, document, and preserve. *Courtesy of Donny Howard, Agent, Oklahoma State Fire Marshal's Office.*

The conditions at the scene and type of examination/testing will determine if large devices, such as furnaces, should be protected in place or removed. If controls or other components must be removed for preservation, investigators should collect enough of the component and related material such as wiring or piping to support their hypothesis regarding the device. The evidence should be maintained in the condition in which it was originally found. Switches and other types of controls should not be operated, and the equipment should be packaged to prevent damage during transport and storage.

Devices, controls, or other equipment may be X-rayed prior to completing a destructive type of examination. This allows the position of controls or conductors to be observed without damaging the components. Often the conductors or controls become embedded in melted plastic. X-rays can reveal what is inside the melted plastic and the condition of these items.

To examine conductors or controls encased in melted plastic, an investigator can use a heat gun to slowly melt the plastic so it can be strategically removed. This technique should be employed in a well-ventilated area and the investigator must wear appropriate PPE. Testing and examination that changes the condition of items should not be initiated until after confirming that the items will not be needed again in their current form.

Glass Pieces

Glass from the scene of a fire or explosion may be collected to determine its composition and application. Examples of glass found at an incident include a bottle that was used as a Molotov cocktail, and a pane of window glass. Analysis of glass fragments can indicate conditions including:

- The direction of the force that resulted in a fracture.

- Soot present on only one side would indicate that the glass was broken after the fire.

- Fracture patterns may have been caused by heat, impact, high-velocity projectiles, or a glass cutter.

If the intent of collecting glass evidence is to determine the direction of force that broke the window glass, use the following steps:

Step 1: Photograph all glass remaining in the broken window (**Figure 9.27**).

Step 2: Using adhesive tape or adhesive-backed note paper, mark the remaining pieces as to their location in the frame — top, bottom, left side, right side, inside surface, or outside surface.

Step 3: Remove and individually wrap each of these pieces before carefully packaging them to prevent further breakage.

Step 4: Identify and collect all available glass fragments from the broken window.

Step 5: Mark each fragment individually to indicate its location — inside or outside the building.

Step 6: Wrap and package each fragment.

Figure 9.27 Photograph glass in context at the scene to show the condition and relative location. *Courtesy of Donny Howard, Agent, Oklahoma State Fire Marshal's Office.*

Packaging techniques for glass evidence depends on the size of the fragment. Tweezers or small pliers should be used to handle fragments. Any thin edges that require preservation can be embedded in modeling clay or putty before the fragment is packaged.

For small glass fragments collected as evidence, use the following precautions:

- Handle small glass fragments carefully to avoid breaking or scratching the item.

- Place the fragments in containers such as pill boxes or small vials (glass or plastic).

- Pack these containers with cotton to prevent movement that could cause chipping during transit.

- Document each container according to the appropriate policies and procedures of the AHJ and laboratory doing the analyses.

The following precautions should be taken to preserve large pieces of glass evidence:

- Document the location and initial condition of the glass according to the appropriate policies and procedures of the AHJ and laboratory doing the analyses (**Figure 9.28**).

- Wrap the glass in cotton or use padded packing material for protection.

- Place the glass in a sturdy box that can be sealed and marked.

NOTE: Package glass comparison samples separately, in the same way as glass evidence.

Figure 9.28 Investigators should mark glass at the scene in place to prevent damage or injury. *Courtesy of Donny Howard, Agent, Oklahoma State Fire Marshal's Office.*

CAUTION: Do not use paper envelopes to package glass fragments because the glass can cut through the paper and cause injury when handled.

When collecting glass evidence, fire investigators should recognize that additional types of physical evidence may be attached to the glass. For example, if the evidence is a broken window, the material used to hold the glass in place (putty, glazing compound, or portions of the surrounding frame) may show tool marks or wood splinters associated with the breakage. Investigators may also find fibers, hair, or blood on the glass. Any additional items discovered should be preserved using the proper technique for that material. Care should be taken to prevent this additional material from being destroyed or contaminated during the collection and packaging process.

If glass is sent to a laboratory for impact analysis, investigators should attempt to collect a comparison sample from as close as possible to the point of impact. Similarly, if the material analyzed is from a bottle that contained flammable liquids, and similar containers are found in the area, an unbroken bottle can be collected for comparison with the sample collected.

Wood Materials

Wood materials may be collected to determine the item's composition or for the analysis of fire patterns or tool marks found on its surface. As with most materials, the collection and preservation of wood evidence depends on the physical size of the item and the intended use of the evidence.

Large pieces of wood evidence can include structural components such as paneling from the interior of a room; trim materials from around doors, windows, or floors; or window or door frames. Entire sheets of wall paneling may be removed from the area of origin to show a particular fire pattern, or a section of a door frame may be collected to preserve a tool mark **(Figure 9.29)**.

In contrast, the evidence may be very small, in the form of wooden matches, splinters, or sawdust **(Figure 9.30)**. Small items may be placed in containers such as vials or pillboxes padded with cotton or other material to prevent damage in transit. An investigator should seal and mark these containers per the AHJ.

The following precautions apply to preserving wood evidence:

* Wrap large items in paper, plastic, or fabric, if possible, to prevent additional damage.

* Tag, log, and transport large items to a secure location; transport may require the use of a larger vehicle.

* Wrap small items in paper or plastic and then seal, tag, and secure until tested.

* Place charred materials on a sheet of cardboard or similar material to prevent additional deterioration during the collection and packaging process **(Figure 9.31)**.

Figure 9.29 A whole door frame may be collected if needed. *Courtesy of Donny Howard, Agent, Oklahoma State Fire Marshal's Office.*

Figure 9.30 The matches used in this incendiary device would be preserved in a small container as evidence.

Figure 9.31 Charred materials tagged and organized for packaging as evidence. *Courtesy of Yates & Associates.*

- Dry items, if necessary, and then wrap or pack them using packaging material or cotton batting to prevent motion during transport.

Wood is a porous material and may contain trace evidence of an ignitable liquid. Wood that will be tested for an ignitable liquid should be placed in a clean metal container or specialty plastic bag suitable for laboratory analysis.

Evidence packaged while wet may form mildew that could potentially destroy the evidentiary value of the item. Wet wood items that are not going to be examined for the presence of ignitable liquid evidence should be air-dried before packaging. For example, the wooden rafters around a bathroom fan that contains a fire pattern needed for examination may have to be dried before a proper analysis can be completed.

NOTE: Air drying may allow ignitable liquid evidence to dissipate.

Paper Materials

Paper-based evidentiary materials may include:

- Paper matches
- Fireworks or dynamite tubing
- Materials used to wrap or hold a device
- Wet and/or burned documents found at the scene
- Paper products used as trailers to spread an intentionally set fire

As with other types of evidence, paper materials should be well-documented before any effort to remove or collect them. The following precautions should be taken when collecting paper evidence:

- Place paper fragments in vials or small boxes.
- Do not fold, staple, or otherwise disturb documents.
- Use cotton batting to protect paper material during transport.
- Take close-up photographs of documents or materials with writing or printing on them.
- Use tweezers to handle dry, intact, and undamaged paper sheets; collect these items as found.
- Place any documents into a protective covering such as a large envelope or plastic document protector.
- Place dry items in a container large enough to hold the materials without damaging them; mark the outside of the container as evidence.
- Collect remaining fragments of documents and package them with the larger item when a document is in numerous pieces; use tweezers to handle small fragments.
- Leave paper materials found in a container, such as a wastebasket, drawer, or safety deposit box, intact in the container; collect, package, and identify the entire container as evidence.
- Collect charred documents and other larger paper items by sliding them onto a cardboard sheet, or flat stiff surface such as a glass or metal plate; place the document and the material used to support it into a cardboard box or other rigid container.

A paper product suspected of being used as a trailer may have ignitable liquids on it. Wet paper products that were not collected as ignitable liquid evidence should be air-dried before packaging to prevent the formation of mildew.

Laboratory analysis can provide information about the type and use of the sample. When comparison samples are available, analysis can indicate whether there is a physical match between samples.

Cigarette Evidence

Cigarettes or related materials found at the incident scene may have value as evidence **(Figure 9.32)**. Cigarettes may have been part of an igniter for an incendiary device, left by an intruder, or involved as the source of the heat of ignition in an accidental fire. Often the paper covering the filter (if equipped) of a cigarette butt will be consumed. The fibers of the filter may be the only evidence remaining. Potential evidence should be documented before it is disturbed.

Cigarette filter fibers may be isolated by sifting debris with multiple size screens. An investigator may need a low power magnifier to determine the difference between a cigarette filter or similar material such as partially consumed polyurethane foam. As with other similar materials, investigators should handle cigarettes and butts carefully so that any evidence that can be obtained is not destroyed.

Figure 9.32 Cigarette at a site flagged as evidence. *Courtesy of Donny Howard, Agent, Oklahoma State Fire Marshal's Office.*

Dry cigarettes and butts should be packaged individually and placed in air-permeable envelopes. The use of plastic bags is not recommended. Investigators should identify and mark each envelope containing cigarettes as evidence.

Laboratory analyses can provide information such as brand, type, flavor, length, and packaging. Analysis cannot determine whether cigarette evidence originated from a specific pack.

Fire-Safe Cigarettes

Fire-safe cigarettes are so named because they are designed to self-extinguish when they are left idle. The words *Fire Standards Compliant (FSC)* on the bottom corner of a cigarette package indicates that the cigarettes meet the national standard.

However, according to laboratory testing, not all of these cigarettes self-extinguish before the threat of being an ignition source. For this reason, many disagree with the designation of any cigarettes as being "fire-safe." They are more accurately described as "reduced ignition propensity" cigarettes.

Adhesive Tape

As individuals often use adhesive tape to construct or secure incendiary or explosive devices, investigators may find tape pieces during scene examination. Adhesive tape should be packaged in a paper bag or box to prevent additional damage. Tape used to hold a device should be handled in a way that will not destroy any trace materials attached to the adhesive side.

A roll of tape suspected of matching the tape found on a device should be packaged to prevent damage or contamination. The roll is then given to the laboratory along with the sample from the device. The analysis and examination may provide evidence such as fingerprints, hair, or fibers.

Laboratory examination of the tape pieces and roll may be able to indicate:

- Tape type
- Tape color
- Tape construction
- Tape manufacturer
- Tape chemical composition
- Whether pieces of tape are from the same roll

Review Questions

1. What types of physical evidence may be present at an incident?

2. How can investigators maintain the security of evidence they collect?

3. What factors determine the method that will be used to collect evidence?

4. Why are evidence identifiers and evidence logs important?

5. What is the purpose of collecting comparison samples?

6. How can an investigator secure evidence during transportation and storage?

7. What are three sources of contamination?

8. What types of containers are appropriate for collecting and transporting evidence?

9. What types of protective equipment should be worn when collecting evidence?

10. What is the difference between physical evidence and trace evidence?

11. What safety precautions must be taken when collecting electrical evidence?

12. How do the collection methods for glass evidence vary depending on the size of the glass to be collected?

13. What precautions should be taken to preserve wood evidence?

14. What precautions should be taken to protect fragile pieces of paper evidence?

15. How should adhesive tape evidence be packaged and collected?

Discussion Questions

1. How might the chain of custody documentation impact legal proceedings?

2. Provide two scenarios in which ignitable liquid evidence would be collected and describe the methods that you would use to collect the samples.

3. Which types of evidence would you expect to be most difficult to collect? What actions can you take to ensure that this type of evidence is properly collected and preserved?

Chapter Reference

Rankin, J. Graham Ph.D., Nicholas Petraco. September 2014. Interpretation of Ignitable Liquid Residues in Fire Debris Analysis: Effects of Competitive Adsorption, Development of an Expert System and Assessment of the False Positive/Incorrect Assignment Rate. U.S. Department of Justice. Award Number: 2010-DN-BX-K272. Accessed online.

Chapter Key Terms

Artifact — Remnant of materials involved in a fire or explosion that are in some way related to the ignition, development, or spread of the fire or explosion.

Chain of Custody — Continuous changes of possession of physical evidence that must be established in court to admit such material into evidence. In order for physical evidence to be admissible in court, there must be an evidence log of accountability that documents each change of possession from the evidence's discovery until it is presented in court.

Comparison Sample — Evidence collected from undamaged areas or materials to offer a comparison to similar materials damaged by a fire.

Cross-Contamination — Material in one location at the scene that is moved to another location at the scene.

Evidence — (1) One of three requirements of evaluation; the information, data, or observation that allows the investigator to compare what was expected to what actually occurred. (2) In law, something legally presented in court that bears on the point in question. (3) Information collected and analyzed by an investigator.

Exclusionary Evidence — Evidence collected to show that a particular device or scenario can be excluded with relation to the ignition or fire spread scenario.

Final Disposition — Point in the chain of custody where a piece of evidence is determined to no longer have value as evidence; options at this point may include permanent storage, return to the owner, or authorized destruction.

Locard's Exchange Principle — Investigative principle that states that whenever someone comes into contact with a scene, they leave something at the scene and take something from the scene.

Physical Evidence — Any physical or tangible item that tends to prove or disprove a particular fact or issue. (Reproduced with permission from NFPA 921-2017, *Guide for Fire and Explosion Investigations*, Copyright 2018, National Fire Protection Association.)

Trailer — (1) Combustible material, such as rolled rags, blankets, newspapers, or flammable liquid, often used in intentionally set fires to connect remote fuel packages (such as pools of ignitable liquid and other combustible materials) in order to spread fire from one point or area to other points or areas; frequently used in conjunction with an incendiary device. (2) Fire pattern left behind after the combustible material has burned.

NOTE: This skill sheet covers basic steps that must be taken for any piece of evidence. Additional steps may need to be taken, depending on the types of evidence at the scene. Always follow SOPs.

Step 1: Document the evidence in place.

NOTE: Evidence should be photographed using overall, mid-range, and close-up photographs **(Figure 9.33)**. Photographs should also be taken during excavation.

Step 2: Don gloves.

NOTE: New gloves should be donned between each sample that is collected.

Step 3: Uncover and expose the evidence.

CAUTION: Care should be taken not to disturb surrounding evidence. Outside help may be required for larger pieces of evidence or for large piles of debris.

Step 4: Select an appropriately sized sample of evidence.

Step 5: Place the evidence in the proper type of collection container.

NOTE: The evidence container should remain sealed until the time of collection.

Step 6: Seal the evidence container.

Step 7: Label the evidence container.

Step 8: Collect comparison samples and document appropriately, if necessary.

Step 9: Complete the evidence log, chain of custody forms, and any other required documentation.

Step 10: Store the evidence in a secure location or prepare it to be transported for testing.

Step 11: Dispose of evidence per local SOPs.

Figure 9.33 *Courtesy of Donny Howard, Agent, Oklahoma State Fire Marshal's Office.*

Photo courtesy of Donny Howard, Agent, Oklahoma State Fire Marshal's Office.

Chapter Contents

NFPA 1033 JPRs addressed in this chapter

This chapter provides information that addresses the following job performance requirements (JPRs) of NFPA 1033, *Standard for Professional Qualifications for Fire Investigator* (2022):

4.1.1	4.2.3	4.2.6	4.4.1	4.6.3
4.2.1	4.2.4	4.2.7	4.4.2	4.6.5
4.2.2	4.2.5	4.2.8	4.4.4	

This chapter provides information that addresses the following course outcomes for the FESHE courses:

Fire Investigation I (C0283)

1. Demonstrate the importance of documentation, evidence collection, and scene security process needed for successful resolution.

6. Explain how the basic elements fire dynamics, construction, and fire protection systems as to how they affect origin and cause determination.

Fire Investigation II (C0284)

5. Evaluate the use of incendiary devices, explosives, and bombs.

Fire Investigation and Analysis (C0285)

2. Document the fire scene in accordance with best practice and legal requirements.

3. Analyze the fire scenario utilizing the scientific method, fire science, and relevant technology.

4. Analyze the legal foundation for conducting a systematic incendiary fire investigation and case preparation.

1. List the conditions an investigator should document during a scene examination. [NFPA 1033, 4.2.1, 4.2.2, 4.4.2]

2. Describe teamwork considerations at an investigation. [NFPA 1033, 4.1.1, 4.2.6, 4.4.1, 4.6.3]

3. Describe the exterior examination process of a fire scene. [NFPA 1033, 4.2.2, 4.2.8]

4. Describe the interior examination process of a fire scene. [NFPA 1033, 4.2.3, 4.4.2]

5. Explain the process of scene investigation at an incident with victims or fatalities. [NFPA 1033, 4.4.1, 4.4.4]

6. Summarize the process of debris removal at a fire scene investigation. [NFPA 1033, 4.2.6, 4.4.2]

7. Summarize the fire scene reconstruction process. [NFPA 1033, 4.2.4, 4.2.5, 4.2.7, 4.6.5]

8. Skill Sheet 10-1: Survey the exterior and interior of the scene. [NFPA 1033, 4.2.2, 4.2.3, 4.2.4, 4.2.5, 4.2.8]

9. Skill Sheet 10-2: Manage victims and fatalities. [NFPA 1033, 4.4.1, 4.4.4]

Chapter 10
Fire and Explosion Scene Examination

The preceding chapters have been focused on the basic knowledge and skills a fire investigator needs in order to perform a complete investigation. This chapter addresses scene examination in more practical terms:

- Scene examination overview
- Teamwork at an investigation
- Exterior examination
- Interior examination
- Scenes with victims and fatalities
- Debris removal
- Scene reconstruction

Scene Examination Overview

NFPA 1033 (2022): 4.2.1, 4.2.2, 4.4.2

As indicated throughout this manual, an investigator must follow a plan throughout the investigation, regardless of the type of fire scene. One model of a **consistent methodology** is the scientific method, as described in Chapter 7, Initial Actions in the Investigative Process. This manual uses the scientific method as the default methodology. **Skill Sheet 10-1** shows the steps for surveying the exterior and interior of the scene. Investigators should perform the following tasks during all examinations:

- Identify and secure potential evidence.
- Identify areas that may need more detailed examination.
- Note the overall condition of the scene including structural condition.
- Maintain site security while performing other actions such as documenting scene conditions.

Documenting Scene Conditions

NFPA 1033 (2022): 4.4.2

Documentation is an essential part of an investigation, as explained throughout this manual. Before entering an incident site, an investigator should confirm that all documentation equipment is intact and functional for the expected scene conditions over the duration of the investigation. Investigators document a fire or explosion scene according to their agency's best practices and legal requirements as detailed in Chapter 8, Scene Documentation.

Scene conditions an investigator documents during a scene examination can include:

- Evidence of an explosion
- Extent of fire or explosion damage
- Evidence of fire department operations
- Evidence of relocated contents
- Complexity of the scene (including scene size, level of safety, number of investigators needed, level of fire department operations, time between fire and investigation)
- Determination of the need for additional resources or equipment
- Prefire conditions of the structure, contents, and area

Investigators should also document the locations and conditions of:

- Building utilities
- Potential evidence
- Remaining contents
- Witnesses who may provide information regarding the fire
- Potential sources of ignition such as heating, ventilating, and air conditioning (HVAC) systems; appliances; and electrical systems **(Figure 10.1)**

Figure 10.1 An investigator's first step after assessing safety is to walk the exterior or perimeter of the scene.

Shared Information

Since 9/11/2001, public safety agencies at every level of the U.S. and Canadian governments have enhanced the quality and quantity of shared information. Multiagency information gathering units now exist to screen and disseminate this information, commonly referred to as *intelligence*.

Fire investigators and dispatchers remain acutely familiar with activities within their jurisdiction, and they may identify trends and patterns. This awareness can aid in identifying acts or activities that bring an ignition source and materials together.

As multiagency intelligence sharing becomes increasingly available and varied, fire investigators may increase the amount of analysis they conduct before arriving on the scene. Investigators can access alerts from various channels and potentially relevant information from dispatch. Alerts may describe specific methods and motives concerning arson, terrorism, and other criminal activities. Dispatch information can range from descriptions of global terrorist incidents to local activities.

Scene Security Zone

NFPA 1033 (2022): 4.2.1

An investigator evaluates the safety of a scene continuously from arrival at the incident site, as discussed in Chapter 2, Safety. Site safety checks include an on-scene security assessment to confirm that unauthorized entry is not permitted, and all potential evidence is protected from further damage, possible contamination, or destruction. If sufficient security is not in place, immediate steps must be taken to enhance existing security or expand the protected zone. Finally, security at the scene should be maintained throughout the investigation. After establishing the safety and security considerations at the scene, the fire investigator begins a scene examination, whether alone or as a team.

During an exterior examination, fire investigators search for potential evidence beyond the secured scene. They also adjust the size of the scene security zone as needed to maintain protection of potential evidence located beyond the immediate scene area.

In an investigation of a scene with an explosion, the investigator or other responder should measure from the epicenter to the farthest point of evidence, and use that measurement plus half again for the outside of the perimeter. When establishing the initial perimeter, the investigator should remember that it is easier to decrease the size of a secured area than to increase it.

When investigators find potential evidence at the exterior of a structure, they should secure the material and protect it from contamination or destruction **(Figure 10.2)**. A fire investigator who is unable to protect the material in its current location should document the findings and take measures to collect the item or material as soon as possible.

Figure 10.2 Evidence outside of the burn structure may need to be protected, or possibly documented/photographed. *Courtesy of Donny Howard, Agent, Oklahoma State Fire Marshal's Office.*

Teamwork at an Investigation

NFPA 1033 (2022): 4.1.1, 4.2.6, 4.4.1, 4.6.3

After establishing the safety and security considerations at the scene, the fire investigator begins a scene examination. Fire investigators often work alone, but they may need to call for assistance at some scenes. For example, law enforcement officers may be present at a fire or explosion incident scene for functions including scene security. Incidents with large debris, and/or human fatalities may require specialized support and are explored at more length later in this chapter.

Debris Removal and Excavation

NFPA 1033 (2022): 4.2.6

Scene evaluation work often requires that an investigator move debris, furniture, and other heavy objects to look at all of the elements present. At a large and/or complex incident, an investigator may need help from heavy equipment operators and additional investigators. The number of team members must remain an appropriate size for their respective tasks. Teams tasked to remove and examine debris should follow directions from a lead investigator who will provide continuous supervision.

Investigative teams using heavy equipment to excavate large areas should include:

- **Lead investigator** — Monitors the process as the primary point of contact. This helps to ensure that equipment operators preserve the integrity of the structure and do not increase the hazards to other personnel or themselves. Assigning one person to this task prevents confusion and improves scene safety.

- **Equipment operator(s)** — Removes large debris in logical layers with clear guidance from the lead investigator.

- **Teams with hand tools** — Removes small debris and examines the scene between heavy equipment excavations of each layer.

Fatality Incident Response

NFPA 1033 (2022): 4.4.1

Teamwork at a fatality response will require an investigator to coordinate with a range of professionals with specialized knowledge and training, including:

- **Law Enforcement Officers** — Every fatal fire scene should be treated as a crime scene. Locard's Exchange Principle (1934) notes that when the perpetrator(s) of a crime comes into contact with the scene, the perpetrator(s) will bring something into the scene and leave something behind **(Figure 10.3a-b). Simply put: every contact leaves a trace.**

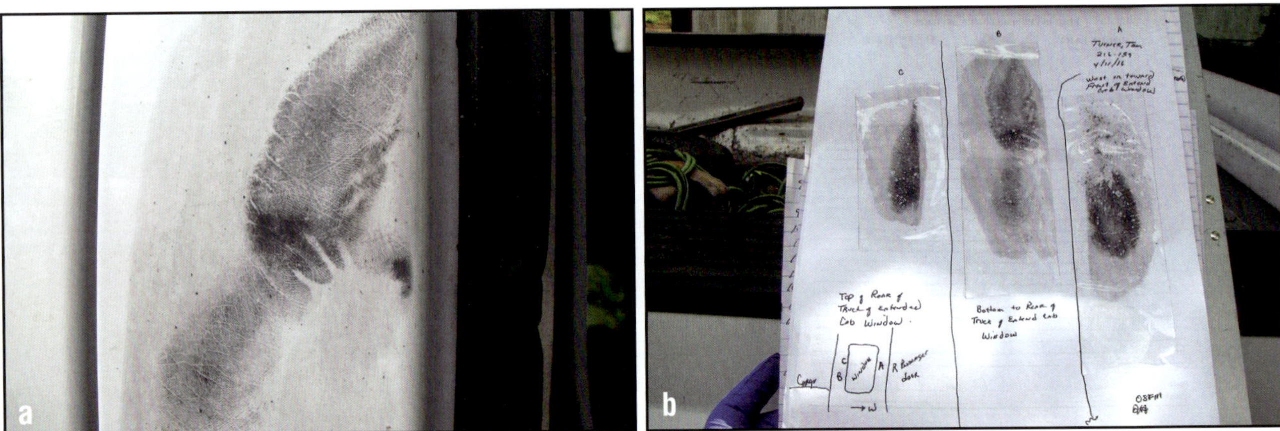

Figure 10.3a-b A photograph of fingerprints, and a photograph of the logged fingerprints, give an investigator some redundancy in the photo record. *Both courtesy of Donny Howard, Agent, Oklahoma State Fire Marshal's Office.*

- **Testing Laboratory Technicians** — Provided samples that comply with their requirements, the testing laboratory team can generate results including the victim's toxicology information at the time of death, including levels of impairment or disease. Toxicological testing can be completed on a very badly burned body.

- **Forensic Pathologist** — Medical professionals such as **medical examiners** and coroners who apply training and skills to determine as much information about a fatality as possible.

- **Forensic Odontologist** — Dentists with specialized knowledge will examine each tooth to determine whether the tooth is clean and solid, decayed, restored, or filled. This information can be compared to available dental records.

- **Forensic Anthropologists** — Investigators often consult forensic anthropologists in criminal cases when the victim's remains are more or less reduced to a skeleton. A forensic anthropologist can also assist in identifying features on remains that are decomposed, burned, mutilated, or otherwise unrecognizable.

Exterior Examination

NFPA 1033 (2022): 4.2.2, 4.2.8

The exterior of the fire scene includes all of the area from the perimeter of the scene to the outside of the structure. Steps for performing an exterior and interior scene examination are included in **Skill Sheet 10-1**.

When conducting the exterior examination, investigators look for any evidence that may assist in determining:

- The area of origin
- Explosion damage
- The cause of the fire
- Damage to doors and windows
- Position of doors and windows
- Areas of collapse where the fire may have burned longer, causing more fire damage

NOTE: Areas of origin and cause are addressed later in this manual.

Evidence found at the exterior of the fire scene can often provide clues about aspects of a fire, such as:

- Possible responsibility for the fire
- Possible location of the area or areas of origin
- Types and extent of fire-suppression activities
- Fire dynamics, including ventilation and possible locations of interior fuel packages

One method of conducting an exterior examination is to first examine the ground around the structure for potential evidence. During this systematic examination of the exterior, the investigator should take field notes and photographs of the entire structure, as well as additional photographs of areas of interest as discussed later in this section.

NOTE: If evidence is not in immediate threat of damage or destruction, it should be collected during the documentation and evidence collection phase per Chapter 9, Evidence Collection and Preservation. If necessary, evidence can be collected during the initial scene examination.

After assessing the security of the fireground and adjacent areas, and the safety of the building, an investigator makes observations regarding the fire and resulting fire-suppression operations. While conducting the exterior examination, a fire investigator determines, identifies, and interprets some or all of the following:

- Fire patterns
- Fire dynamics
- Potential evidence
- Explosion dynamics
- Building construction types and systems

If contents were removed from the interior, investigators must determine who removed them and where the contents were originally located **(Figure 10.4)**.

Figure 10.4 When building contents have been removed from the scene, investigators should ask fire suppression personnel what the original position of the contents was before they were moved. *Courtesy of Yates & Associates.*

Investigators must determine whether these items were removed before, during, or after the fire or explosion. This information will be needed if the contents must be returned to their original position to reconstruct the fire scene.

Broken Glass

NFPA 1033 (2022): 4.2.2

Investigators should examine broken glass found outside of the structure. During this examination, investigators should note any soot or cracks in the glass. Glass found with soot on one side is an indication that the glass was broken during or after the fire. Glass broken before the fire will not have soot deposits on either side.

Common causes of cracks in glass include:

- **Mechanical force such as a hammer or rock** — Concentric cracks (conchoidal fracture lines), or spiderweb type fractures.
- **Thermal stress** — Cracks start from the edge of the pane and intersect, or appear as small crescent or straight cracks within the glass (crazing). Crazing is caused by a temperature difference on both sides of the glass and typically does not extend through the entire thickness of the glass. Rapid cooling by a fire stream on one side of the glass is typically responsible for crazing.

Exterior Fire Department Operations

During the exterior scene examination, a fire investigator should document all evidence of fire department operations on the exterior of the structure, such as:

- Use and location of power tools
- Areas where overhaul operations were conducted
- Opened doors and windows, for entry and/or ventilation
- Identifiable tool marks or boot prints on structural openings
- Location of debris removed from the interior of the building and the avenue used to transport the material

Building Utilities

NFPA 1033 (2022): 4.2.8

During the exterior scene examination, the locations and condition of all utilities entering the building should be examined and documented. A fire investigator should determine whether the utilities are currently operational, and whether they were operational at the time of the fire. An investigator should also document readings on any utility meters on the outside of the structure, and any signs of tampering or meter removal.

Electricity service is a significant consideration during an investigation. Utility companies may add a disconnect sleeve (**meter booting**) to temporarily isolate a meter. This practice adds insulators to disconnect electrical service to empty buildings or structures with unpaid accounts.

After ensuring the electricity has been disconnected, the investigator should examine:

- Electrical meters **(Figure 10.5)**
- Utility distribution systems
- Pole mounted transformers
- Interior of the meter base, specifically the lugs and conductors
- Service drop connectors (triplex or quadraplex) extending to the structure

Figure 10.5 Exterior scene with an electrical meter. *Courtesy of Donny Howard, Agent, Oklahoma State Fire Marshal's Office.*

These components should be assessed for signs of fire or other damage. In the analysis stage of an investigation, this information assists the fire investigator in excluding potential sources of ignition and may also indicate whether the building was occupied at the time of the fire.

Building Systems

NFPA 1033 (2022): 4.2.8

Exterior building system components should be examined and documented. An investigator should consider whether to exclude or confirm the involvement of the following building systems in the ignition or spread of the fire:

- Vents
- Chimneys
- Fire department connections
- Exterior alarm signaling devices
- Fire protection system control valves
- Heat pump and air conditioning condensing units

Interior Examination

NFPA 1033 (2022): 4.2.3, 4.4.2

After completing the exterior examination of the fire scene structure or vehicle, the fire investigator moves towards the interior of the scene. The interior examination allows the fire investigator to build on the data collected from prior intelligence, interviews, and the exterior examination. The most important factor is that the investigator follows a systematic approach to the investigation. Investigators should document the process they use for all aspects of an investigation, and be prepared to explain the decisions they make.

As with an exterior examination, the fire investigator conducting the interior examination uses the scientific method to collect data and develop hypotheses. Inside a burned structure, the investigator will develop hypotheses based on:

- Fire patterns
- Fire dynamics
- Analyzed evidence
- Explosion dynamics
- Building construction and systems

A fire investigator at the scene is responsible for determining the level of detail that is required for the interior examination. This determination is based on the complexity (size, extent of damage, level of necessary reconstruction) of the incident scene and the circumstances surrounding the investigation. For example, in very large or complex structures, the fire investigator determines the extent of the examination based on the incident and the information available at the time the interior examination is conducted.

When documenting observations, an investigator should be careful to substantiate any opinions and analysis offered. Without data, the recorded observation may be interpreted differently by someone else. The fire investigator should:

- Distinguish observations from interpretations
- Indicate a level of certainty for any assertions
- Offer other hypotheses that could explain the anatomic data

Initial Interior Walkthrough

NFPA 1033 (2022): 4.2.3

Investigators may find it helpful to conduct an informal walkthrough of the structure to better understand the construction, fire spread, and degree of damage. Observations made during this walkthrough will assist the investigator in developing a fire spread hypothesis and determining the magnitude of the investigation. The investigator should also consider fire dynamics and movement.

During the initial walkthrough, the investigator should note hypotheses without making a determination of the area of origin. A walkthrough may not match the order of fire suppression actions or the detailed examination. For example, suppression efforts may have started from the front door, the walk-through may start from the garage, and investigation may start at the back door.

When a fire investigator determines that the incident merits a detailed examination of the interior of a structure, the investigation should include the building's interior, the remaining contents, and any other available evidence. An investigator often conducts the detailed examination after completing an initial walkthrough of the structure.

Systematic, Detailed Examination

NFPA 1033 (2022): 4.2.3

Investigators should begin the interior examination by conducting an initial assessment of the building and then moving to a more detailed examination of specific rooms, locations, or areas. This detailed examination usually progresses from the area of least damage toward the area of most damage **(Figure 10.6)**.

Figure 10.6 Investigators begin a scene examination working from an area of less damage and progressing toward an area with more damage. *Courtesy of Donny Howard, Agent, Oklahoma State Fire Marshal's Office.*

During the detailed interior examination, a fire investigator should:

- Analyze the spread and direction of fire movement within the structure.
- Analyze the force and direction of overpressure created during an explosion.
- Correlate observations with what is known about the fire and fire-suppression efforts.
- Attempt to locate the area(s) of origin of the fire or explosion.

Similar to the exterior examination, the detailed internal examination should document items including:

- Locations and conditions of building contents
- Positions of doors and windows (open or closed) **(Figure 10.7)**
- Locations of appliances and positions of controls
- Locations and conditions of building systems within the structure, including fire protection systems
- Locations and conditions of building utilities within the structure, including the electrical panel board containing the circuit breakers or fuses for the portions of the building involved in the fire

Figure 10.7 Position of doors. *Courtesy of Donny Howard, Agent, Oklahoma State Fire Marshal's Office.*

Beginning at the chosen entry point, the investigator should also document findings, such as:

- Broken windows
- Heat or fire damage
- Ventilation openings
- Forcible entry damage **(Figure 10.8)**
- Items of evidentiary value
- Identification of fire effects
- Damage that may be attributed to ventilation
- Mirror effects on door hinges, protection patterns on door jambs
- Position and condition of all doors and windows (closed, opened, locked, unlocked, forced, intact)

Figure 10.8 A broken door frame can indicate forcible entry. *Courtesy of Donny Howard, Agent, Oklahoma State Fire Marshal's Office.*

NOTE: Investigators should recognize that increased damage may be the result of the effects of ventilation and may not necessarily indicate the origin of the fire.

Recognizing Evidence

NFPA 1033 (2022): 4.4.2

Throughout the interior examination, investigators must recognize potential evidence so that it can be preserved and documented as found. Potential evidence may relate to the area of origin and the cause of the fire. Investigators must be able to recognize and correlate the evidence. For example, forced entry damage does not indicate how the fire began, but it may have evidentiary value and should be documented when discovered.

An investigator must secure and protect any material that may become evidence. If a fire investigator is unable to protect the material in its current location, the findings should be documented, and measures taken to collect the item or material as soon as possible.

Scenes with Victims and Fatalities

NFPA 1033 (2022): 4.4.1, 4.4.4

In terms of using a systematic approach at an incident, investigators should approach the scene of a fatal fire as they would any other structure or vehicle fire. Because the presence of a fatality will require additional team members in the response, scene protocols or standard operating procedures (SOPs) should be in place. Autopsy data is essential to corroborate fire scene findings (McAllister, 2014).

Discovering and Managing Fatalities

NFPA 1033 (2022): 4.4.1

As soon as a body is discovered, and it is determined that the victim is deceased, firefighters should minimize the effects of firefighting in the area where the victim was discovered. Fire suppression personnel should not touch or remove the body from the scene until both documentation and examination of the body and the area surrounding the body have been completed. Premature removal of the body often results in contamination of the body and a potential loss of essential evidence.

Local jurisdictional policies should always be followed when dealing with a fire fatality. The incident commander should understand the importance of leaving bodies in place and securing the scene prior to documentation and examination occurring. A body and its surroundings should be considered evidence and be treated as such.

The ONLY exceptions to leaving the body in place as found include:

- Imminent collapse of the building
- Uncontrollable fire approaching the vicinity of the body

Documenting Fatality Scenes

NFPA 1033 (2022): 4.4.1, 4.4.4

In addition to normal fire scene documentation, investigators at a fatal fire scene need to show the entire scene, including the surrounding area. To accomplish this, investigators may perform the following sequence of tasks:

- Photograph the body and body parts and their surroundings as soon as possible. All exposed surfaces of the body should be photographed prior to debris removal.
- Create a video record of the entire scene, per SOPs.
- Diagram the entire scene (multiple diagrams).
- Sketch the entire scene (multiple sketches).
- Sketch the scene showing the position of the body relative to its surroundings.
- Complete an inventory of recovered remains and artifacts.
- Photograph the completed inventory.
- Document findings.

Skill Sheet 10-2 shows the steps for managing victims and fatalities. Additional photographs should be taken during the debris removal process, when additional evidence may be uncovered. Photographs should be taken of the body as it is being moved within the scene and also during the removal process. All burns and injuries should be photographed during the examination, with an indicator of the photographed items' scale. The body may show important fire effects, fire patterns, and/or blast effects. This evidence may be a crucial component of the scene examination.

Fatality Scene Debris Search

As indicated earlier in this chapter, one method of conducting debris search involves dividing a scene into a grid. An investigator can use string, surveyor ribbon, or a pre-made grid tool to divide the area for systematic search.

An investigator can place a tarp away from the scene, on a flat surface, and use tape to draw a corresponding scene grid onto the tarp. As team members remove debris from the scene, they place it in the corresponding area on the tarp, then sift the debris **(Figure 10.9)**.

A grid system consists of two sets of parallel lines running at right angles to each other. When utilizing this search technique, the investigator should:

- Establish a point of reference

- Establish a baseline

- Establish the grid within the scene

- Number each square within the grid and draw a corresponding grid over the scene plan

The investigator should collect all debris surrounding the body **(Figure 10.10)**. The debris may be placed in containers to be removed to the exterior or sifted in place. Each container should be identified by the grid location from which the debris was removed. This on-scene documentation may be evaluated again at a laboratory.

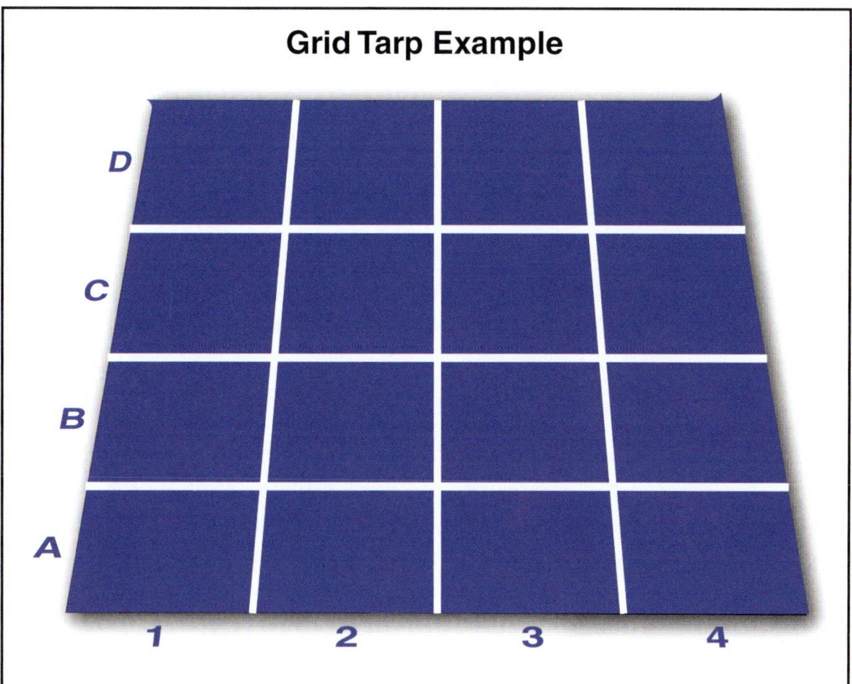

Figure 10.9 One technique for organizing debris removal is to assign a grid to a search area and repeat that grid sequence on a tarp.

Figure 10.10 At a fatality incident, the investigator should collect all debris surrounding the body to screen for significant findings. *Courtesy of Donny Howard, Agent, Oklahoma State Fire Marshal's Office.*

Investigators can use small battery-operated vacuums to collect debris. After collection, the debris should be sifted using multiple sized screens. Sifting with a ½ inch screen followed by a ¼ inch screen is normally sufficient to collect any evidence found within the debris. The fire debris may need to be sifted while wet. Any debris that could contain bone fragments should be sifted when dry. Gypsum wallboard and human bone have a consistency similar to chalk which can disintegrate when handled wet.

Biological Evidence

The severity of the thermal trauma on tissues may make it more difficult to differentiate between human, animal, or commingled remains. Careful and complete sifting using multiple size screens is often necessary. Investigators should carefully search the fatality scene for clues as to the cause and **manner of death**, and for evidence of the origin and cause of the fire.

Critical evidence is often found within arms-length of the body. Smaller bone fragments or teeth may be present within the fire debris underneath or in the area surrounding the body. Smaller masses, such as fingers, feet or entire limbs, may be consumed during the fire. The remains of a smaller person, such as a child, may be completely destroyed during a fire.

NOTE: Investigators should seek timely assistance from a forensic anthropologist when they are uncertain if debris that looks like gypsum wallboard also contains bone fragments.

Fire investigators should be aware that human body fat may be rendered (melted) by the heat of a small flame, resulting in self-sustaining flaming combustion. The fat can continue to burn as it is absorbed into the victim's clothing or onto furnishings where the victim was sitting or lying down at the time of the fire. Often, a protection pattern will be present under the body where the fire did not consume the clothing.

Depending on the circumstances surrounding the fire, the investigator may also have the clothing tested for an ignitable liquid, blood, or explosive residue. Identification of fire death victims can include reliable sources of information such as:

- DNA
- Fingerprints
- Dental comparison
- Skeletal comparison
- Identifying marks such as scars and tattoos

Less reliable sources of identification include:

- Clothing
- Other possessions
- Identifying documents on the person
- Visual identification of poorly preserved remains

Quantifying the Incident

Incident scenes where fatalities or injuries have occurred are more complicated than scenes without those considerations. In addition to the types of observations used at non-fatal fire scenes, such as fire effects and fire patterns, investigators at a fatality scene must also answer questions such as:

- Does the scene itself make sense?
- Does the position of the body appear to fit with the rest of the scene?
- Does unburned clothing on the body indicate that a victim was appropriately dressed at the time of the fire?

The following situations describe fires in which fatalities are likely to occur:

- Accidental fires where the victim is subjected to heat, flame, smoke, or other toxic gasses.
- Accidental fires caused by other conditions, such as a heart attack that causes a smoker to drop a cigarette.
- Incendiary fires that result in intended or unintended deaths, such as a homicide or a firefighter fatality.
- Incendiary fires where the intent is to conceal a homicide or suicide scene.

Response Teams

The following questions will help an investigator determine important facts at a fatality fire scene:

- Who is the victim?

- How did the victim die?

- When did the victim die (before or after)?

- Why did the victim die — asphyxiation, burns, falling debris?

- What was the **cause of death** (result of fire or unrelated cause)?

- Where did the victim die? Does the location of the body provide clues to the fire's origin?

The questions will help an investigator present an informed and substantiated opinion on the circumstances around the victim's condition **(Figure 10.11)**. The cause of death is the injury or disease that started an uninterrupted chain of events that resulted in death. The manner of death reflects the circumstances of death. There are five manners of death:

- Natural

- Homicide

- Suicide

- Accident

- Undetermined

The investigator may not be able to establish a cause and manner of death until after all three points of the so-called Golden Triangle of Death Investigation (Bowes) are finalized:

- The history

- The scene

- The autopsy (**postmortem examination**)

Figure 10.11 An investigator has additional responsibilities at an incident with a fatality. *Courtesy of Donny Howard, Agent, Oklahoma State Fire Marshal's Office.*

Forensic Pathology

Forensic pathology applies medical science to legal questions (DiMaio, 2001). Forensic pathologists apply their training and skills to determine as much information about a fatality as possible.

The pathologist may attend the fatality scene and work with the fire investigator to gather information about the victim, including:

- Photographs

- Fingerprints

- Physical description

- Medical and dental records including x-rays

- DNA from victim and relatives if necessary

The ability to collect and test DNA can yield precise information indicating the identity of the deceased or other persons who were present at the fatality scene. A forensic pathologist will often get the best results from testing with whole blood. Other sources for DNA include bone marrow, tissue from brain, skin, or muscle, and hair with roots.

Should only a portion of the body remain the documentation should follow standard procedures such as noting skin color, anatomic parts present, scars, and tattoos. If only bone or teeth remain, those materials may also be tested.

During a postmortem examination (autopsy), a forensic pathologist can establish more details that may indicate the victim's identity, cause of death, and manner of death. Factors considered during a postmortem examination include results from X-rays of the victim and effects. X-rays can show:

- Indicators of age and injuries
- Indicators of identity such as unique physical attributes
- Potential hazards, such as sharp objects including bones and metal parts

Depending on local jurisdictional policies, the fire investigator may attend the postmortem examination to brief the pathologist about the scene and answer any questions the pathologist may have. The postmortem examination in fire-related deaths may determine if the victim was alive at the time of the fire. Fire causes are discussed later in this chapter.

At a postmortem examination, an investigator can perform or document the following tasks:

- Photograph the post mortem examination
- Collect and dry items removed from the victim
- Describe jewelry, clothing, and other possessions
- Document height, weight, apparent ethnicity, hair and eye color, scars, tattoos

Common tests include:

- Sexual assault
- Carbon monoxide levels
- Drugs, poisons, alcohol, toxins

Other samples taken during the postmortem examination test for characteristics and condition of:

- Blood
- Airways
- Internal tissue
- External tissue
- Stomach contents

Forensic Dentistry

An examination of the victim's teeth and jaw may assist with age determination. To that end, an investigator may ask a forensic odontologist to aid in identifying a victim through dental records and examinations.

Dental work and records accumulate over time. They may be completed and/or stored at multiple locations. Incomplete or inaccurate dental records further complicate the ability of a researcher to find definitive information.

Forensic Dental Examination

Dental records often include enough unique detail to serve as an identification method. A forensic odontologist, a dentist with additional training, can offer assistance in identifying human remains from a fatal fire. An investigator can aid a forensic odontologist by understanding and providing information from dental artifacts.

For example, an average adult human mouth has 28 teeth, excluding wisdom teeth which are often removed. Each tooth has five exposed surfaces:

- Cheek side
- Tongue side
- Biting surface
- Mesial (towards front of mouth)
- Distal (towards rear of mouth)

Forensic Anthropology

Forensic anthropology is the study of the human skeleton applied to a legal setting. To identify skeletal remains, forensic anthropologists look for:

- Indicators that show the victim's sex and age
- Skeletal fragments that are recognizable as human
- Any material that looks like human bone, in the context of the fire scene
- Indicators that show if a bone or fragment belonged to a human or an animal
- Anatomical characteristics that indicate where the bone or fragment belongs in the skeletal system

Causes of Death

NFPA 1033 (2022): 4.4.1

Fire deaths may occur days or weeks after exposure. More fire victims die of asphyxiation than as a result of burns sustained during the fire. Investigators should research whether their jurisdiction's incident history shows trends in common causes of fire deaths.

Carbon Monoxide (CO)

Carbon monoxide is a colorless, odorless, tasteless, non-irritating poison gas with a relative density a little lighter than air. During combustion, organic materials possessing an adequate supply of oxygen can produce CO, which displaces oxygen concentrations in air.

Carbon monoxide (CO) produces hypoxia by limiting a person's ability to use oxygen, because carbon monoxide (CO) attaches more readily to hemoglobin (Hb) than oxygen (O^2) does. Once CO attaches to the blood's hemoglobin, the blood carboxyhemoglobin (COHb) level rises rapidly.

Carboxyhemoglobin is not a definitive measure of the conditions the victim experienced. For example, healthy older adults tend to present lower saturation levels; children tend to present higher levels. Furthermore, COHb levels immediately decrease when a respirating victim receives supplemental oxygen.

Inhalation Injuries

Inhalation of hot gases may result in severe injury or death. Burns or inflammation from inhaling smoke or other chemical irritants causes rapid swelling of the tongue or pharynx which results in asphyxiation. In addition to burn patterns, the presence of soot in the trachea past the vocal cords is conclusive proof of active respiration in fire conditions **(Figure 10.12)**. Facial burns will normally be present when a victim has also sustained inhalation injuries.

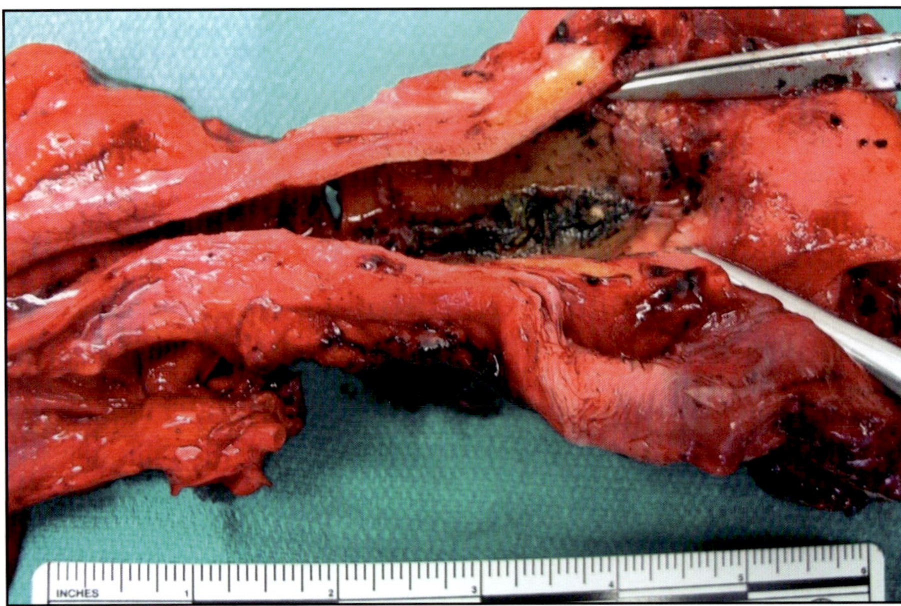

Figure 10.12 A photograph of a dissected trachea shows evidence of smoke inhalation. *Courtesy of Wayne Chapdelaine, Metro-Rural Fire Forensics.*

Burn Injuries

Burn injuries may indicate that a victim had been exposed to specific scene conditions ahead of death, such as fuel-limited burning. An investigator must ensure proper documentation of burn injuries and investigate all scenes where potentially fatal burn injuries were sustained. For example, burns with blisters may be evidence that the victim was alive at the time of the fire, because it indicates that blood was still circulating at the time of that damage.

The investigator should note the position of limbs. Limb positions can indicate whether the body was exposed to intense heat conditions that coagulated and dehydrated body tissue and muscle. For example, a burned body may show a posture with arms raised and clenched fists, known as *pugilistic attitude*. This arrangement may occur after death, as opposed to a living victim's reflexive attempt to defend from a threat.

Documenting the burn patterns may be helpful to determine the victim's activities and location during the fire. A body diagram should be used to document any wounds or trauma. Damage including burns should include an indication of severity. Burn severity is measured in two ways:

- Degrees indicate a measurement of injury depth
- Percentages indicate an estimated surface area affected

When describing a burn depth, using a standardized scale to indicate severity will increase the precision of the record. For example, the National Institute of General Medical Sciences defines 6 degrees of burns:

- **1st degree** — Outer layer of skin (epidermis); usually heals within a week.
- **2nd degree** — Epidermis and second layer (dermis) damage and blistering; may need a skin graft and may leave a scar.
- **3rd degree** — Destruction of both layers of skin; will require skin grafts.
- **4th degree** — Damage extending into the fat under the skin layers.
- **5th degree** — Damage extending into muscle.
- **6th degree** — Damage extending into bone.

To describe burn surface area, the Rule of Nines is a standardized system to categorize the surface area of a human body. For example, on an adult, both legs (18% x 2 = 36%), the groin (1%) and the front chest and abdomen involve approximately 55% of the body, when calculated as follows:

- Head = 9% **(Figure 10.13)**
- Each arm = 9% x2
- Chest (front) = 9%
- Abdomen (front) = 9%
- Groin = 1%
- Upper back = 9%
- Mid/lower back and buttocks = 9%
- Each leg, front and back = 18% x2

Human Behavior in Fire Conditions

An investigator must understand common basic human behaviors in fire conditions, to effectively determine whether the responses seen at the scene match those expectations. Several factors will affect how a person behaves in fire conditions.

Human behavior is affected by fire in many ways. Vulnerabilities and characteristics that complicate reactions in a fire situation can make rescue, recovery, and survival much more difficult. Physical limitations or vulnerabilities that complicate a fire scene can include:

- Intoxication
- Physical disabilities
- Sensitivity to toxins
- Circumstances that limit mobility
- Incapacitating injuries or medical conditions
- Compromised or inefficient circulatory or respiratory systems

Cognitive comprehension limitations:

- Mental illness
- Comprehension
- Developmental disabilities
- Person's level of rest/fatigue
- Alcohol, drugs, and medication
- Age related mental development
- Inhalation of smoke or toxic gases **(Table 10.1, p. 302)**

Physical setting characteristics and familiarity:

- Number and accessibility of means of egress
- Inadequate number of exits, blocked escape routes
- More likely to accurately assess a fire's progression at home than in an unfamiliar setting

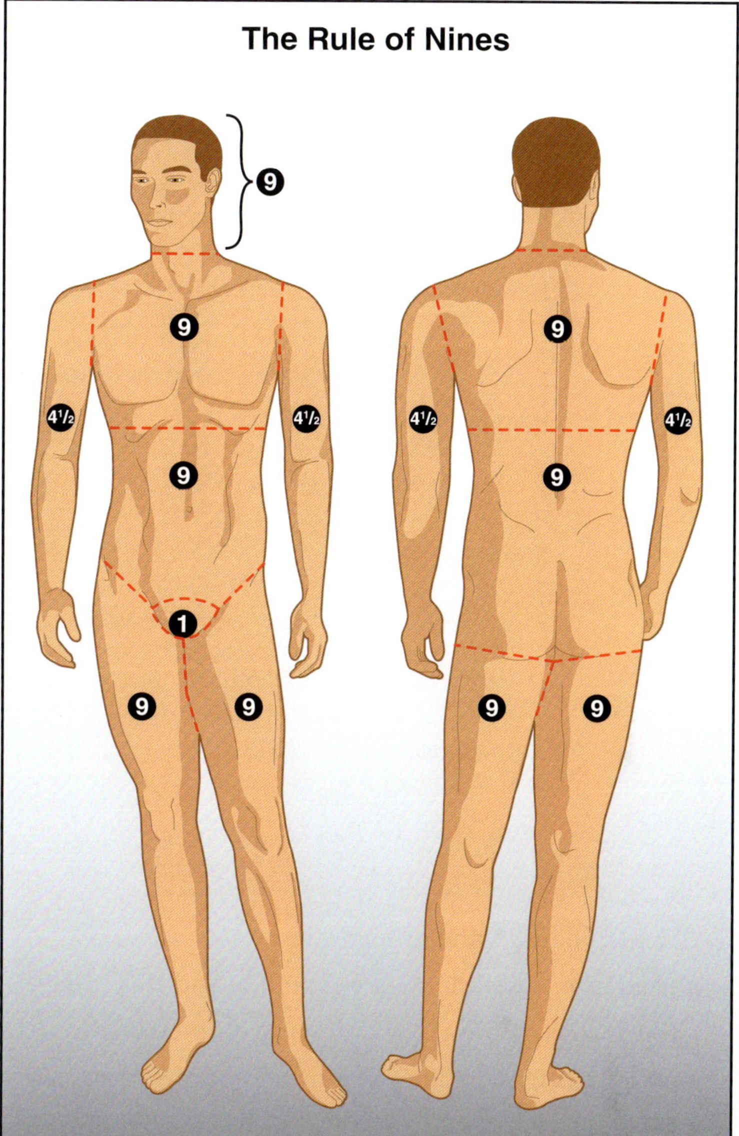

Figure 10.13 Calculating the surface area of burns on a victim or remains can be generalized by estimating the area of body zones.

Table 10.1
Common Products of Combustion and Their Toxic Effects

Carbon Monoxide	Colorless, odorless gas. Inhalation of carbon monoxide causes headache, dizziness, weakness, confusion, nausea, unconsciousness, and death. Exposure to as little as 0.2 percent carbon monoxide can result in unconsciousness within 30 minutes. Inhalation of high concentration can result in immediate collapse and unconsciousness.
Formaldehyde	Colorless gas with a pungent, irritating odor that is highly irritating to the nose. 50-100 ppm can cause severe irritation to the respiratory tract and serious injury. Exposure to high concentrations can cause injury to the skin. Formaldehyde is a suspected carcinogen.
Hydrogen Cyanide	Colorless, toxic, and flammable liquid below 79°F (26°C) produced by the combustion of nitrogen-bearing substances. It is a chemical asphyxiant that acts to prevent the body from using oxygen. It is commonly encountered in smoke in concentrations lower than carbon monoxide.
Nitrogen Dioxide	Reddish-brown gas or yellowish-brown liquid, which is highly toxic and corrosive.
Particulates	Small particles that can be inhaled and deposited in the mouth, trachea, or the lungs. Exposure to particulates can cause eye irritation, respiratory distress (in addition to health hazards specifically related to the particular substances involved).
Sulfur Dioxide	Colorless gas with a choking or suffocating odor. Sulfur dioxide is toxic and corrosive, and can irritate the eyes and mucous membranes.

Source: *Computer Aided Management of Emergency Operations (CAMEO) and Toxicological Profile for Polycyclic Aromatic Hydrocarbons.*

Group dynamics and group size:

- Larger groups take longer to mobilize and clear an area.
- Groups with a formal structure and defined leaders usually react more quickly.
- Individuals tend to delay response to initial signs of fire until others in the group begin to react.

Debris Removal

NFPA 1033 (2022): 4.2.6, 4.4.2

Debris at a scene may obstruct important evidence. As a result, debris removal is often required for a complete and thorough scene investigation. Debris removal processes must be methodical and planned.

Examples of debris include:

- Burned structural members
- Collapsed walls and ceilings
- Partially burned building contents

The key objectives of debris removal are to:

- Locate physical evidence.
- Discover fuel packages and contents.
- Facilitate locating the point or area of origin.
- Reveal potential ignition sources and the material first ignited.
- Reveal hidden fire effects as well as intensity and movement patterns.

Excavation Process

NFPA 1033 (2022): 4.2.6

Before debris removal and reconstruction begins, the fire investigator should have developed one or more preliminary ignition sequence hypotheses. During the excavation process, the investigator tests those hypotheses against the findings at the scene. Should the evidence indicate that a certain sequence is not probable, the investigator should consider other hypotheses. If no probable sequence can be determined, the investigator may need to remove and analyze debris in other areas of the scene.

As with all operations during the scene examination, the fire investigator must proceed with caution to avoid damaging, destroying, or contaminating potential evidence. Some debris may support other debris or structural components. The removal of key pieces of debris may result in structural collapse.

What This Means to You

The process used by investigators to excavate a fire scene resembles the tasks performed by an archeologist. In both cases, the layers of debris represent periods of time. For an archeologist, the layers can represent many years; for a fire investigator, the layers represent the history of a fire before ignition through extinguishment.

An archaeologist examines materials that have changed due to age and decay. The fire investigator examines materials that have changed due to heat and flame exposure and/or fire suppression activities.

Excavation Locations

NFPA 1033 (2022): 4.2.6

The investigator must determine where to start examining and excavating, based on analyses of fire movement and intensity patterns. To determine this location, the investigator should consider:

- Size of the total area damaged

- Type, degree, and location of damage

- Number of rooms or compartments involved

- Amount of debris and its composition (**layering**) (**Figure 10.14**)

- Patterns present that seem to indicate an origin area

The investigator must select an area that is large enough to:

- Examine nearby patterns and their correlation

- Identify ignition sources and first material ignited

- Support documentation of initial fire development and spread patterns

Figure 10.14 Debris from a collapse zone includes structural members and contents from higher floors. *Courtesy of Donny Howard, Agent, Oklahoma State Fire Marshal's Office.*

Debris Layers

In addition to providing clues as to the area(s) of origin, damage configuration helps the investigator assess the hazards present during debris removal. The damage configuration also provides information regarding what likely exists in the upper layers of the debris. For example, broken ceiling joists or rafters are typically discovered during the early stages of layer removal. These structural items usually fall on top of other contents that have already burned.

Layering is a systematic process that investigators can use to examine and remove fire debris. At a fire scene, the top layer of debris generally comes from the highest point in the structure while the bottom layer of debris comes from the lowest point in the structure. Layering at an explosion scene may not be the same as layering at a fire scene. At an explosion scene, the debris may become mixed during the positive and negative pressure phases of the explosion.

Systematic analysis of debris layers can provide the investigator with potential evidence. For example, finding plasterboard with the ceiling side unburned and the attic side burned on a layer near the floor could indicate that the fire began in the attic, rather than in the room below. If the fire had started in the room, the ceiling side would likely demonstrate greater damage than the attic side and the plasterboard would likely appear among upper layers of debris rather than near the floor.

Regardless of the size of the scene, the investigator should thoroughly document the removal process with photographs and grid diagrams. Establishing a grid promotes systematic examination, helps to document the process, and aids in proper reconstruction.

When using a grid system, the scene is subdivided (without disturbing debris) into manageable segments that are marked using string, rope, barrier tape, or another marking system. Once the space is subdivided, an investigator can begin the debris removal process one grid at a time. An investigator should document each grid-search sequence in notes, sketches, and photographs. If using a portable electronic device, the investigator may be able to annotate the photographs with grid lines.

Protected Areas

Protected areas should also be examined and documented as part of origin determination **(Figure 10.15)**. If possible, large items found in the debris, such as furniture or appliances, should be left intact. If the items must be removed to facilitate the examination of the area, they should be placed in an accessible location for additional examination and potential reconstruction.

Figure 10.15 Protected areas may not seem significant, but they still indicate placement of items during fire conditions. *Courtesy of Donny Howard, Agent, Oklahoma State Fire Marshal's Office.*

The fire investigator may also locate structural components, such as ceiling materials or wall-framing components. The investigator should carefully examine and document these components. If the investigator can determine the prefire location of a component, the patterns found on the component may assist in determining the direction of fire movement.

Debris Management

NFPA 1033 (2022): 4.2.6, 4.4.2

Debris removal plans may change as the investigator collects new information. For example:

- When fire development patterns indicate a localized area, the debris removal area can be confined to that localized area.

- An investigator may need to expand the debris removal area as localized areas of origin are ruled out.

- In some cases, the removal can encompass the entire damaged area or confines of the structural foundation.

When choosing an area to excavate, the investigator should prioritize evaluation of origin area patterns and damage. An investigator who only focuses on assessing a possible ignition source may report inaccuracies in:

- Classification

- Cause determination

- Alleged responsibility

Care in a Fire Scene

Investigators must consider the potential importance of all fire scene debris and evidence. Items discarded as simple debris may later prove to be:

- Evidence

- Otherwise needed to help an investigator understand a scene

- Necessary for determining the original locations of other items

Layering Process Questions

During the layering process, the fire investigator should answer the following questions:

- Are there directional fire effects or patterns on the material?

- What was the prefire location of the material being examined?

- Does the damage and debris indicate the site may have included an explosion?

- Was the debris layer created in fire conditions or as a result of fire-suppression activities such as overhaul?

The investigator should take photos of each item as it was found and note its location on a floor plan or drawing to document the removal process. After indicating where items were found in the excavation area, the investigator may separate items from the other layers for further examination. For example, the positions and locations of door or window locks, window glass, and interior finish materials can indicate site conditions before, during, and after the incident.

As the layers are removed, additional fire patterns may become visible. For example, investigators may find unburned material below a burned layer, which could indicate the level at which the ignition occurred. Also, a layer of white/gray ash may indicate areas of more complete burning, possibly resulting from lack of extinguishment during suppression efforts.

Debris Removal Strategy

After establishing where to excavate, where to take debris, and how to move materials, the investigator formulates a debris removal strategy. Regardless of scene size, the investigator must continually examine the debris being moved.

The investigator must consider:

- Debris should be removed in such a manner that the investigator can accurately determine the debris layering sequence.

- Debris should be examined in a given area, then remove the top-most layer before examining the next area **(Figure 10.16)**.

- Debris removal at a relatively small incident requires fewer work hours than at a larger incident.

- At larger incidents, teams may be tasked to aid a lead investigator.

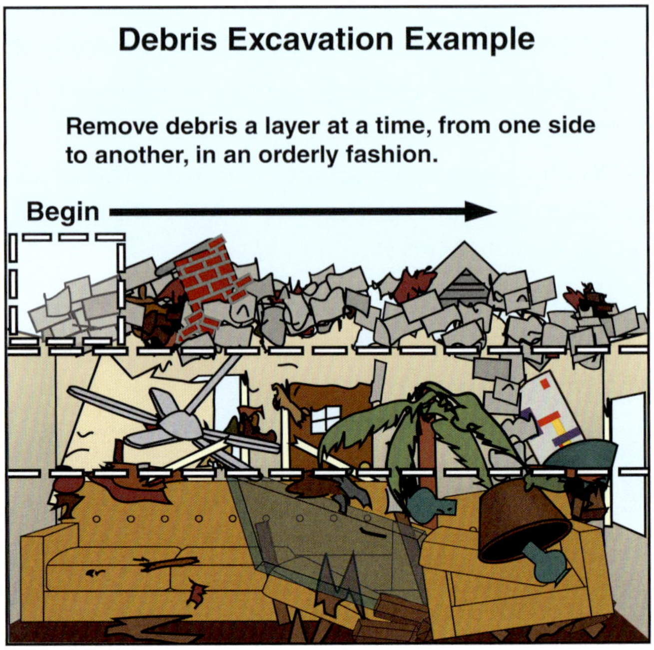

Before starting debris movement, investigators must also identify the path of travel that will be used to move materials away from the excavation site. An ideal examination location:

- Prevents contamination or destruction of potential evidence in other areas of the building

- Provides a debris staging area that sufficiently allows for detailed examination of removed debris

Figure 10.16 Investigators should remove debris in a logical pattern.

Layering Process Tools

In the immediate origin area, debris removal and layering require careful attention to detail. Hand tools are useful in origin areas with extensive damage. Tools used to reveal, identify, and preserve small pieces of evidence include:

- Trowels **(Figure 10.17)**

- Whisk brooms

- Magnets to locate and isolate pieces of ferrous (iron-based) metal

- Small sieves or screens with different-sized openings

Debris may be collected in plastic tubs. Sifting is usually performed before collecting debris in the tubs.

Figure 10.17 Near the origin area, investigators may use hand tools to remove small debris and locate small pieces of evidence. *Courtesy of Nicole Brewer, Portland Fire Investigations.*

Debris that has melted into a large, solid object may be examined at the scene if the material is soft enough for an investigator to melt it with a heat gun **(Figure 10.18)**. Otherwise, solidified debris may be examined with X-rays and CT scans to determine whether evidence is present in the melted material **(Figure 10.19 a-b)**.

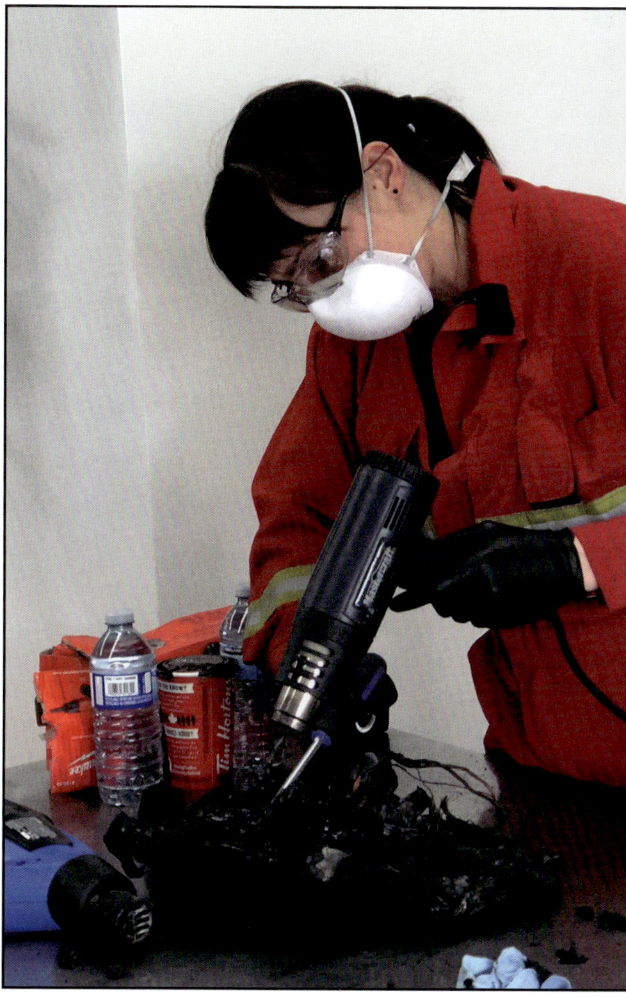

Figure 10.18 Investigators can use a heat gun to separate fused items. *Courtesy of Wayne Chapdelaine, Metro-Rural Fire Forensics.*

Heavy Machinery

The use of heavy machinery may be especially helpful at large area scenes with extensive damage including roof and structural support collapse. Concerns with using heavy equipment during excavation and debris removal include:

- Equipment such as front-end loaders, excavators, or backhoes may cause extensive damage to floor structures.

- Heavy machinery is usually fueled with ignitable liquids, which may contaminate the scene.

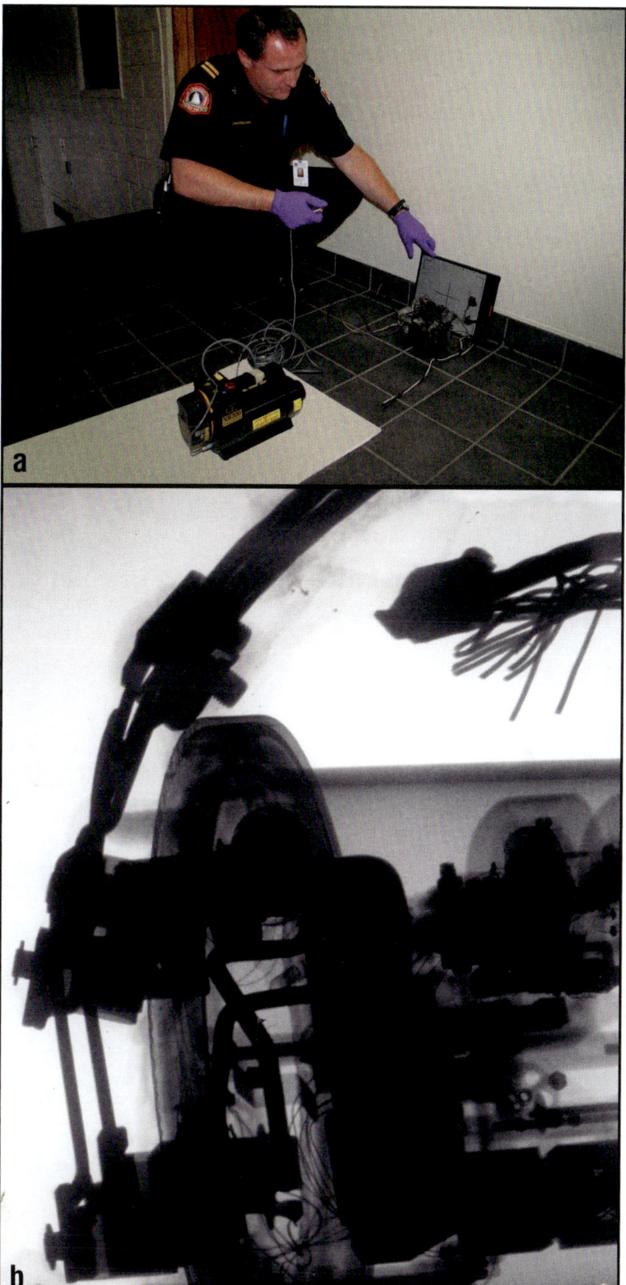

Figure 10.19 a-b Portable X-ray machines can provide investigators an image of the interior of larger pieces of debris. *Courtesy of Wayne Chapdelaine, Metro-Rural Fire Forensics.*

Scene Reconstruction

NFPA 1033 (2022): 4.2.4, 4.2.5, 4.2.7, 4.6.5

As part of incident analysis, fire investigators attempt to rearrange the fire scene to match prefire conditions. Functionally, fire scene **reconstruction** involves the relocation of burned materials and contents. Structural elements (wiring), and utility services (piping) are returned to their prefire positions after the fire debris has been removed. A reconstruction can help investigators visualize fire patterns and conduct a more thorough analysis of the fire scene.

Reconstructing the Scene

NFPA 1033 (2022): 4.2.7

Reconstructed scenes allow a fire investigator to consider fire patterns within the context of their development. Fire pattern analysis is discussed in Chapter 12, Fire Origin.

To successfully reconstruct a scene, fire investigators should:

- Recreate the scene with contents in their prefire locations.
- Document fire growth patterns.
- Identify and document potential ignition sources.
- Identify and document first material ignited.
- Document and collect evidence.

Resources that can assist investigators in scene reconstruction include:

- Thorough notes and close attention to detail during debris removal
- Information from interviews or prefire photographs
- Diagrams or notes with measurements that indicate the exact location where the materials were discovered during debris removal
- Presence of fire patterns and protected area patterns

Fire patterns may not always indicate an item's exact orientation, position, or location. Some items may have been moved or removed from their expected position. This is an example of why the reconstructed scene may be more definitive for determining the origin area than the fire patterns themselves.

After completing the fire scene reconstruction, the investigator should sketch and photograph the entire reconstructed scene. At this stage, analysis of the scene may reveal patterns and items that may be significant as evidence.

Hypothesis Testing

NFPA 1033 (2022): 4.6.5

Reconstructing the fire scene allows the investigator to directly observe some hypothesis-testing considerations in the precise space where the fire happened, such as:

- Protected areas
- Fire effects and patterns on adjacent materials
- Indications of missing items based on disruptions in patterns
- Areas that look different than expected from the initial walkthrough, for example, an unexpected indication of a possible area of origin

All potential sources of ignition in the area of fire origin should be identified and tested against the working hypothesis. If one ignition sequence hypothesis emerges as the most plausible in this review, all evidence that supports or negates that hypothesis should be collected and preserved. If more than one hypothesis remains plausible, more evidence should be evaluated and collected.

Review Questions

1. What should the fire investigator confirm when performing an on-scene security assessment?

2. What should the investigative teams include when using heavy equipment to excavate large areas?

3. What professional support will an investigator require when responding to a fatality incident?

4. What types of clues can be found at the exterior of a fire scene?

5. While conducting the exterior examination, what can a fire investigator determine, identify, and interpret?

6. What should the fire investigator do to substantiate any opinions and analysis offered?

7. When a fire investigator determines that the incident merits a detailed examination of the interior of a structure, what should the investigation include?

8. What should fire investigators do if they are unable to protect the material in its current location?

9. What are the only exceptions to leaving the body in place as found?

10. What should the investigator do when utilizing the grid system search technique?

11. What are the three points of the golden triangle of death investigation?

12. What should the fire investigator have developed before debris removal and reconstruction begins?

13. What are the key objectives of debris removal?

14. What should the fire investigator consider to determine where to start examining and excavating?

15. What is layering?

16. What is a fire scene reconstruction?

17. What should be done if one ignition sequence hypothesis emerges as the most plausible?

Discussion Questions

1. What should the fire investigator do when finding potential evidence at the exterior of a structure?

2. What kind of building systems can be found in your jurisdiction?

3. What should fire investigators base their hypotheses on?

4. What is the trend in common causes of fire deaths in your jurisdiction?

Chapter References

Bowes MD, Matt. Forensic Pathology Review. Accessed online.

DiMaio, Vincent, and Dominick DiMaio. 2001. Forensic Pathology (Practical Aspects of Criminal and Forensic Investigations) 2nd Edition. CRC Press; New York.

Locard, Edmond. 1934. "La police et les méthodes scientifiques." Bibliothèque illustrée, number 24.

McAllister, J.L., Carpenter, D.J., Roby, R.J. et al. 2014. The Importance of Autopsy and Injury Data in the Investigation of Fires. Fire Technol 50, 1357–1377. Springer Link. Accessed online.

Chapter Key Terms

Cause of Death — Specific injury or disease that leads to death.

Consistent Methodology — A system of broad principles or rules that define procedures and methods to help investigators work predictably and efficiently within the scope of a fire and/or explosion scene. A common example is the Scientific Method.

Forensic Pathologist — Science of determining the cause and manner of death.

Layering — Deposition of fire debris in identifiable layers, such as above or below a floor assembly, ceiling materials, or roof assembly.

Manner of Death — Determination of how the injury or disease leads to death.

Medical Examiner — Medically qualified government officer whose duty is to investigate deaths and injuries that occur under unusual or suspicious circumstances, to perform post-mortem examinations, and in some jurisdictions to initiate inquests. *Also known as* Coroner.

Meter Booting — The addition of insulators to an electrical meter by the utility company.

Postmortem Examination — A surgical examination of a dead body to determine the cause of death. *Also known as* Autopsy.

Protected Area — Undamaged surface within an otherwise fire damaged area, possibly resulting from objects shielding the surface from the effects of the fire; generally used to refer the fire investigator to where large objects such as furniture were positioned before the fire.

Reconstruction — Process of excavating and rearranging a fire scene with the intent to rebuild all or part of an area to its prefire configuration.

NOTE: The scope of examination will be dependent upon the characteristics of the particular fire or explosion scene. The steps in this skill may be accomplished in a different order and repeated as many times as necessary, depending on the situation at the scene. The fire investigator should work alongside other professionals, such as a coroner or medical examiner, as the scene requires it.

CAUTION: Use caution when moving debris to ensure that it does not affect scene safety.

Step 1: Assess scene safety and security before entering the perimeter and then continually throughout the investigation.

Step 2: Move systematically around the exterior of the scene, documenting observations along the way.

    a. Field notes

    b. Diagrams and drawings

    c. Photographs/video

Step 3: Record observations and information gathered from firefighters regarding fire suppression operations.

Step 4: Record observations about bystanders and other individuals present at the scene.

Step 5: Identify and manage potential evidence on the exterior. Remove debris as necessary to reveal the evidence.

    a. Secure the evidence in place, if possible.

    b. Photograph and document the evidence.

    c. Collect the evidence, if applicable.

Step 6: Conduct a preliminary walkthrough, if needed.

Step 7: Move systematically throughout the interior, documenting observations along the way.

    a. Field notes

    b. Diagrams and drawings

    c. Photographs/video

Step 8: Conduct a more detailed examination in areas that require it.

Step 9: Identify and manage potential evidence on the interior **(Figure 10.20)**. Remove debris as necessary to reveal evidence.

    a. Secure the evidence in place, if possible.

    b. Photograph and document the evidence.

    c. Collect the evidence, if applicable.

Figure 10.20 *Courtesy of Donny Howard, Agent, Oklahoma State Fire Marshal's Office.*

Step 10: Reconstruct the scene, if applicable.

Step 11: Interpret and record fire patterns and fire effects.

Step 12: Inspect building utilities and systems. Document the results.

Step 13: Analyze and document explosion effects, if applicable.

    a. Effects on glass, walls, foundations, and building materials

    b. Low- and high-order effects

    c. Blast zone and origin

NOTE: This skill sheet provides general guidelines for ensuring that victims and fatalities are protected. Students should follow the SOPs of the jurisdiction. In addition, each investigation may present special circumstances that should be addressed individually as they arise.

Step 1: Contact any specialized personnel that may be needed to assist with examination or recovery of the victim. Work with these personnel as needed throughout the investigation.

Step 2: Protect the area where the victim was found.

   a. Place a barrier to the room or location to prevent unauthorized entry.

   b. Establish a path of entry and search the area prior to approaching the victim.

NOTE: Bodies should not be covered with fire fighting tarps.

Step 3: Systematically search for and mark additional evidence near the victim.

Step 4: Document the area.

   a. Take photographs of the victim.

   b. Record observations such as location and position of the body, clothing, and damage.

   c. Sketch the area surrounding the victim.

NOTE: Do not move the victim until the surrounding area has been thoroughly documented.

Step 5: Maintain the chain of custody throughout the investigation.

Photo courtesy of Donny Howard, Agent, Oklahoma State Fire Marshal's Office.

Chapter Contents

NFPA 1033 JPRs addressed in this chapter

This chapter provides information that addresses the following job performance requirements (JPRs) of NFPA 1033, *Standard for Professional Qualifications for Fire Investigator* (2022):

4.1.7   4.2.1   4.2.9   4.4.2

This chapter provides information that addresses the following course outcomes for the FESHE courses:

Fire Investigation I (C0283)

7. Discuss the basic principles and identify cause and origin of fires.

Fire Investigation II (C0284)

5. Evaluate the use of incendiary devices, explosives, and bombs.

Fire Investigation and Analysis (C0285)

3. Analyze the fire scenario utilizing the scientific method, fire science, and relevant technology.
4. Analyze the legal foundation for conducting a systematic incendiary fire investigation and case preparation.

Learning Objectives

1. Differentiate the damage typically caused by different kinds of explosions. [NFPA 1033, 4.1.7, 4.2.9]

2. Describe explosion effects. [NFPA 1033, 4.1.7, 4.2.9]

3. Describe the characteristics of seated explosions. [NFPA 1033, 4.1.7, 4.2.9]

4. Describe the characteristics of combustion explosions. [NFPA 1033, 4.1.7, 4.2.9]

5. Explain how to approach an explosion scene. [NFPA 1033, 4.1.7, 4.2.1, 4.2.9, 4.4.2]

6. Explain methods for investigating explosion scenes. [NFPA 1033, 4.1.7, 4.2.1, 4.2.9, 4.4.2]

Chapter 11
Explosion Dynamics and Investigation

Courtesy of the U.S. Air Force, photo by Senior Airman Sean Worrell.

A fire investigator must understand the basic science of explosions and be able to recognize the damage caused. This chapter addresses the following topics:

- Characterization of explosion damage
- Explosion effects
- Seated explosions
- Combustion explosions
- Approaching an explosion scene
- Explosion scene examination

Characterization of Explosion Damage

NFPA 1033 (2022): 4.1.7, 4.2.9

Explosions are a physical or chemical conversion of energy that results in the rapid release of high pressure gas into the environment. This is most often a chemical or mechanical conversion of potential energy into kinetic energy. Nuclear and electrical explosions are beyond the scope of this manual. For more information on chemical explosive materials, refer to the IFSTA manuals, **Hazardous Materials for First Responders**, and **Hazardous Materials Technician**.

In general terms, **explosives** are the materials, either a pure single substance or a mixture, that undergo a very rapid chemical or physical change, releasing large volumes of gas(es). The **chemical explosion** process is driven by the explosive material(s).

Mechanical Explosions

Mechanical explosions result from pressure increase within a confined container. Examples of mechanical explosions include dry ice bombs, steam boilers, and pressure cooker failures.

The increase and release of this pressure causes various effects, including:

- Damage to confining compartments and/or containers
- Secondary fires some distance from the point of ignition
- Extensive damage to the area surrounding the explosion

NOTE: Physical confinement is a necessary factor of a mechanical explosion.

In a mechanical explosion, the materials do not undergo a chemical change. As the pressure increases beyond the structural capacity of the container, a rupture forms. The violence of the rupture may depend upon the speed with which the pressure in the container rises and the strength of the container.

Combustion or an ignition source is not necessary to trigger a mechanical explosion; however, there must be some force that increases the pressure within the container. **Boiling liquid expanding vapor explosions (BLEVEs)**, the most common mechanical explosions that investigators will encounter, are discussed in more detail later in this chapter.

Chemical Explosions

Chemical explosions involve exothermic reactions in which an ignition source initiates (or an increase in temperature self-initiates) the explosion, and combustion **propagates** along the **blast-pressure front** of the reaction in all directions. The reaction chemically changes the fuel.

Chemical explosions can result from solid fuels such as explosives; however, investigators will more frequently encounter chemical explosions involving gases, vapors, or dusts suspended in air. Chemical explosive materials are addressed extensively in the IFSTA manual, **Hazardous Materials Technician**.

Deflagrations

Deflagrations are combustion reactions that are generally fueled by diffused gases or dust suspended in air under normal pressure. The fuel may be a pure product or mixture. The velocity of the reaction front is subsonic (slower than the speed of sound). As the pressure on the fuel increases, the speed of the explosion reaction increases. For example, a given amount of natural gas mixed with air will propagate more quickly in a smaller compartment because of its higher degree of confinement, than in a larger compartment.

Deflagrations can also be associated with **low explosives** such as smokeless powder, flash powder and black powder **(Figure 11.1)**. The flame front moves through the fuel at a speed limited by the concentration (density) of the fuel and the ease with which heat can be transferred through the fuel.

Figure 11.1 Debris with fireworks residue contains low explosives. *Courtesy of Donny Howard, Agent, Oklahoma State Fire Marshal's Office.*

Detonations

As the pressure on a fuel increases, it becomes more likely that an explosion in that fuel will reach the level of **detonation**. A detonation may be ignited by a shock and/or a flame.

Detonations are most commonly associated with **high explosive** compounds in which the solid fuel and the oxidizer for the fuel are under high pressure (such as in a stick of dynamite). Detonations are rapid and powerful, with supersonic (faster than the speed of sound) reaction velocities. Also associated with detonations is an intense **shock wave** that causes much greater damage than the blast-pressure front associated with deflagrations. Confinement is not a requirement for a detonation.

While detonations are a subcategory of chemical explosions, detonations can also occur in mechanical explosions involving flammable liquids such as liquid propane gas tanks used to heat homes **(Figure 11.2)**. Detonations may occur with diffused gases when the gases are under great pressure such as in pipes or in pressurized gas canisters.

Stoichiometric Ratio

The **stoichiometric ratio** is a mathematical ratio of fuel and air necessary for complete combustion to occur, leaving no byproducts of combustion, such as ash. Put another way, this is the ratio at which all of the fuel is converted from matter to energy.

Figure 11.2 Liquid propane tanks exposed to high temperatures for sufficient duration can detonate. *Courtesy of Donny Howard, Agent, Oklahoma State Fire Marshal's Office.*

This idealized ratio does not occur at incidents. However, it is helpful to understand the concept of complete combustion to predict and define similar, more common conditions. For example, the most powerful explosions occur in ratios slightly richer than the stoichiometric ratio.

Fuel vapor to air ratio is introduced in Chapter 3, Fire Dynamics. The closer a **fuel-to-air ratio** gets to the stoichiometric ratio, the greater the pressure rise in the explosion. Deflagrations near or slightly richer than stoichiometric can create high-order explosion damage associated with detonations if the compartment is not properly ventilated.

When the air/fuel mixture is slightly richer than stoichiometric, the speed of the flame front is greatest and the ignition of common combustibles less likely because they are not sufficiently heated for a long enough period of time. When the air/fuel mixture increases in richness, the speed of the flame front slows, and allows more time for the combustion of available materials.

Order of Damage

Low- and high-order of damage should not be confused with low and high explosives. Explosion damage is classified by its magnitude:

- **Low-Order Damage** — Generally associated with **nonseated explosions** and deflagrations. Walls may bulge away from the explosion or fall outward, intact from the structure; roofs may be lifted slightly; windows may be dislodged and/ or broken; debris is large and only moved short distances **(Figure 11.3)**.

- **High-Order Damage** — Generally associated with **seated explosions** and detonations. Structure or structural members are shattered or splintered; debris is small, plentiful, and widespread; structure may be completely destroyed **(Figure 11.4, p. 318)**.

Figure 11.3 Walls fallen intact, away from a structure, may indicate low-order-explosion damage. *Courtesy of Donny Howard, Agent, Oklahoma State Fire Marshal's Office.*

Investigators may find that the use of the terms low-order damage and high-order damage may not always be appropriate. The damage present at some explosion scenes may encompass both types of damage. Seated and nonseated explosions are described later in this chapter.

Figure 11.4 High-order-explosive damage, such as at the 1995 bombing of the Murrah Federal Building, shows small debris and shattered structural members. *Courtesy of AP /WIDE WORLD PHOTOS.*

 **Explosive Materials Situational Awareness**

Investigators should not enter a scene if they suspect explosive materials to be present on site. Fire investigators can ensure their own safety by not attempting to exceed limitations in the experience, resources, and certifications they hold.

Explosive materials may create effects that can be seen at a fire investigation scene. Investigators should be aware of the arrangement and presence of materials, common camouflage techniques, and device components that may indicate the presence of an explosive device.

NOTE: Incendiary devices are discussed in Chapter 13, Fire Cause.

Investigators can also expect to encounter the presence of explosives intended for the following targets:

- Mass transit
- Mail rooms
- Religious facilities
- Government buildings
- High profile individuals
- Heavily populated areas
- Places that have previously received threats
- Social conditions including protests and civil unrest

The ATF's Bomb Arson Tracking System (BATS) identified the following four locations as the most frequently threatened types of location targets in 2017 (USBDC):

- Education (high school, middle school, junior high)
- Office/business (department/discount store)
- Residential (single family or two-family dwelling)
- Assembly (restaurants and courthouses)

Should an investigator suspect that explosive materials may be present at a scene, resources that may be called for assistance include:

- ATF
- Local LEO
- Ordnance Disposal

Explosion Effects

NFPA 1033 (2022): 4.1.7, 4.2.9

Explosions, under ideal conditions, project heat and pressure in a relatively spherical shape, expanding in all directions **(Figure 11.5)**. The blast-pressure front created is followed by a flame front that ignites the expanding gases as they move away from the center of the explosion.

The blast-pressure front tends to do the majority of the damage in an explosion. The flame front moves so quickly that there is not sufficient time for heat to transfer to combustible structural members for ignition to occur. In addition, the blast-pressure front may carry objects, pieces of a structure, or pieces of a container (**shrapnel**) all of which can cause additional damage. The sections that follow discuss in greater detail the possible effects of explosions and the factors that control these effects **(Figure 11.6)**.

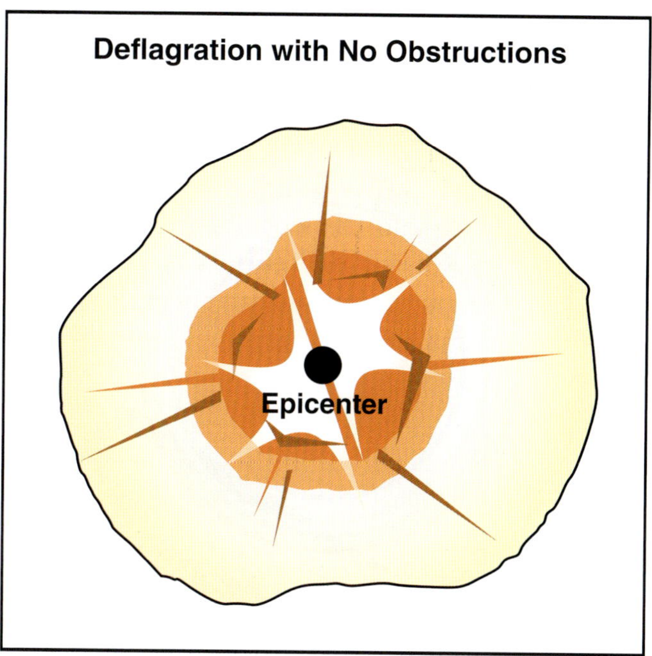

Figure 11.5 Under ideal conditions, a deflagration expands outward in all directions.

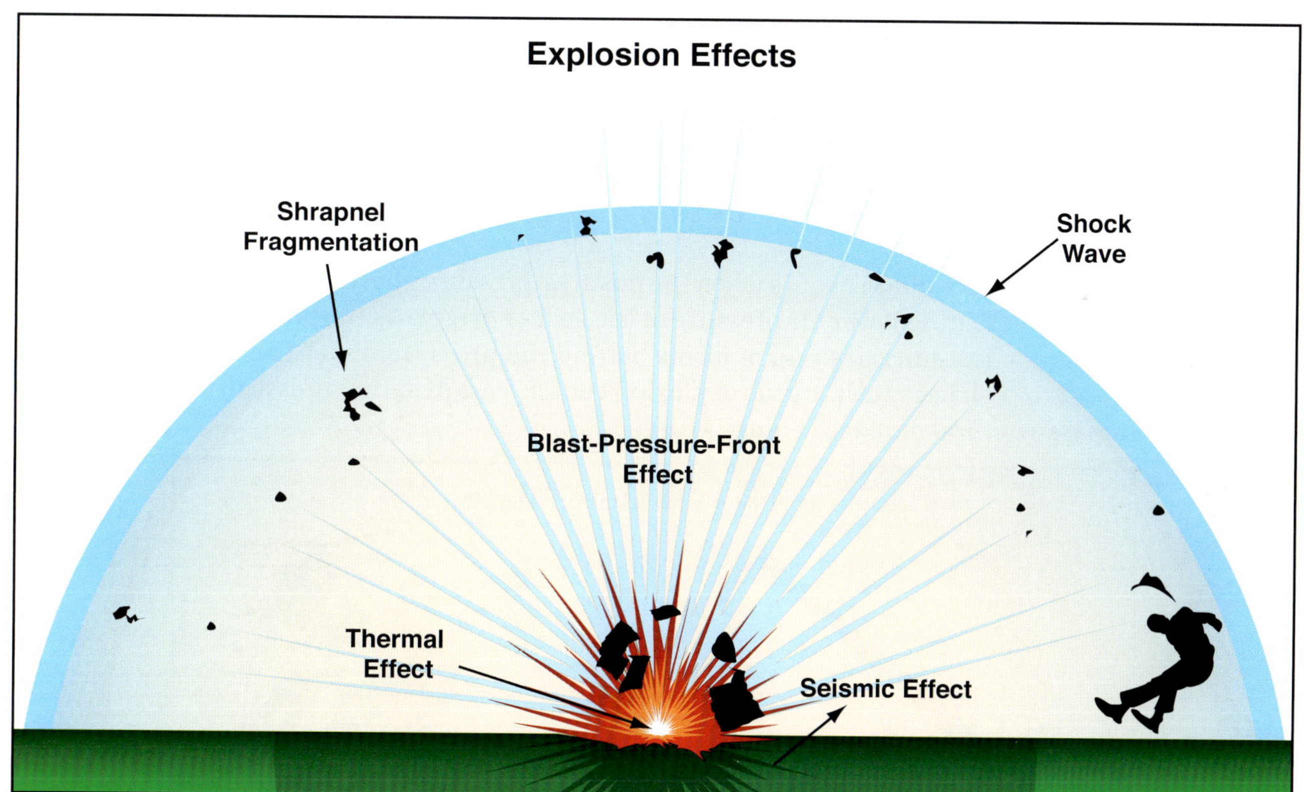

Figure 11.6 Explosion effects include considerations from pressure, heat, and debris.

Blast-Pressure Front Effect

Detonations produce large quantities of gases, which expand very rapidly from the point of origin. These rapidly released gases create a blast-pressure front that travels outward from the center. As the explosion reaction consumes the available fuel, the front moves farther away from the center and the strength of the front decreases. This blast-pressure front is the primary cause of injuries and damage.

Deflagrations also produce a blast-pressure front, in which a slower rate of pressure rise and a lower order of damage do not cause the gases involved to explode. The blast-pressure front in a deflagration moves through and expands the gas/air mixture surrounding the ignition source.

The blast-pressure-front effect has two phases:

1. The **positive-pressure phase** results from the blast wave moving away from the center of the blast as gases expand. The positive-pressure phase is responsible for the majority of damage because of the force it exerts **(Figure 11.7)**.

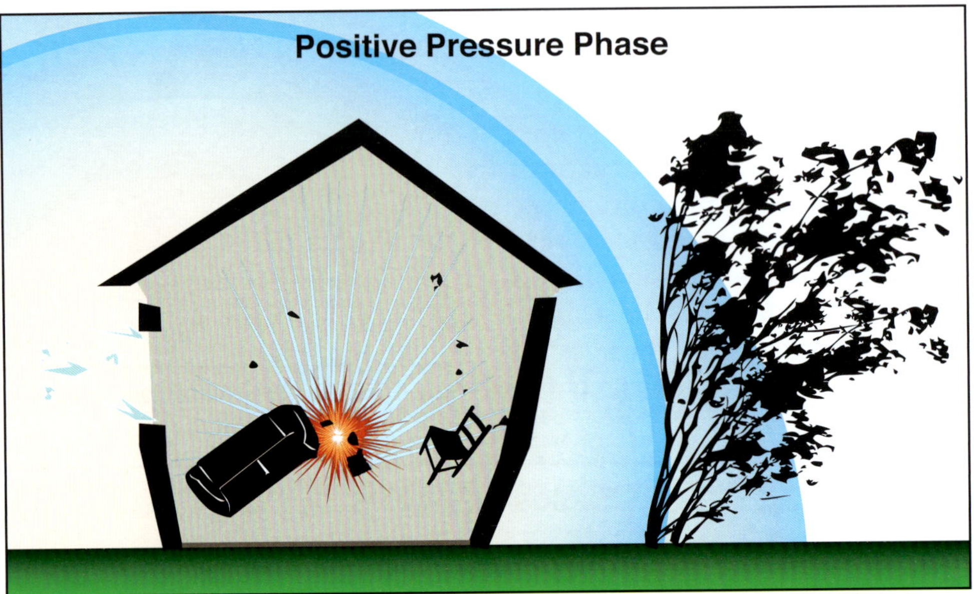

Figure 11.7 The positive-pressure phase is when blast pressure expands outward, causing most of the damage in an explosion.

2. The **negative-pressure phase** occurs when the positive-pressure phase heats the surrounding air, displacing it and creating lower air pressure at the seat or origin. As the positive-pressure phase moves outward, a low-pressure region forms behind the pressure wave. The negative pressure causes the air to rush back towards the explosion center. This negative-pressure wave may result in additional structural damage **(Figure 11.8)**.

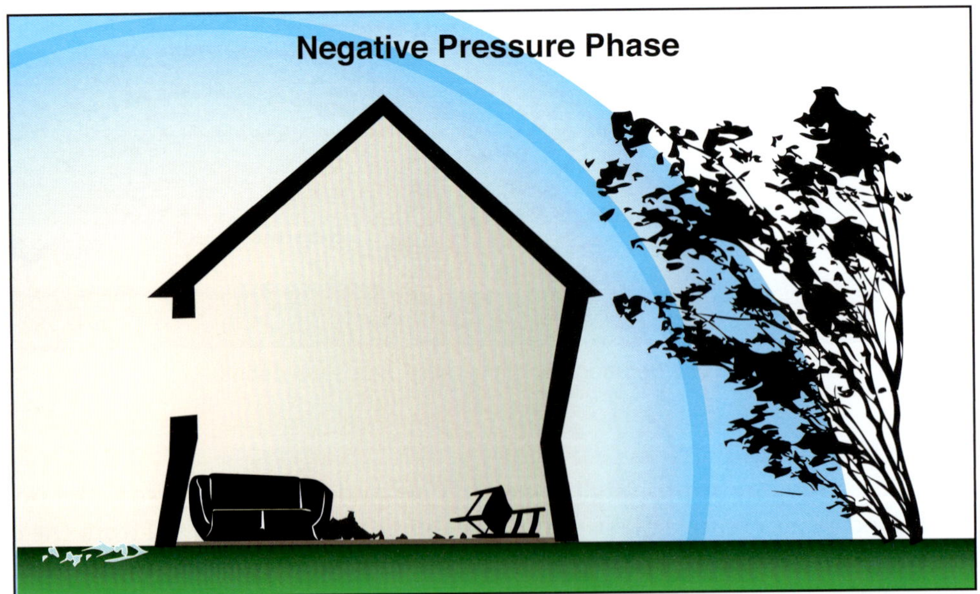

Figure 11.8 The low pressure behind the positive-pressure phase causes negative pressure moving back toward the epicenter of the explosion.

Ventilation openings and the shape of compartments or ruptures in a containing vessel can affect the shape of a blast-pressure front. The force of the front may be strong enough to destroy some structural members. Structural members or parts of a vessel that can contain the pressure will "reflect" the pressure. Reflected pressure can result in greater damage on one side of an explosion site than on the other. For example, a dust explosion in a granary might have the force to destroy wooden members but not steel members. The blast would reflect off the steel members, increasing damage to the opposite side of the structure or pushing the front toward ventilation areas such as doorways and windows.

Fragmentation Effect

Fragmentation is debris moved during the explosion, whether by the shrapnel or the explosion itself **(Figure 11.9)**. These pieces may be thrown over a wide area and great distances, causing personal injury and other types of damage to surrounding structures or objects.

The **fragmentation effect** can be used to map the explosion scene after the fact. Damage can include:

- Bodily harm such as bruises, punctures, or tears.

- Structural damage that destabilizes buildings and other resources.

- Severed utilities such as electrical wiring or gas lines.

Figure 11.9 Shrapnel are small pieces of debris moving with great speed and force that can cause damage far away from an explosion. *Courtesy of Captain James A. Sobota, Fairfax County, Virginia, Fire and Rescue Department.*

Seismic Effect

When a blast occurs at or near ground level, the air blast creates a ground shock or crater – which is the seat of the explosion. A seismic disturbance (an earth vibration similar to an earthquake) is formed as shock waves move across or through the ground. The distance the shock wave travels depends on the type and size of the explosion and type of soil. Seismic effects are generally associated with large, seated explosions. The collapse of a large structure after an explosion can also contribute to the seismic effect.

Thermal Effect

All types of explosions release thermal energy, which has the potential to ignite combustible gases and heat the surrounding air. This heat can also ignite other nearby thermally thin combustibles. Additional fires caused by these explosions results in damage that the fire investigator must evaluate.

Detonations produce the greatest amount of heat, generating high temperatures that only exist for a brief duration. While the initial detonation can cause fires, the blast-pressure front and any resulting shrapnel generally cause the damage to the surrounding structure. Shrapnel in the form of firebrands may cause secondary fires some distance from the seat of the explosion.

Deflagrations are more likely to spread fire to surrounding combustibles because high temperatures from the flame front move more slowly away from the ignition source. Therefore, the flame front in a deflagration has a longer duration to transfer heat to other combustibles. When examining portions of an explosion scene, investigators should attempt to determine which came first, the deflagration or the fire.

In addition to heat in the flame front of a detonation or deflagration, thermal heat energy in the form of a **fireball** may result from burning combustible gases or flammable vapors and ambient air. Fireballs are especially prevalent in BLEVEs that involve flammable vapors. The fireball is present for a limited time during or after the explosive event. Radiated heat from the fireball may be extreme and can contribute to secondary fires and bodily injuries.

Seated Explosions

NFPA 1033 (2022): 4.1.7, 4.2.9

Seated explosions leave an identifiable **epicenter** (the seat) in the form of a crater or area of greater damage. The presence of a seat indicates an explosion occurred in a concentrated fuel source at or near the seat. The seat of an explosion is the remaining crater or concentrated area of damage. Sometimes material is thrown from the seat into other areas.

Characteristics of Seated Explosions

Seated explosions are generated by specific types or configurations of explosive fuels. The energy of the explosion creates the seat; the size of the seat of these explosions depends upon the size of the explosive blast at the epicenter **(Figure 11.10)**. The seat may be circular or spherical in shape.

Seated explosions may involve:

- BLEVEs

- Explosives

- Steam boilers

- Tightly confined gaseous fuels or liquid fuel vapors

NOTE: Low pressure containers can generate a high-velocity heat induced tear. This is most often seen in high volume hazardous materials containers.

The pressure involved in seated explosions is very high, and the rate of pressure release is both rapid and powerful. The high rate of pressure release is due to the concentrated nature of the fuel source (explosive or confined space) at the center of the explosion. Seated explosions' reaction velocities are supersonic (faster than the speed of sound) and cause loud, explosive noises.

Figure 11.10 Detonation leaves recognizable seats/craters. Large craters like this one are the result of extremely powerful detonations. *Courtesy of the U.S. Air Force, photo by Senior Airman Sean Worrell.*

 **Butane Hash Oil (BHO)**

Recent legalization of marijuana in several U.S. states has led to an increased prevalence of labs that pose a fuel gas explosion risk. Fuel gas explosions in these labs are capable of causing injuries to civilians, fire personnel, and hash oil manufacturers, as well as damage to structures. Some manufacturers expedite the tetrahydrocannabinol (THC) extraction process by placing the material in an oven, on a stove, or heating it with a hair dryer or heat gun. In an unventilated, enclosed area, this process can create significant explosion hazards.

Fire investigators responding to incidents should be aware of any evidence of the butane hash oil (BHO) manufacturing process, such as:

- Marijuana shake (sometimes in large quantities)

- Tempered glass or specimen dishes (especially with yellow or brown residue)

- Butane canisters (may be in large quantities or in cylinders)

- Extraction tube (may be made of metal, PVC, glass pipe, plastic beverage or even prescription bottles)

BLEVEs

One of the most common seated explosions is a boiling liquid expanding vapor explosion (BLEVE). These mechanical explosions can cause high-order explosion damage, similar to the damage associated with high explosives. BLEVEs are deflagrations that may reach the speed and destructiveness of detonations. They often occur as a result of:

- An exposure to a fire
- Mechanical damage that has weakened a container's walls
- The heating of a container above the level that allows its contents to remain in a liquid state

When a liquid inside a vessel begins to boil, the additional pressure created cannot be relieved through the vessel's normal means of venting the gas. As a result, the liquid's temperature continues to rise beyond the boiling temperature, increasing the pressure inside the pressurized vessel. When the pressure exceeds the container's structural limitations, the vessel fails and the liquid turns into a gas. The energy release of the liquid rapidly expanding to a gaseous state results in a violent explosion. Any liquid under pressure has the potential to produce this type of explosion.

The most violent BLEVEs occur in liquefied compressed gases such as LPG (propane). Gases can be stored as compressed liquids through pressure. Because the pressurized liquid is being actively prevented from returning to its natural state, it takes less time for the liquid to boil and create enough excess pressure to destroy its canister or vessel. This faster reaction time results in a much more violent explosion. In addition, when flammable gases are released, these materials can result in a fireball and secondary fires. The damage resulting from BLEVEs involving liquefied gases are often described as total or catastrophic.

NOTE: BLEVEs can occur in very small vessels like aerosol cans and cigarette lighters. Also, water and other nonflammable materials can explode in a BLEVE.

Combustion Explosions

NFPA 1033 (2022): 4.1.7, 4.2.9

Combustion explosions are a type of chemical explosions. Combustion explosions occur within fuels diffused or dispersed in air. The pressure release rates in these explosions travel at subsonic speeds.

Nonseated explosions may be powerful enough to do structural damage similar to seated explosions, but generally nonseated explosions result in lesser damage. Features of nonseated explosions include:

- No crater
- Deflagrations
- Wide-spread (diffuse) fuels
- Moderate rates of pressure rise

Combustion explosions are categorized by their primary material, such as:

- Dust
- Smoke
- Fuel gas
- Aerosols
- Particulates
- Mists or mixtures

Fuel Gas Explosions

Normally, **gas/vapor explosions** occur in compartments or structures where ignitable vapors have mixed with air in a ratio suitable for ignition. In some rare instances, such as ignition of gases inside pipes, gas/vapor explosions may reach the level of a detonation.

Gas/vapor explosions mostly produce low-order damage. They involve liquid vapors and fuel gases. In contrast, gas/vapor explosions producing high-order damage involve natural gas and lighter than air gases.

When air in a compartment mixes with an ignitable vapor in a ratio suitable for ignition, and a competent ignition source is present, conditions for a deflagration occur. If the ratio is within the flammable range, the introduction of an ignition source of sufficient duration and temperature will begin a chain reaction in which the vapors rapidly expand and combust, creating a blast-pressure front followed by a flame front. In a confined area, this pressure can cause great damage to a surrounding structure including collapse of structural members. The force and extent of the blast-pressure front is determined by a number of factors.

The sections that follow examine important factors present in a gas/vapor explosion. Taken as a whole, these topics describe the reaction that occurs during this type of deflagration.

Fuel-to-Air Ratio

Similar to fire conditions introduced in Chapter 3, the fuel-to-air ratio in a compartment is perhaps the most important factor in determining the violence of a gas/vapor explosion. Explosions near the lower explosive (flammable) level (LEL) ratio tend to consume almost all of the available fuel, thus reducing the chance for postexplosion fires. Ratios near the upper explosive (flammable) level (UEL) tend to create postexplosion fires, because unburned fuel gases remain after the reaction and can remix with available air in the ventilation phase or the negative-pressure phase of an explosion. For more information about upper and lower explosive (flammable) limits, refer to the IFSTA manual, **Hazardous Materials for First Responders**.

Low order explosions may have fuel-to-air ratios near the UEL or LEL. These explosions burn inefficiently.

High order explosions may have a fuel-to-air ratio near or just above the stoichiometric values. These explosions burn efficiently.

Vapor Density

When considering the fuel-to-air ratio, it is also necessary to understand how vapor density affects the way gases mix. Vapor density was introduced in Chapter 3. Generally speaking, lighter-than-air gases (less dense than air) tend to collect near the ceiling of compartments while heavier-than-air (more dense than air) gases tend to collect near the floor of a compartment.

Propane, being a heavier-than-air gas, will tend to stay near the floor or ground. A heavier-than-air gas acts like syrup poured into water. The syrup will remain at the bottom unless an outside force, like someone stirring the liquid with a spoon, forces the syrup to mix with the water **(Figure 11.11)**. Left to itself, the solution will separate out again. Likewise, if propane or another heavier-than-air gas has settled to the floor of a compartment, then some sort of airflow is required to agitate the propane. Agitation may occur as a result of air movement within the compartment or the stream produced from a leak in a gas container or pipe or a rise in the ambient temperature.

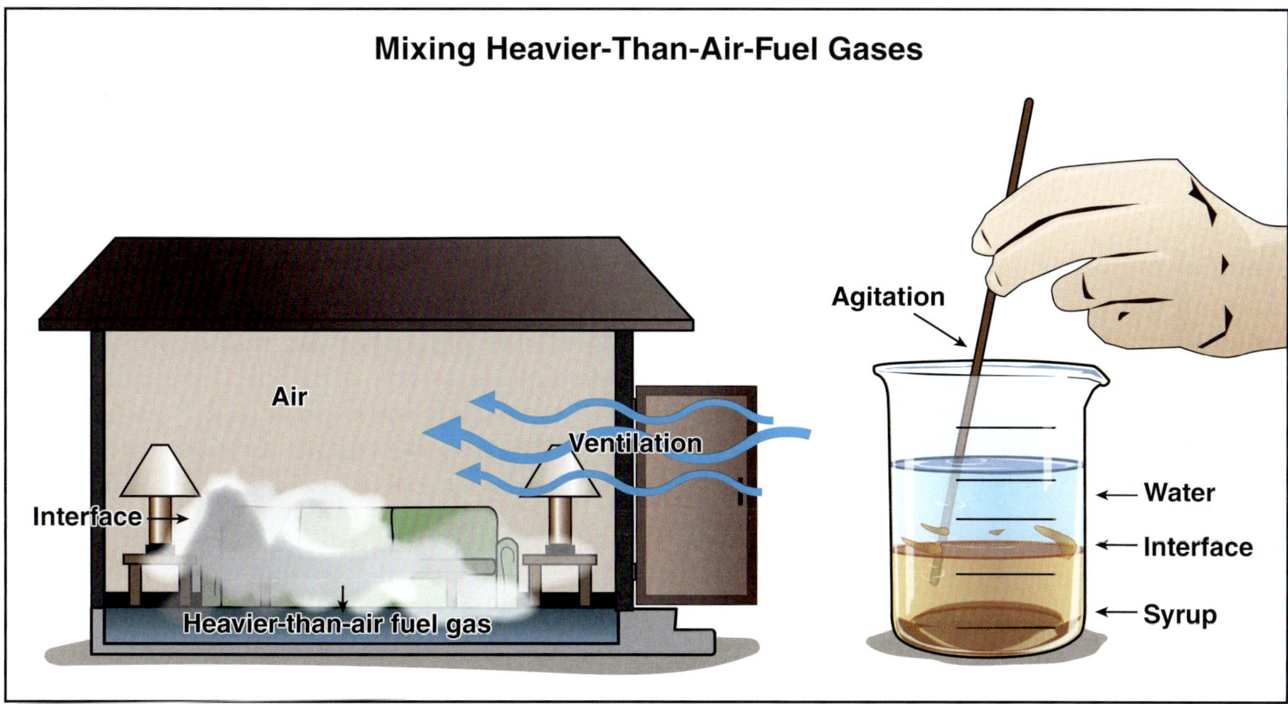

Figure 11.11 Heavier-than-air gases require some form of agitation, such as ventilation, to mix with the air above them.

In buildings that have basements, heavier-than-air gases can sink into sublevels. The mixing as the gases move downward through compartments is similar to what happens as lighter-than-air gases rise through a compartment.

Because of varying types of construction, the area most affected by a deflagration such as the roof or floor does not necessarily indicate the vapor density of the exploding gas or vapor. The area between the fuel-rich area and the air-rich area is known as the **interface** and is where the two are mixing **(Figure 11.12)**. Many other factors can affect the mixing of gases such as people walking through rooms, HVAC systems, fans, gases introduced under pressure, and other ventilation sources.

Multiple Gas/Vapor Explosions

Deflagrations may spark several secondary explosions in one structure, especially if fuel has been leaking and mixing with air for an extended period of time. As the fuel moves away from its source, it begins to fill that compartment and others. Fuel gases can move through vents and small spaces around doors. The new pockets of fuel will be at different fuel-to-air ratios than in the source compartment. When an ignition source causes an explosion in one compartment, the ensuing heat may ignite gases in other rooms or stories of a structure. Turbulence from the initial explosion can result in mixing of the gases that may have been above their flammable limits. Once this mixing occurs and gases are within the flammable limits, secondary explosions can result. The explosions are separate incidents with the initial explosion acting as the ignition source for the others.

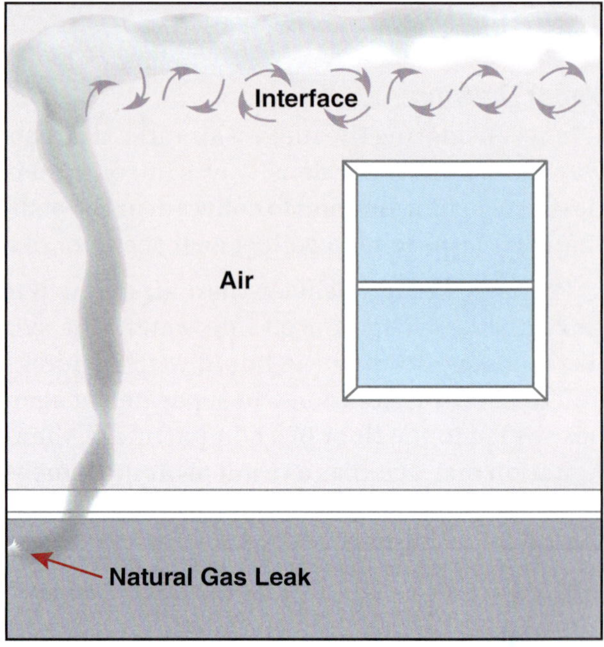

Figure 11.12 The area in which a fuel gas mixes with ambient air is known as the interface.

Fuel-Gas Migration

Fires in gases may follow the fuel into pipes, or fuel gases may seep downward into the ground and provide a new path for ignition. Fuel gases may also migrate into pipes or underground channels such as sewer tunnels. The fuel gases may even spread into surrounding structures. Fuel gases encountering an ignition source may cause explosions or fires some distance from the original leak. Because of the distances fuel gases can travel and still remain a hazard, investigators should determine where fuel gases first originated and map their travel whenever examining a possible gas/vapor explosion scene.

Representatives from utility services should be consulted to assist investigators in tracking the source of a **fuel-gas migration**. Utility companies have equipment that creates holes in the ground to allow the insertion of monitoring equipment to check subterranean gas levels. This process can assist the investigator in mapping the travel of the gas and arriving at the source of the leak.

Investigators should also be aware that until the source of the leak has been identified and the gas isolated, subsequent explosions may occur as a result of continued migration. Continued migration of the fuel gas may also appear in a location other than the site of the original explosion.

Added Odorants

Typically, fuel gases that are used in residences or businesses have an odorant added to the gas so that the gas can be detected. However, detection of the odorant should not be solely relied upon to track the migration of a fuel. As the fuel gas migrates, it may lose this quality as it moves through the ground.

Flame Speed and Burning Velocity

The **flame speed** is the velocity of the flame front relative to the center of the explosion. For example, the maximum flame speed of propane is 11.5 ft/s (3.5 m/s) assuming a steady flow. Flame speed can vary significantly, based on a number of factors.

Burning velocity is the velocity of the flame front relative to the unburned gases ahead of it. The burning velocity describes how rapidly the heat of the explosion is combusting available fuel and generating the flame front. The faster the combustion, the more agitated the gas/air mixture becomes, and the greater the pressure increase as the fuels are pushed outward and consumed.

Burning velocity is dependent upon the composition, temperature, and pressure of the unburned gases. The fundamental burning velocities of various fuel/air mixtures can be calculated and are provided in NFPA 68, *Standard on Explosion Protection by Deflagration Venting*. This standard provides examples of calculating burning velocities based upon standard temperature, pressure, and composition of unburned gases. Variables such as lower temperatures away from the ignition source, lean gas mixtures, or tightly confined spaces would all reflect variations in the expected burning velocity.

The blast-pressure front created during a gas/vapor explosion is a direct result of the explosion's flame speed and burning velocity. The faster the speed and velocity, the greater the increase in pressure and therefore, the more powerful the deflagration **(Figure 11.13)**.

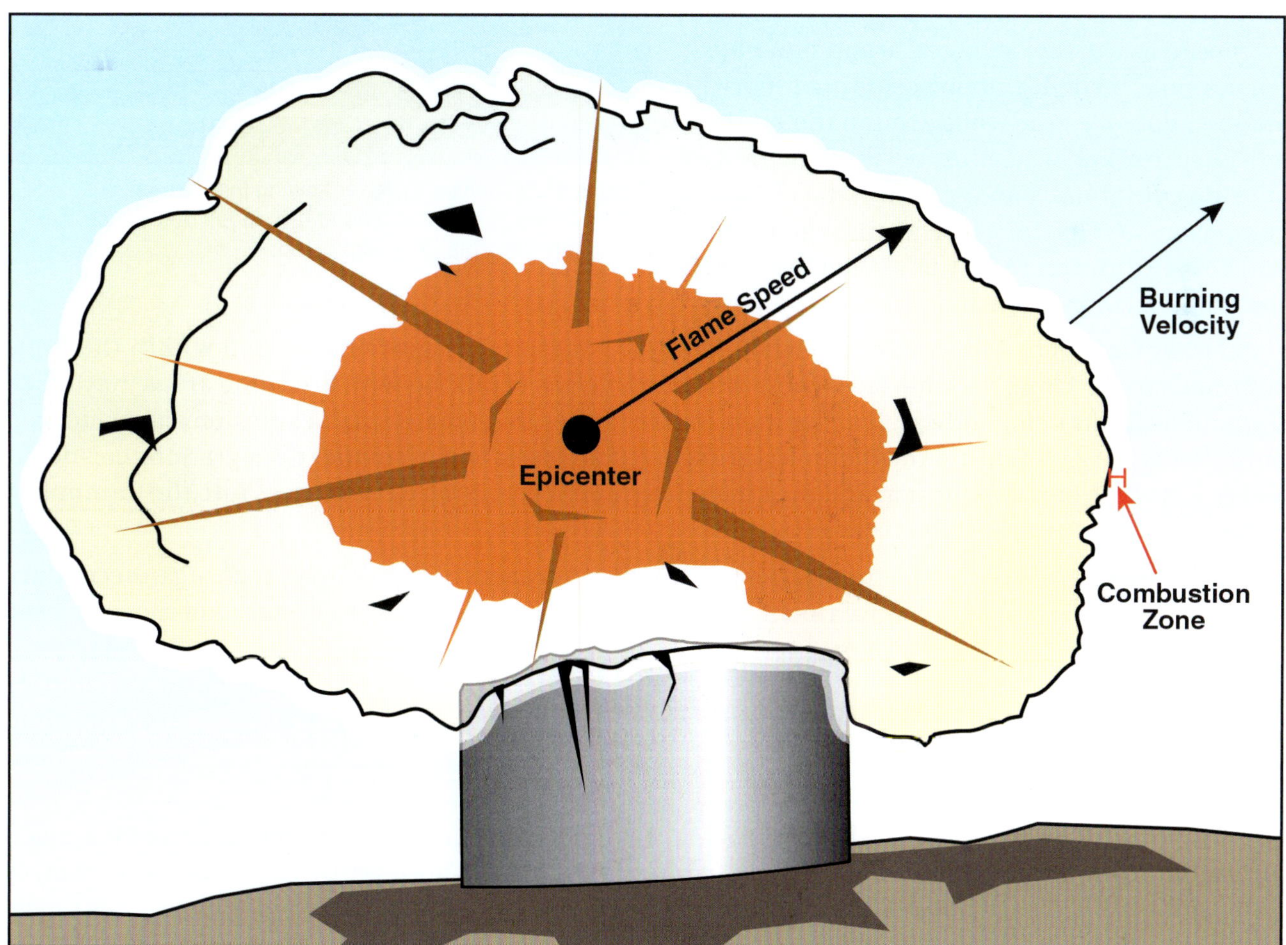

Figure 11.13 Depiction of the epicenter, flame speed, combustion zone, and burning velocity in a dust explosion.

Turbulence

Provided ideal circumstances, any explosion will produce a nearly spherical shape, expanding in all directions until the available fuel has been consumed. Very rarely do explosions occur in which there are no obstructions to this expansion. Obstructions such as furniture, walls, ceilings, or appliances, create turbulence within the expanding pressure front of a deflagration. **Turbulence** changes the shape of the explosion, alters the movement of the gases involved, and increases burning velocity.

A good analogy for understanding turbulence is water flowing in a stream. When the water is deep and there are relatively few obstructions, the water moves at approximately the same speed and volume. When the water reaches an obstruction, it is pushed around the obstruction, temporarily increasing the velocity of the water and creating eddies in the flow **(Figure 11.14)**.

Gases in a deflagration react the same way to obstructions. The burning velocity (flow) is increased as the fuel is forced around an object. For example, if a deflagration occurs in a room with an open doorway, the walls around the doorway will force the flow of burning fuel through the door causing additional damage within the adjoining compartment. The front moving through the door will be faster and more powerful than the front in the original compartment.

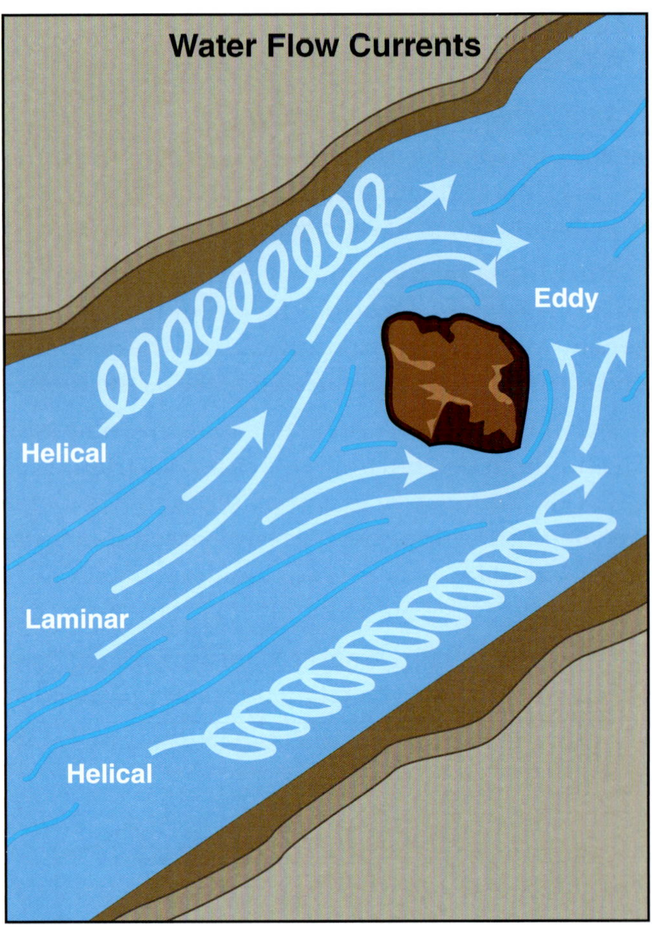

Figure 11.14 Similar to water flowing in a stream, obstructions in a space, and at the edges of a space, can affect gas currents in an explosion.

Another form of turbulence occurs when multiple, smaller deflagrations happen within the same confined space. Along the interface where a flammable gas is mixing with air, some areas may reach flammable limits before others. If an ignition source ignites this area, a small explosion can occur that only affects the gases within the flammable range. A chain reaction of smaller deflagrations can occur as a result of the initial explosion and the created turbulence, mixing nearby gases into the flammable range.

Turbulence can also enhance the fuel-to-air ratio in other areas of a structure resulting in secondary explosions outside the initial compartment. Fuel that may have seeped into other stories or rooms may be above the flammable range. Turbulence from the original deflagration can quickly return the mixtures to within the range by actively forcing air into the mixture. The heat from the initial explosion or a secondary fire can then serve as an ignition source for the newly mixed gases resulting in a secondary explosion.

Confining Space

Generally speaking, the smaller a confining space is, the greater the rate of pressure rise for a given amount of fuel-air mixture. For example, the same amount of fuel-air mixture in a canister, a room, or a structure will all react very differently **(Figure 11.15)**. The more the gas is diffused (larger confining space), the longer it takes flame to propagate through the available fuel. The burning of fuel releases the energy that causes the rise in pressure in an explosion. In smaller confining spaces, the more concentrated fuel burns more quickly and releases greater pressure more rapidly.

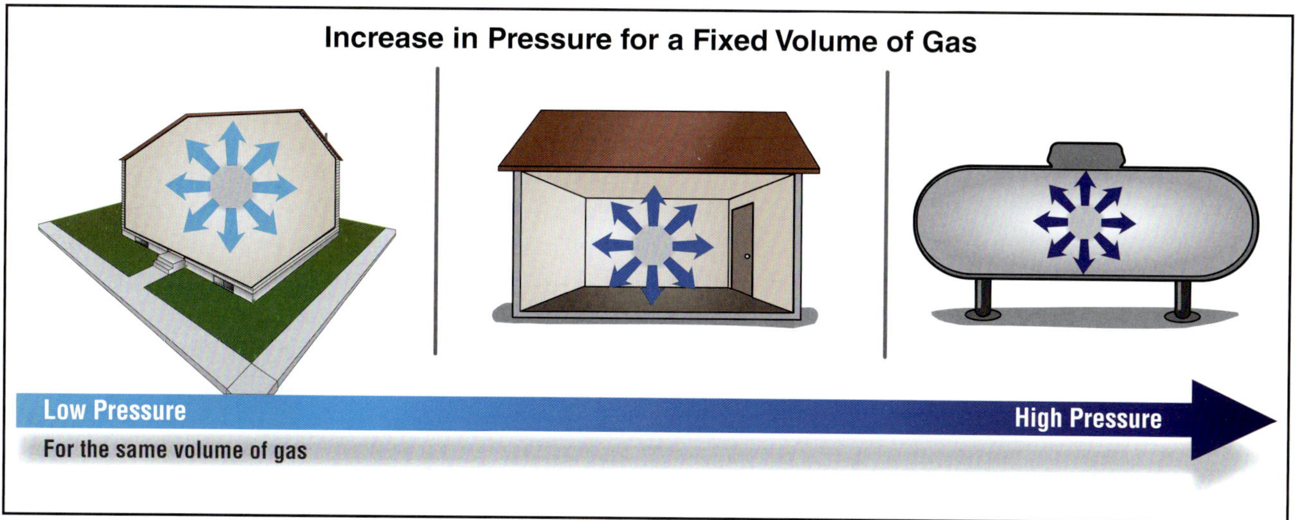

Figure 11.15 For a fixed volume of gas, pressure increases as the size of the container decreases.

As the flame propagates through the fuel, the flame front expands until it comes into contact with the confining vessel. At this point, the vessel will either reflect the blast-pressure front, absorb the blast pressure (become damaged), or completely fail. Any expanding blast front will attempt to reach a ventilation point. If the confining space is totally enclosed, failure of the containment allows the explosion to ventilate. In a compartment, this failure might be broken windows or walls that have ruptured or collapsed. In small vessels such as LPG tanks, the vessel itself fails and releases much higher levels of pressure.

Ignition Source

For an ignition source to initiate a gas/vapor explosion, the following criteria must be met:

- Fuel/air mixture is within the flammable range.
- The fuel/air mixture and the ignition source make contact.
- Fuel and ignition source are in contact long enough for effective heat transfer.
- Ignition source must be capable of reaching a higher temperature than the autoignition temperature of the fuel/air mixture.

Possible ignition sources for explosions are similar to those for fires: heated objects, open flames, electrical arc, etc. Contact and duration of heat transfer are key to the explosive reaction. Direct contact is necessary for an ignition source to be the initiator of a gas/vapor explosion.

The location of the ignition source within a compartment has an effect on the power of an explosion. A deflagration that ignites near the center of compartment will be more powerful than one beginning near a wall or corner. A centralized ignition allows the explosion to gain heat and increased pressure before coming in contact with structural members. When the ignition source is near a wall, for example, much of the initial heat is transferred to the wall which causes the blast-pressure front not to develop in the direction of the wall. This lack of development results in an overall weaker explosion.

Venting

Pressure, by definition, wants to be released. A blast-pressure front will continue to expand until the flame front has reached its maximum distance (consumed all available fuel) and the differences in pressure between the excited gases and the atmosphere return to equilibrium. If the pressure cannot be diverted, it will attempt to move through anything in its path causing structural damage, moving/throwing objects around (fragmentation) or creating ventilation in the form of failed structural members. Venting offers a direction or path of least resistance for the blast-pressure front in an explosion. The greater the ventilation in a compartment confining an explosion, the lower the order of damage from the explosion.

As an explosion ventilates, it spreads damage to areas not initially exposed to an explosion. Investigators may see damage immediately opposite a doorway, for example, when the rest of the compartment shows no damage. The flame front that ventilates along with the blast-pressure front could create secondary fires in adjoining compartments as well.

Industrial facilities that are in danger of gas/vapor explosions and **dust explosions** may be equipped with pressure vents (blast walls) designed to open or detach when a blast-pressure front moves through a structure **(Figure 11.16)**. The vents ensure that the pressure released from an explosion can expand without destroying the surrounding structure. The structure will sustain damage, but the damage will not be catastrophic and injuries or loss of life may be kept to a minimum.

Figure 11.16 The panels on this roof can detach easily in an overpressure situation. *Courtesy of David M. Smith, Associated Fire Consultants, Inc.*

NOTE: Ventilation has little or no effect on mitigating the damage from detonations because detonations move so quickly that the explosion has no time to ventilate.

Dust Explosions

Dust explosions are chemical, nonseated, deflagrations similar to gas/vapor explosions. Dust explosions occur when solid matter in very fine particulate form suspended in air ignite within the flammable range. Dust generated from any dry commodity operations can also create conditions for a dust explosion, regardless of the size of the space. These explosions can occur in industrial facilities such as lumber mills, granaries, mines, or metal fabrication facilities, and also smaller spaces such as garages and grain bins (Triangles, 2018).

Dust explosions can even occur when the fuel would not normally be flammable. Materials that may be fuel for dust explosions include:

- Coal
- Sawdust
- Charcoal
- Plastics and resins
- Grain dust and flour
- Metals (aluminum, magnesium, titanium, iron)

A dust explosion propagates similarly to a gas/vapor explosion. The blast-pressure front, flame speed, and burning velocity also behave very similarly. The main difference between the two is the nature of the fuel involved. Dust explosions are solid-fuel explosions. The fire in the explosion uses the ambient air as an oxidizer and the suspended dust particles as a fuel. The energy released when the solid fuels ignite is greater than when gas fuels ignite. As a result, the rate of pressure increase is much higher in dust explosions, and the expansion rate much faster and more powerful. Dust explosions are comparable to gas/vapor explosions as explained in the following sections.

Particle Size

The smaller the particle size, the greater the surface area and the more violent the explosion. Similarly, gases or vapors with a lower molecular weight will have a lower flash point compared to gases/vapors with a higher molecular weight (**Figure 11.17**).

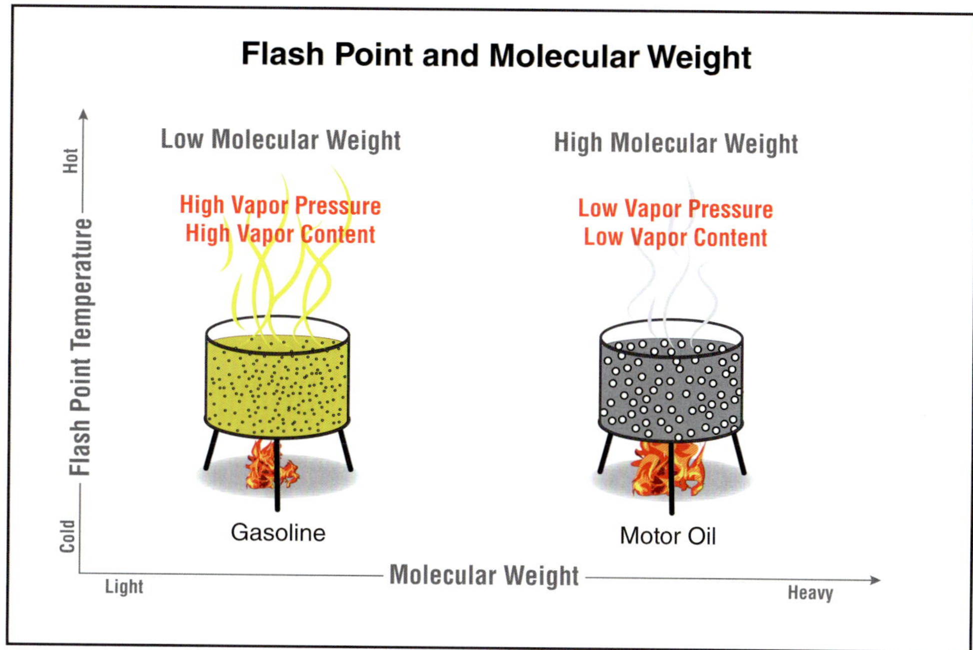

Figure 11.17 Gasoline has a low flash point and low molecular weight. Motor oil has a high flash point and high molecular weight.

Concentration

Similar to the fuel-to-air ratio in gas/vapor explosions, dust concentration has a lower flammable limit, but not an upper flammable limit. Concentration is measured in ounces per cubic foot (grams per cubic meter).

Types of dusts have different flammable ranges and optimum (stoichiometric) ratios. Most common dusts have a lower flammable limit concentration of less than 1.0 oz./ft^3 (1kg/m^3). As with gas/vapor explosions, the rate of pressure rise for mixtures near the optimum ratio are the fastest. Similarly, dust concentrations can be described as too lean or too rich to burn. For dust concentrations that are very lean or very rich, explosive pressure is very low.

Turbulence

The same as gas/vapor explosions except in one aspect; turbulence in dust explosions can render dust that has settled on horizontal surfaces and floors airborne. The additional fuel can strengthen the explosive reaction and increase its duration.

Moisture

The moisture concentration of dust has an effect on the amount of energy required for ignition. The higher the moisture content, the more energy required for ignition. The opposite is also true: dryer dust requires less energy to ignite.

Ignition Energy

More ignition energy is required to cause a dust explosion than is required to cause a gas/vapor explosion. Temperatures ranging from 600° F to 1,100° F (320° C to 600°C) are required for ignition of particulate material.

Explosive Series

The blast-pressure front in a dust explosion entrains dust into the reaction as it moves ahead of the flame front. As a result of this phenomena, dust explosions frequently occur in a series or cascade as each successive reaction meets with new, properly mixed fuel.

Dust Combustion Considerations

Dangerous on its own, dust fires can destroy property and injure workers. The following sections address combustible dust hazard models (Triangles, 2018).

Combustible Dust Fire Triangle. For a combustible dust hazard to cause a fire, a fuel, oxidant, and an ignition source must be present. In the case of combustible dust, the fuel is the powder or finely-divided material and the oxidizing agent (oxidant) is usually oxygen in air. The ignition source can be direct heat, open flame, friction, sparks, self-combustion, or any other process that adds energy to the system.

Combustible Dust Flash Fire Square. Dust suspended in the air creates a flash-fire hazard. A **flash fire** describes a flame front that travels rapidly through a diffuse fuel source without producing pressure. The flash fire square shows that the addition of a dispersed material creates the conditions for a flash fire to occur.

Dust Explosion Pentagon. For a combustible dust fire to produce an explosion, there must be confinement of the flame propagating through the dust-air mixture. Confinement is the final element required to create the potential for a dust explosion, represented as the dust explosion pentagon. With confinement of the propagating flame front, pressure will rise which may cause rupture of the enclosure.

Approaching an Explosion Scene

NFPA 1033 (2022): 4.1.7, 4.2.1, 4.2.9, 4.4.2

To accurately and effectively examine the scene of an explosion, investigators follow a systematic approach similar to the methods used to investigate fire scenes. The fire investigator must determine the explosion's:

- Cause
- Origin
- Fuel source
- Ignition source
- Ignition sequence

Many times the explosion scene will be larger than that of a post fire scene. Investigators may notice that the damage associated with an explosion scene is also greater than that of a fire scene. Utilizing a methodology that incorporates a systematic approach to examining the scene will ensure that the scene is processed correctly and efficiently.

Establish a Secure Perimeter

As explained in Chapter 2, Safety, the investigator must secure the explosion scene and prevent unauthorized entry. As indicated at the beginning of this chapter, the presence of undetonated explosive materials may require wider scene perimeters and additional resources. The outside boundary of the scene perimeter cannot be established before the furthest piece of debris is found.

The scene perimeter should be 1.5 times larger than the furthest distance of debris from the explosion's suspected epicenter. An adequately sized scene perimeter allows an investigator to conduct a thorough examination and minimizes the likelihood of overlooking debris that could be relevant. The scene can be made smaller after an investigator has located, collected, and examined debris and the scene for evidence.

Gather Information

Prior to conducting a search of the scene for evidence, the investigator should gather information relating to the event. This information should include witness information and electronic data such as video surveillance. Witness interviews are discussed later in this manual.

Origin Determination

To begin the search for evidence and to locate the origin of the explosion, the investigator should establish the search areas and select the search pattern. The boundaries of each search area assigned should overlap to ensure the searcher(s) find(s) all evidence. Search patterns may include:

- Line
- Grid
- Spiral
- Circular

Search from the scene perimeter and move inward toward the area of greatest damage. Investigators may consider tasking their local ground search and rescue team or other responders to aid in the work of searching for evidence.

As debris that may potentially be evidence is identified, it should be labeled per the investigator's instructions, which may include **(Figure 11.18)**:

- Assigning identification numbers
- Documenting damage characteristics
- Documenting the global positioning system (GPS) location
- Marking the item location with a flag for further examination

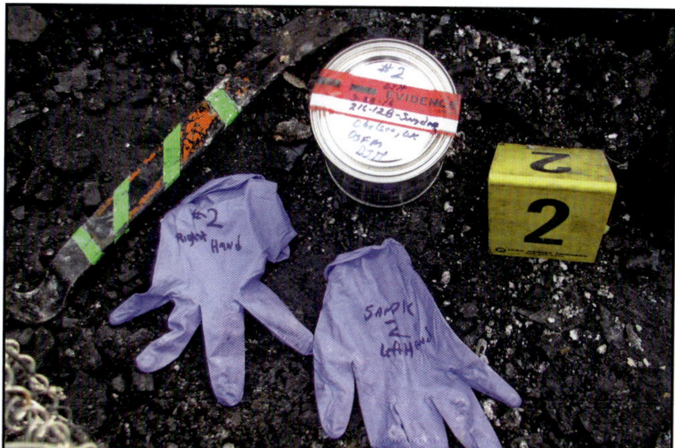

Figure 11.18 Labeled evidence should be distinct from other evidence, and identifiable per the incident. *Courtesy of Donny Howard, Agent, Oklahoma State Fire Marshal's Office.*

Flags may also assist an investigator during **explosion-dynamics analysis**. Orienting flags in the direction of debris travel may assist the investigator with identifying force-vectors and the epicenter of the explosion. Photographing flags from above via drone or other aerial photography may better show the direction and distance the debris travelled.

Explosion Scene Examination

NFPA 1033 (2022): 4.1.7, 4.2.1, 4.2.9, 4.4.2

Explosion scene examination is similar to fire scene examination in many ways. Explosion scenes will often have more complexity and prevalent debris. The investigator should follow the scientific method or other systematic approach in all investigations.

Explosion Scene Examination

Investigators perform various analyses to conduct a thorough explosion scene examination. As at a fire scene, investigators adhere to all AHJ and laboratory procedures for evidence collection, preservation, and documentation, and apply techniques outlined in this manual. An investigator examines an explosion scene to:

- Ensure scene safety
- Identify and collect potential evidence
- Identify the ignition source and sequence
- Establish a hypothesis about the origin of the explosion

Safety Assessment

In addition to hazards found at fire scenes, explosion scenes have unique hazards that must be assessed before evidence can be identified, documented, and collected. The possibility of structural collapse is much more likely at an explosion scene because structural members may be more severely weakened than at fire scenes. Also, fuel-gas mixtures and dust-air mixtures may still exist in the flammable range at the scene. These mixtures could result in secondary explosions.

Presence of a seated explosion, or evidence of an explosive device, may be an indicator of a secondary device on the scene. Secondary devices may have been left intentionally to do further damage to responders after an initial device. Investigators should follow agency or AHJ SOPs.

Investigators should:

- Evacuate and deny entry to the explosion area.
- Notify resources that can help identify and mitigate devices.
- Seek and isolate any additional explosive devices.

Explosions may introduce toxic gases into the air or cause leaks in fuel lines. This damage may introduce hazardous materials to the scene. A safety assessment for these and other fire scene related hazards is necessary to ensure the safety of investigators and should be the first step in any scene assessment.

Fire Versus Explosion Recognition

Following a safety examination, the first responsibility for an investigator examining a potential explosion scene is to determine what damage is from fire and what damage is from an explosion. In addition, an investigator should determine whether an explosion at the scene caused a fire or if a fire at the scene initiated an explosion.

To make these determinations, an investigator should survey damage at the scene. If this initial survey reveals damage more commonly found at fire scenes, it is less likely that an explosion occurred. If an investigator finds signs of overpressure, such as bulging of walls or window glass shattered outward some distance away from the structure, then further investigation seeking explosion evidence is recommended **(Figure 11.19)**.

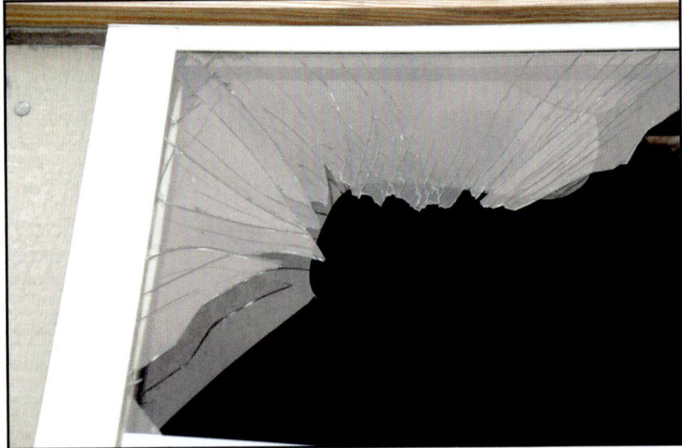

Figure 11.19 Shattered windows with no evidence of fire damage may indicate overpressurization damage or shrapnel damage from explosions. *Courtesy of Yates & Associates.*

Fuel Identification

Damage at the scene may indicate possible fuels that could have caused an explosion. Determining the fuel involved is a critical step in later determining the area of origin and ignition source.

Investigators should locate and identify the utilities and services at the scene as a first step in identifying the fuel source for a possible explosion. Homes heated by natural gas, for example, would provide lighter-than-air fuels while those heated by propane or fuel oil would provide heavier-than-air fuels.

During the early stages of a scene examination, an investigator will have a general idea of damage at the scene. Further investigation will provide more detail, but an investigator needs to know when to conduct a more detailed analysis. The damage first found at the scene can indicate the type of fuel involved, which may guide an investigator to conduct more detailed analyses beyond the initial assessment. Scene characteristics associated with fuel types include:

- **Lighter-than-air gases** — Almost always result in deflagrations; typically produce low-order damage, secondary explosions, and postexplosion fires.

- **Heavier-than-air gases** — Almost always result in deflagrations; typically produce low-order damage and can contribute to BLEVEs, secondary explosions, and postexplosion fires.

- **Liquid vapors** — Almost always result in deflagrations accompanied by low-order damage. Postexplosion fires almost always result where liquid vapor fuels are present. Flammable liquids in pressurized containers escaping as vapors, and subsequent ignition, are always associated with BLEVEs.

- **Dusts** — Almost always result in deflagrations and often show evidence of secondary explosions. Preexplosion and postexplosion fires are common with dust explosions.

- **Explosives** — Almost always result in detonations and seated explosions. High-order damage is likely. Explosives never result in secondary explosions at a scene and do not contribute to BLEVEs.

- **Backdrafts** — Always result in deflagrations that leave low-order damage. Because backdraft explosions occur in products of combustion suspended in air, preexplosion fires are always present. Postexplosion fires should also always be expected after backdraft explosions **(Figure 11.20)**.

- **BLEVEs** — Result in deflagrations that can rise to the level of detonations. Preexplosion fires are almost always present and act as the heating source to raise the pressure in the confining vessel surrounding a liquid. Postexplosion fires may result from the fireball when flammable liquids are involved in the explosion. BLEVEs are singular incidents and do not result in secondary explosions.

Figure 11.20 Backdraft explosions occur when the flammable gases produced in the fire reach the flammable range, ignite, and cause a deflagration. *Courtesy of Wayne Chapdelaine, Metro-Rural Fire Forensics.*

Explosion Damage Identification

NFPA 1033 (2022): 4.2.9

After determining an explosion's fuel source, an investigator should begin a more detailed analysis of damage at the scene. The investigator should describe and document the damage by answering the following questions:

- Were walls shattered or did they bulge?

- Was masonry broken?

- Were steel members bent?

Investigators should also review how damage to a structure and its contents compares to typical effects seen in explosions. Explosion effects and related damage include:

- **Blast-pressure-front positive phase** — This phase produces the majority of damage in an explosion as overpressure from the blast-pressure front shatters windows, bulges walls, and creates greater damage in ventilation locations.

- **Blast-pressure-front negative phase** — The rush of gases back toward the center of an explosion can break glass, move debris, and further twist structural members but does far less damage than the positive-pressure phase.

- **Shrapnel impact** — Generally associated with detonations in explosives, shrapnel can cause bodily harm to victims, sever power lines or gas lines, or ignite causing fires some distance from an explosion.

- **Thermal energy** — In deflagrations, thermal energy may last for a long enough duration to produce postexplosion fires. Near the epicenter of the explosion, thermal energy may have no effect after the detonation. However, thermal energy may have some effect a distance away from the epicenter after the blast-pressure front has lost much of its expansive power.

- **Seismic energy** — Typically associated with very powerful, seated explosions, seismic shock waves can shake and damage adjacent structures as well as down power lines and rupture fuel gas lines, resulting in secondary fires and explosions. Seismic energy can also cause the collapse of structures damaged by the initial explosion.

Determining the approximate overpressure levels is subjective but possible through the identification of damage. This determination can be very complex and difficult.

If investigators observe damage such as shattered unreinforced concrete, bent or deformed steel structural members, or near/total destruction of a structure, then the explosion is likely the result of high explosives or extremely powerful mechanical explosions. In these situations, an investigator may solicit technical expertise to complete a thorough examination of the scene. Typically, agencies within the U.S. Department of Homeland Security (or similar agencies in other countries) also investigate suspected high-order explosion scenes.

Overpressure Damage

Overpressure is measured in pounds per square inch (psi) or in kilopascals (kPa) in metric measurements. Normally, investigators will not encounter overpressures of more than 1.0 or 2.0 psi (7 kPa or 14 kPa). These overpressure levels can cause minor to severe damage to materials in common construction:

- Below 1.0 psi (7 kPa), minor damage such as breaking of glass, bulging of walls, and damage to ceilings may occur **(Figure 11.21)**.

- Between 1.0 and 2.0 psi (7 kPa and 14 kPa) most lightweight construction will suffer great damage and may become uninhabitable.

- Near 2.0 psi (14 kPa), partial collapse of a structure may occur, including the failure of thin, steel members such as corrugated siding.

- Above 2.0 psi (14 kPa), damage becomes progressively more extensive.

Figure 11.21 Overpressurization can move walls and ceilings, leaving cracks or ceilings misaligned with walls. *Courtesy of Nicole Brewer, Portland Fire Investigations.*

Preexplosion and Postexplosion Fire Damage

Debris, especially glass, can aid an investigator in determining whether observed damage is the result of preexplosion or postexplosion combustion. The thermal effect of an explosion moves very rapidly, even in deflagrations. Debris moved or thrown by the blast-pressure front would not be affected by the thermal energy of the explosion because the duration of heat exposure is very brief. If debris appears to have been burned or shows evidence of the products of combustion, such as soot, then the debris more than likely was subjected to a fire *before* it became part of the explosive reaction.

Window glass is an especially important type of evidence for determining preexplosion or postexplosion damage. Even in explosions that are near 0.25 psi (1.725 kPa) overpressure, glass will shatter and be pushed away from a structure even if structural damage is very minor. Because glass evidence is common at almost all explosions, glass may be the only debris that investigators have to analyze.

Shards of glass scattered some distance away from a structure generally indicate an explosion. The relative size of the glass shards likewise provides an indication of the explosion force or rate of pressure rise. If the interior side of the shards show evidence of soot or products of combustion, then it is very likely that a preexplosion fire was present **(Figure 11.22)**. The absence of this evidence would indicate that the window shattered as the result of an overpressure alone.

Figure 11.22 Soot-covered shards or crazed shards of glass typically indicate that a fire was present before an explosion broke the glass. *Courtesy of Hugh W. Graham, Yates & Associates.*

Potential Evidence

The evidence at an explosion scene is similar to that found at fire scenes; however, explosion evidence tends to be more widely dispersed and more greatly damaged. Explosion evidence may also be embedded in other portions of the scene or buried beneath rubble. As a result, explosion evidence may be more difficult to identify, which makes documenting and collecting the evidence of even greater importance. In addition to evidence that would be collected at a fire scene, investigators at an explosion scene should document and/or collect any of the following pieces of potential evidence:

- Clothing and injuries of individuals at the scene
- Damaged or displaced structural components
- Damaged contents including the position of the contents and their original positions, as possible
- Condition and position of utilities, especially utilities that could contribute fuel or an explosion ignition source
- Injuries of victims
- Evidence removed from victims

NOTE: Should the cause of the explosion be from the detonation of an explosive device, device reconstruction will most likely take place remote from the scene at a forensic lab.

Other techniques and factors to implement while collecting evidence at the scene of an explosion include:

- Establishing a grid in the secure area to allow for an organized and well documented scene examination. The grids can then be carefully examined by marking each piece of evidence with flags, cones, or markers to allow for accurate documentation of each item.

- Not allowing vehicles to operate within the secure area to prevent tires from picking up evidence.

- Identifying evidence by conducting a line search. In a line search, personnel spread out at arm's length from one another and walk through an area marking each piece of evidence with flags, cones, or makers as they walk.

Total Stations

Some departments use a piece of equipment known as a **total station** when investigating large fire scenes, explosion scenes, and traffic accidents. A total station consists of surveying equipment attached to a computer-imaging system. Investigators align examination poles to the central surveying unit to mark the piece of evidence digitally. The surveying unit generates a computer model of the scene that maps each piece of evidence's location. The examination equipment can also mark specific points in a scene such as the corner of a collapsed wall. In this way, the total station can also help recreate the scene as a digital model for future reference.

Force-Vector Analysis

Investigators can use **force-vector analysis**, a graphical tool, to determine blast-pressure direction in an explosion. Similar to a diagram that shows the **vectors** of fire movement at a fire scene, a force-vector analysis uses arrows to depict the direction a blast-pressure front moved in an explosion **(Figure 11.23)**. Damage will often show an investigator the directions of forces applied in an area. An investigator can map those forces by:

- Drawing an arrow on the diagram in the direction of a bulge in a wall.

- Documenting the direction of shattered windows and damage through doorways.

- Documenting any debris travel or movement.

- Noting the trajectory of debris and the distance travelled.

- Noting anything the debris damaged along its trajectory.

While documenting damage force-vectors for the scene, an explosion pattern should begin to form. The completed diagram should also indicate the estimated force needed to cause the observed damage.

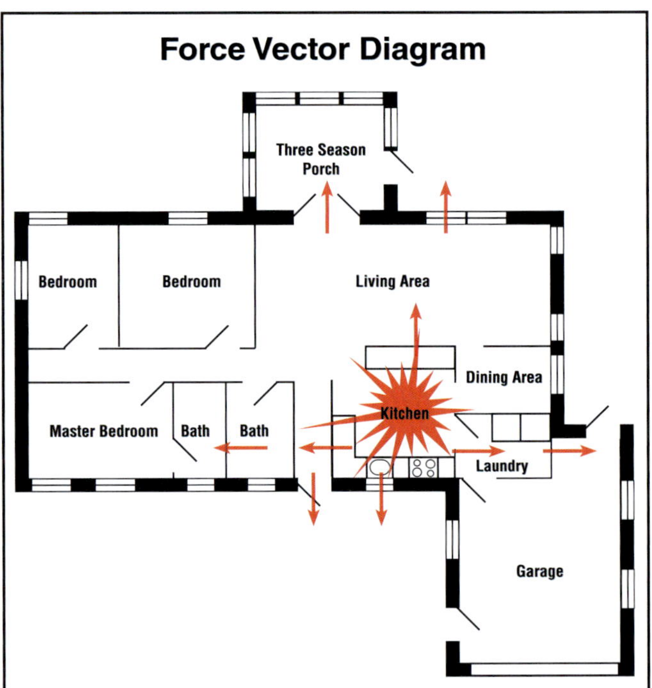

Figure 11.23 A force vector diagram depicts the direction of force damage at the scene of a deflagration.

Review Questions

1. What is the difference between a mechanical explosion and a chemical explosion?
2. Compare and contrast low-order damage and high-order damage.
3. How do the effects of blast-pressure fronts differ in detonations and deflagrations?
4. Describe what happens during the positive-pressure and negative-pressure phases of a blast-pressure front.
5. How does the fragmentation effect cause damage?
6. When can a seismic effect occur?
7. What are the differences between deflagration and detonation?
8. What are the characteristics of a seated explosion?
9. How does a BLEVE occur?
10. How can you differentiate a nonseated explosion from a seated explosion?
11. List the characteristics of a fuel/gas explosion.
12. How do dust explosions differ from fuel/gas explosions?
13. What may include labeling of potentially identified evidence?
14. How can glass aid an investigator in determining whether observed damage is the result of preexplosion or postexplosion combustion?
15. What is a force-vector analysis?

Discusson Questions

1. What categories of explosion damage have occurred in your jurisdiction?
2. How does the size of the dust particle influence the magnitude of an explosion?

Chapter References

Firefighter Nation Content Directors. 2009. "Smoke Explosion Nearly Kills 3 Firefighers." Firefighter Nation. Accessed online.

"Triangles, Squares, and Pentagons: The Anatomy of a Combustible Dust Hazard." 2018. Dust Safety Science. Accessed online.

United States Bomb Data Center (USBDC). 2017. Explosives Incident report (eir). Bureau of Alcohol, Tobacco, Firearms and Explosives. Accessed online.

Chapter Key Terms

Blast-Pressure Front — Expanding edge of the pressure in a detonation or deflagration that causes the majority of damage in an explosion.

Boiling Liquid Expanding Vapor Explosion (BLEVE) — Rapid vaporization of a liquid stored under pressure upon release to the atmosphere following major failure of its containing vessel. Failure is the result of over-pressurization caused by an external heat source, which causes the vessel to explode into two or more pieces when the temperature of the liquid is well above its boiling point at normal atmospheric pressure.

Burning Velocity — Velocity of the flame front in an explosion relative to the unburned product ahead of it.

Chemical Explosion — Rapid, exothermic reactions in which an ignition source initiates the explosion (or an increase in temperature self-initiates it), and combustion propagates along the blast-pressure front of the reaction in all directions.

Deflagration — Chemical reaction producing vigorous heat and sparks or flame and moving through the material (as black or smokeless powder) at less than the speed of sound. A major difference among explosives is the speed of this reaction.

Detonation — (1) Supersonic thermal decomposition, which is accompanied by a shock wave in the decomposing material. (2) Explosion with an energy front that travels faster than the speed of sound.

Dust Explosion — Rapid burning (deflagration), with explosive force, of any combustible dust. Dust explosions generally consist of two explosions: a small explosion or shock wave creates additional dust in an atmosphere, causing the second and larger explosion.

Epicenter — The center of a seated explosion, located at the center point of the seismic shock wave created by the explosion.

Explosion — Physical or chemical conversion of energy that results in the rapid release of high pressure gas into the environment.

Explosion-Dynamics Analysis — Process of following force vectors backwards toward the initial compartment where an explosion occurred.

Explosive — (1) Any material or mixture that will undergo an extremely fast self-propagation reaction when subjected to some form of energy. (2) Materials capable of burning or bursting suddenly and violently.

Fireball — Brief, roughly spherical fire suspended in the air resulting from ignitable gases released during an explosion; may occur during or after an explosive reaction.

Flame Speed — Velocity of the flame front relative to the center of the explosion.

Flash Fire — Type of fire that spreads rapidly through a vapor environment.

Force-Vector Analysis — Graphical tool investigators can use to determine the direction of blast pressure in an explosion.

Fragmentation — In terms of explosions, describes the process of a confining vessel losing its structural integrity and becoming shrapnel. Low-order damage involves large pieces of debris and shrapnel due to fragmentation, while high-order damage typically involves small, more widespread debris. *See* Shrapnel.

Fragmentation Effect — Process in which bomb parts and components are blown outwards by the explosion in the form of fragments, shards, or shrapnel.

Fuel-Gas Migration — Tendency of leaking fuel gases to move through pipes, sewer systems, permeable soil, or HVAC ducts, spreading flammable gases to areas often great distances away from the source.

Fuel-to-Air Ratio — Percentage of a fuel-gas suspended in air; ratios within the flammable range can sustain combustion when met with a competent ignition source.

Gas/Vapor Explosion — Chemical, nonseated deflagration that occurs when a fuel-gas mixed with air in the proper ratio rapidly ignites; may reach the level of a detonation if the fuel-gas mixture ignites in a confined space, such as in a gas pipe.

High Explosive — Chemical explosive material that decomposes extremely rapidly (almost instantaneously) and has a detonation velocity faster than the speed of sound.

High-Order Damage — A characterization of damage that includes bulging of walls, walls fallen intact away from a structure, roofs lifted and dislodged, and large debris moved only short distance.

Interface — Area between the fuel-rich area and the air-rich area where the two are mixing.

Low Explosive — (1) Explosive that decomposes or burns rapidly, but does not produce an explosive effect unless it is confined. (2) Explosive material that deflagrates, producing a reaction slower than the speed of sound.

Low-Order Damage — A characterization of explosion damage including small shattered debris, widespread damage, and near or total destruction of the confining vessel.

Mechanical Explosion — Explosion that is the result of physical change to a product in a closed container causing the buildup of pressure overcoming the structural resistance of the container.

Negative-Pressure Phase — Portion of an explosion in which air rushes back toward the center; caused by the lower pressure created from the positive pressure phase.

Nonseated Explosion — Explosion having no identifiable epicenter or seat; typically associated with deflagrations in fuel-air mixtures and dust clouds.

Positive-Pressure Phase — Portion of an explosion in which gases are expanding outward from the center.

Propagation — Spread of combustion through a solid, gas, or vapor, or the spread of fire from one combustible to another.

Seated Explosion — Explosion with a clearly defined epicenter or seat, often a crater; typically associated with boiling liquid expanding vapor explosions (BLEVEs) and high explosives.

Shock Wave — Blast-pressure front moving faster than the speed of sound; becomes an amplitude wave that travels through solid objects and the ground.

Shrapnel — Descriptor of debris, large and small, carried by the blast-pressure front of an explosion.

Stoichiometric Ratio — Ideal fuel-to-air ratio at which complete combustion of fuels occurs without byproducts of combustion; does not naturally occur. *Also known as* Optimum Ratio.

Total Station — Surveying equipment attached to a computer-imaging system, used to create computer models of incident scenes.

Turbulence — In terms of explosions, the changes in shape of the blast-pressure front as the expanding pressure is forced around objects and/or toward areas of ventilation; occurs in deflagrations but not in detonations.

Vector — Quantity in mathematics that has magnitude and direction and that is commonly represented by a directed line segment or arrow whose length represents the magnitude and whose orientation in space represents the direction; common quantities with vector representations include force, pressure, and velocity.

Photo courtesy of Donny Howard, Agent, Oklahoma State Fire Marshal's Office.

Chapter Contents

NFPA 1033 JPRs addressed in this chapter

This chapter provides information that addresses the following job performance requirements (JPRs) of NFPA 1033, *Standard for Professional Qualifications for Fire Investigator* (2022):

4.1.7	4.2.3	4.2.5	4.2.8	4.6.5
4.2.2	4.2.4	4.2.7	4.2.9	

This chapter provides information that addresses the following course outcomes for the FESHE courses:

Fire Investigation I (C0283)

7. Discuss the basic principles and identify cause and origin of fires.

Fire Investigation II (C0284)

5. Evaluate the use of incendiary devices, explosives, and bombs.

Fire Investigation and Analysis (C0285)

2. Document the fire scene in accordance with best practice and legal requirements.

3. Analyze the fire scenario utilizing the scientific method, fire science, and relevant technology.

4. Analyze the legal foundation for conducting a systematic incendiary fire investigation and case preparation.

1. Describe analytical methods and procedures used for investigation hypotheses. [NFPA 1033, 4.6.5]

2. Analyze information to determine potential area of origin. [NFPA 1033, 4.1.7, 4.2.2, 4.2.3, 4.6.5]

3. Correlate data from the fire to determine potential area of origin. [NFPA 1033, 4.2.5]

4. Interpret fire patterns and fire effects. [NFPA 1033, 4.2.2, 4.2.3, 4.2.4, 4.2.5, 4.2.7, 4.2.8]

5. Interpret fire patterns produced away from the fire plume. [NFPA 1033, 4.2.2, 4.2.3, 4.2.4, 4.2.5, 4.2.7, 4.2.8]

6. Describe analytical methods and procedures used for potential origin area evaluation of an explosion. [NFPA 1033, 4.2.7, 4.2.9, 4.6.5]

7. Skill Sheet 12-1: Determine the potential area of origin. [NFPA 1033, 4.2.4, 4.2.5, 4.2.7, 4.2.8, 4.2.9, 4.6.5]

Chapter 12
Potential Area of Origin Determination

Photo courtesy of Donny Howard, Agent, Oklahoma State Fire Marshal's Office.

A fire investigator uses observation data to determine the area where the ignition source and material first ignited came together for the first time. The accuracy of the information that goes into the origin determination will affect the accuracy of the cause determination. This chapter addresses the following:

- Investigation hypotheses
- Potential origin determination considerations
- Fire scene data
- Fire pattern development
- Other fire patterns
- Explosion origin determination

Investigation Hypotheses

NFPA 1033 (2022): 4.6.5

Fire investigation is a forensic science where evidence is simultaneously destroyed or altered as it is generated (Houck and Siegel, 2015). As such, investigators' requirements to increase stringency in evaluation and documentation will continue over time. To present a compelling case for their determination of an origin area, investigators must understand how fire changes an environment, how to safeguard against interpretation fallacies, and how to evaluate conclusions developed during investigation.

Analytical methods and procedures include:

- Timelines
- Link analysis
- Systems analysis
- Fault tree analysis
- Data reduction matrixing
- Hypothesis development and testing

Hypothesis Development and Testing

Using a consistent methodology, an investigator develops one or more working hypotheses regarding the origin of the fire. A common methodology starts with a general survey, to a more focused examination of specific portions, while testing hypotheses about the patterns and other conditions discovered.

Hypothesis testing should follow a set procedure. When evaluating a hypothesis regarding an area of origin, investigators should consider:

- Ventilation
- Drop down/fall down
- Fuel packages ignited
- High heat-release-rate fuels
- Lack of access for extinguishment

One method of hypothesis testing focuses on identifying locations within the structure that appear to merit additional study and attempting to correlate the patterns found in these areas with the fuel packages that were located in the structure or compartment before the fire. If a pattern cannot be explained after this correlation, then additional analysis of that area is warranted. During the analysis of fire patterns, investigators also consider protected areas as markers that may assist in scene reconstruction.

The analysis of adjacent patterns and the correlation with available fuel packages can assist in the determination of the area or areas of origin. Steps in determining the potential area of origin are found in **Skill Sheet 12-1**.

While examining a fire scene, investigators should consider:

- What does fire spread/fire modeling show?

- Did primary or secondary burning cause the flame plume damage?

- Is the ignition source a competent ignition source for fuels present **(Figure 12.1)**?

- Can a fire that starts at this hypothetical area of origin result in the fire that evolved?

- Does the fire scenario from this hypothetical area of fire origin result in the totality of the damage?

- How do other area(s)/cluster(s) of damage, that can be attributed to other causes, affect the current hypothesis?

- Does damage from a cluster of fire effects/patterns indicate flame spread to another cluster of effects/patterns?

Figure 12.1 An investigator may be able to recreate ignition conditions using materials similar to those involved in the incident. *Courtesy of Nicole Brewer, Portland Fire Investigations.*

Bias

Everyone is susceptible to **cognitive biases**. Investigators must be aware of how bias may influence their judgement and decision making during investigations (Dror, 2010). The careful use of knowledge, experience, and training will help an investigator evaluate the basis of their conclusions during an investigation (Bieber, 2014). For example, an investigator can practice reflexively self-questioning perceptions. Actively trying to disprove a hypothesis will yield a hypothesis with more support than detraction.

An unbiased opinion is supported by the data, facts, records, reports, documents, scientific references, and evidence. When using a consistent methodology to evaluate an incident scene, an investigator should carefully consider new information that shifts the working hypothesis (Taroni, 2010). For example, the presence of irregular burn patterns on carpet may resemble intentionally added fuel trailers, but may have been caused by a different fire effect such as melted plastics.

Combating Bias with the Scientific Method

Asking questions and consulting a checklist can help you follow the scientific method and make better decisions during an investigation. Nicole Brewer (2019) offers the following series of suggestions:

- Check yourself at multiple points during your investigation. Make sure you are not focusing on collecting data on one particular hypothesis and neglecting others (data collection and hypothesis testing).

- Expand your line of interview questions to include multiple potential ignition scenarios (data collection).

- Consider multiple explanations for fire patterns and effects that you observe (analyze data).

- Consider all potential sources of ignition in an area of origin and remember to think of your area of origin in three dimensions (data collection and analysis). Your ignition source may have come from above.

Potential Origin Determination Considerations

NFPA 1033 (2022): 4.1.7, 4.2.2, 4.2.3, 4.6.5

Fire investigators must always use a consistent methodology when determining the origin of a fire. The scientific method can be applied to every aspect of the work. The scientific method as an investigative process was introduced in Chapter 7, Initial Actions in the Investigative Process. **Skill Sheet 12-1** shows the steps for determining the area of origin.

An investigator should use as many procedures and methodologies as are reasonable to determine the area of fire origin. Regardless of the model, or combination of models, investigators compare information from different sources to gain a clearer overview of the overall incident. One way to compare data is to indicate findings on a floor plan as a Venn diagram where areas of overlap demonstrate where the information sources corroborate each other. A properly constructed Venn diagram can serve as the origin matrix. Venn diagrams were introduced in Chapter 8, Scene Documentation.

NFPA 921 states that determination of the fire origin area is coordinated from one or more considerations, which will be explored in the following sections. Investigators consider three primary categories of information while determining the area of origin:

- Fire dynamics

- Fire patterns and fire effects

- Witness information and electronic data

Fire Dynamics

NFPA 1033 (2022): 4.2.2

Evidence of damage from an incipient-phase fire may persist for a significant time as the fire develops further. As fire conditions intensify, visible fire effects and patterns can change the initial patterns. In addition, ventilation in a compartment may cause increased damage at or near openings, or across from these ventilation openings. A fire investigator must understand fire dynamics in order to interpret fire patterns and test fire origin hypotheses. An overview of fire dynamics was presented in Chapter 3. The two factors most significant in the investigator's analysis, especially while conducting heat and flame vector analysis, are oxygen and ventilation, and **radiant heat flux**.

Oxygen and Ventilation

Investigators carefully consider the availability of oxygen in the compartment or structure over the course of the fire. In an oxygen deficient environment, such as a compartment that generated full-room involvement, it is possible that less damage may be present or persist around the origin of the fire then elsewhere in the compartment. The most damaged areas may create ventilation openings.

In an incident with full-room involvement, basing the origin area on the degree of damage alone will likely lead to an incorrect hypothesis. Ventilation can cause fire movement from the area of origin to other locations. This can create greater fire damage in other areas of the building at ventilation openings or due to fuel load. Also, damage near ventilation openings may have occurred much later in the fire as a result of increased air flow in these areas. New fire patterns will be created in this instance.

Radiant Heat Flux

The amount of heat transferred to a surface is termed *radiant heat flux*. Higher temperatures result in increased heat flux, which then has more capacity to generate damage including **fire patterns**.

In the early stages of a fire, the fire plume and the heat flux rates are the primary means of pattern production. As a fire develops, a substantial upper layer begins to form and starts transferring heat to the wall and ceiling surfaces. As the temperature in the upper layer increases and the duration of contact between the upper layer and the wall/ceiling surfaces increases, the radiant heat flux to these surfaces reaches a critical threshold that begins damaging the material and creating patterns.

A ceiling jet formed by the intersection of the plume and a ceiling causes greater heat to transfer first to the ceiling surface and then to wall surfaces. The temperature of the affected surface is hottest near the plume centerline and cooler proportionally to the distance from the plume centerline. As a result, patterns are more distinct closer to the centerline of the plume **(Figure 12.2)**.

NOTE: The ceiling jet and the gases from the upper layer radiate heat to all exposed surfaces below them.

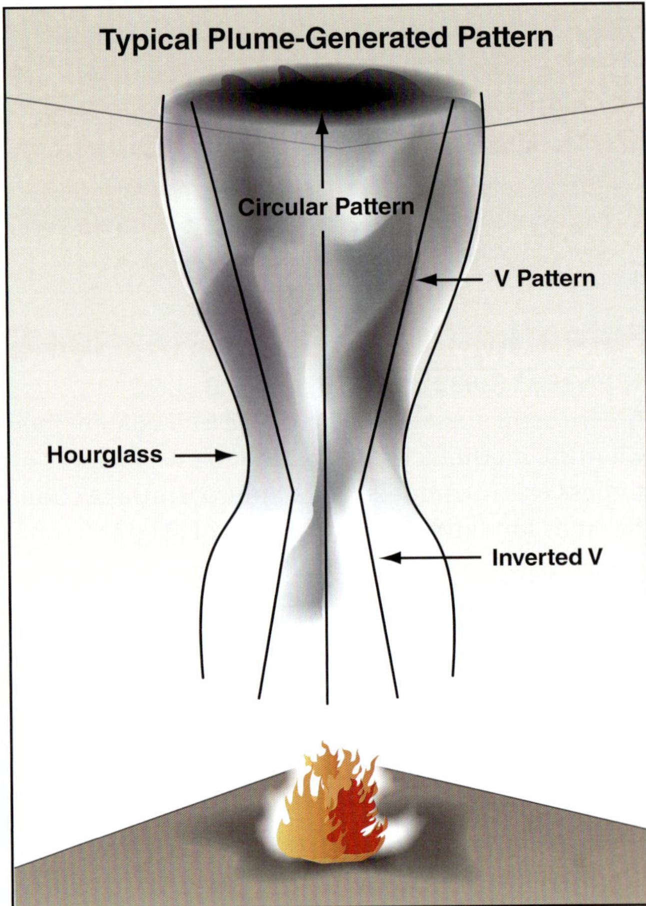

Figure 12.2 A plume-generated pattern shows stronger effects along the centerline of the plume.

As the compartment transitions through flashover and into full-room involvement, the upper layer expands toward the floor and encompasses nearly the entire volume of the compartment. At this stage of development, the walls, ceiling, and floor surfaces receive a similar magnitude of heat flux for a short period. Because of this uniformly applied heat, the initial plume patterns may appear to have been destroyed or obscured.

As a fire transitions to full-room involvement, it also changes from being fuel-limited to being ventilation-limited. Therefore, the most intense combustion and highest temperatures occur in ventilated areas. As is true in other stages of fire development, the new locations of the highest temperatures are also areas where the greatest damage occurs and most pronounced patterns may emerge.

Fire Effects and Fire Patterns

NFPA 1033 (2022): 4.1.7, 4.2.2, 4.2.3

Recognition and analysis of **fire effects** and fire patterns serve an integral role in determining the area of origin at a fire. Fire effects include physical and chemical changes made to individual materials when exposed to products of combustion. Fire patterns are the overall collection of fire effects **(Figure 12.3)**.

Each fuel package that burns generates its own fire effects and fire patterns. Fire investigators may choose to recreate a scene to evaluate the effects and patterns, to correlate each to a specific fuel package. Scene reconstruction, and correlation of damage with fuel sources, can be difficult when numerous fuel packages create patterns in an area. During scene evaluation, debris removal, and scene reconstruction, an investigator should develop a hypothesis as to the potential area or areas of origin of the fire.

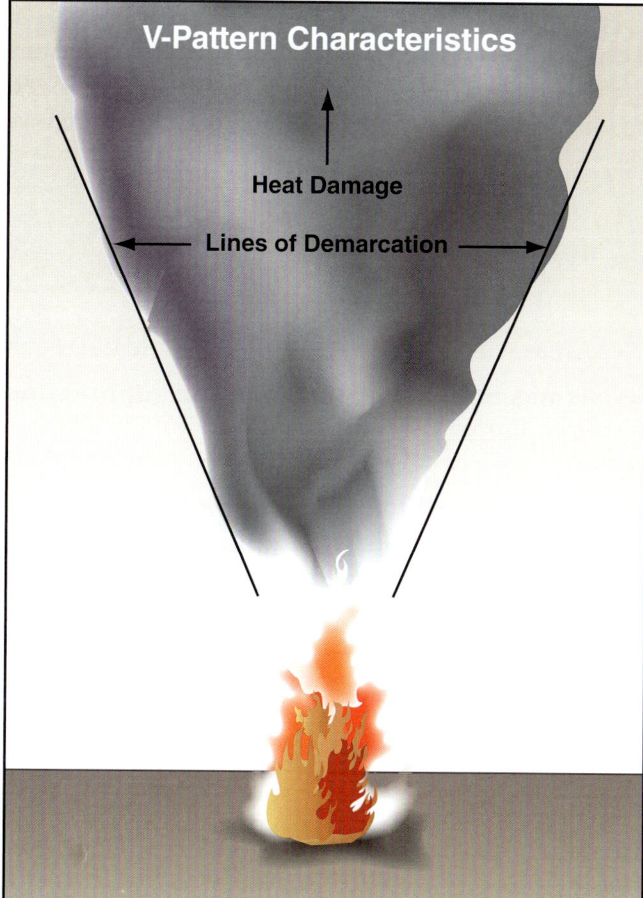

Figure 12.3 Fire effects include lines of demarcation that an investigator can track to form hypotheses.

Heat Effects

An investigator should understand the temperatures and conditions that can produce physical effects in a compartment. For example, when a fire ignites within a compartment, the products of combustion soon begin to interact with the surfaces and materials in the compartment, such as the walls, floor, and ceiling, found within the compartment. The effect on each of these materials will depend on several variables, including the characteristics of the material, the temperature of the products of combustion, and how long the materials were exposed to the products of combustion. Effects form patterns, and patterns can show how conditions changed over time in that space.

Fire Patterns

The analysis of fire patterns involves identifying damage to materials (fire effects) and then comparing those findings to analysis of other materials and their damage. Fire effects are caused by one or a combination of the following: heat, **deposition**, and/or **consumption**.

The investigator should identify and evaluate all of the fire effects present and cluster or trend these fire effects into fire patterns. And, the investigator may use each effect as an indicator to initially classify the cause of the observed damage (i.e. clean burn = heat).

Fire patterns are formed on surfaces (walls, floors, ceilings, and contents) of a structure as a result of direct flame contact or the exposure to heat generated by a fire. When heat causes a pattern, the method of transfer is by conduction, convection, or radiation. Fire patterns have visible boundaries or borders where the fire or products of combustion affected a surface, leaving adjacent surfaces less affected or intact. These borders or lines of demarcation are what a fire investigator searches for when looking for fire effects and fire patterns.

Fire Pattern Analysis History

Fire patterns have long formed the basis of investigators' understanding how a fire has moved within an environment (Gorbett, 2015). Historically, the practice of determining fire origin from fire patterns has been highly subjective, and later reviews determined the interpretations to be incorrect. The increased use of consistent methodologies, such as the scientific method, has significantly improved the reliability of results.

Witness Information and Electronic Data

As introduced in Chapter 8, Scene Documentation, an investigator obtains data from witnesses to the fire or witnesses aware of the conditions within the fire scene prior to the fire. Investigators should interview all possible witnesses and correlate their information to determine if the information is useful to determine the fire origin.

Witnesses

Investigators should obtain statements from employees, owners, and occupants regarding the locations of fire, smoke and flames, and witness vantage points. First responders, including firefighters involved in suppression efforts, are also considered witnesses with more technical perspectives. Investigators should obtain statements from firefighters regarding:

- Security
- Overhaul
- Rekindles
- Exposures
- Ventilation
- Suppression efforts
- Location of burning
- Type of extinguishment
- Method/location of entry
- Areas where suppression efforts could not reach
- Times of dispatch, arrival, and suppression completion

An investigator should corroborate witness information with conditions at the scene. For example, if a witness describes exiting the structure through a window and leaving it open, the investigator should check the window itself and any other notes related to that window. When conducting an analysis of the witness statements, answer: "Do the witness statements confirm or refute the hypothesized area of fire origin?"

Investigators may wish to use the following process with witness information:

1. Organize witness information in a manner conducive to analysis.
2. Analyze the content of witness interviews.
3. Present the results.
4. Study witness statements to test fire origin hypotheses.

Electronic Data

Where possible, investigators collect the electronic data from the scene and nearby locations. Fire and security alarm data provide potentially valuable information such as the location and activation sequence of devices such as alarm system components. Analyzing alarm data can help an investigator to develop a hypothesis regarding the origin of a fire. Useful alarm data includes:

- Door/window openings
- Smoke alarms/detectors
- Fire alarm activation or security cameras **(Figure 12.4)**
- Location and activation of motion sensors

Figure 12.4 Security cameras may have recorded information an investigator needs. *Courtesy of Donny Howard, Agent, Oklahoma State Fire Marshal's Office.*

Correlating Data

Data must be collected in a way to be comparable to other data. For example, observations from witnesses and electronic data should be ordered chronologically, synchronized, and added to the incident timeline. In some cases, the witnessed and documented sequence of the fire growth may indicate a hypothetical area of origin.

Fire Scene Data

NFPA 1033 (2022): 4.2.5

Fire effects and patterns represent individual pieces of data at a fire scene. Taken together, the measurable and visible damage from the fire, combined with the patterns created from the various heat sources, can show a sequence of events that occurred during the fire.

NOTE: Weather is included in Chapter 7, Initial Actions in the Investigative Process, as a factor that investigators should consider during scene evaluation and witness interviews.

Fire investigators should be able to reconstruct a hypothetical sequence of events, beginning with identifying a plausible area(s) of origin. Throughout the process, the investigator should not assume that the most damaged area indicates the area of origin. This assumption is a form of bias and must be strongly guarded against. Investigators must test each hypothetical area of origin.

Figure 12.5 Investigators determine fire origin by evaluating competent ignition sources, fuel packages, and ventilation sources.

For example, at one fire scene, a candle left burning on a bedroom dresser ignited nearby combustibles resulting in an initial V-shaped pattern extending from the top of the dresser **(Figure 12.5)**. The room eventually reached full room involvement and the initial V-shaped fire pattern was no longer visible without a visual aid. A window (ventilation opening) in the room had failed during the fire. Greater damage was present around the window than at the top of the dresser.

Determining the Sequence

Each fire effect and fire pattern found at a scene is the product of an individual fuel source, including from the upper gas layer. Understanding the order in which those fuels ignited can indicate an origin area to an investigator. However, one of an investigator's most challenging tasks is determining this sequence of events because many factors can complicate the analysis.

Patterns tend to change over the course of a fire. Examples:

- Larger patterns may not necessarily indicate an area of longer or hotter burning.

- Patterns from rekindled areas can be misinterpreted as having occurred during the fire.

- Full involvement in a compartment will change the patterns that existed before that phase.

- A triangular or inverted cone pattern may evolve into a columnar or V-shaped fire pattern.

Only with all of the data can investigators create an accurate history of the fire and, as a result, confidently identify the area of origin. Fire pattern analysis data collection includes:

- Analysis of sequential events
- Arc surveys or arc mapping
- Depth of calcination survey
- Depth of **char** analysis
- Fire dynamics
- Fire pattern analysis
- Heat and flame vector analysis
- Origin matrix analysis

An investigator should avoid forming any conclusions until all the data has been gathered and analyzed. The following sections describe several of these data points and their significance.

Fire Damage Assessment

An investigator who is reconstructing the history of the fire must understand how each pattern at a fire scene was generated and, if possible, its sequence. From that knowledge, investigators develop a hypothesis for each likely scenario and test those hypotheses.

Fire Damage

Fire patterns can indicate fire movement, fire intensity, or both. It is very rare that one fire pattern, no matter how large or dominant, can individually point investigators to the area of origin. Fire patterns must be examined as a whole picture, accounting for all fire patterns or fire effects. Furthermore, an investigator must determine how the various fire patterns at the scene give collective evidence about what happened.

An investigator should be able to account for every fuel package and the effects of ventilation. For example, when a couch has ignited, the investigator should have an idea as to whether the fire spread to or from the couch.

A fuel load's proximity to walls and corners affect the patterns generated from its burning **(Figure 12.6)**. For example, identical fuel packages may burn differently because of their different locations in the compartments. The investigator must consider all such information when analyzing fire patterns to determine the area of origin.

Heat Damage

Heat damage is normally the most visible and measurable indicator of a fire's effects. As fire investigators continue

Figure 12.6 The position of furnishings and location of the fire can change the development of patterns. *Courtesy of Iowa State Fire Training Bureau.*

the examination of the building, this information assists in the determination of fire growth, development, and spread.

Damage to a building can be caused by:

- Fire-suppression efforts
- Fire and products of combustion
- Ignition points exterior to the building **(Figure 12.7)**
- Fire spreading from one compartment to another
- Fire spreading from the interior of the building to the exterior

NFPA 921 contains a list of fire effects and whether these effects are visible only or measurable data can be obtained from the effect:

- Calcination
- Char
- Clean burn
- Collapsed furniture springs
- Color changes
- Deposition of smoke on surfaces
- Distorted light bulbs
- Mass loss of material
- Melting of materials
- Oxidation
- Rainbow effect
- Spalling
- Thermal expansion and deformation of materials
- Victim injuries
- Window glass

Figure 12.7 An investigator should determine whether heat effects on the structure's exterior could have resulted from ventilation from the interior. *Courtesy of Donny Howard, Agent, Oklahoma State Fire Marshal's Office.*

Investigators should carefully examine all points of low burning. These patterns may be the result of burning debris that drops from the building and causes a secondary ignition, or they may be an indicator of an exterior area of origin **(Figure 12.8)**. If the pattern indicates that the fire originated lower than the base of the structure, the investigator should look for structure openings in this area, and additional indications that the fire originated outside of the structure.

Fire investigators must determine whether these patterns are distinct and separate areas of fire origin or result from the natural progression of the fire. In addition to ventilation and low-burning indicators, an investigator should analyze all patterns on both the exterior and interior of a structure.

As with all potential evidence that fire investigators use in the analysis of the origin and cause of a fire, they should properly document and analyze all fire effects and fire patterns. Investigators should collect and preserve any physical evidence according to the procedures explained in Chapter 9, Evidence Collection and Preservation.

Figure 12.8 Indicators of low burning sometimes indicates the origin area, especially when they affect the structure's exterior. *Courtesy of Donny Howard, Agent, Oklahoma State Fire Marshal's Office.*

Depth of Char

The solid by-product of burning wood is referred to as *char*. Investigators can examine qualities of char to inform a hypothesis regarding how fire spread. For example, char depth decreases the farther away the char is from a heat source.

NOTE: Depth of char is NOT a reliable indicator of specific fire duration.

Investigators can take measurements of char depth and the extent of char across a damaged surface using a **char gauge**; blunt-ended, thin probes such as dial calipers; or tire-tread gauges. For example, a V-shaped pattern, with a greater extent of charring on one side than the other, can indicate that the fire moved in the direction of more damage. This movement may be attributed to increased ventilation or the introduction of another fuel package on the more damaged side of the pattern.

Collecting Char Data

Using a consistent pressure for each measurement, each depth is taken by inserting the char gauge or other tool into the center of char blisters at various points across a fire pattern. The probe should be inserted into the char until it meets significant resistance. These measurements should only be compared with similar woods at similar elevations.

Investigators must remember that wood may be missing from the area being measured. This missing wood may have broken due to weakening from exposure to heat or complete destruction in the fire. Investigators should indicate any missing wood in the depth-of-char measurement **(Figure 12.9)**.

Figure 12.9 The cross-sectional area of a wood member gradually decreases as it burns. *Courtesy of Donny Howard, Agent, Oklahoma State Fire Marshal's Office.*

When measuring char, investigators should consider variables including:

- **Number of heat sources that created the char pattern** — Depth of char measurements can indicate whether multiple heat sources created a pattern.

- **Char measurements from the same material** — Softer materials will show deeper char than harder materials, which can affect conclusions from measurements.

- **Ventilation increasing the rate of burning** — An increased mixture of gases and oxygen within the compartment supports increased combustion and can also produce greater char depth. Wood adjacent to or near a ventilation source can show greater char depth due to hot fire gases escaping from the compartment.

- **Consistent measuring technique** — Each char depth being compared should be measured using the same equipment and method.

Using Char Data

The ultimate outcome of depth-of-char analyses is an objective and measurable determination of the movement or intensity of a fire. These measurements can aid an investigator in making comparisons about which portions of the surface burned longer than others.

Investigators develop an **isochar map** using a floor plan of the investigated space. As depth of char measurements are collected and recorded, their locations are indicated on the plan. Similar depths are connected, and the resulting diagram shows **intensity patterns** and lines of demarcation in the area **(Figure 12.10)**. Isochar maps provide objective evidence of areas with greater heat flux.

Figure 12.10 An isochar diagram can indicate places with intense burning that may not be otherwise visible in a severely damaged space.

Depth of Calcination

Gypsum wallboard changes texture as it is heated at fire temperatures **(Figure 12.11)**. This process and the result are known as **calcination**. Investigators can measure calcination using the same process and probes used in char depth measurements: tire tread depth gauges, insulation thickness gauges, or calipers.

Figure 12.11 Calcination is the result of evaporating water from gypsum wallboard. *Courtesy of Donny Howard, Agent, Oklahoma State Fire Marshal's Office.*

NOTE: Wet wallboard often yields no reliable measurements of calcination.

Wallboard is measured same as char depth, but with additional considerations:

- Finishes, such as paint, wallpaper, or paneling, can affect patterns left on wallboard. Such finishes are combustible even when the wallboard is not.

- Damage to gypsum wallboard can occur at any time during fire-scene activities such as fire suppression, overhaul, or investigation.

Wallboard Calcination

As indicated in Chapter 10, Fire and Explosion Scene Examination, an investigator will interact with significant quantities of gypsum wallboard in various conditions at a fire scene. An investigator should understand how this material reacts under various conditions.

Water in gypsum wallboard is chemically bound in a crystal lattice (calcium sulfate dihydrate). As gypsum wallboard heats, water molecules dissociate from the crystal lattice. When one water molecule departs, gypsum becomes calcium sulfate hemihydrate (also known as *plaster of paris*), which has one water molecule per two calcium sulfate molecules. At higher temperatures, the remaining water disassociates, and the material becomes anhydrous calcium sulfate. This dehydration process is known as *calcination*.

Gypsum wallboard is commonly used to protect building structural components. It is effective for this purpose over relatively long spans of time because heat energy delivered to the gypsum drives off the water. The mineral temperature rises slowly until the water is evaporated, then increases more rapidly. In fire conditions, the materials behind the gypsum wallboard will remain relatively cool as water disassociates from the gypsum.

Collecting Calcination Data

Investigators should use a consistent methodology to examine gypsum wallboard to determine the amount and the significance of the damage. Regardless of the method used, accuracy is maintained if one person completes the probe measurements with consistent pressure through the process.

Investigators complete a gypsum wallboard calcination survey via the following steps:

1. A grid pattern is established on the surface of the gypsum wallboard using chalk line or laser. The typical grid size is 1 x 1 foot areas **(Figure 12.12)**.

2. A blunt-ended probe is inserted in each of the grids at the same relative position. For example, an investigator may insert the probe into the top left corner of each grid. The investigator inserts the probe until resistance is felt.

3. The investigator removes the probe from the gypsum wallboard taking care to collect an accurate measurement. To create a visual representation of the findings, investigators may insert a roofing nail at each collection point with the measurement written on it.

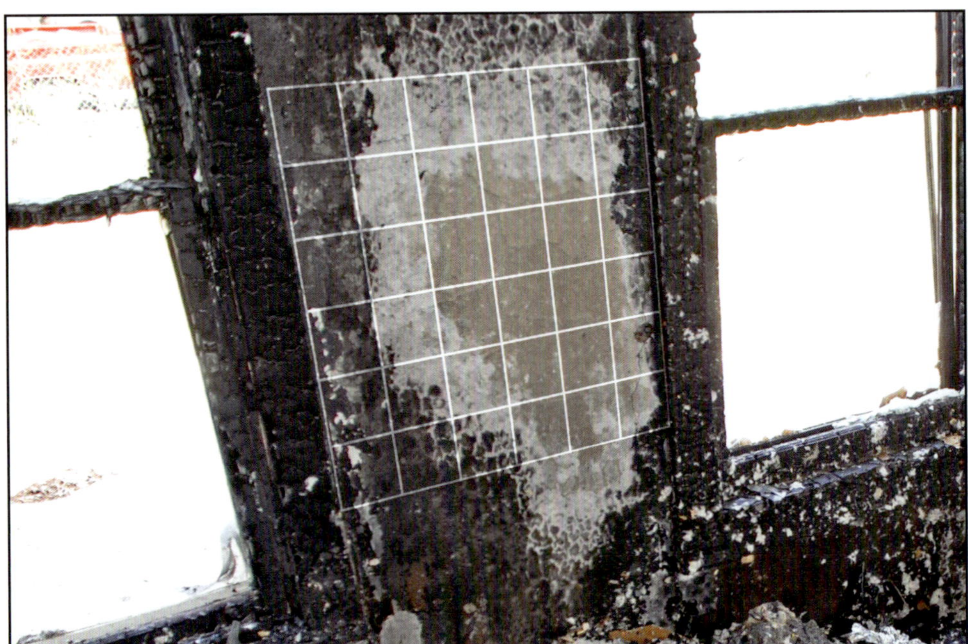

Figure 12.12 An investigator can draw or superimpose a grid on gypsum wallboard to help survey calcination depth.

While collecting probe measurements, an investigator may chose to use a pre-established scale to indicate relative measurements. For example, on one scale intended for assigning a degree of visible fire damage to gypsum wallboard, the degree of fire-damage scale ranges from 0, indicating no damage, to 6, indicating complete consumption.

Wallboard Exposure Data

Current exposure studies demonstrate that fire damage to wallboard occurs progressively. As cited in Gorbett (May 2015), Mann and Putaansuu (2010) identified the following progression:

1. Soot coating of undamaged facing paper;

2. Discoloration/degradation of facing paper;

3. Facing paper burns away;

4. Partial dehydration and discoloration/soot staining of surface layer of gypsum;

5. Formation of anhydrous and hemihydrate layers with layers progressing through the cross section;

6. Complete conversion of dihydrate to anhydrous and hemihydrate;

7. Anhydrous extends through the entire cross section;

8. Wallboard becomes catastrophically heat damaged and lacks structural integrity.

Using Calcination Data

As with depth of char measurements, investigators should record depth of calcination readings to a format that allows for additional pattern comparisons. Char depth and calcination depth are both an objective and measurable determination of the movement and intensity of a fire. In both cases, greater depth indicates greater heat exposure.

Investigators record the measurements to establish trends in the depth of calcination, which will indicate the overall amount of calcination that has occurred to the gypsum wallboard. The numerical values may then be assigned a color or contour plot to visually represent the fire patterns that are present.

Vector Analysis Diagrams

Investigators test hypotheses with an origin matrix heat and a flame vector analysis. Vector diagrams show scene examination findings drawn over a scale diagram of an incident scene or compartment. The vector diagram is especially useful when it can be added or removed from a diagram (overlay) showing the prefire condition of the room or area.

The diagram should show a top-down view of the fire scene, and can illustrate fire patterns in environments such as:

- Vehicles
- Wildland environments
- Structures or compartments
- Appliance or other equipment casings

An accompanying legend explains the symbols used for each vector, material, and fire pattern. The vectors or arrows can either point away from the area of fire origin or toward it. The fire investigator must indicate in the legend the method used, and remain consistent throughout the diagram. Investigators may also include photographic references within the legend to further illustrate the observed damage. Each fire pattern is assigned a vector style (format, color) and a number. This identifier represents the approximate length of the fire pattern to give an idea of its relation to the size of the space.

The prefire diagram should indicate the position and condition of:

- Walls
- Windows
- Doorways **(Figure 12.13)**
- Contents of the structure or vehicle

An overlay can show the locations of potential fuel packages in the structure or vehicle. Investigators may then add another overlay to show the direction of heat or flame spread within the space using arrows. Investigators can then use the vector analysis diagram for heat and flame pattern analysis.

Investigators should use vectors with solid lines for horizontal fire spread, and vectors with dashed lines for vertical fire spread. Analysis of the damage

Figure 12.13 Door positions marked with a durable tape can provide context later in the incident evaluation. *Courtesy of Donny Howard, Agent, Oklahoma State Fire Marshal's Office.*

is also included. For example, the arrows can help direct the investigator's attention to the sources of the most intense burning fuel packages. Similarly, investigators should explore locations that do not show a clear cause for their condition.

The combination of prefire features plus focus on fuel packages, plus vectors indicating fire dynamics can show the investigator several important considerations in the incident such as:

- Which burned areas do not match a fuel source

- Whether a given fuel package was at or near the area of origin

- Whether a particular fuel package was ignited after the initial fire package

Arc Mapping

Investigators must first identify where arcing occurred before they can see the overall arc fire pattern. To determine arcing, an investigator surveys and maps conductors for damage. Types of conductors commonly found in structures include outlets, wall conduits, junction boxes, and appliance cords. During **arc mapping** analysis, fire investigators consider:

- Do arc sites match a hypothetical area of origin?

- Are there electrical arc sites anywhere within the scene?

- Is there any trend within the geographical location of arc sites?

- Are there electrical arc sites within or near the proposed or hypothetical area of fire origin?

Examining Possible Arc Sites

Investigators should have available a low-power magnification device, such as a jewelers loup, a lighted magnifying glass with scale, or an optician's headset, to examine the conductors in the field. Investigators should have some method, such as colored zip ties, string, ribbon, or tape, for marking arc sites at the scene **(Figure 12.14)**. Investigators photograph and document evidence of a potential electrical arc.

When circumstances indicate that a conductor may have arced, the investigator examines the entire conductor. Each arc site on the conductor is its own event, and investigators document it as such. For example, if an arc site is located on a cord to a floor lamp near the base of the lamp, the investigator examines the cord along its full distance to the socket into which it was plugged.

Figure 12.14 This wiring system has been tagged with uniquely numbered tags at each location where an arc has occurred. *Courtesy of Yates & Associates.*

As with any evidence, investigators should minimize handling delicate items. Any component that appears as though it would disintegrate or otherwise sustain more damage as a result of being moved should not be moved. Other means of assessing arcing on fragile physical evidence is discussed later in this section.

CAUTION: If a conductor appears too fragile to touch, then it probably is.

The investigator may remove a section of conductor and/or electrical components to examine them properly. When collecting a conductor for analysis, the investigator cuts the wires at some distance away from the arc, and protects all pieces of electrical evidence from further damage such as breaking or mechanical abrasion.

Types of Arcing Damage

After finding imperfections, the investigator determines whether the damage resulted from an arc or from other processes. The following characteristics of an arc site may be visible without aid or, in some cases, low power magnification:

- Localized damage

- Localized point of contact

- Localized round depressions

- Round, smooth shape of artifact

- Small beads and divots over a localized area

- Identifiable corresponding damage on opposing conductor

- A sharp line of demarcation between the more damaged and less damaged portions of the conductor

- Spatter (melted copper or aluminum depending on the type of conductor) may be present on the conductor as a result of the molten metal produced during the arc to have been ejected from the arc site and resolidifying on the conductor or nearby adjacent surfaces of other materials

Concave or convex changes in the conductor's surface may indicate the presence of an arc site **(Figure 12.15)**. Dirt accumulation, char, rust, frayed insulation, or cut insulation are the results of other fire damage and do not indicate an arc site. The investigator should examine the conductors carefully to ensure the damage is not the result of tool marks or mechanical fracture. Melting damage may present as elongated damage (necking) or the presence of blisters on the surface of the conductor.

The presence of melting itself does not indicate an arc site. Conductors can melt through exposure to heat without creating an arc. Melt damage on conductors may occur in ways including:

- Alloying

- Extraneous melting

- Melting caused by fire temperatures

Types of Arcs

An arc may occur through charred insulation. Investigators should look for damage on the conductor beneath charred insulation. Checking the conductor establishes whether the area was just charred or charred and arced **(Figure 12.16)**.

Figure 12.15 Arc damage may affect noncombustible surfaces.

Figure 12.16 Examination of electrical outlets can reveal the nature of the damage. *Courtesy of Donny Howard, Agent, Oklahoma State Fire Marshal's Office.*

Other types of arcs may also occur leading to ignition of various materials:

- **Parting Arcs** — Occur when a cord from an appliance is removed from a receptacle while current is still flowing (appliance is on). May ignite combustible dust and vapors.

- **Arcing Through Char** — Occurs when the insulation protecting a conductor breaks down and becomes somewhat conductive, allowing current to flow through the insulation.

Examining and Collecting Arcing Evidence

The presence of an arc site cannot always be observed in the field. When an investigator suspects that a conductor may have been involved with an arc, a physical examination of that conductor can reveal damage. Investigators can pass the conductor through their fingers while feeling for imperfections. An investigator may also remove dirt and scale from the conductor allowing observation of any imperfections. Investigators can use cotton balls or cotton swabs to remove char and damaged insulation. Larger items can be cleaned with a dry brush or place the conductor in an ultrasonic bath.

The investigator should have access to a stereo microscope, preferably with a camera, to examine and photograph damage under higher magnification. The use of scanning electron microscopy (SEM) and computed tomography (CT scan) may also be required to determine if an arc site is present as some of the characteristics of an arc site are not visible without aid or under low power magnification. An X-ray can capture the arcing for examination and may eliminate the need to conduct destructive testing.

The following characteristics of an arc site are not visible without aid or with low power magnification, and the use of scanning electron microscopy (SEM) must be employed to detect them:

- Resolidification waves

- Locally enlarged grain size

- High internal porosity when viewed in a cross section

- Copper drawing lines visible outside the damaged area

Using Arcing Data

Creating an arc diagram begins with an accurate sketch of the compartment under investigation. An investigator should follow a consistent methodology for this process. NFPA 921 provides one possible procedure for completing the diagram after an initial sketch. This data can be added to other data for comparison, such as an additional overlay on the vector analysis diagram.

Any arc-map diagram should include the following information:

1. Identification of all conductors with the following information included whenever possible:

    — Load on each circuit **(Figure 12.17)**

    — Direction in which power was flowing

Figure 12.17 The load on a circuit can indicate whether the circuit was overloaded or improperly loaded. *Courtesy of Donny Howard, Agent, Oklahoma State Fire Marshal's Office.*

— Locations of junction boxes, switches, outlets, or other control mechanisms in the electrical system

2. Physical descriptions of examinations at arc sites:

   — Grounded conduit

   — Site faulted to another cable

   — Conductor completely severed

   — Building element that is conductive

   — Site faulted to another portion of the same cable

Tagging arc sites helps investigators create a sketched, arc-survey diagram. The diagram assists investigators in determining the sequence of events at the fire and establishing any pattern to electrical movement at the scene. For example, an investigator observes that a power strip or power bar with multiple appliances in it has arced as have the appliance cords plugged into it. The farthest cord from the electrical source would have arced first, indicating that it was first exposed to heat, which may further indicate that it was nearest to the area of origin. By diagramming the scene, investigators document these patterns. Investigators can also make diagrams for small zones such as one section of a wall around a junction box or an area of a room that had many cords plugged into the same outlet.

Arc Mapping Procedure

Previous editions of NFPA 921 have enumerated a multi-step procedure for arc mapping. The 2021 edition of 921 puts more focus on the investigator's ability to locate, recognize, and identify patterns throughout the course of a fire investigation. In 921 (2021) 6.3.21.10, the arc mapping procedure is simply as follows:

"Arc mapping involves several processes including arc identification, as discussed in in 6.3.21.10.2. The arc sites first need to be found and then documented. There are a number of ways to do this work, but all seek to develop relationships between the location of the arc sites themselves as well as the relationship of the arc sites to other evidence, such as ignition sources, in the fire scene. Depending upon the requirements, some fire scenes may only be partially arc mapped or arc mapping may only be applied to a particular piece of equipment."

Fire Pattern Development

NFPA 1033 (2022): 4.2.2, 4.2.3, 4.2.4, 4.2.5, 4.2.7, 4.2.8

The visible and measurable fire patterns are caused by one or a combination of the following physical effects to the material: heat, deposition, and/or consumption. These fire effects will assist investigators in better identifying the cause of the resulting damage.

The patterns found during the scene examination assist investigators in determining the sequence of the fire from ignition to extinguishment. In many cases, they lead investigators to the area or areas of origin. The patterns also provide information about how the fire spread and about the fuels involved during the progression of the fire. A fire investigator must be able to identify and interpret fire patterns found during the examination of a fire scene. Additional data that may aid in determining origin is discussed later in this chapter.

Fire Pattern Characteristics

Fire patterns evolve and change during the progression of a fire. When examining fire patterns, fire investigators should consider at what point the fire pattern was developed. For example, a basic stand-alone, V-shaped pattern is likely to have been created in the early stages of the fire from a single fuel package **(Figure 12.18)**. Should the room or area reach full-room involvement, the V-shaped pattern continues to evolve beyond its initial configuration. The earlier pattern may still exist, but may not be readily or visibly discernible.

The material that the pattern is displayed upon has a significant effect on what is observed after the fire. Combustible surfaces, such as wood paneling used as a wall covering, may actually be consumed and add to the overall fuel load available to the fire. When exposed to less energy from the fire, the damage to combustible surfaces may only cause charring or discoloration. On noncombustible surfaces, such as brick, plaster, or metals, the patterns may cause surface discoloration, spalling, melting, or distortion. The physical characteristics of a surface can also affect the type of pattern found. For example, rough surfaces are often more heavily damaged than smooth surfaces of the same material. The reason for this difference is that a rough surface presents more surface area for hot gases or flames than a smooth surface. The increased surface area and turbulence created as fire gases flow over rough surfaces result in additional damage.

Figure 12.18 V-patterns can develop after an electrical fire in a circuit breaker. *Courtesy of Wayne Chapdelaine, Metro-Rural Fire Forensics.*

Fire Damage Evaluation

Investigators should use a consistent metholodgy for determining and area of origin. Gorbett (Fire Science Reviews, 2015) identifies seven steps that an investigator can use while evaluating fire damage:

1. Identifying the value in further analysis of a surface or compartment;

2. Identification of the varying degrees of fire damage (DOFD) along the surfaces of the compartment and contents;

3. Identifying clusters and trends of damage (fire patterns);

4. Interpreting the causal factors for the generation of the fire patterns;

5. Developing area(s) of origin hypotheses;

6. Testing the hypothetical area(s) of origin; and,

7. Selecting a final area of origin hypothesis.

Movement versus Intensity Patterns

NFPA 1033 (2022): 4.2.5, 4.2.7

Fire patterns can indicate both movement of the fire and the intensity of its burning. Investigators should be able to recognize indications of both varieties. Fire development is discussed in Chapter 3, Fire Dynamics.

NOTE: Investigators must remember that fire is three dimensional and therefore fire patterns may be present on multiple surfaces.

Movement patterns result from the application of heat, flame, and smoke, which are produced as a result of the fire's growth and spread from the initial heat source. For example, a stud wall with greater char to the north side of each stud than to the south side of each stud may indicate fire movement from the north to the south **(Figure 12.19)**. Analysis of movement patterns assists investigators in tracing damage back to the initial heat source that produced it.

Intensity patterns are produced by the flame and hot-gas zones of a fire plume (3D conical shape of the plume). Intensity patterns leave their physical marks, such as lines of demarcation, surface melting, or oxidation marks, on a surface. For example, a vehicle surface exposed to a large structure fire may have an oxidation pattern on the side of the vehicle, showing greater fire intensity directed at one end of the vehicle than at another point on the same surface **(Figure 12.20)**. An investigator examining the pattern on the refrigerator could conclude that an isolated fuel source burned in that particular spot. Analysis of intensity patterns assists investigators in determining the characteristics (heat release rate, size, orientation, etc.) of the fuel package that produced them.

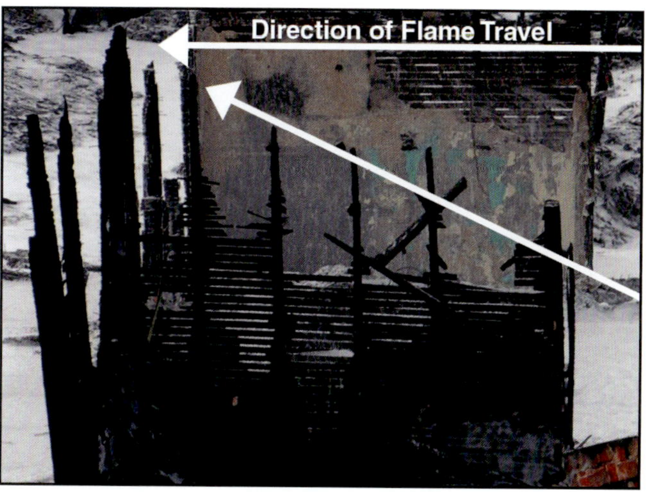

Figure 12.19 Partial studs may point toward the area of greatest heat or fire duration. *Courtesy of Yates & Associates.*

Figure 12.20 An intensity pattern can reflect on a surface some distance away from the fuel package. *Courtesy of Donny Howard, Agent, Oklahoma State Fire Marshal's Office.*

Common Fire Patterns

Fire investigators should make every effort to determine how a fire pattern was generated. If a fire pattern present within a compartment cannot be conclusively determined as to how it occurred, investigators should note it as being undetermined as to its generation. This information should also be recorded. Fire patterns with clear definitions include:

- Clean burns
- Saddle burns
- Protected areas
- Plume-generated
- Pointer and arrows
- Ventilation-generated
- Upper-layer generated
- Suppression-generated
- Full-room-involvement
- Irregular shaped patterns on floors

Lines of Demarcation

Lines of demarcation are the visible borders on the edges of fire patterns. These lines appear as a difference between areas with greater and lesser fire damage. Lines of demarcation may be formed by fire, heat, or smoke **(Figure 12.21)**.

When evaluating lines of demarcation and damage within the patterns, investigators consider:

- Is there increased magnitude of damage near the fuel item?
- Are the observed lines of demarcation angled and extending from the fuel package?
- Is the elevation of the line of demarcation consistent with the height of the fuel package?
- Is the loss of mass to the fuel package consistent with the damage to the affected surface?
- Is the width of the base of the observed damage approximately the width of the fuel package and not greater than two times the width of the fuel package?

Figure 12.21 Lines of demarcation include heat and smoke lines in a room affected by thermal layering. *Courtesy of Donny Howard, Agent, Oklahoma State Fire Marshal's Office.*

Plume-Generated Fire Patterns

NFPA 1033 (2022): 4.2.3

An investigator can use plume-generated patterns to identify potential areas of origin or the location of significant fuel packages that became involved after ignition. After observing plume-generated patterns, the investigator should consider how the pattern formed. For example, plume shape is affected by the relative distances of:

- The resulting plume from the vertical surface
- The burning fuel package from the vertical surface
- The height of the ceiling or intersecting horizontal surface

Investigators should closely examine plume patterns located in areas without any significant fuel packages. For example, a plume-shaped pattern may appear on a wall on the main level of a residence from a fire in a basement if there is an opening such as a grate in the floor.

Investigators categorize plume-generated patterns by the geometrical shapes formed by the lines of demarcation:

- **Triangular or inverted cone** — At the start of a fire, the flames have not yet reached the ceiling.
- **Conical shaped pattern** — The next phase of fire pattern development results from a burning fuel package that is close to a vertical surface.
- **Columnar pattern** — The second phase of a fire pattern occurs when the flame height has reached the ceiling, resulting in a vertical pattern on the surface.

Not all fire plumes will generate V-shaped patterns, U-shaped patterns, or circular-shaped patterns because of factors including:

- Ventilation within the space
- Surface on which the pattern is created
- Shape, orientation, and location of the fuel as the pattern is made

Further, plume-shaped patterns may be altered or disguised by factors such as combustible surface material in the plume area burning simultaneously with the initial fuel package. For example, wood paneling on a wall may ignite when heated by another fire.

The following patterns are generally the result of plumes and are discussed in greater detail in the sections that follow:

- Columnar
- V-shaped (conical)
- U-shaped
- Truncated cone
- Inverted cone
- Hourglass
- Circular-shaped
- Pointer and arrow

Columnar

A columnar fire pattern is found on vertical surfaces when the flames (plume) from a burning fuel package that are near a vertical surface, such as a wall, and reach an intersecting horizontal surface such as a ceiling early in the effect development. At this point, the by-products of combustion spread outwards in all directions (ceiling jet). The resulting effect on the vertical surface is a fire pattern resembling a column with relatively vertical lines of demarcation that are normally close to the width of the fuel package that was responsible for the pattern. Should the fire continue to burn, the pattern will evolve into a "V" shaped or conical fire pattern.

V-Shaped (Conical)

A V-shaped pattern is frequently found in compartment fires **(Figure 12.22)**. As a fire develops next to a vertical surface, the flames and hot gases widen into a V-shaped or conical pattern on the surface. The fuel package must be very near the vertical surface for a "V" shaped pattern to occur.

The width of the V is a function of the size of the flames that create the plume. For example, a fire in a small trash container close to a wall results in a pattern on the wall that is relatively narrow, while a sofa burning next to a wall generates a wide pattern on the wall.

The closer the package is to the vertical surface, the more sharply angled the lines of demarcation appear. The V-pattern extends as high as the plume reaches and, in many cases, extends to the ceiling of the space in which the pattern was found. The bottom of the V points toward the flame source as the pattern was made.

The angle formed by the edges of the V depends on several variables, including:

- Heat release rate of the fuel

- Geometry of the fuel package that causes the plume

- Effects of ventilation on the flow of the plume (direction of entrainment)

- Physical characteristics of the surface on which the pattern appears, its combustibility, and how easily it ignites

- Any horizontal surfaces, such as building overhangs or another horizontal surface such as a table, that would interrupt the flow of the fire plume.

When a plume intersects a horizontal surface lower than the ceiling, the pattern changes. Depending on the height of the surface, the pattern may not appear as a V, or the angles evidenced by the lines of demarcation may be altered **(Figure 12.23)**.

U-Shaped

U-shaped patterns appear higher on vertical surfaces than do V-patterns **(Figure 12.24)**. U-shaped patterns appear when the burning fuel package is further away from the vertical surface. As a result, the plume generated by the fire has more room to spread before reaching surrounding walls **(Figure 12.25)**.

Truncated Cone

Truncated cone patterns are three-dimensional fire patterns created by a fire plume that are observed on both horizontal and vertical surfaces. A conical-shaped fire plume may create patterns on both the wall and the ceiling at the same time **(Figure 12.26)**.

Figure 12.22 A V-pattern indicates a fire burned near this vertical gypsum wallboard. *Courtesy of Nicole Brewer, Portland Fire Investigations.*

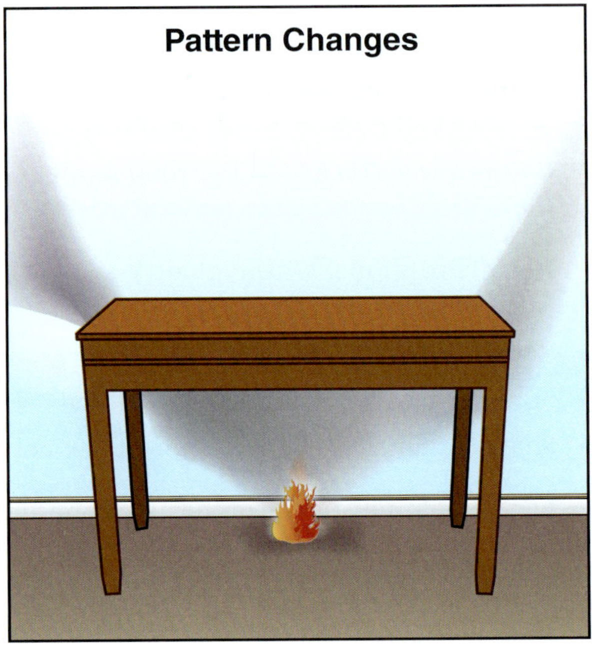

Figure 12.23 An intervening horizontal surface, such as the underside of a table, may alter a V-shaped pattern.

Figure 12.24 U-shaped patterns appear relatively high on vertical surfaces compared to V-shaped patterns. *Courtesy of Wayne Chapdelaine, Metro-Rural Fire Forensics.*

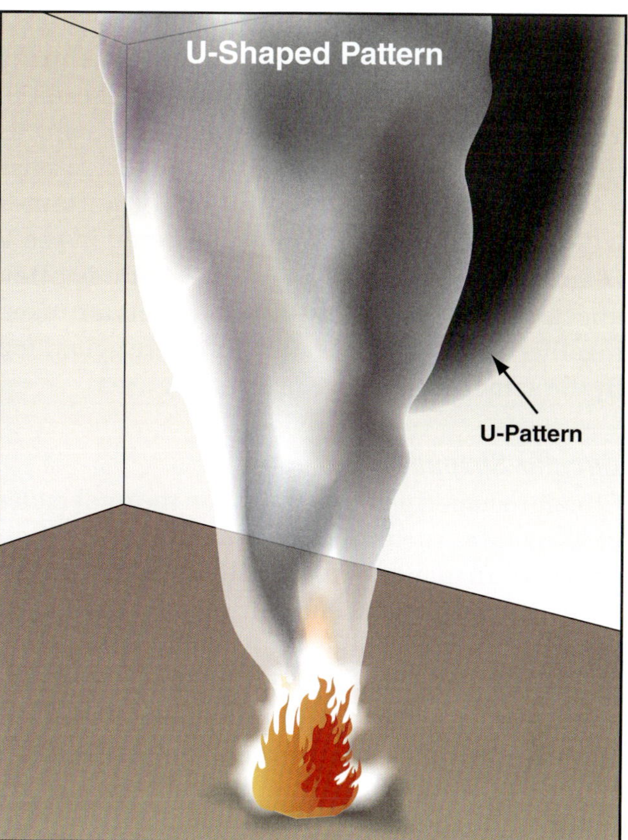

Figure 12.25 U-shaped patterns are typically formed by fires that are some distance from the wall on which the pattern appears.

Inverted Cone

Inverted cone patterns are triangular in shape, with a wide base that narrows to a point centered over the fire **(Figure 12.27)**. An inverted-cone pattern can develop from a number of situations including:

- Fires that extinguish before the flame height reaches the ceiling

- Very small fires or fires with a low heat-release rate, located near a vertical surface or wall

- Fires involving fuel gases, such as propane or natural gas, especially from leaks in pipes below the fire floor; ignited gases can escape through ventilation points or openings

Figure 12.26 A truncated-cone pattern forms on the wall and ceiling at the same time.

Figure 12.27 Inverted-cone patterns are formed by fires that are very small or have a low heat release rate. *Courtesy of Donny Howard, Agent, Oklahoma State Fire Marshal's Office.*

Hourglass

An hourglass pattern forms when a fire burns uninterrupted, close to a vertical surface. The initial V-shaped pattern has a bottom width the same as the fire package, and the top expands until it reaches a horizontal surface. The height of the flames will create an inverted-V effect centered in the V **(Figure 12.28)**. The top of the V-shape is often affected by the hot gas layer. These effects are taken together as a pattern because they were generated by the same fuel package in the same area.

Circular-Shaped

Circular-shaped patterns appear on the undersides of horizontal surfaces such as ceilings or tables. This pattern may be a full circle shape, or it may be cut or distorted by discontinuities in the horizontal surface above the plume. For example, a fire under a table may burn the entire underside of the table and also show circular patterns on the ceiling where the plume escaped around the edges of the table.

Circular-shaped patterns appear directly above a burning fuel package, assuming that ventilation or other forces do not push the plume away from the fire **(Figure 12.29)**. The center of the circular-shaped pattern should show greater damage from the fire by having a greater depth of char or depth of calcination or other measurable damage effect. As a result, an investigator can use these circular-shaped patterns to establish where individual fuels were located.

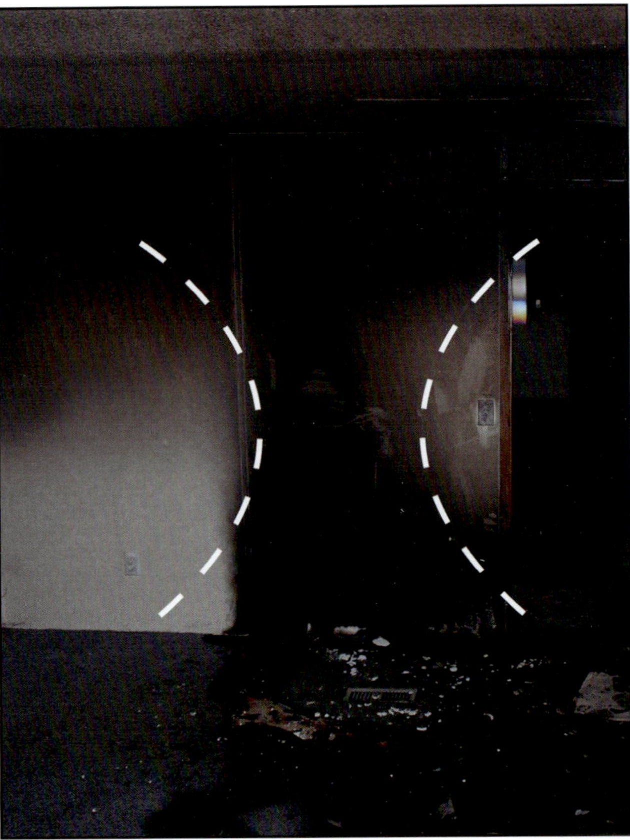

Figure 12.28 A flammable liquid fire on the floor can form an hourglass pattern. *Courtesy of Donny Howard, Agent, Oklahoma State Fire Marshal's Office.*

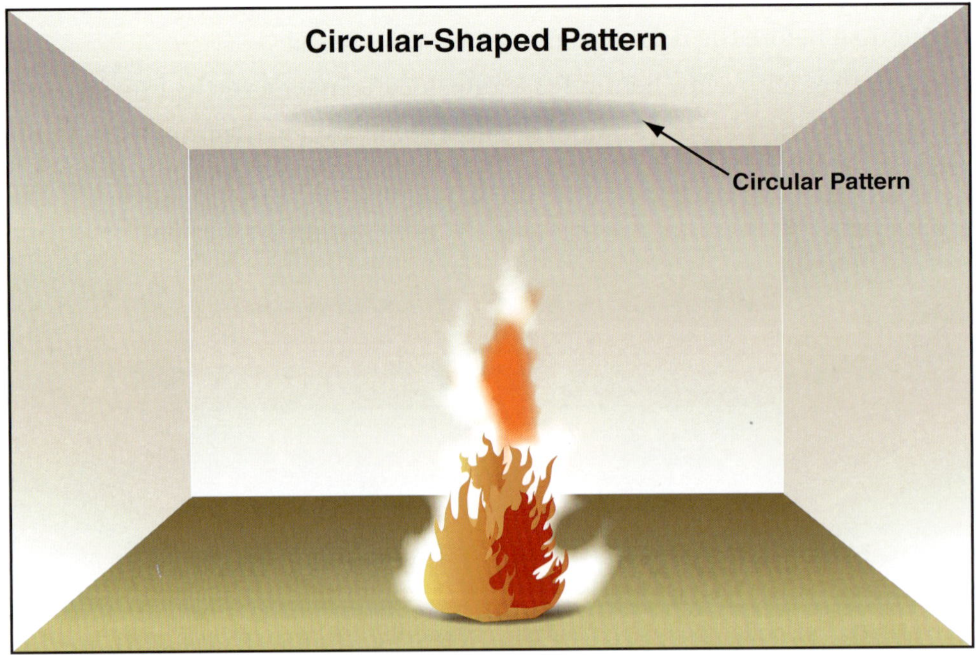

Figure 12.29 A circular pattern, without other fire effects, forms directly above a fire.

Pointer and Arrow Patterns

Pointer and arrow patterns occur within the wall assembly instead of on the wall surface. Pointer and arrow patterns are often helpful in tracking the source and path of travel of flame and hot gases during the fire. When analyzing pointer and arrow patterns, investigators should check to make sure that the damage is fire-related and not failure or breakage caused by fire-suppression efforts.

Fire damage will be most severe on the side of surfaces and components facing the heat source, especially at unprotected edges. When the component is part of a series, such as wall studs, often the tops of the studs will sustain mass loss or be consumed **(Figure 12.30)**. This mass loss will be greater closer to the fire that produced this pattern, with the shortest remaining stud indicating the most severe exposure.

Figure 12.30 The shortest stud indicates the area that burned the longest with exposure lessening in proportion to the length of the studs. *Courtesy of Donny Howard, Agent, Oklahoma State Fire Marshal's Office.*

Other Fire Patterns

NFPA 1033 (2022): 4.2.2, 4.2.3, 4.2.4, 4.2.5, 4.2.7, 4.2.8

Other fire patterns can be produced away from the fire plume. The fire investigator must be able to recognize and analyse these patterns, determine how they formed, and define their significance. The fire investigator must develop a thorough understanding of the fire dynamics in the compartment, especially as those dynamics changed due to increased ventilation.

Fire patterns created by increased ventilation may alter or completely destroy the patterns related to the fire's origin **(Figure 12.31)**. The original

Figure 12.31 Given sufficient fuel, oxygen, and time, fire patterns can destroy patterns formed when fire entered a space. *Courtesy of Donny Howard, Agent, Oklahoma State Fire Marshal's Office.*

patterns may persist if they are remote from the ventilation openings. Visual observations of the fire patterns alone in a compartment that experienced a long duration post-flashover environment will in many cases be insufficient to correctly determine the area of fire origin. Obtaining verifiable and measurable data obtained by depth of char, depth of calcination, or other measurable data may assist the fire investigator in determining the area of origin.

Knowing what form ventilation took and when it was introduced to the fire are key pieces of evidence that an investigator should not overlook. For example, when a room has been subjected to a flashover or has transitioned to full-room involvement, many patterns that might have been visible before flashover or full-room involvement may no longer be readily visible. However, an investigator can use depth of char or depth of calcination analysis to determine where earlier fire patterns were before the room reached full involvement.

Hot Gas or Upper Layer Patterns

As discussed in Chapter 3, a hot gas layer develops at the highest horizontal surface in a compartment with sufficient fuel and ventilation. Radiant energy from the gases within this layer causes patterns to develop on the ceiling and wall surfaces, as well as on the compartment's contents and surfaces **(Figure 12.32)**. Combustible surfaces are charred or consumed while noncombustible surfaces are discolored or distorted.

When the fire is extinguished before flashover, these patterns may be distinct. These patterns are often located on horizontal surfaces, such as floors and exposed surfaces of contents located in the compartment, and may be more significant based on their relative distance from the radiant heat layer. For example, the top portion of a sofa's backrest may sustain greater thermal damage than the seat cushions. Nonexposed surfaces, such as the undersides of furnishings, usually sustain no damage from this source.

Figure 12.32 Heat effects include patterns that indicate the height of the hot-gas layer during the fire.

Ventilation Effects

Ventilation effects on a compartment include patterns formed via conduction when hot gases flow past vertical surfaces such as walls and open doors. Ventilation effects can develop in both fuel and ventilation controlled compartments, but are especially prevalent where significant air flow is present. Ventilation-controlled fires burn intensely near air sources.

Because of the variety of effects that ventilation sources can have on a fire, investigators must consider all ventilation activities when analyzing fire patterns. Examples:

- Broken windows
- Fire-suppression activities
- Forced ventilation using building systems
- Openings, such as windows, doors, and vents

Ventilation directly affects the heat release rate of burning fuels, any analysis of fire patterns must include how and when compartments were ventilated. For example, an increase in ventilation to a ventilation-limited compartment can cause full-room involvement. When a fire reaches full-room involvement, patterns created during the incipient and growth stages of the fire may be altered or destroyed.

Ventilation can also force trapped, unburned products of combustion outside the compartment through duct systems, broken windows, or open doors. These products of combustion may generate fire patterns outside the compartment as the flames or by-products of combustion escape the compartment to another compartment or to the exterior. Ventilation patterns may appear to originate outside of the compartment in these conditions.

Ventilation-Generated Patterns

Ventilation patterns result from the movement of air in the compartment or building during the fire. The first configurations that an investigator should consider include a combination of air movement and glowing embers. For example, embers on a combustible floor surface can burn a hole in or through the floor, which can create a new airflow source and additional damage **(Figure 12.33)**. A fire investigator should examine patterns that may have multiple explanations, and collect evidence to determine whether the fire source was in fact an area of origin or the result of ventilation.

Development of Ventilation Patterns

Investigators frequently observe ventilation-generated patterns on and around doors and windows of compartments involved in fire **(Figure 12.34)**. When air flows from under the door into the compartment while hot fire gases flow over the top of the door, the portion of the door facing the fire shows evidence of fire damage at the upper portion and over the top where fire gases flowed. In the case of a metal, hollow-core door, the interior

Figure 12.33 Burn-through effects show that combustible materials contacted flooring for some time and may have caused a ventilation increase. *Courtesy of Donny Howard, Agent, Oklahoma State Fire Marshal's Office.*

Figure 12.34 Damage around the edges of a door is an example of a ventilation pattern.

portion will normally fail from the top and "peel" downwards. Investigators should examine the door for evidence of whether it was open/closed and also examine the locking mechanism in an attempt to determine if the door was locked.

When the hot-gas layer fills the entire compartment, an investigator is likely to find evidence of the gases flowing from the compartment from both the top and bottom of the door opening. The fire compartment side of these doors often show damage or charring from the top to the bottom on the vertical surface as well as where the gases flowed over and under the door. Evidence of the flow may also be seen on the opposite side of the door at both the top and bottom.

When fire investigators examine patterns found on a door, they should also consider whether hot debris falling against or near the lower portion of the door contributed to the development of patterns being observed. Air movement under a door can result in increased damage should hot or glowing debris be deposited there during the fire.

Fire investigators may find other patterns or effects caused by ventilation when fire gases flow through structural openings such as doors and windows. Any time a gas flows through a restricted opening, its velocity is increased. An increase in the velocity of hot gasses may result in damage patterns that are different from adjacent spaces or surfaces. As with any fire pattern, the fire investigator has to examine the area next to the pattern and attempt to determine the factors that contributed to the development of what is observed.

Using Ventilation Data

Understanding these patterns and how they are created assists investigators in correlating them with other evidence obtained at the scene. This is especially true in short- and long-duration post-flashover fires. In the case of long duration post-flashover fires, the most damage will certainly be found at the ventilation openings.

While examining ventilation-generated patterns in a structure, fire investigators should remember that increased ventilation in an area can result in:

- Longer periods of burning
- Greater damage in exposed locations
- Higher rates of heat release from fuel packages

During ventilation-limited conditions, incomplete combustion causes fuel to be suspended in the upper layer of the compartment in the form of hot gases. This suspension creates a fuel-rich atmosphere. Combustion occurs at locations where the flammable limits are met, which will typically be near the areas of open ventilation. This ventilation pattern occurs where no fuel package appears to be present. Therefore, the greatest area of structural damage may be related solely to ventilation.

Full-Room-Involvement Patterns

In compartments where full-room involvement takes place, investigators will likely observe fire damage extending from the ceiling to the floor. Fire investigators can expect to find damage to almost every surface in the compartment **(Figure 12.35)**. Carpeting or combustible floor coverings will show heavy damage and the floor may have burn holes. When the fire burns for long periods of time or when the space has a heavy fuel load, the damage is more extensive. Because a full-room involvement fire includes burning at all levels, the undersides of furnishings may be burned or damaged. The extent of the destruction is determined by:

- Fire duration
- Fuel configuration
- Available ventilation
- Amount of fuel in the compartment

Figure 12.35 An item showing heat damage on all exposed sides was probably surrounded by fire or sufficient temperatures to sustain fire. *Courtesy of Donny Howard, Agent, Oklahoma State Fire Marshal's Office.*

Suppression-Generated Patterns

NFPA 1033 (2022): 4.2.2, 4.2.3, 4.2.5, 4.2.8

Fire-suppression efforts may also affect the patterns that fire investigators observe during the examination. Water streams used to suppress a fire may have the effect of spreading the fire to areas that it would not normally have reached. Extinguishing agents applied during the fire are capable of creating and altering existing fire patterns.

In addition, firefighters may use passive or mechanical ventilation to strategically move air and hot gases through the building to assist in fire-suppression activities. These ventilation practices can cause fire patterns that are altered versions of typical patterns. Investigators must develop a thorough knowledge of all fire-suppression activities that occurred at the scene to interpret these patterns.

Clean-Burn Patterns

Clean-burn patterns are found on noncombustible surfaces where there has been direct flame contact with or intense radiant heat on the surface. Soot or smoke deposits or other items, such as wallpaper and paint, on the surface are burned away by the flame, leaving an appearance of a clean or lighter shaded area between the area of flame contact or intense radiant heat and the soot- or smoke-darkened area surrounding it **(Figure 12.36)**. Sometimes this occurs when the surface of the materials was too hot for the products of combustion to adhere to it. A clean-burn pattern (inverted V) can indicate the flame area at the base of an hourglass pattern created by a plume on a noncombustible wall surface.

Figure 12.36 Areas that have burned clean show that combustible materials in that space were heated sufficiently to be consumed. *Courtesy of Donny Howard, Agent, Oklahoma State Fire Marshal's Office.*

Irregular Patterns on Floors

NFPA 1033 (2022): 4.2.5

Fire investigators often observe irregularly shaped patterns on floor surfaces. The lines of demarcation observed with these patterns depend on the material on which the pattern is formed, such as the type of surface and the amount of heat exposure the material receives.

Irregular patterns may be formed on floor surfaces in the following ways:

- Flashover
- Protected areas
- Wear or traffic patterns
- Pooling of ignitable liquids
- Hot gases flowing on the floor surface
- Smoldering in an area that was difficult to extinguish
- Flaming debris dropping onto the surface during the fire
- Polyurethane foam melting and pooling around furniture items **(Figure 12.37)**
- Plastic contents melting, flowing, and then burning onto the surface

Figure 12.37 Polyurethane foam from the cushions in this burned-away couch has left an irregular pattern on the floor. *Courtesy of Yates & Associates.*

- Upper-layer radiant heating that creates damage on exposed surfaces
- Fuel packages (such as clothing, paper, corrugated cardboard) burning on the floor

The investigator can determine the cause of the pattern by correlating nearby fuels with the pattern and determining whether there is a plausible explanation of what is observed. The investigator should consider the method of generation because there are a number of ways that irregular patterns on floors can be formed. Should an ignitable liquid be suspected as to the cause of the irregularly shaped pattern, laboratory testing is necessary to confirm that an ignitable liquid was present. The following sections describe two particular irregular patterns, trailers and doughnut-shaped patterns, that investigators should be able to identify at a fire scene.

Fuel Trailers

A trailer of ignitable liquid or solid fuel or a mixture of both causes a type of irregular pattern. Trailers are often used in intentionally set fires to connect remote fuel packages (combustible materials, pools of ignitable liquid, etc.) with each other or to move fire from one area to another **(Figure 12.38)**. The pattern found is normally narrow and extends between burned areas. Trailers may appear to be linear in nature on horizontal surfaces.

When ignitable liquids are used to form a trailer on a porous surface, investigators may also observe a **wicking**

Figure 12.38 The burned areas of this carpet indicate where ignitable liquids were poured to create trailers. *Courtesy of Donny Howard, Agent, Oklahoma Fire Marshal's Office.*

effect at the edges of the pattern. Wicking occurs when quantities of the ignitable liquid are absorbed by the material on which the liquid is poured. The liquid seeps (wicks) away from the edge of the actual pour. Wicking may appear as small fingers of damage perpendicular to the major pattern. As with other irregular patterns, the pattern observed by investigators depends on the type and composition of the surface on which the trailer was made and the material used to form the trailer.

NOTE: Heavy traffic areas and flashover can cause damage patterns that are similar to those caused by trailers.

Doughnut-Shaped Patterns

A doughnut-shaped pattern forms when a pool of liquid ignites on a floor surface. The pooled liquid cools the surfaces beneath it. The lines of demarcation are at the edges of the pool while the center portion of the pool shows little or no damage.

As the fire continues to burn and vaporize the liquid, the pool becomes smaller, providing less protection. This process continues until the liquid is mostly consumed. The center area of the liquid will remain protected the longest and therefore will sustain less damage, making it an excellent area to sample. Fire investigators should not collect samples from the charred area as any trace amounts of ignitable liquid will be most likely consumed.

Saddle Burns

Saddle burns are most commonly found on the top edges of floor joists **(Figure 12.39)**. The pattern forms deep charring of the wooden joist, and is normally very localized.

Saddle burns can be caused by:

- Use of ignitable liquids
- Fire burning downward through the floor surface above the joist
- A significant fuel package, such as a sofa, in the area of the pattern
- Pooling of polyurethane foam from furniture located above the burn
- May be observed on wood roof trusses as a result of burning asphalt shingles
- Lightning strikes to a cable in an attic may create saddle burns across ceiling joists
- Intense burning on the floor surface above the joist and could result from ventilation factors

Figure 12.39 Partially burned floor joists show evidence of being burned from above, leaving a concave saddle burn in the wood. *Courtesy of Donny Howard, Agent, Oklahoma State Fire Marshal's Office.*

Protected Areas

NFPA 1033 (2022): 4.2.3

Protected areas are not themselves a fire pattern, but they can indicate conditions within a space during fire conditions. Undamaged areas on surfaces within the fire-damaged area may have been caused by objects that shielded the surface from the effects of the fire and hot gases that damaged other portions of the surface **(Figure 12.40)**. This phenomenon is known as **heat shadowing** and can obscure other patterns in the shadowed area. Heat shadowing is often found alongside protected areas.

For example, investigators may find circular-shaped patterns that are the result of trash receptacles or furniture located in the area of the pattern. Bottoms of cardboard boxes that contained combustible materials, such as paper, may remain intact and protect the floor surface even though the box and its contents are burned completely to ash.

During scene reconstruction, investigators can use protected areas to place the contents back into their locations at the time of the fire **(Figure 12.41)**. Investigators can correlate these patterns with the fuel packages and the fire dynamics that occurred within the compartment to understand what caused the patterns.

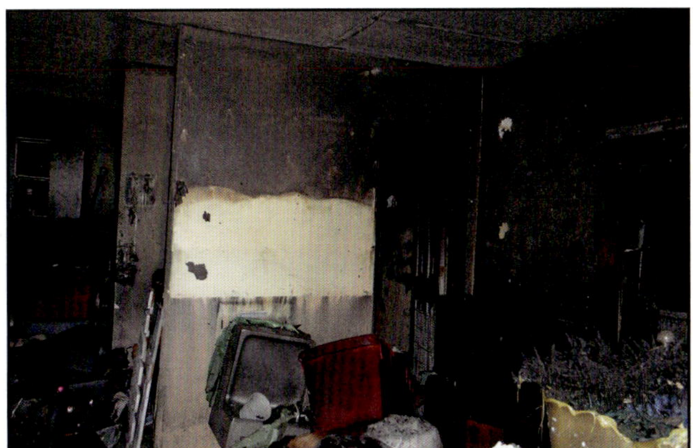

Figure 12.40 Undamaged surfaces show that an item may have been between that space and the fire/heat conditions that affected adjacent spaces. *Courtesy of Donny Howard, Agent, Oklahoma State Fire Marshal's Office.*

Figure 12.41 A sofa protected this area of wallboard, leaving an area of less damage surrounded by an area of greater damage on the wall. *Courtesy of Hugh W. Graham, Yates & Associates.*

Investigators should examine the underside of furniture or other components in the compartment to further understanding the development of the fire in the compartment. Heavily charred surfaces with uncharred protected areas beneath them indicates that the exposure was from the area on the other side of the protected surface. Uniform charring on an item indicates heat flux from both the areas above and below the item.

Explosion Origin Determination

NFPA 1033 (2022): 4.2.7, 4.2.9, 4.6.5

Depending upon the type of explosion, determining the area of origin of the explosion may be relatively easy or very difficult. In the case of seated explosions, finding the area of origin involves finding the area of greatest damage. Where explosives are involved, the area of origin is the seat of the explosion often accompanied by a crater or extreme splintering and destruction of structural members. BLEVEs originate in confining vessels. When the vessel or its remains are located, so is the area of origin.

The origins of deflagrations are more difficult to determine. Often, an investigator may only be able to identify the room in which the deflagration occurred. For example, when one gas/vapor explosion mixes gases into the flammable range, subsequent explosions will add effects that complicate the determination of the order of detonation. Generally speaking, the area of greatest damage is the area of origin in gas/vapor explosions.

Where dust explosions are suspected, secondary explosions are the norm and not the exception. Further, the secondary explosions fueled by dust are generally more powerful than the initial explosion.

Because of the possible complexity of determining and analyzing the area of origin, investigators must use organized and scientific methods. The two sections that follow describe systematic methods investigators can use to help identify and analyze the origin of an explosion.

Explosion-Dynamics Analysis

NFPA 1033 (2022): 4.2.9

Explosion-dynamics analysis is the process of following force vectors backwards toward the initial compartment where an explosion occurred. Tracing the paths of debris is a part of this analysis. When a pattern emerges suggesting a room of origin, investigators may create a more focused force-vector diagram of the room of origin to further pinpoint the center of an explosion.

Explosion-dynamics analysis does not always clearly indicate a single explosion. When secondary explosions have occurred at a scene, investigators must take these into account when determining the area of origin. The investigator can diagram the damage from each explosion and make a determination about origin after documenting all explosions. Secondary explosions tend to have less force and do less damage to structural members. Another key piece of evidence to identifying the initial explosion is that the initial explosion tends to show the most evidence of ventilation. Investigators should trace ventilation damage around doorways and in adjacent compartments, which can point to the initial explosion.

Fuel-Source Analysis

Once the epicenter or area of origin has been identified, investigators should identify the fuel for the explosion and its source. As mentioned previously, comparing the type of damage observed with possible available fuels provides the basis for a hypothesis about the explosive fuel. An investigator should do the following to confirm the hypotheses about the fuel source for an explosion:

- Perform chemical tests on collected debris samples in the area of origin.

- Take air samples especially when a gas/vapor explosion is suspected.

- Perform laboratory analysis of collected samples.

In the event of a gas/vapor explosion, an investigator should examine all gas utilities at the scene. If a leak is not readily apparent, then the entire gas piping system should be leak tested to determine from where the fuel originated. This testing may require additional technical expertise from the utility company or other experts.

NOTE: Any pressure test should be run at the normal operating pressure associated with the system being tested. If the leak is underground, the pressure and flow tests should be performed prior to exposing a leak. The debris that surrounds the leak location will influence the test and should be in place during these tests.

Review Questions

1. List analytical methods and procedures for potential origin area evaluation.
2. What should the investigator consider when evaluating hypothesis regarding potential area of origin?
3. How can an investigator avoid bias during potential origin area evaluation?
4. While determining the area of origin, what three categories of information should the investigator consider?
5. In an incident with full-room involvement, what will likely lead to an incorrect hypothesis?
6. What is the difference between fire effects and fire patterns?
7. What does involve the analysis of fire patterns?
8. What is an example that patterns tend to change over the course of a fire?
9. Which is normally the most visible and measurable indicator of a fire effect?
10. Why should the fire investigator carefully examine all points of low burning?
11. What should the fire investigator use to measure char depth?
12. What variables should a fire investigator consider when measuring char?
13. How is an isochar map developed?
14. What is shown in a vector analysis diagram?
15. What should a fire investigator consider during an arc mapping analysis?
16. How does melt damage on conductor occurs?
17. What causes a fire pattern?
18. How does movement patterns differs from intensity patterns?
19. List pattern types that have a clear definition.
20. Which considerations should a fire investigator take when evaluating lines of demarcation?
21. What geometrical shapes fire investigators use to categorize plume-generated patterns?
22. List patterns that generally are the result of a plume.
23. Where are columnar patterns found?
24. Which are the key pieces of evidence that an investigator should not overlook?
25. Where does a hot gas layer develop?
26. Where does ventilation-controlled fire burn intensely?
27. What are ventilation-generated patterns the result of?

28. Where does the investigator frequently observe ventilation-generated patterns?

29. In case of a long duration post flashover, where will the most damage be found?

30. What will determine the extent of the destruction on a full-involved room fire?

31. When are doughnut-shaped patterns formed?

32. Where explosives are involved, what is the area of origin?

33. What kind of explosion has secondary explosions as a norm?

34. What is an explosion-dynamics analysis?

35. What should an investigator do to confirm the hypotheses about the fuel source for an explosion?

Discussion Questions

1. How does the scientific method help investigators to avoid bias?

2. How is electronic data collected, analyzed, and correlated in your jurisdiction?

Chapter References

Bieber, Paul. December 2014. "Fire Investigation and Cognitive Bias." Wiley Encyclopedia of Forensic Science. Accessed online.

Dror, Itiel E, and Cole, Simon A. 2010. "The Vision in 'Blind' Justice: Expert Perception, Judgement, and Visual Cognition in Forensic Pattern Recognition." *Psychonomic Bulletin & Review*; 17 (2), 161-167.

Gorbett, Gregory Edward, et al. 2015 -05-28. "Use of damage in fire investigation: a review of fire patterns analysis, research and future direction." Fire Science Reviews 4:4 DOI 10.1186/s40038-015-0008-4. Accessed online.

Gorbett, Gregory Edward. 2015-08-07. "Development and Assessment of a Decision Support Framework for Enhancing the Forensic Analysis and Interpretation of Fire Patterns." Worcester Polytechnic Institute; Digital WPI. Accessed online.

Houck, Max M, and Jay A Siegel. 2015. *Fundamentals of Forensic Science, Third Edition*. Elsevier; New York.

Taroni F, et al. 2010. Data Analysis in Forensic Science: A Bayesian Decision Perspective. John Wiley & Sons; Chichester (UK).

Chapter Key Terms

Arc Mapping — Visual documentation of the path of electrical arcs at a scene.

Calcination — Process of driving free and chemically bound water out of gypsum; also describes chemical and physical changes to the gypsum component itself.

Char — Carbonaceous material formed by incomplete combustion of an organic material, commonly wood; the remains of burned materials.

Char Gauge — Blunt-ended, thin probe similar to dial calipers or tire-tread gauges that is inserted into blistered char to measure the depth of char.

Cognitive Bias — Systematic errors in thinking, frequently based on social influences, memory, belief and distorted impressions of probability, which cause people to deviate from true objectivity in their decision-making.

Consumption — The loss of mass from a material when exposed to heat. This is based on heat exposure and material properties.

Deposition — Process in which smoke decreases in temperature and/or reaches cooler surfaces, and smoke particulates, liquid aerosols, and gases accumulate on surfaces.

Fire Effect — Visible or measurable physical changes, or identifiable shapes, formed by a fire effect or group of fire effects. *Similar to* Fire Pattern. (Reproduced with permission from NFPA 921-2017, *Guide for Fire and Explosion Investigations*, Copyright 2018, National Fire Protection Association.)

Fire Pattern — The apparent and obvious design of burned material and the burning path of travel from a point of fire origin. A collection of fire effects that, taken together, show a progression in fire development.

Heat Shadowing — Fire pattern left when an object blocks a fire's radiant heat from reaching a combustible surface behind the object.

Intensity Pattern — Damage from flame and hot-gas zones of a fire plume that shows areas of higher temperatures.

Isochar Map — Graphical depiction of char depths, with connections between equal measurements.

Movement Pattern — Damage from fire or combustion products during the course of the fire's growth and spread from the initial heat source.

Radiant Heat Flux — Measure of the rate of heat transfer to a surface, expressed in kilowatts/m^2 per second, or Btu/ft^2 per second.

Wicking — Pattern that occurs when quantities of ignitable liquid are absorbed by the material onto which the liquid is poured.

NOTE: The scope of the examination will depend upon the characteristics of the particular fire or explosion scene staged. The steps in this skill can follow a different order and repeat as many times as necessary to match the staged scene. The fire investigator should work alongside other professionals as the scene requires and complete appropriate documentation during each step of the investigative process.

Step 1: Conduct interviews.

    a. Witnesses

    b. Owner, occupants, and employees

    c. Firefighters and other first responders

Step 2: Observe fire pattern damage on the structure and contents.

Step 3: Conduct arc mapping.

Step 4: Collect any evidence specific to area of origin determination **(Figure 12.42)**. Take the necessary steps to maintain the chain of custody throughout the investigation.

Step 5: Analyze all data and evidence that has been gathered, utilizing outside expert resources, as necessary.

Step 6: Reconstruct the fire scene, as necessary.

Step 7: Develop hypotheses regarding the area of origin.

Step 8: Test area of origin hypotheses.

 NOTE: Test hypotheses to disprove, rather than prove them.

Step 9: Select a final hypothesis regarding the area of origin based on analysis and testing.

Figure 12.42 *Courtesy of Donny Howard, Agent, Oklahoma State Fire Marshal's Office.*

Photo courtesy of Donny Howard, Agent, Oklahoma State Fire Marshal's Office.

Chapter Contents

NFPA 1033 JPRs addressed in this chapter

This chapter provides information that addresses the following job performance requirements (JPRs) of NFPA 1033, *Standard for Professional Qualifications for Fire Investigator* (2022):

4.1.7	4.2.6	4.3.1	4.4.4	4.6.1	4.6.5
4.2.3	4.2.7	4.3.2	4.5.1	4.6.2	
4.2.4	4.2.8	4.4.2	4.5.2	4.6.3	
4.2.5	4.2.9	4.4.3	4.5.3	4.6.4	

This chapter provides information that addresses the following course outcomes for the FESHE courses:

Fire Investigation I (C0283)

2. Understand and demonstrate the process of conducting fire origin and cause investigation.

7. Discuss the basic principles and identify cause and origin of fires.

Fire Investigation and Analysis (C0285)

2. Document the fire scene in accordance with best practice and legal requirements.

3. Analyze the fire scenario utilizing the scientific method, fire science, and relevant technology.

4. Analyze the legal foundation for conducting a systematic incendiary fire investigation and case preparation.

Learning Objectives

1. Infer an opinion concerning the cause of a fire or explosion. [NFPA 1033, 4.6.5]

2. Identify the material first ignited on a fire scene. [NFPA 1033, 4.2.7]

3. Identify potential ignition sources at the fire scene. [NFPA 1033, 4.2.6]

4. List common ignition sources at a fire scene. [NFPA 1033, 4.2.6, 4.2.7, 4.2.8]

5. Identify oxidizing agents on a fire scene. [NFPA 1033, 4.2.6]

6. Evaluate circumstances that allow combustion to occur on the fire scene. [NFPA 1033, 4.1.7, 4.2.4, 4.2.6]

7. Determine competent ignition sources at the fire or explosion scene. [NFPA 1033, 4.6.5]

8. Determine possible and probable cause hypotheses. [NFPA 1033, 4.2.5, 4.4.3, 4.6.1, 4.6.3, 4.6.5]

9. Explain steps for the post-scene investigation process. [NFPA 1033, 4.2.5, 4.4.3, 4.6.1, 4.6.3, 4.6.5]

10. Describe the incendiary fire analysis process. [NFPA 1033, 4.6.4]

11. Skill Sheet 13-1: Determine the cause of the fire. [NFPA 1033, 4.2.3, 4.2.4, 4.2.5, 4.2.6, 4.2.8, 4.2.9, 4.3.1, 4.3.2, 4.4.2, 4.4.3, 4.4.4, 4.5.1, 4.5.2, 4.5.3, 4.6.1, 4.6.2, 4.6.3, 4.6.4, 4.6.5]

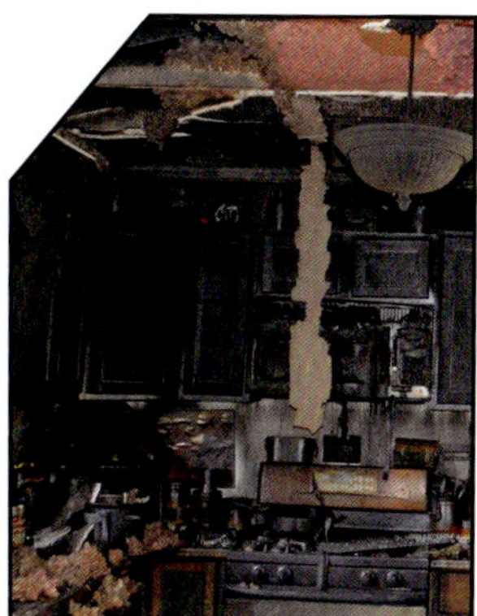

Chapter 13
Cause Determination

Courtesy of Donny Howard, Agent, Oklahoma State Fire Marshal's Office.

After determining the potential origin, the investigator next determines the cause of the fire or explosion. The process to evaluate and determine the fire or explosion cause includes:

- Introduction to cause determination
- Material first ignited
- Heat (ignition) source
- Common ignition sources
- Oxidizing agent
- Ignition sequence
- Developing cause hypotheses
- Testing the cause hypothesis
- Post-scene investigation
- Cause classifications
- Incendiary fires analysis

Introduction to Cause Determination

NFPA 1033 (2022): 4.6.5

With few exceptions, the origin area will include the **fire cause**. **Skill Sheet 13-1** shows the steps for determining the cause of the fire. During the cause determination process, investigators use a consistent methodology, such as the scientific method, to collect data that allows the investigator to accurately identify the:

- **Material first ignited**
- Heat (ignition) source
- Oxidizing agent
- **Ignition sequence**

Material First Ignited

NFPA 1033 (2022): 4.2.7

A key component in making a fire cause determination is identifying the material first ignited. Characteristics of a first material that complement the ignition source include:

- Available heat energy must be sufficient to ignite the material
- The ignition source must be near enough to the material to transfer sufficient energy
- Material must be able to absorb and retain sufficient heat energy from the ignition source to begin combustion

    In addition to the material's compatibility with an ignition source, an investigator should consider:

- **Fuel State** — An ignition source may be competent to ignite fuel at one state of matter but not another. For example, a spark can ignite vapors within the flammable range, but may not be able to ignite a pool of the same material as liquid. Examination of the origin area may not show an apparent source of ignition.

- **Surface-to-mass ratio** — A wood match is a competent ignition source for wood shavings. In contrast, a wood match may not be competent to ignite a 2 foot x 4 foot (0.6 meters by 1.2 meters) wood structural member.
- **Orientation of the fuel to material** — A competent heat source applied near the bottom of a vertical 1 x 6-inch ((0.025 m to .15 m) wood board will be more likely ignite the board than the same heat source applied to the middle of a horizontal wood board.

An investigator should compile a list of the fuels found or known to have been in the origin area. The list of likely first fuels ignited can be narrowed down based on:

- Fire effects
- Fire patterns
- Fire dynamics
- Witness statements or other factors

Investigators must consider that fuels and first materials ignited may have been completely destroyed in fire conditions. For example, a small electrical arc may have the capacity to ignite natural gas or propane within flammable ranges. In this case, the flammable gas would be burned away, and the only indicator of the ignition source would be evidence of arcing.

Heat (Ignition) Source

NFPA 1033 (2022): 4.2.6

The use of a consistent methodology to determine the cause of a fire requires data collection. The data collection process includes identifying all potential ignition sources within the origin area. An investigator also analyzes individual heat (ignition) sources to determine whether they are competent to ignite any of the fuels under the conditions found at the scene.

A careful examination of surfaces in the origin area may yield fire patterns that could assist in determining the ignition source. Heat sources may have any or none of the following qualities:

- May have been removed or altered.
- Recognizable components, such as an electric space heater.
- May be unrecognizable, or completely destroyed by the fire.

A competent ignition source must have the three following qualities:

- **Sufficient temperature** — To ignite the first material
- **Sufficient time** — To transfer the required heat to the first material
- **Sufficient heat energy transfer** — To result in the ignition of the first material

Sufficient Temperature

To determine whether any potential heat source could ignite the first fuel ignited, an investigator must know the ignition temperature of the first fuel ignited. The investigator can use that information to determine whether the object had heated to the necessary ignition temperature.

Rejecting Potential Ignition Sources

An investigator collects a list of potential ignition sources during the investigation. To narrow that list, an investigator may remove any potential ignition sources at or near the origin area that could not have started the fire. The fire investigator should document, collect, and preserve evidence that supports removal.

Solid Fuels

Most solids must pyrolize before they will ignite. One or more heat sources must supply enough energy for pyrolysis and also to ignite resulting vapors **(Figure 13.1)**.

Liquid Fuels

Liquid fuels must vaporize before they will ignite. A common example of vaporization are the gases on the surface of a liquid. A heat source must supply sufficient energy to cause a state change from liquid to vapor. The ignition source must provide a temperature high enough to cause ignition of the vapors. The investigator must remember that vapors from the fuels may travel some distance before encountering an ignition source capable of igniting the vapors.

Figure 13.1 Heavier structural members will not ignite from intermittent sparks. *Courtesy of Yates & Associates.*

Gaseous Fuels

Gaseous fuels do not have to change physical state before ignition; therefore, gaseous fuels require the least amount of energy for ignition. Conditions that must be met for ignition include:

- Gaseous vapors must be concentrated within the flammable range.
- The ignition source must contact the gaseous vapors for sufficient duration.
- The ignition source must provide a high enough temperature to ignite the fuel.

Vapor concentrations above or below the flammable range will not ignite, even given a competent ignition source. The ignition source and gaseous fuels may be remote from the fire origin area.

Sufficient Time to Transfer

Higher temperatures usually do not require as much time to transfer sufficient heat. Heat sources operating correctly may ignite a fuel package if they are in direct contact with a fuel for sufficient duration. In other cases, a heat source may operate outside normal parameters.

Analysis of possible ignition sources should include consideration of whether a utility or tool was used correctly or incorrectly. Damage to or bypassing of components intended to regulate a heat source's range of temperature and prevent malfunctions can result in temperatures that exceed the anticipated range for that heat source. Regulation features include:

- Safety switches
- Thermal cutoffs
- Temperature controls

Sufficient Heat Energy Transfer

As introduced in Chapter 3, Fire Dynamics, the fire investigator knows the methods for heat energy transfer are conduction, convection, or radiation. During debris removal, the investigator evaluates each potential material first burned against each potential competent ignition source. The fire investigator should consider how the heat energy transferred from an ignition source to the fuel.

Common Ignition Sources

NFPA 1033 (2022): 4.2.3, 4.2.6, 4.2.8

Investigators will sometimes determine the ignition source was destroyed, altered, or removed from the scene. Supporting data that can indicate the presence of an ignition source includes:

- Witness interviews
- Fire dynamics analysis
- Weighing of forensic analysis
- Analysis of fire effects and fire patterns
- An examination of the electrical system
- Evidence such as a fuel container or ignitable liquid residue

For example, an investigator can reasonably infer that an open flame device (lighter or match) was the ignition source given the presence of an ignitable liquid in the debris of a sofa in the living room. The following sections outline a number of common ignition sources.

Inferring Conclusions Based on Evidence

Investigators make logical **inferences** in accordance with NFPA 921, *Guide for Fire and Explosion Investigations*:

There are times when there is no physical evidence of the ignition source found at the origin, but where an ignition sequence can logically be inferred using other data.

The following are examples of situations that lend themselves to formulating an ignition scenario when the ignition source is not found during the examination of the fire scene. This list is not exclusive, and the fire investigator is cautioned not to hypothesize an ignition sequence without data that logically supports the hypothesis.

(A) Diffuse fuel explosions and flash fires.

(B) When an ignitable liquid residue (confirmed by laboratory analysis) is found at one or more locations within the fire scene and its presence at that location(s) does not have an innocent explanation.

(C) When there are multiple fires.

(D) When trailers are observed.

(E) The fire was observed or recorded at or near the time of inception or before it spread to a secondary fuel.

Sparks, Embers, or Flames from Outside Fires

NFPA 1033 (2022): 4.2.3

Open fires outside a building can generate sparks or embers capable of igniting combustible materials in structural components or adjacent fuel packages. For example, a flaming wood ember can reach temperatures of approximately 1,880°F (1030°C). While this temperature is high enough to ignite a wood shingle or other combustible material, the fire investigator should also evaluate the distance the ember had to travel before landing on the combustible surface because the ember loses heat energy as it travels. If the outside fire is close to the point of origin, the burning ember may serve as a competent ignition source. In this example, the method of ignition is conduction.

Outside fires can also result in the ignition of combustible surfaces as a result of heat energy radiated from the fire to the target fuel. When evaluating this potential ignition source, the fire investigator must consider the distance from the original fire to the target fuel as well as the heat release rate of the original fire. As the distance from the source to the target increases, the potential for ignition decreases.

Fuel-Fired or Fuel-Powered Equipment (Gas, Liquid, or Solid Fuel)

Fuel-fired or fuel-powered equipment can serve as competent ignition sources under certain conditions. Equipment in this category include:

- Gas-fueled equipment
- Liquid-fueled equipment
- Solid fuel-powered equipment

Gas-Fueled Equipment

Operations using a gas-fueled torch, such as Mapp or oxyacetylene as fuel, present known significant ignition hazards. Two sides of the fire tetrahedron, heat and oxygen, are always present during welding, grinding, and cutting operations. Because the intent of these operations is to separate or join metals using high temperatures, the flame temperatures of an oxygen/gas mixture are always above the ignition temperature of common combustibles.

Cutting operations can generate thousands of potential ignition sources in the form of hot molten metal (slag). Sparks and slag from grinding operations can travel many feet (meters) from the work area. Also, the torch flame itself is competent to ignite most fuels in the immediate work area.

In cases where sparks or slag ignite a material, investigators may not find the competent ignition source during the debris removal and origin area examination process. However, the investigator's knowledge that hot work was conducted in the area before the ignition of the fire can inform an inference of the ignition source.

NFPA 51B, *Standard for Fire Prevention During Welding, Cutting, and Other Hot Work*, requires that an area of 35 feet (11 m) around the operation be free of combustibles or cover combustibles in the area with flame-resistant materials or guards. A fire watch must also be posted for at least 30 minutes after completing the operation.

Soldering operations can also ignite nearby combustible materials. The pipe may conduct enough heat from the torch to ignite adjacent combustible materials. Solder is less likely an ignition source since the melting temperature is at or below the ignition temperature of wood and paper **(Table 13.1)**.

	Table 13.1							
	Time Required to Ignite Wood Specimens							
Wood 1¼ in. × 1¼ in. × 4 in. (32 mm × 32 mm × 100 mm)	No Ignition in 40 Min		Exposure Before Ignition, by Pilot Flame, Minutes					
	°F	°C	356°F (180°C)	392°F (200°C)	437°F (225°C)	482°F (250°C)	572°F (300°C)	662°F (350°C)
Long leaf pine	315	157	14.3	11.8	8.7	6.0	2.3	1.4
Red oak	315	157	20.0	13.3	8.1	4.7	1.6	1.2
Western larch	315	157	30.8	25.0	17.0	9.5	3.5	1.5
Noble fir	369	187	—	—	15.8	9.3	2.3	1.2
Redwood	315	157	28.5	18.5	10.4	6.0	1.9	0.8
Sitka spruce	315	157	40.0	19.6	8.3	5.3	2.1	1.0
Basswood	334	167	—	14.5	9.6	6.0	1.6	1.2

Liquid-Fueled Equipment

Equipment that use liquid fuel includes propane fueled stoves and gasoline or diesel powered generators. During debris removal, the investigator should protect any such device plausibly competent to ignite the first material, and the area around it **(Figure 13.2, p. 388)**. The investigator should photograph and document the equipment before its removal. Investigators should note the positions of knobs, switches, and valves. The investigator may solicit additional opinions about the unit from other experts.

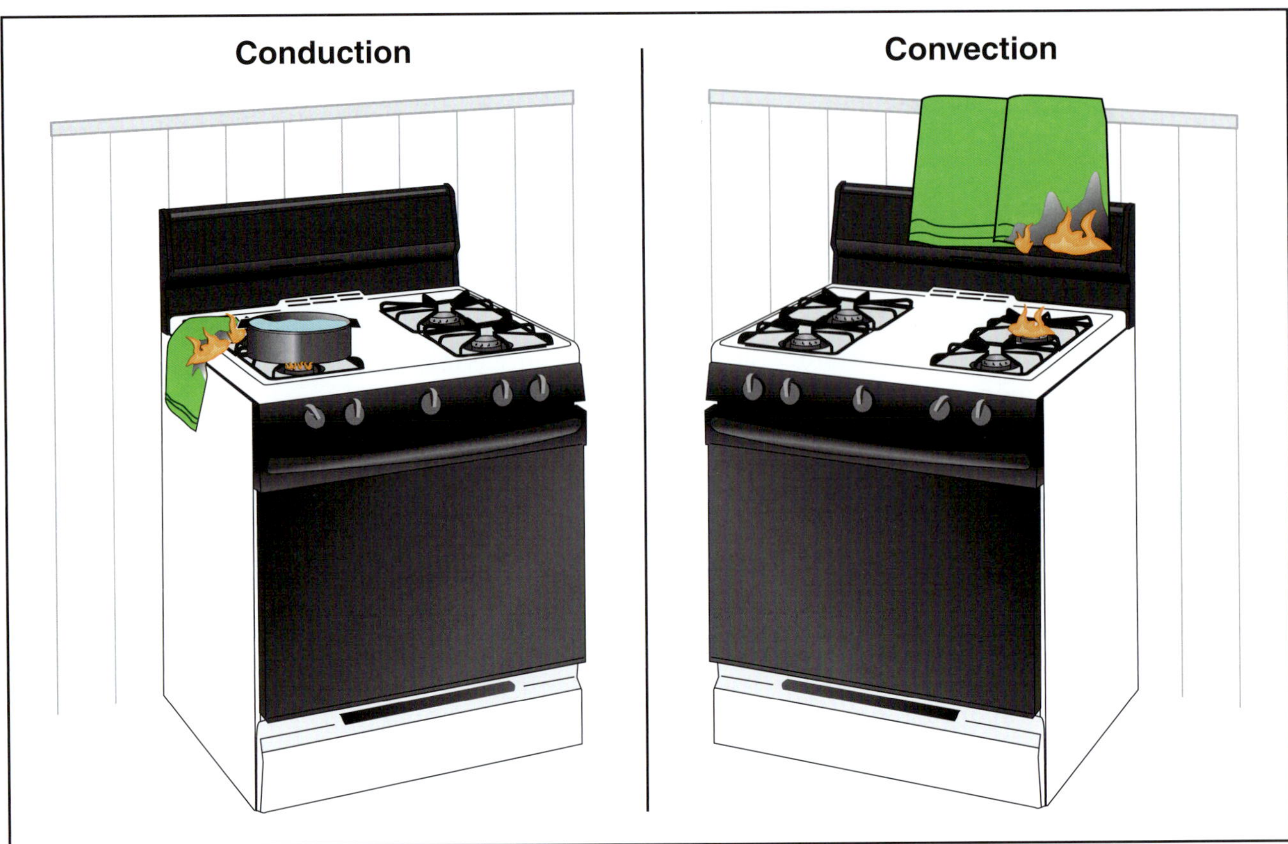

Figure 13.2 The location and method of heat transfer are two important components of the ignition sequence.

Solid Fuel-Powered Equipment

Solid fuel-powered equipment uses radiant heat to serve a purpose, such as heating air or water **(Figure 13.3)**. In normal operation, solid fuel-powered equipment is the ignition source for a controlled fire. The usual method of heat transfer is conduction though radiated heat from the equipment. Combustible materials in direct contact with the hot surfaces of the device or its chimney will heat by conduction. Burning or glowing embers of solid fuel may fall from the unit onto nearby combustibles.

Investigators must understand the clearance requirements of the equipment. For example, most appliances include the clearance requirements on the appliance itself or in the operating manual. Heat sources from solid fuel-powered equipment include:

Figure 13.3 Combustibles in contact with wood-burning stoves may ignite from conduction to heat. *Courtesy of Hugh W. Graham, Yates & Associates.*

- Wood and coal heating equipment

- Associated piping if sufficiently heated

- Chimneys and vents serving those devices

- Sparks or embers escaping from the device or chimney

- Hot embers in ashes may remain competent to ignite some materials for a significant length of time

The improper installation of a device or its chimney that places a hot surface (device or chimney) too close to combustible materials may be a competent ignition source through radiant heat transfer. Heat transfer may also occur via convection, in which hot gases escape due to improper flue or chimney construction.

Investigators should evaluate the appliance's ignition method. In some cases, the unit will show use of fuels that it was not designed to burn. An incompatible fuel may cause overheating and/or appliance failure, particularly at the floor where the radiant heat transfer can affect surrounding combustible surfaces. For example, coal in a wood stove may burn hotter than the stove can contain. Similarly, a unit may not contain the rapid heat release rate common with ignitable liquid use.

The investigator must document the device's condition and the position of any controls and openings. This information can assist investigators in determining whether to include or reject the heat producing device as a possible source of ignition.

Electrical Equipment

NFPA 1033 (2022): 4.2.8

Heating from purpose-built electrical appliances such as stoves, heaters, and lightbulbs, at ranges within their respective normal operation, is expected and would not be considered an electrical source of ignition. These devices, and others, feature current levels capable of ignition in specific conditions usually outside of normal usage. See Chapter 7, Electrical Analysis, for basic information on this topic.

> **WARNING:** Confirm deactivation of electricity service before inspecting electrical components.

Analyzing Components

When analyzing electrical components involved in a fire, the investigator first determines whether the structure received power at the time of the fire **(Figure 13.4)**. Then, the investigator looks at plugs and connections for indications of conditions before and after the fire. Components that may show evidence of electrical service and/or failure include:

- The transformer

- Service connections

- Overcurrent devices

An electrical system or component may be a competent ignition source for a particular fire if energized at the time of the fire. The electrical event itself must also meet the following conditions:

- Occurred within the origin area of the fire

- Transferred heat energy to the first fuel ignited

- Persisted for sufficient duration to ignite the first fuel

- Generated sufficient heat energy to overcome the thermal inertia of the first fuel ignited

Figure 13.4 Inspecting damage around plugs near the origin area may lead investigators to the ignition source. *Courtesy of Hugh W. Graham, Yates & Associates.*

Common Failures

Common failures found on electrical conductors or at electrical connections include **(Figure 13.5)**:

- **Loose connections** — May result in resistance heating, as an oxide interface forms on the terminal screw, conductor, and metal plate. Resistance heating may cause the terminal screw and associated components to eventually glow red hot and fail.

- **Damaged insulation on conductors** — Common damage results from manufacturing defects, rodent activity, and pinching, stretching, or penetration from an overdriven nail or screw. Conductors that touch can result in a short circuit or electrical arc.

Figure 13.5 Damaged electrical cords can create a competent ignition source. *Courtesy of Donny Howard, Agent, Oklahoma State Fire Marshal's Office.*

- **Damaged electrical appliance cords** — Occurs due to wear or mechanical fracture. Signs of damage include microarcing on the male plug blades and damage where the cord attaches to the appliance.

Generally, these systems and their components only generate uncontrolled heat energy as a result of one of the following common incidents:

- Sparks
- Short circuits
- Electrical arcs
- Resistance heating

An investigator may consult an electrical engineer regarding the electrical system and wiring. A forensic examination can help determine whether damage resulted from arcing, melting, or alloying. An electrical arc may occur if the conductors come in contact with each other. Arcing of the conductors through char (insulation) is also possible. Insulation may melt as the conductor overloads. Nearby combustible materials may ignite if the overload condition is severe.

Alloys Created from Melting

Metals melting at fire temperatures may combine to form alloys with different characteristics than the original materials. For example, copper melts at 1,985°F (1085°C). Aluminum melts at 1,220°F (660°C). Depending on the combination, an alloy of copper and aluminum can melt at 1,020°F (585°C).

Zinc and other so called **pot metals** can have the same effect. Severed branch circuit conductors that have a silvery appearance most likely have been subjected to **eutectic alloying**, not electrical arcing. Final determination of eutectic alloying may have to be verified through metallurgical testing.

Short Circuits

A short circuit is an abnormal path of current in a circuit that normally leads to an overcurrent condition. For example, the following sequence can occur:

1. A short circuit resulting from contact between metal objects can melt metal.

2. Metal heated to vaporization creates an arc.

3. During the brief arcing period, droplets of molten metal (sparks) separate from the site.

4. Depending on the fuels at or near the point of the arcing, an ignition can occur.

The ability of metal droplets to cause ignition depends on the available fuel and the size and heat content of the molten particles. Most droplets will only be capable of igniting fine fuels with a high surface-to-mass ratio (dust, thin paper, or grass), or those already heated almost to the point of ignition. The type of metal may affect droplets' ability to ignite any fuel. For example, copper conductors seldom ignite a fuel because the metal droplets cool very rapidly as they travel through the air. In contrast, molten droplets from aluminum conductors burn as they travel through the air and have a greater ability to cause the ignition of combustible materials on which they land **(Figure 13.6)**.

Figure 13.6 Melted aluminum wiring may pose an additional ignition source if it lands on easily combustible material. *Courtesy of Yates & Associates.*

Resistance Heating

Poor electrical connections commonly cause electrical fires. The flow of current through a conductor normally produces some heat due to the resistance of the material used to form the conductor. Normally, the heat dissipates to the surrounding air. If an excessive amount of current flows through the conductor, it may produce enough heat to ignite adjacent combustibles or damage electrical components or the wiring in appliances.

For example, a 100-foot (30 m) heavy-duty extension cord can safely provide electrical energy to a window air conditioner when it is fully extended **(Figure 13.7)**. The same cord tightly coiled on the floor near an upholstered chair could generate sufficient thermal energy to damage the extension cord and insulation, and pyrolyze the fabric.

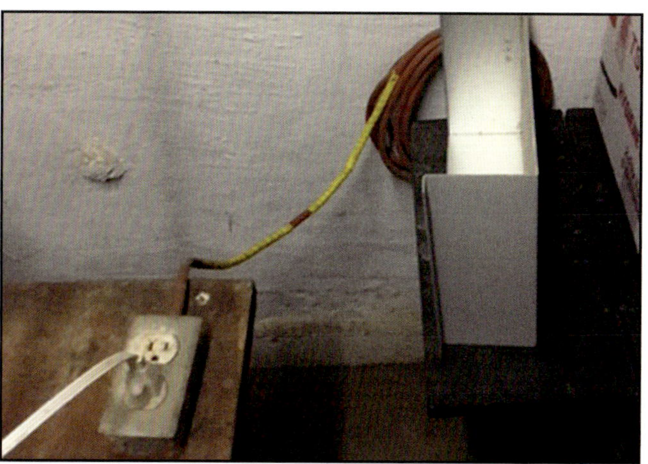

Figure 13.7 Coiled electrical wires may not cause problems if they are not near combustible materials. *Courtesy of Donny Howard, Agent, Oklahoma State Fire Marshal's Office.*

Localized high temperature conditions atypical in a fire environment can be generated by factors including:

* Mismatch between wiring and conductors

* Current flow through aluminum conductors with high-resistance oxide formation.

* Shorting and/or arcing in a branch circuit when fire burns away the wiring insulation.

* Oxide formation on copper wiring connection in loose duplex receptacles or switches.

* Loose connections, such as partially melted terminal screw heads, copper conductors melted under a wire nut, damage to connection plates, or eroded steel lug screws in the wiring

Batteries

The ongoing proliferation of battery-operated resources, such as electronic tablets, smart phones, electronic smoking devices, along with portable power sources, have increased the prevalence of batteries in most places. Damage to the individual battery cells can cause bulging and other failures. Batteries can become a competent heat source when mishandled, such as through:

* Damage
* Overheating
* Overcharging

Hot Objects

Hot objects commonly result in ignition. Examples of hot objects include:

- Improperly discarded ashes
- Metal fragments ejected from machinery
- Electrical devices, such as an electric heater or lightbulb
- Friction-heated objects such as overheated fan belts, hot boxes on railcars, or tires

The fire investigator should evaluate the heat transfer method for the ignition source to the material first burned. For example, a fuel in direct contact with the hot object (heat source) heats by conduction. A fuel at some distance from the heat source more likely heats by radiation and requires more energy than conduction.

Evaluating Electrical Lighting

When evaluating electric lighting as a potential source of ignition, the fire investigator should note the type of bulb involved and how the bulb was oriented to the fuel. A low-wattage incandescent lightbulb can generate surface temperatures of up to approximately 500°F (260°C). The surface temperature of an incandescent bulb depends on its wattage and position (for example, base up or base down). The normal surface temperature of an incandescent bulb (depending on its wattage and position) typically cannot serve as a competent ignition source for most fuels. However, if insulated while in use, the bulb can generate temperatures high enough to ignite paper or cloth. This is why many building codes include requirements that prevent accidental ignition of materials should they fall from a shelf onto the light **(Figure 13.8)**.

High-intensity lamps with quartz halogen bulbs are a competent ignition source for most materials, as the bulbs generate surface temperatures of up to 1,650°F (900°C). A quartz halogen bulb in contact with or in close proximity to combustible materials could result in flaming ignition.

Metal halide lamps operate under high pressures and temperatures (1,830°F [1000°C]). Failure may result in extremely hot pieces of glass being ejected and presenting a potential source of ignition.

Figure 13.8 A light bulb wrapped in a blanket may be a competent ignition source because the cloth insulates the bulb, trapping its heat. *Courtesy of Nicole Brewer, Portland Fire Investigations.*

Explosive or Fireworks

Explosive materials can be a competent ignition source for low-mass fuels. Items in this category include:

- Fireworks
- Munitions
- Model rockets
- Blasting agents
- Incendiary devices
- Tracer ammunition

Investigators removing debris from an explosive or fireworks fire scene should attempt to locate portions of the ignition source. If an incendiary device was used, the investigator should take the samples at and around the origin area for laboratory analysis to determine the type of material used in the device. Laboratory analysis can identify trace explosive or firework materials.

Incendiary Devices

Some improvised incendiary devices use materials commonly found in a fire scene. For example, a person making an incendiary device can rig a competent ignition source from a high-wattage lightbulb, matches, and an insulation material such as shredded paper or cloth. An investigator must thoroughly document findings, especially when evidence indicates that the fire was intentionally set.

Other Open Flame, Sparks, or Smoking Materials

Common ignition sources include items or devices intentionally designed with an open flame. These devices may have engineered limitations to minimize unintentional heat transfer to other materials.

Smoking Materials

Materials designed to ignite and burn are common ignition sources. Smoking materials include cigarettes, cigars, and pipes. Current information on this topic may be found in NFPA 260, *Standard Methods of Tests and Classification System for Cigarette Ignition Resistance of Components of Upholstered Furniture,* and NFPA 261, *Standard Method of Test for Determining Resistance of Mock-Up Upholstered Furniture Material Assemblies to Ignition by Smoldering Cigarettes.*

Generally speaking, a cigarette most often ignites upholstered furniture when located partially between the cushions with a sufficient air supply to allow it to burn **(Figure 13.9)**. Then, the cushions and padding absorb the generated heat over time (Babrauskas, 1985). The preignition smoldering process can generate a significant amount of smoke and harm a structure's occupants.

Laboratory burn tests show that smoldering ignition requires durations from two minutes to several hours. In addition to the position of the smoking device, the furniture materials affect the time required to ignite.

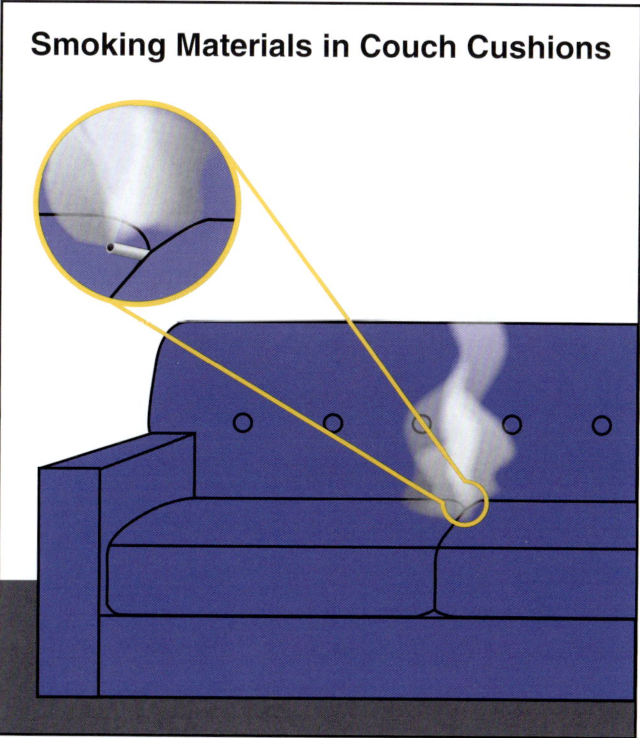

Figure 13.9 A lit cigarette allowed to smolder in the cushions of a couch can become a competent ignition source.

Self-Extinguishing Cigarettes

As of 2011, commercial cigarettes include measures that cause the cigarette to self-extinguish under conditions including when the fire does not intensify within a relatively short period of time. Smoking materials that do not have any such regulation include cigarette wrappers, marijuana items, and materials without wraps such as pipes.

Open Flame Materials and Devices

Materials designed to ignite and sustain an open flame include:

- Candles
- Matches
- Lighters
- Warning flares

Methods of heat transfer with open-flame items include conduction, where fuel is in contact with the flame source, or convection where the heat from the burning item comes into contact with combustibles above the flame.

Candles commonly factor in fires. Candles produce sufficient energy to ignite most combustible materials, and can sustain that energy for a substantial period of time. For a candle flame to ignite materials from radiation, the fuel would have to be located in close proximity to the flame.

An investigator evaluating an area with candle evidence should indicate plausible materials first ignited, and any associated burn patterns. Remnants of candle use, whether as an incendiary device or as part of the overall scene, can include:

- Wax
- Wick
- Metal wick plate
- Broken candle holder at or near the point of origin

Natural Sources

Natural sources of heat energy do not require human intervention for ignition. Though relatively rare, scenarios in which natural heat energy sources serve as competent ignition sources include:

- Solar energy
- Static discharge
- Lightning strike
- **Spontaneous heating**

Solar Energy

Under normal conditions, heat energy from the sun on any given combustible material will not cause ignition. However, solar energy focused onto a small area can provide enough heat for pyrolysis and ignition.

For example, sunlight focused with a magnifying mirror onto a combustible surface, such as draperies, wood interior finish, or paper, for a sufficient period of time, can be a competent ignition source. In that scenario, an investigator examining the point of origin would identify the item that caused the energy focus, and document the availability of sunlight to that item.

Static Discharge

Static electricity describes energy trapped or prevented from escaping from the surface of an object. A body or object with an accumulation of trapped electrical energy is "charged."

For ignition to occur as a result of a static discharge:

- Static electricity must generate.
- An ignitable mixture of fuel at the discharge arc must exist.
- A discharge arc of sufficient energy to cause ignition of the fuel must exist.
- Separate static charges must accumulate, and the electrical potential maintained.

Some physical evidence may identify a source of static electricity. For example, a belt-driven machine may generate sufficient sparking to ignite materials in some conditions. The investigator may collect circumstantial evidence that supports the findings from interviews and other data collected after the detailed scene examination.

The types of fuels most susceptible to ignition from static include:

- Ignitable gases
- Combustible dusts
- Ignitable liquid mists or vapors
- Solid fuels with a high surface-to-mass ratio
- Gases or vapors that are within their flammable limits

Lightning Strike

Lightning is a natural static discharge capable of delivering very high levels of electrical energy to localized areas. Lightning commonly strikes the tallest object in its path to the ground and can enter a structure along paths such as:

- The structure itself
- Power, telephone, or cable lines via the service connection
- Tall objects (trees, other structures) near enough for horizontal travel
- Metal objects or other components of the structure that extend higher than the roof, including antennas, masts, rooftop HVAC units

The location of the strike may not be able to safely absorb or redirect the energy delivered in a short amount of time. This excess energy may ignite nearby combustible materials, and result in explosive destruction of objects or structural components in the path of travel. Damage to conductors can include:

- Melting
- Breaking
- Vaporization of the conductors at several locations along the path of travel

Spontaneous Heating

Spontaneous heating is a process through which a material increases in temperature (oxidizes) without an external source of heat energy. References such as the NFPA® *Fire Protection Handbook* list materials that may self-heat (**Table 13.2**). In addition to organic materials, some metal shavings or powders generate heat as a result of oxidization. In isolated instances, these materials may generate sufficient temperatures to self-ignite.

Factors that influence the ignition of combustibles via self-heating include:

- **Rate of heat generation** — The material must generate heat at a rate greater than the heat dissipation rate. Material stored warm or hot will self-heat faster than cooler materials.

Table 13.2 Examples of Materials Subject to Self-Heating	
Material	**Potential for Self-Heating**
Charcoal	High
Linseed oil rags	High
Foam rubber	Moderate
Hay	Moderate
Manure	Moderate
Sawdust	Low
Grain	Low

- **Effects of ventilation** — The material must have sufficient air supply to support oxidation. However, too much air supply will dissipate heat via convection.

- **Insulation properties of the immediate surroundings** — The material must be insulated so that the heat generated by the oxidation process does not dissipate.

For example, a linseed-oil soaked rag, crumpled and placed in a container, has high potential to self-ignite. The same rag spread out and placed in open air so that any heat generated can dissipate with the help of air movement will not normally generate sufficient heat to self-ignite.

The unsealed container does not limit ventilation, while blocking most air currents. The container also holds the rag in a way that enables the rag itself to serve as insulation. Similarly, insulation may be provided by piles of materials, when sufficient air supply is available.

Heat Spreading from Separate Fire Sources

Fires often ignite secondary fuel packages. The investigator must determine whether multiple fires have separate origins or resulted from natural fire propagation. Sources of heat energy in this category include:

- Radiated heat

- Convected heat

- Conducted heat

- Direct flame impingement

- Flying brands, embers, or sparks

Fire exposure is a viable source of heat energy to ignite additional fuels at some distance from the material first ignited. When fire spread is apparent, investigators hypothesize the order of ignition. The fire investigator documents the method and path of heat transfer back to the point of fire origin.

Oxidizing Agents

NFPA 1033 (2022): 4.2.6

While using a consistent methodology, such as the scientific method, to determine fire cause, investigators must evaluate all available oxidants or oxidizing agents. For example, investigators evaluate the oxidants' form, and how oxidants interacted with the fuel and heat (ignition) source to cause combustion. Oxidants, or oxidizing agents, found or hypothesized to be present within the fire origin area have special significance.

The oxygen contained within an atmosphere is the most common oxidant for a fire. Increased oxygen may be present within the fire scene and may lead to increased combustion. For example, a leaking tank or oxygen generator may affect the overall oxygen concentration in an atmosphere.

In addition to oxygen, a fire investigator should consider the involvement of other oxidants in the structure. For example, chlorine is commonly found near swimming pools.

Ignition Sequence

NFPA 1033 (2022): 4.1.7, 4.2.4, 4.2.6

An ignition sequence brings a competent ignition source, a fuel package, and an oxidant together under circumstances that allow combustion to occur. Circumstances include sufficient quantities of each factor plus sufficient time. Fire investigators must remember that an ignition source and a fuel found within the origin area may not have started the fire. To determine the ignition sequence, investigators must evaluate each fuel, ignition source, and oxidant in combination as part of the cause determination process.

Determining the Fire Cause

NFPA 921 list several factors and questions that investigators should consider when determining the cause of a fire. These include:

(1) How and sequentially when the first fuel ignited came to be present in the appropriate shape, phase, configuration, and condition to be capable of being ignited

(2) How and sequentially when the oxidant came to be present in the right form and quantity to interact with the first fuel ignited and ignition source and allow the combustion reaction

(3) How and sequentially when the competent ignition source came to be present and interact with the fuel

(4) How and sequentially when the competent ignition source transferred its heat energy to the fuel, causing ignition

(5) How safety devices and features designed to prevent fire from occurring operated or failed to operate

(6) How and sequentially when any acts, omissions, outside agencies, or conditions brought the fuel, oxidant, and competent ignition source together at the time and place for ignition to occur

(7) How the first fuel subsequently ignited any secondary, tertiary and successive fuels that resulted in any fire spread. If the hypothesized ignition location is not within the main area of fire destruction, then, for the hypothesis to be valid, the investigator should be able to demonstrate that there was a path of fire propagation along which fire would have been able to propagate.

Fire Ignition-Source Identification

When evaluating ignition sequences, investigators consider whether someone's action or omission (failure to act) brought together the competent ignition source with the first material ignited **(Figure 13.10)**. For example, an unattended lit candle (omission) on a windowsill some distance from curtains, with the window closed, may not result in an ignition sequence. If someone opens the window and the candle flame ignites the curtains, that action is part of the ignition sequence. In this case, the investigator determines whether the window was open or closed to identify this ignition sequence **(Figure 13.11, p. 398)**.

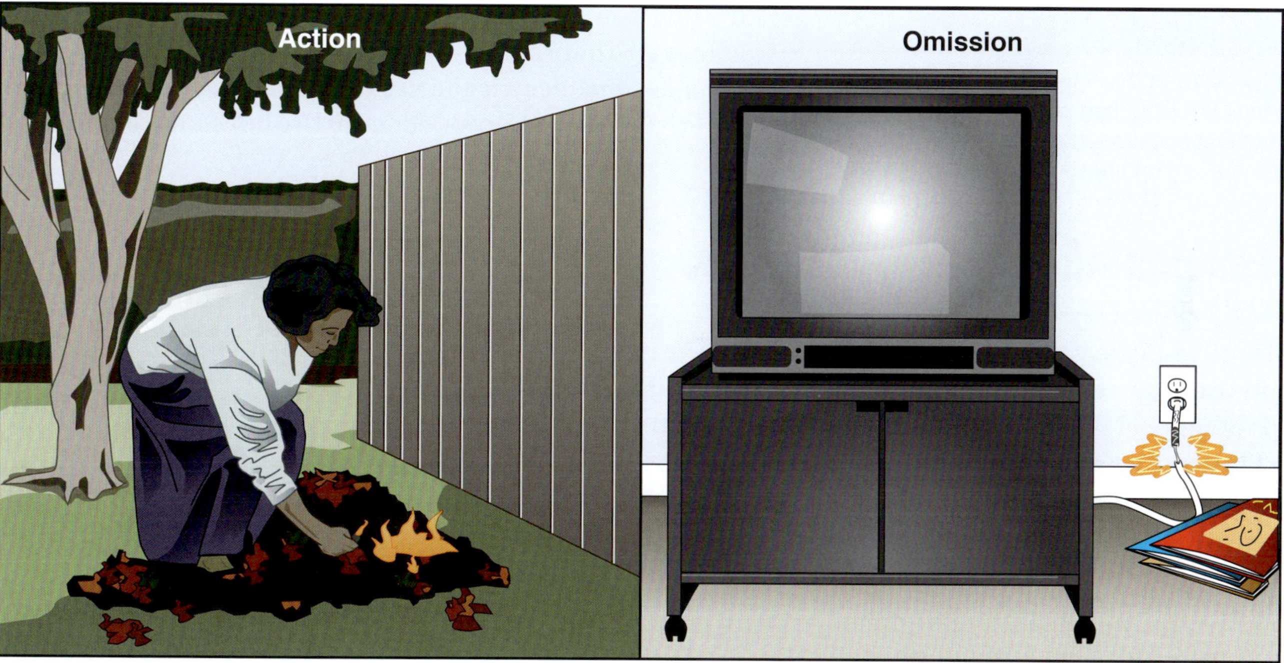

Figure 13.10 Human action or inaction can have an influence on the fire cause determination.

To determine the ignition sequence, an investigator should develop a series of hypotheses and test each one by asking questions and gathering more information if necessary.

Questions:

• Was the candle lit (a competent ignition source)?

• Was the window open or closed (part of the ignition sequence)?

• Was there a wind or breeze that day (part of the ignition sequence)?

Figure 13.11 The ignition sequence describes the first fuel ignited and the ignition sources meeting to cause the fire. For example, a candle is a competent ignition source for lightweight curtains, in close enough proximity.

- What was the distance from the candle to the curtains (part of the ignition sequence)?

- Was the curtain material capable of igniting and sustaining a flame (first material ignited)?

- Were other windows open that may have contributed to cross-ventilation or did some other event occur, such as the slamming of a door, that created air current capable of moving the curtain and the candle?

Explosion Ignition-Source Identification

NFPA 1033 (2022): 4.1.7, 4.2.4, 4.2.6

As with other types of fire investigations, an investigator may need to evaluate multiple ignition sources in an explosion investigation. An investigator can eliminate possible ignition sources via analysis of:

- Fuel

- Origin area

- Information from other sources

For example, an investigator considering natural gas as the fuel would recognize the different height properties than a heavier-than-air gas. Similarly, the investigator should consider the electrical arc as a possible ignition source if the origin area has no evidence of fire but shows evidence of electrical failure and arcing.

Developing Cause Hypotheses

NFPA 1033 (2022): 4.6.5

The investigator should research the activities occurring in the prefire origin area to determine the equipment, appliances, or other heat producing devices that were present or in use **(Figure 13.12)**. With that information, the investigator evaluates how those factors may be a competent ignition source either as a result of a failure or misuse.

Identifying Ignition Sources

For each potential ignition source found within the origin area, the investigator develops a separate cause hypothesis. When determining whether an appliance could serve as a competent ignition source, an investigator considers:

- Repair history

- Age of the item

- Previous damage

- Manufacturer recalls

- Problems with the equipment

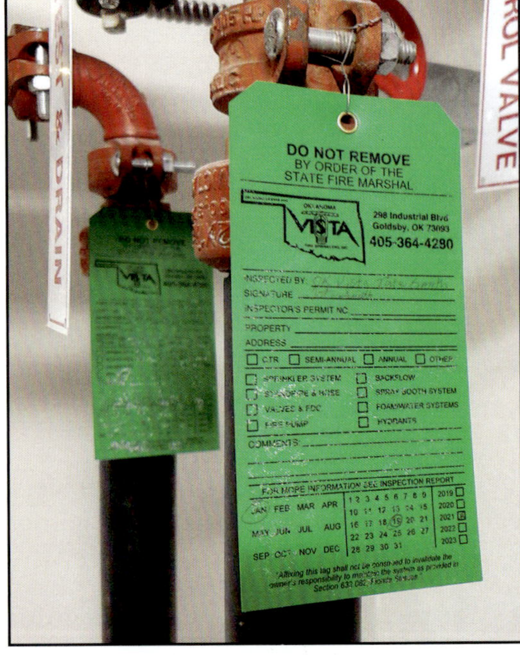

Figure 13.12 Maintenance tags indicate the most recent inspection and the responsible party. *Courtesy of Donny Howard, Agent, Oklahoma State Fire Marshal's Office.*

The investigator should also consider ignition sources that may have been present at the time of the fire but are absent at the time of the scene examination, such as:

- Lightning
- Static electricity
- Lighters/matches

NOTE: A hypothesis without evidence is not valid.

Example of Witness Testimony Use

A fire originated in the corner of a bedroom. In that area, a lamp and an alarm clock plugged into power bar, which in turn connected to a wall outlet.

The fire damaged the wall, the power bar, and the power cords of both the alarm clock and the lamp. The homeowners said they purchased the power bar two-years-old at a local retail store. In the days before the fire, the homeowner noticed that the alarm clock and lamp worked intermittently and the power bar's switch light no longer lit.

The witness's information provides evidence that the power bar was not performing as expected and may have failed, resulting in a fire. Even with that information, the investigator must also examine physical items as part of the investigation to determine cause, such as:

- Lamp
- Alarm clock
- Electrical distribution panel
- Wall outlet including associated wiring **(Figure 13.13)**
- Position of the breaker (ON / OFF / TRIPPED)
- Condition of the fuse (BLOWN / NOT BLOWN / CORRECT SIZE)

Figure 13.13 An investigator should consider all associated wiring when evaluating whether an outlet was a factor in an incident's development. *Courtesy of Donny Howard, Agent, Oklahoma State Fire Marshal's Office.*

Evaluating Ignition Sources

Investigators evaluate each ignition source found within the origin area against the fuels within the origin area. For example, an investigator may list each of the fuels and each of the ignition sources and determine the possible interactions between the fuel and the ignition source.

While completing this analysis, an investigator considers:

- Is there any evidence that ignition occurred?
- Is there a competent ignition source for the fuel?
- Is a pathway present from the first fuel ignited to any secondary fuels?
- Is the fuel close enough to the potential ignition source to be capable of ignition?

This analysis will ensure the investigator identifies each fuel and ignition source found within the origin area and considers it as a potential cause. Investigators complete this analysis to reduce the number of possible ignition sequences. This analysis also tests cause hypotheses.

Testing the Cause Hypothesis

NFPA 1033 (2022): 4.2.5, 4.4.3, 4.6.1, 4.6.3, 4.6.5

Testing the fire cause hypothesis is a significant step in applying a consistent methodology, such as one based on the scientific method. In this phase of the investigation, the investigator develops one or more final data-based hypotheses.

At this point in evaluation, the investigator must decide which hypotheses are possible versus which are probable. In order to make this determination, the investigator should:

- **Test the cause hypothesis** — Conduct experiments that improve the investigator's understanding of the incident.

- **Review literature** — Evaluate whether available research matches the circumstances and conditions of the current fire under investigation.

- **Follow basic scientific principles** — Consider whether the hypothesis adheres to accepted scientific principles of chemistry, physics, or thermodynamics.

An investigator may use data collected from the scene to test a hypothesis in a cognitive (thought) experiment. For example, after leaving a scene the investigator hypothesized that an open window may have influenced the *flow path*. Considering other evaluation notes, the investigator reviews the frame and window for indications of mechanical failure. The window was closed at the time of the investigation and the glass had thermal fractures. This evidence disproves the investigator's hypothesis that the window had been open during the incident **(Figure 13.14)**.

Figure 13.14 Broken glass external to an incident scene can indicate the direction of the window break, and whether the fire had been a factor in the break. *Courtesy of Donny Howard, Agent, Oklahoma State Fire Marshal's Office.*

Post-Scene Investigation

NFPA 1033 (2022): 4.2.5, 4.4.3, 4.6.1, 4.6.3, 4.6.5

The post-scene investigation occurs between when the investigator leaves the site and before the incident scene is released. Some organizations may refer to the post-scene investigation as a post-incident or follow-up investigation, or completion of the fire scene exam. Up to this point, the fire investigator has focused on evaluating the fire scene itself. The following sections discuss the investigators next steps:

- Collecting and organizing investigative information
- Evaluating investigative information
- Analyzing data
- Determining responsibility

Collecting and Organizing Investigative Information
NFPA 1033 (2022): 4.6.1

The process of collecting investigative information begins when the investigator is first assigned to an incident. Investigators gather information in formats including:

- Notes
- Reports
- Sketches
- Photographs
- Evidence packages
- Video or audio recordings of interviews

All evidence must be documented and secured throughout all processes, for use in the investigation and subsequent court proceedings. As discussed in Chapter 11, Evidence Collection and Preservation, the chain of custody must be maintained.

While reviewing the available information, the investigator considers whether sufficient information is present to complete the investigation. Investigators then organize the information into a format accessible for all parties working on the case. This may include digitizing, organizing, and storing data, per the AHJ.

The investigator may compile the collected information on printed or digitized forms as per the authority having jurisdiction. In the absence of such a form/format requirement, the investigator may compile a list documenting the contents of the investigative file, and identify information or reports needed for further case evaluation.

The investigator also identifies and obtains information from other sources, such as:

- Insurance reports
- Fire department incident reports
- Police reports regarding the incident
- Reports developed by private investigators
- Photos and videos obtained from investigative and other sources, such as security cameras, television stations or bystanders at the fire scene

A systematic approach to data organization allows the fire investigator to:

- Identify relevant information
- Obtain a better overview of the available data
- Use that information to make informed decisions related to the investigation

As the complexity of an investigation increases, so does the investigator's task of sorting through the information and making decisions regarding its relevance. The investigator must organize the data so it remains manageable.

Evaluating Investigative Information

NFPA 1033 (2022): 4.4.3

With the available investigative information organized and cataloged, the investigator begins to evaluate the investigative file to develop an understanding of the current information and the potential scenarios.

Investigators add information obtained during the post-scene investigation to the investigative file for use in incident analysis and the cause determination. The investigator must continually retest the current hypothetical ignition scenarios, and remain willing to change the scenario based on new or conflicting information.

At this stage, the investigator continues to develop and test hypotheses regarding potential ignition scenarios. This work may identify special expertise needed to support the investigation including laboratory work to analyze evidence collected at the fire scene.

Depending on the incident complexity, other elements of the post-scene investigation may include:

- Developing data necessary to assist in the determination of the fire cause
- Submitting evidence for laboratory analysis and relating the results to the fire scene
- Seeking experts to assist in the development and/or analysis of investigative information
- Reviewing and analyzing records and documents related to the incident and/or the involved parties
- Focused interviews with involved parties or individuals who might possess information regarding the investigation

Expert Resources

Professionals with training and experience will be able to offer a different perspective on a case (NIST). Investigators may solicit assistance from experts qualified in a range of fields, including:

- Physicians and forensic pathologists
- Psychologists or behavioral disorder specialists
- Attorneys for guidance in specialized legal areas
- Polygraph operators able to administer and interpret polygraph tests
- Forensic accountants for assistance in the acquisition and analysis of financial data
- Insurance representatives to research information such as coverage, prior losses, and claims
- Engineers with a background in a topical discipline such as electrical, mechanical, chemical, civil, structural, or fire protection
- Industry experts associated with trade associations or companies that produce machinery and processing equipment, for assistance in incidents that involve a specialized industry, process, or piece of equipment

Forensic Laboratories

An investigator may have a choice in laboratories. If possible, before working with a forensic laboratory, the fire investigator should check the references provided by the laboratory as well as confer with any peers who may have had dealings with the facility in the past.

When choosing, an investigator should consider criteria such as:

- **State-of the-art techniques** — Use the most current techniques and technology

- **Proper evidence handling** — History of avoiding evidence destruction or contamination

- **Reputation** — History of clients who were pleased with the work the laboratory provided to them

- **Certification and accreditations** — Correct professional certifications and accreditations to perform the forensic work needed

When selecting a laboratory to perform forensic analysis of evidence collected as part of a fire investigation, the investigator primarily focuses on the reliability of the results. Laboratories demonstrate commitment to providing high-quality results by:

- Following national standards

- Using recognized procedures

- Employing certified forensic scientists

The forensic science community has developed standards and certifications that can assist fire investigators in the evaluation process. Examples:

- **American Board of Criminalistics (ABC)** — National peer-review group that certifies forensic scientists in specific disciplines including fire debris analysis

- **The American Society of Testing and Materials (ASTM)** — Maintains a committee on Forensic Sciences (E-30) that establishes standard methods, practices, and guidelines for the forensic field

- **Accreditation Board of Engineering and Technology (ABET)** — Provides accreditation, promotion, and advancement of education in applied science, computing, engineering, and technology

- **Association of Crime Laboratory Directors (ASCLD)** — Provides criteria used to judge whether a facility and its practices provide an atmosphere conducive to the quality of work necessary in the forensic field

Joint Examinations

Public and private sector laboratories frequently host joint examinations of appliances, vehicles, and other equipment. This type of examination requires cooperation of the parties and reduces the likelihood of spoliation claims, and is generally cost effective for multiple parties.

In joint examinations, interested parties meet and examine, dismantle, and evaluate items from the scene. Typically, parties follow an agreed-upon examination protocol to examine the evidence.

Forensic Testing

A forensic laboratory is a resource that can assist in fire cause determination. The fire investigator typically requests the following types of analyses:

- **Fire debris analysis** — Identifies trace materials found at the fire scene, such as ignitable liquids

- **Microanalysis** — Examines physical evidence that may be too small to differentiate without magnification. Examples: damaged electrical wiring, tool marks, impressions from tires and shoes, broken glass, smoking materials and matches, and hair or fibers **(Figure 13.15)**

Figure 13.15 Smoking materials found at the scene may require microanalysis in a laboratory during the post-scene investigation. *Courtesy of Donny Howard, Agent, Oklahoma State Fire Marshal's Office.*

- **Latent fingerprint identification** — Compares fingerprints found at a fire scene to those of a suspect

Depending on fatalities, an investigator may also request:

- **Forensic anthropology** — Identifies deceased victims
- **Forensic pathology** — Determines the cause and manner of death
- **Serology** — Analyzes body fluids such as saliva, blood, urine, or semen
- **Firearms analysis** — Examines firearms or projectiles for classification and/or identification

Analyzing Data

NFPA 1033 (2022): 4.2.5, 4.4.3, 4.6.5

The investigator must have the ability to analyze investigative information. For the results to be meaningful, the investigator must follow a formal analysis process. Systematic analysis is especially important where the incident complexity and/or size may affect the investigation's speed.

The quality of the fire investigator's work product and the time required to perform essential tasks may be improved through the use of analytical investigative techniques such as:

- Link analysis
- Financial profiling
- Graphs and timelines
- Inference development
- Event and commodity flowcharting

Link Analysis

The U.S. Bureau of Alcohol, Tobacco, Firearms, and Explosives (ATF) *Fraud Investigation Manual* describes *link analysis* as a method of computing, organizing, and utilizing data relating to an investigation. Link analysis presents complex data in clear graphics showing relationships through data such as:

- Street addresses
- Telephone numbers
- Vehicle license plates

Financial Profiling

Financial profiling is an investigative tool that allows the investigator to organize and display the financial data of an individual or organization onto a graph or chart. Therefore, it is easier to detect any deviations that may establish a pattern of financial behavior.

Graphs and Timelines

Investigators may draw high-quality graphs on paper or generate them in a computer software program. Graphs can be set-up to efficiently represent numerical data and relationships such as **(Figure 13.16)**:

- Time versus temperature for a fire
- Water flow over time for an automatic sprinkler system
- Profit/loss in dollars over a length of time at a business

Graphing may also be useful in the analysis and presentation of the information to other investigators, supervisors, or a jury. One type of graph, timelines, depicts the occurrence of events over a span of time with intervals such as seconds and minutes. Timelines with longer time periods, such as building ownership and changes in insurance or financial status of individuals or businesses, can also clarify topics in the analysis phase of the investigation.

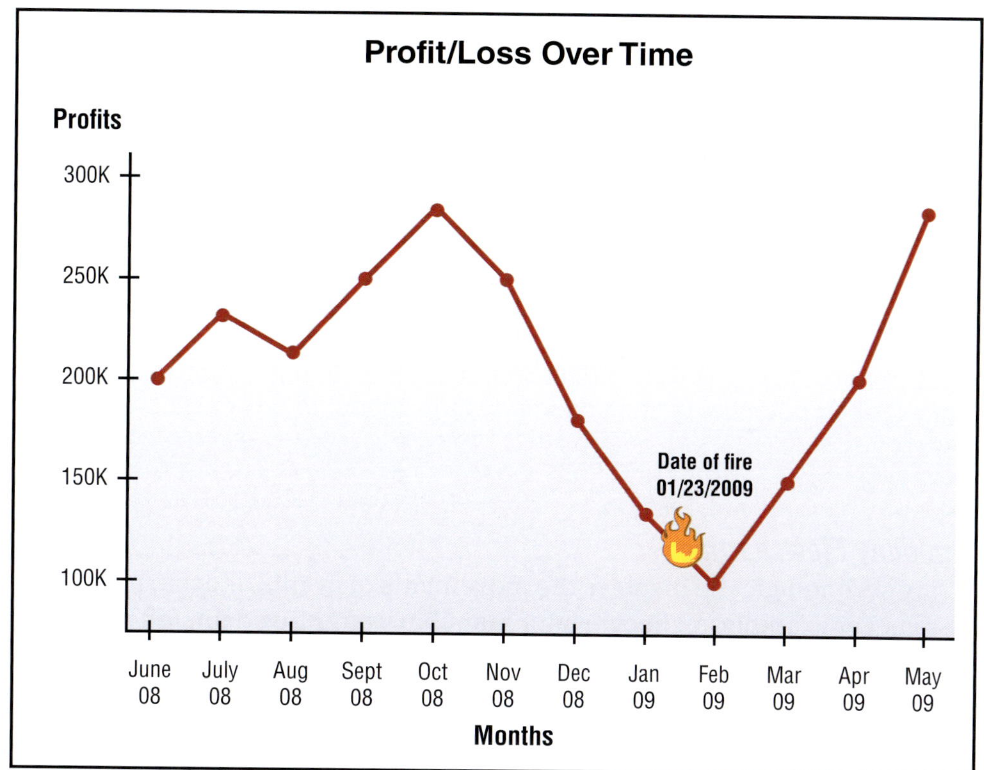

Figure 13.16 Graphing a business' finances over a period of time may help an investigator discover the motive behind an incendiary fire.

After selecting the time scale, an investigator decides which reference times to use. A well-designed timeline with carefully correlated times and events provides an understandable visual representation of known events and potential scenarios that led up to those events.

For example, to develop a timeline from fire ignition through extinguishment, an investigator may include points during fire progression, such as when a witness first saw fire or when the fire department received the report of the fire. Times recorded by different sources of information may not corroborate each other. When reconciling discrepancies, the investigator may select the source with the most information or credibility, such as the fire department dispatch center.

Inference Development

An inference is a conclusion derived from a set of **premises**. Inferences may take the form of a hypothesis, conclusion, prediction, or estimate. Premises are based on an investigator's:

- Knowledge
- Experience
- Expertise

Investigators use inference to:

- Test the validity of information
- Give direction to the investigation
- Recognize the absence of needed data

The value of this very subjective process depends on the level of confidence the investigator has in the inference. The investigator uses data to test all hypotheses.

Testing Inferred Hypotheses

While investigating a fire in a single-family residence, the investigator determined the origin area was in the living room. The investigator observed irregular fire patterns on the floor, and hypothesized that an ignitable liquid was the first fuel.

No ignition source was recovered from the scene. The investigator hypothesized that a cigarette lighter or another open-flame device ignited the vapors of the liquid fuel and was then removed.

Two months after the fire, a laboratory provided test results from flooring samples. The results were negative for the presence of an ignitable liquid.

An investigator should have also collected and analyzed other data, such as flashover conditions within the room or other potential ignition sources at the scene, to be consistent with the use of the scientific method.

Event and Commodity Flowcharting

Event flowcharting chronologically displays the movements of events or occurrences either over time or through a system. For example, an investigator may chart gambling debt, followed by the loss of job, marriage breakup, and fire.

Commodity flowcharting portrays the movement of a tangible item, such as money or stolen property, through a system. A good example of commodity flowcharting would be the portrayal of the movement of items removed from a structure before a fire and placed in storage, pawned, or sold at a flea market.

Determining Responsibility

Given all available information on a fire or explosion incident, and evaluation and analysis of evidence, the fire investigator may be involved in determining who or what was responsible for:

- Cause of the fire
- Human acts or omissions
- Cause of bodily injury or loss of life
- Cause of damage to property at the incident

  NOTE: Different parties/conditions may have caused individual aspects of the incident.

If the origin and cause are easily determined, the investigator may complete the analysis phase during the scene examination. When investigators cannot readily determine the origin and cause during the examination of the scene, the postincident investigation can be complex and time-consuming.

To make a final determination of the origin and cause of a fire, the fire investigator considers:

- What was the spread of the fire once it was ignited?
- What human factors were involved in the ignition of this fire?
- Were the fire protection systems adequate? Did they function as designed?
- What factors were responsible for injuries or fatalities resulting from the fire?
- What factors or circumstances brought the components of the fire tetrahedron together?
- Did the building construction and interior finish or contents contribute to the spread of the fire?

Incendiary Fires Analysis

NFPA 1033 (2022): 4.6.4

Incendiary fires are deliberately set with malicious intent. In addition to the physical evidence necessary to classify a fire as incendiary, a fire investigator may have to identify a party who is responsible for having set the fire. This identification may require research, using a consistent methodology, to determine which individuals may have had both a motive and an opportunity to commit the crime.

Incendiary Fire Indicators

Investigators must systematically test all possible indicators of incendiary fires. Categories of evidence, and their tests, include:

- **Multiple points of origin** — The investigator must prove that the multiple separate and distinct points of origin, such as lightning strikes, electrical failures, flying brands, or the flow of hot gasses from the upper layer, did not result from normal fire dynamics.

- **Trailer** — Ignitable materials used to spread fire usually leaves char or burn patterns and may be used with incendiary devices. An investigator should consider whether patterns that resemble trailers were instead caused by factors related to heavy traffic areas or flashover.

- **Incendiary devices** — Most incendiary devices leave evidence of their use, especially the metal parts of electrical or mechanical devices. More than one device may have been used, and sometimes a faulty device can be found **(Figure 13.17)**.

- **Ignitable liquid evidence** — An investigator may test for this evidence after observing patterns that could have been caused by pooled ignitable liquid foreign to a particular location.

After an investigator classifies a fire as incendiary based upon the physical evidence at the scene, answering the following questions may assist in suspect identification:

- Were there signs of forced entry?

- Are contents removed or out of place?

- Is there an absence of personal items?

- Was the fire department access blocked?

- Did the fire occur at an unusual time of day?

- Was there structural damage before the fire?

- Was a second fire started in the same structure?

- Is there evidence of other crimes in the structure?

- Does the owner's or occupant's story match the circumstances?

- Did the fires occur during inclement weather, causing a delayed response or providing a possible reason for the fire?

Figure 13.17 Components often remain as evidence after an incendiary device causes a fire.

Investigators may face legal ramifications should they indicate that factors are incendiary when they may not be. In cases where an investigator cannot determine sufficient evidence for a particular fire cause classification, the correct classification should be "undetermined." The term "suspicious" should not be used when referring to the cause classification.

Motives

NFPA 1033 (2022): 4.6.4

Identifying the motive may aid an investigator in determining the party responsible for setting an incendiary fire. Common motives, and motive indicators, associated with incendiary fires include:

- Revenge
- Vandalism
- Profit (fraud)
- Crime concealment
- Excitement (vanity)
- Extremism (terrorism)

Revenge

Fires set because of personal or professional revenge comprise the largest category of arson fires. They also account for fifty percent of arson cases.

Often a history of domestic disputes precedes this type of fire. In most cases, revenge fires occur as an impulsive reaction to some other situation. Generally, the victim can provide information regarding the suspect's identity.

Ignitable liquids are seldom used when the fire is not planned in advance. In other cases, the arsonists may utilize Molotov cocktails or "firebombs." Fires set to homes or businesses are often set to the exterior or through a broken window, targeting property such as:

- Fences
- Vehicles
- Outbuildings
- Storage rooms

Vandalism Fires

Vandalism fires are usually set by two or more individuals (usually juveniles), often with no given or apparent reason. Before setting the fire, vandals usually damage property or paint graffiti. Vandalism on the interior of a scene may indicate that entry to the building was forced. Prime targets commonly include:

- Schools
- Vegetation
- Trash containers
- Vacant buildings

Profit (Fraud)

Monetary gain is the primary motivator for profit driven incendiary fires. Firesetters motivated by fraud often attempt to cause the most possible damage in the least possible amount of time. Firesetters with a motive for profit or fraud generally have unlimited access to the interior of a structure without fear of discovery.

Fires intended to defraud often feature multiple fires and ignitable liquids. Holes broken in walls or ceilings and/or trailers are often used to assist in the fire spread; time-delay ignition devices are also common. Property owners often set elaborate fires that require significant preparation.

Personal property that cannot be easily replaced is often removed before the fire. The property owner is frequently absent from the building, and doors are found locked. Fraud fires are classically set because of poor financial status; however, the motivation for arson for profit may be quite abstract and as varied as the imagination of the firesetter. An example might include a profitable and well-established business set on fire for alternative fraud reasons such as providing a reason for a remodel of the interior. If the owner were to close for the week to complete the remodeling, revenue would be lost. If a fire were set to a business with adequate insurance, the overhead expenses, lost revenues, and remodeling fees would be paid through business interruption insurance (if the fire were not properly investigated).

Other reasons for fraud fires do not involve the property owners. The fires are generally not elaborate and if set to the interior, require the firesetter to force entry. In addition to owners or insured parties, the following individuals may set fires for economic gain:

- Firefighters to obtain overtime or call-out pay
- Competitors to drive the victim out of business
- Contractors to secure a contract for rebuilding the loss
- Insurance agents to sell insurance to uninsured persons in the area
- Individuals intending to devalue the property so that they can purchase it at a lower price

Financial Analysis

When investigating a fire set in a home or business, the investigator reviews and evaluates the owners' current financial condition. The evaluation starts with a review of the most basic factors: assets and liabilities. Any anomalies that emerge from this review should be investigated further. For example, motivating factors may include extraordinary liabilities, such as loans coming due, or diminishing assets, such as a series of significant losses **(Figure 13.18)**.

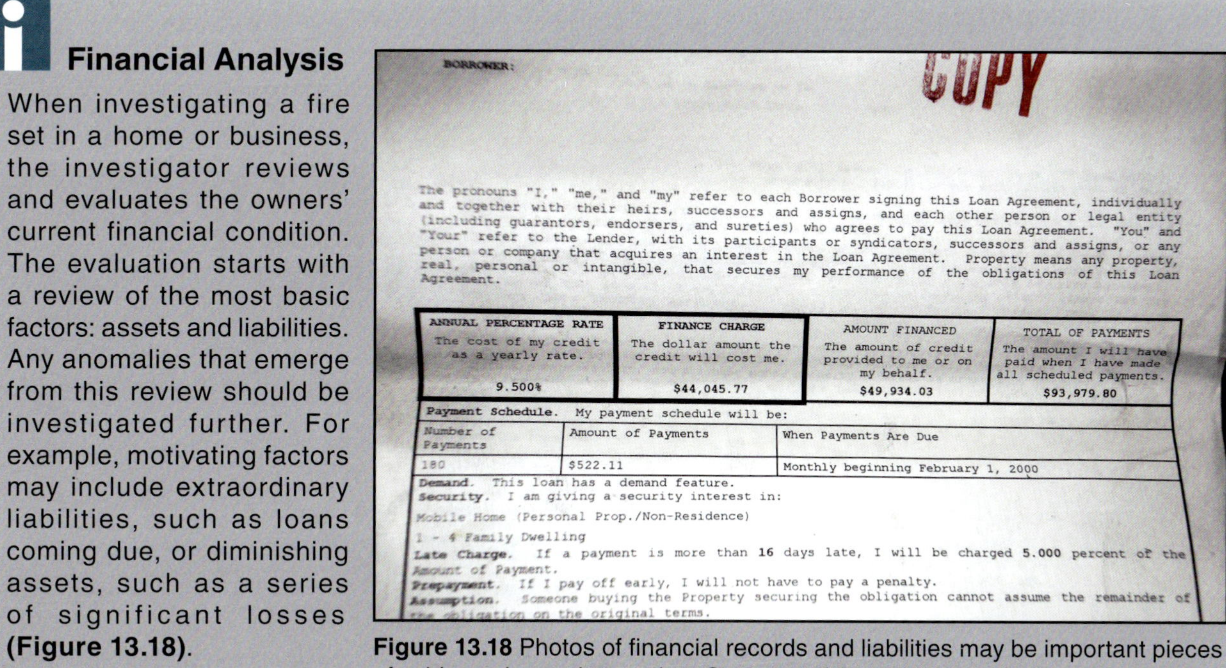

The pronouns "I," "me," and "my" refer to each Borrower signing this Loan Agreement, individually and together with their heirs, successors and assigns, and each other person or legal entity (including guarantors, endorsers, and sureties) who agrees to pay this Loan Agreement. "You" and "Your" refer to the Lender, with its participants or syndicators, successors and assigns, or any person or company that acquires an interest in the Loan Agreement. Property means any property, real, personal or intangible, that secures my performance of the obligations of this Loan Agreement.

ANNUAL PERCENTAGE RATE The cost of my credit as a yearly rate.	FINANCE CHARGE The dollar amount the credit will cost me.	AMOUNT FINANCED The amount of credit provided to me or on my behalf.	TOTAL OF PAYMENTS The amount I will have paid when I have made all scheduled payments.
9.500%	$44,045.77	$49,934.03	$93,979.80

Payment Schedule. My payment schedule will be:

Number of Payments	Amount of Payments	When Payments Are Due
180	$522.11	Monthly beginning February 1, 2000

Demand. This loan has a demand feature.
Security. I am giving a security interest in:
Mobile Home (Personal Prop./Non-Residence)
1 - 4 Family Dwelling
Late Charge. If a payment is more than **16** days late, I will be charged **5.000** percent of the Amount of Payment.
Prepayment. If I pay off early, I will not have to pay a penalty.
Assumption. Someone buying the Property securing the obligation cannot assume the remainder of the obligation on the original terms.

Figure 13.18 Photos of financial records and liabilities may be important pieces of evidence in proving motive. *Courtesy of Donny Howard, Agent, Oklahoma State Fire Marshal's Office.*

Crime Concealment

Arsonists set fires as a mechanism to destroy evidence of a crime. The three most common types of crime concealment fires are:

- Burglary concealment
- Embezzlement concealment
- Homicide concealment

Burglary Concealment. These types of fires rarely involves ignitable liquids because a burglar usually enters a structure with the intent to steal. In most cases, the arsonist sets a fire after deciding that incriminating evidence was left behind. To conceal or destroy evidence, the arsonist chooses a location where they believe evidence, such as fingerprints or blood, may have been left. Arsonists often set burglary concealment fires using combustibles on hand.

Embezzlement Concealment. Arsonists may set a fire to erase or destroy evidence of embezzlement with the origin area located near the "paper trail." Often the paperwork itself is used as the fuel with ignitable liquids sometimes used to assist in the destruction of the documents **(Figure 13.19)**. These fires may also conceal theft or destruction of computers or other repositories of digital accounting records.

Homicide Concealment. In homicide concealment fires, arsonists often use ignitable liquids in an attempt to destroy the body and the evidence of the manner and cause of death. Arsonists generally set fires on and around the body.

Figure 13.19 Financial paperwork or other evidence may be intentionally burned to make it more difficult to assign motive for a fire. *Courtesy of Donny Howard, Agent, Oklahoma State Fire Marshal's Office.*

Excitement (Vanity)

Action and excitement accompany fires and fire-suppression activities. Some firesetters commit arson to have the private satisfaction of instigating a fire service and law enforcement response.

The spur-of-the-moment fires, however, develop as a recognizable pattern over a period of time. Examples of patterns in excitement arsons include:

- **Dates and day of the week** — The pattern may follow paydays, normal work days/times, or days spent consuming alcohol.

- **Time of day** — Most arsonists set excitement fires during the hours of darkness. Times may coincide with travel to and from work or other activity.

- **Type of structure** — An arson pattern may show a conscious or unconscious focus on a certain type of structure such as schools, churches, or vacant structures.

- **How the fire is set** — Arsonists typically ignite these unplanned excitement fires with readily available combustibles. The arsonist may repeat the same ignition method due to familiarity/comfort with the process.

- **Where the fire is set** — Repeat arsonists may choose a concealed location that yielded the intended level of emergency response and lack of detection in previous experience, such as under a crawl space.

Arsonists who seek recognition, or wish to be viewed as heroes, may set and "discover" fires. These individuals are consistently present at the fire scene and often attempt to assist in fire fighting activities. They may have any employment background, but trends indicate that some have experience as security guards, volunteer firefighters, and reserve law enforcement officers. When the presence of individuals behaving in this way is observed, the investigator should check their background for past incidences of firesetting behavior.

Pyromania

Pyromania is a mental state and a recognized psychological disorder characterized by setting fires as a release of tension or to produce a feeling of euphoria. True pyromaniacs are few in number. *Pyromania* is not a motive.

Pyromaniacs seldom set fires with ignitable liquids. They typically light paper products in vehicles, in alleys, or behind buildings. Over a short period of time, it is common for a pyromaniac to set multiple small fires within several blocks of each other.

Extremism (Terrorism)

Social protest by an individual or group may target a government, an ethnic or religious group, or a facility that operates in opposition to their cause. Protestors may create fires or explosions to advertise or advance the arsonist's purpose. Although the arsonist often intends to remain unknown, the arsonist often identifies the group or cause as the responsible party by leaving graffiti or signs at the scene, and/or contacting media services. Arsonists usually set fires and explosives to the exterior of buildings or propel incendiaries (such as a Molotov cocktail) into a structure through broken windows or doorways.

Opportunity

NFPA 1033 (2022): 4.6.4

An opportunity is a set of circumstances that make an action possible. In terms of the three elements of a crime, a suspect having opportunity was available to set the fire under the conditions that occurred, or work with another individual to set the fire. Opportunity alone does not indicate guilt.

Types of information and evidence that may assist in establishing opportunity include:

- Impressions
- Latent prints
- Phone records
- Traffic citations
- Building security
- Witness statements
- Climactic conditions
- Alarm system printout
- Professional knowledge
- Time of day/day of week
- Familiarity with the facility
- Access to keys to the property
- Security or surveillance cameras
- Personal items found at the scene
- Isolation or remoteness of facility
- Access to security codes for alarm or security systems

Review Questions

1. What are the characteristics of a first material that complement an ignition source?

2. What qualities must a competent ignition source have?

3. What should the fire investigator remove to narrow the list of potential ignition sources?

4. List supporting data that can indicate the presence of an ignition source.

5. When evaluating sparks, embers, or flames from outside fires, what must the fire investigator consider?

6. List heat sources from solid fuel-powered equipment.

7. When analyzing electrical components involved in a fire, what should a fire investigator do?

8. What are common failures found in electrical conductors?

9. What metals melting at fire temperatures may combine to form alloys?

10. What is a short circuit?

11. What factors can generate localized high temperatures conditions atypical in a fire environment?

12. List examples of hot objects.

13. List materials that are considered explosive.

14. What is a smoking material?

15. What is an open flame material?

16. Why is solar energy considered a competent ignition source?

17. List sources of heat energy from separate fire sources.

18. What should an investigator do to determine the ignition source?

19. When identifying explosion ignition sources, how can an investigator eliminate possible ignition sources?

20. When determining whether an appliance could serve as a competent ignition source, what should an investigator consider?

21. To determine which hypotheses are possible versus which are probable, what should the fire investigator do?

22. What other sources are used for an investigator to obtain information?

23. From what experts can a fire investigator request assistance?

24. What is a link analysis?

25. What kind of data can be efficiently represented by a graph?

26. What does a fire investigator use inference for?

27. What is a commodity flowchart?

28. To make a final determination of the origin and cause of a fire, what should the fire investigator consider?

29. What are the categories of evidence for incendiary fires?

30. List common motives associated with incendiary fires.

31. What are the three most common types of crime concealment fires?

32. List types of information and evidence that may assist in establishing opportunity.

Chapter References

Babrauskas, Vytenis & Krasny, John. (1985). Fire Behavior of Upholstered Furniture. Final Report National Bureau of Standards, Washington, DC.. -1. Accessed online.

NIST. Organization of Scientific Area Committees for Forensic Science (OSAC). Accessed online.

Chapter Key Terms

Eutectic Alloying — When two metals form an alloy that has a lower melting temperature than either of the original metals.

Fire Cause — (1) Agency or circumstance that started a fire or set the stage for one to start; source of a fire's ignition. (2) The sequence of events that allows the ignition source of ignition and the material first ignited to come together. (3) The combination of fuel supply, heat source, and a hazardous act that results in a fire.

Ignition Sequence — History of the fire, beginning when the ignition source and the first fuel ignited meet at the origin area, and proceeding through the entire duration of fire spread through the scene.

Inference — A conclusion that is derived from a set of premises.

Material First Ignited — Fuel that is first set on fire by the heat of ignition. To be meaningful, both a type of material and a form of material should be identified.

Pot Metal — Slang term that refers to alloys that consist of inexpensive, low-melting point metals that are normally used for inexpensive castings; examples include zinc, lead, copper, and aluminum.

Premise — Proposition from which inferences may be drawn to reach conclusions about a question or problem.

Spontaneous Heating — Heating resulting from chemical or bacterial action in combustible materials that may lead to spontaneous ignition.

Static Electricity — Accumulation of electrical charges on opposing surfaces created by the separation of unlike materials or by the movement of surfaces.

NOTE: The characteristics of the fire or explosion scene will determine the scope of the examination. Responding to the particulars of the scene, learners may complete steps in this skill in a different order and repeat them as many times as necessary. The fire investigator should work alongside other professionals as the scene requires it. Complete the appropriate documentation during each step of the investigation process.

Step 1: Conduct interviews.

    a. Witnesses

    b. Owners, occupants, and employees

    c. Firefighters and other first responders

Step 2: Gather any outside reports or records that may help determine fire cause or responsibility.

Step 3: Identify the origin area.

Step 4: Identify fuels in the origin area.

Step 5: Identify potential ignition sources.

Step 6: Identify the oxidizing agent.

Step 7: Collect any evidence specific to cause determination. Take the necessary steps to maintain the chain of custody throughout the investigation.

Step 8: Observe fire pattern damage on the structure and contents.

Step 9: Analyze all of the data and evidence gathered, utilizing outside expert resources, as necessary.

Step 10: Identify potential ignition sequence(s).

Step 11: Develop hypotheses regarding the cause.

Step 12: Test cause hypotheses.

  NOTE: Test hypotheses to disprove, rather than prove them.

Step 13: Select a final hypothesis regarding the cause.

Chapter Contents

NFPA 1033 JPRs addressed in this chapter

This chapter provides information that addresses the following job performance requirements (JPRs) of NFPA 1033, *Standard for Professional Qualifications for Fire Investigator* (2022):

4.1.5	4.3.2	4.4.2	4.6.3	4.7.2
4.3.1	4.3.3	4.6.2	4.7.1	4.7.3

This chapter provides information that addresses the following course outcomes for the FESHE courses:

Fire Investigation I (C0283)

4. Describe the implications of constitutional amendments as they apply to fire investigations to include case law decisions that have affected fire investigations.

Fire Investigation II (C0284)

2. Explain the rule of law as it pertains to arrest, search, and seizure.

7. Explain the role of the fire investigator in courtroom demeanor and testifying.

Fire Investigation and Analysis (C0285)

2. Document the fire scene in accordance with best practice and legal requirements.

3. Analyze the fire scenario utilizing the scientific method, fire science, and relevant technology.

4. Analyze the legal foundation for conducting a systematic incendiary fire investigation and case preparation.

Learning Objectives

1. Describe how the fire investigator interacts with the media. [NFPA 1033, 4.7.2]

2. Explain the requirements that apply to written reports. [NFPA 1033, 4.3.1, 4.3.2, 4.3.3, 4.4.2, 4.6.2, 4.7.1]

3. Explain the litigation process in which a fire investigator will eventually participate. [NFPA 1033, 4.1.5, 4.3.3, 4.7.3]

4. Describe witness information categories at court proceedings. [NFPA 1033, 4.6.3, 4.7.3]

5. Skill Sheet 14-1: Prepare a written report of investigative findings and present those findings publicly. [NFPA 1033, 4.6.3, 4.7.1, 4.7.2, 4.7.3]

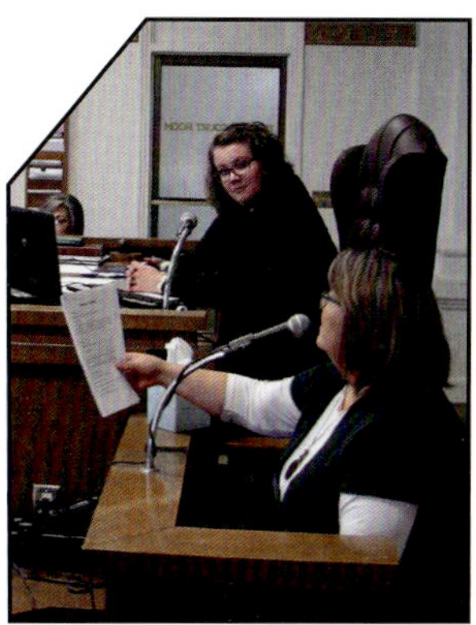

Presentation of Investigative Findings

The investigator's job does not end with the termination of investigation procedures at the fire scene. Most investigators must present some type of formal summary and presentation of their investigative findings. This chapter offers guidance to investigators on the following topics:

- Media relations
- Presentation types
- Report writing
- Legal proceedings
- Testimony at court proceedings

Media Relations

NFPA 1033 (2022): 4.7.2

Positive media relations bolster the public image of fire investigators and investigative agencies. The following sections highlight three primary ways in which interaction between the media and the fire investigator occur.

NOTE: Media interactions including interviews, conferences, and releases are not required to provide all the details of the investigation.

Media Interviews

A media interview may be planned and prearranged, or it may be spontaneous at an incident scene. Some departments have a designated **public information officer (PIO)** who maintains training and formal contacts with local news agencies **(Figure 14.1)**. The PIO may be assigned to answer media questions regarding the incident or its investigation. When a PIO is available and on scene, investigators should politely direct questions and interview requests to the PIO.

News outlets can support or complicate the work of an investigation. If a PIO is not assigned or available, investigators should follow established policies and procedures. An investigator must prepare for the challenge of interacting with media representatives when formally speaking for the response department or company **(Figure 14.2, p. 418)**.

Investigators should provide information as necessary without divulging confidential information about victims' identities or details about the ignition scenario, particularly with incendiary fires. A good practice is to include an agency's fire prevention messages in the media statement, for example, "Smoke detectors provide early notification of a fire."

Figure 14.1 The media and fire and emergency services personnel should establish a task group that follows incident scene protocols during crisis situations. *Courtesy of Ron Jeffers, Union City, NJ.*

Interview Good Practices

Good practices an investigator can use to prepare for an interview include:

- Listen to the questions carefully and thoughtfully. Only answer "yes" or "that's correct" if the facts in the questions are completely accurate, and no inaccurate conclusions have been drawn.

- Avoid arguing with reporters. A defensive attitude may suggest that information is being withheld from the media.

Figure 14.2 In organizations without a full-time PIO, a chief officer may communicate with the media. *Courtesy of Ron Jeffers, Union City, NJ.*

- Take time to prepare thoughts for the interview. Bring a clear message ready to present.

- Do not answer questions beyond your area of expertise or knowledge. Refer such questions to those who have the necessary information or offer to find the answer — and *always* follow through.

- Avoid using fire service terminology. Explain any technical terms as they are used. For example, when describing the difference between an incendiary fire and an arson, be sure to also explain the significant factors in each determination.

- Rehearse your interview technique, and try to improve delivery. Repeat key points so the message is clear.

- Prepare phrases that can be repeated as necessary, such as: "The fire is under investigation," and "If you have any information, please call."

Interview Preparation

An investigator may not be able to prepare for all situations that occur during an interview. Preparing responses to scenarios such as the following will help an investigator maintain control of the message:

- There is no such thing as "off the record." Expect that anything said to a reporter will be quoted.

- Hypothetical ("what if") questions invite unproven facts. Explain that fire investigator(s) are actively working and answers may not be available at that time.

- Speculation may include phrases such as, "The fire is suspicious." A rebuttal can be, "The fire is under investigation," especially if the fire cause has not been determined.

- Listen for false or misleading information in reporters' questions. If a question contains false information, politely identify the errors and correct the facts.

- Beware of forced-choice questions. If the answer is "none of the above," that phrase can precede an explanation of the facts.

Press Conferences

Unlike a media interview, a press conference is a scheduled event. When possible, press conferences should be timed to reach the public efficiently. For example, scheduling a press conference at 10 a.m. will usually ensure that the story meets deadlines for the evening news and the next day's print media. A best practice is to start a press conference five minutes after the scheduled time. This slight delay helps to ensure that everyone is present and prepared to begin.

Before the press conference begins, the investigator should decide whether to take questions at the end of the conference. When an investigator expects to answer questions, best practices in this type of media opportunity include:

- Where possible, ask members of the media to submit questions prior to the press conference.

- Provide full names and positions (or ranks) of individuals presenting the information or mentioned in the conference, in writing, to the media.

- Provide a written media release with pertinent dates, times, names, completed activities, pending activities, and future activities. (See the section on Media Releases for additional guidance.)

- Arrange for knowledgeable people to answer questions related to the incident or the investigation.

Questions Regarding Incendiary or Cause Not Determined Fires

Before releasing any information, the investigator should be aware of how the information may be used. For example, an investigator may discover evidence that a fire was intentionally set. The decision to release this information depends on factors including the circumstances surrounding the fire.

In most cases, investigators are advised to indicate that the investigation is ongoing until the final report is complete. More valuable information may be obtained during the course of interviews when the public believes that the investigation is still underway. Further, releasing an indication of arson evidence may not benefit the public and may embolden a firesetter.

In other situations, informing the public of an incendiary fire may assist in identifying a firesetter or limiting risk. For example, if a serial arsonist is in the area, the public must be notified to reduce the risk of additional fire sets.

Regardless of any circumstance, investigators should avoid using the word *suspicious*. Indication of "suspicion" may suggest "suspicion of arson" to the media and the public. This can have negative unintended consequences, such as fostering an unwarranted belief that the victim is a suspect, even if the fire cause is ultimately ruled accidental or undetermined.

Media Releases

Media/press releases present information in a ready-to-use news story format. The investigator may be asked to provide information regarding the investigation to a PIO or company officer who may be preparing a media release on the incident. In other cases, the investigator may be tasked to generate the media release. When releasing information regarding an investigation, the investigator must consider the legal and ongoing investigation issues associated with the incident. The following sections describe best practices with media releases.

Parts of the Media Release

The media release generally consists of the heading and the body. The body or content of the media release contains the facts that the investigator wants the public to know. The heading includes:

- Subject of the media release

- Name and telephone number of a contact person who can respond to questions

- Date the media release is being disseminated and when the media release can be used

Improving Media Releases

After crafting a release, the investigator should confirm that the document meets the following metrics per the authority having jurisdiction:

- **Acceptable** — The release should meet the needs and preferences of the publishing media.

- **Targeted** — The information should be kept to need-to-know information for a specific audience.

- **Appropriate** — Confirm that all data is accurate factual and free of grammatical and spelling errors.

- **Formal** — The use of company or department letterhead will convey that the release is authorized by the indicated organization.

- **Readable** — Ask at least one other person to proofread the media release to ensure transitions are logical and conclusions match the evidence.

Style guidelines to improve the readability and "human interest" qualities of the release include:

- Write in a narrative style.

- Ask the media service to ask for a follow-up interview or submit written questions as interest merits.

- Include quotes from decision makers and those involved in the incidents whenever possible. Multiple perspectives provide a wider overview of the situation.

When the release is ready for distribution, the investigator should consider guidelines to maximize the reach and effectiveness of the release:

- Include appropriate internal stakeholders, such as chief officers, administration, and elected officials, in the distribution of media releases.

- Send the media release to all local media outlets. Research non-English language media options, as appropriate. Information distribution patterns indicate interest in the included communities, which can open dialogue with those communities.

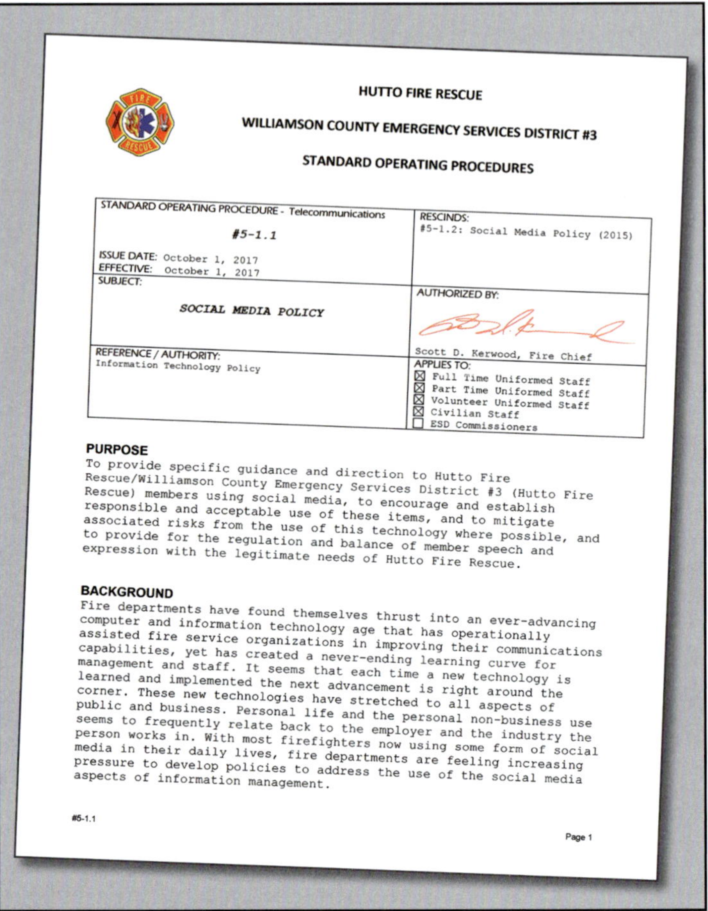

Figure 14.3 A social media SOP is an internal document that can communicate social media expectations.

- Distribute media releases well ahead of news deadlines, via an outlet approved by all parties **(Figure 14.3)**. Consider whether postal mail will arrive quickly enough to remain relevant. Alternative examples may include electronic mail (email), social media, or hand delivery.

Report Writing

NFPA 1033 (2022): 4.3.1, 4.3.2, 4.3.3, 4.4.2, 4.6.2, 4.7.1

Investigators should understand and comply with the requirements that apply to written reports. **Skill Sheet 14-1** shows the steps for preparing a written report of investigative findings and presenting those findings publicly. Investigative files, reports, and analysis inform other data analysis and interventions such as:

- Involvement of products, appliances, or occupancies

- Fire, building, or other codes, standards, and safety practices

- Fire litigations or prosecutions

- Fire statistics and analysis

Investigative Files

NFPA 1033 (2022): 4.6.2

Compiling an investigative file should be viewed as the logical extension and conclusion of each fire or explosion investigation. A complete investigative file contains data that will form the basis of many decisions over time, especially when that information is collected and presented in a way that is accessible and reliable.

Fire and explosion investigation and data analysis is an integral part of the fire protection community. The accuracy of the data and information contained within it is paramount to any other concern.

Investigative File Components

The investigative file is a compilation of all unedited data related to the incident. This normally includes all media in all formats, spanning from the incident occurrence, through response, and ending with the conclusion of the investigation. Media can include:

- Typed or handwritten documents, statements, and notes
- Electronic documents
- Sketches/diagrams
- Photographs
- Videotapes
- Audio recordings
- Magnetic or digital recordings

NOTE: The investigative file must include all unedited photographs taken on scene. Not all photographs from the file may be needed in the report.

Within each investigative file, the investigator should include:

- Nature of the incident scene
- Consequences of the occurrence of the incident
- Any known associated investigations and analysis
- Findings and conclusions drawn from the investigation and analysis of evidence

Policy for Collecting Data at an Investigation

During an investigation, investigators are advised to follow a consistent methodology, possibly based on the scientific method. As a best practice, that methodology should be documented in the investigation notes. Any relevant written laws, ordinances, policies, and procedures should also be noted. The role and responsibility of investigators and their employers may have bearing on the specifics of the written policy and procedures.

Policy for Reviewing Investigative Files

When an investigative report is complete, an investigator should solicit a review to ensure the accuracy and completeness of the findings and conclusions. This accountability provides a level of scrutiny to ensure that the report is understandable by someone other than the author.

Peer review can be accomplished through informal collaboration with other investigators, formalized review processes by supervisory personnel, or the intended audience/recipient of the investigative file. When this step is complete, the report should be defensible and withstand scrutiny from other interested parties. The investigator should be able to use this report to recall all aspects of the investigation and analysis at an indeterminate future date. The peer review concept initially appears in Chapter 1, Organization, Responsibilities, Authority.

Investigative files may include data that can later aid functions such as:

- Reduction of property damage
- Prevention of similar incidents in the future
- Detection and mitigation of product safety hazards
- Discovery and assessment of responsibility for criminal activities
- Development, evaluation, and enforcement of fire safety and life safety codes

Investigative Reports

NFPA 1033 (2022): 4.7.1

As discussed later in this chapter, a thorough written report is necessary for the investigator's qualification as an expert able to give **opinion testimony** at trial in many jurisdictions, including the U.S. federal court system **(Figure 14.4)**. Written reports may also be required from any person wishing to be qualified as a public or private **expert witness**.

Figure 14.4 A prosecutor may call an investigator to serve as an expert witness to present an opinion as credible testimony.

NOTE: In Canada, many provincial, Supreme and/or Courts of Queen's Bench require a written report from witnesses at a trial.

Purpose of Reports

The primary purpose for writing a report is to relay all relevant information obtained during a particular fire investigation. A thorough and professional investigative report may increase the chance of a successful disposition of the case.

The report's style should be clear and concise, focused on the details necessary to continue the investigation. The investigator should maintain a neutral focus when determining what information should be included. Any opinions must tie directly to established facts.

Reasons to Write Reports

Although a written report may not be requested at the beginning of the fire investigation, one may be required later, sometimes years after the investigation. One reason an investigator should use a consistent methodology across investigations is to aid in writing the report at some remove.

Reports are generally required of investigators employed by public agencies, such as fire departments. Public agency reports are often considered public record. Parties that may read these documents with special interest include victims, supervisors, attorneys, and insurance representatives.

Investigators in the private sector are not always required to write reports. This necessity is generally defined by agency or company policy, the entity on whose behalf the fire is being investigated, or judicial requirements. Reports may also be required due to a court decision.

In addition to court requirements, departmental and company policies may dictate the style and format of the report. Local policy dictates to whom, if anyone, the report will be addressed. Some reports are addressed to the fire chief, the fire marshal, law enforcement personnel, or any other person as directed by standard operating procedures. Other reports are a simple narrative that is not addressed to anyone. In either case, the report should be a factual representation of the investigation.

Required Report Components

Because the report itself may be challenged by an outside agency, it should be able to stand alone. As indicated throughout this manual, the investigator should document the use of a consistent methodology (systematic approach). Standards and the scientific method can be used as a template to build the report. The investigative report should follow a consistent methodology, and refer to that methodology to show compliance, especially if the methodology is based on a standard. The report should explain the importance of the methodology and how it was applied to the investigation.

NOTE: Investigators may not know when their reports are challenged.

When an authority having jurisdiction requires an investigation report, that authority will generally set parameters for the report. One widely recognized model for report requirements is included in NFPA 1033. An investigator should follow the requirements of NFPA 1033 if report requirements are not otherwise made available from the AHJ.

NOTE: NFPA 1033 A.4.7.1. cites ASTM International (formerly American Society for Testing and Materials) documents that guide an investigator in reporting opinions of scientific or technical experts and reporting incidents that may involve criminal or civil litigation.

NFPA 1033 requires that a written report contain:

- Concise, error-free writing
- Investigator's opinions, clearly expressed
- Accurate reflection of the investigative findings
- Facts and data from which the investigator derived conclusions
- An indication of the intended audience and meets their needs or requirements
- An explanation of the reasoning and methodology behind the investigator's opinions

The written report should communicate the *observations*, *analyses*, and *conclusions* of the investigator. Instead of suggesting a specific format, NFPA 921 emphasizes the content of the written report should contain:

- Date the report was submitted
- Date and location of the examination(s)
- Date, time, and location of the incident
- Name of the person or entity requesting the report
- Scope of the investigation including the tasks completed
- Name of the person preparing the report
- Nature of the report (preliminary, interim, final, summary, supplemental)

In addition, peer review notes and participants should be noted per the AHJ. After meeting report requirements, the remainder of the report is meant to provide facts and explanations that support the investigator's opinion. The investigator will already have compiled extensive notes, sketches, photos, and other forms of scene documentation. The challenge of writing a finalized report about the scene is to take from that documentation the information that best illustrates the findings. The report should include enough information from the documentation to both:

1. Show how the investigator came to form conclusions and opinions about the origin of, cause of, and responsibility for the fire.

2. Provide enough information that a second party or investigator can evaluate conclusions for their accuracy.

Documented information that may be relevant to a report includes some or all of the following:

- Photographs
- Fire patterns
- Fire dynamics
- Items examined
- Laboratory reports
- Evidence collected
- Witness information

- Computer fire models
- Fire scene description
- Diagrams and sketches
- Investigative methodology
- Potential ignition sources examined, considered, and eliminated

- Observations and information relevant to any opinions expressed
- Other incident or investigative reports, such as separate or previous investigations

Photographs

NFPA 1033 (2022): 4.3.2

Well-chosen photographs contribute context in a way no other medium can **(Figure 14.5)**. In a report, photographs can provide context or otherwise support opinions and conclusions. The documentation of evidence usually includes sequential photographs from the time and location of an item's discovery through its preservation and complete examination.

When writing a report, investigators should select photos that illustrate the in-text conclusions and opinions. These photos may be included as attachments or appendices in an investigative report.

When adding photos, an investigator should consider adding arrows, lines, circles or blocks of text to the photos to assist the reader in understanding what is being depicted. In addition, descriptions of the photographs themselves can provide further clarity.

Figure 14.5 A photo can convey an immersive sense of the scene. *Courtesy of Dan Madrzykowski, NIST.*

As discussed earlier in this manual, the field notes photo log should include unique references and descriptions for each photo. When compiling the report, the investigator should include a report-specific photo log with the field note references. The usage order of report photos does not need to match the chronological order of the field notes.

Scene Sketches and Diagrams

NFPA 1033 (2022): 4.3.1

As discussed earlier in this manual, sketches and diagrams in a report are used to illustrate a concept from the scene.

Examples:

- Explosion dynamics
- Directional fire spread and the fire's origin
- Location from which evidence was removed
- Room identifications and interior dimensions
- Location of contents, appliances, and utilities
- Location and position of victims or witnesses

A field sketch from scene documentation can be turned into a diagram. In some cases, multiple sketches can be compiled onto one diagram. Diagrams can be created using a number of readily available, inexpensive computer programs.

Diagrams included in a report can be used to show context including several different factors **(Figure 14.6 a and b)**. Large scene diagrams, such as those from explosion scenes, may also include GPS coordinates of evidence. Other diagrams used to support report findings may include:

- Arc surveys or mapping
- Depth of char measurements

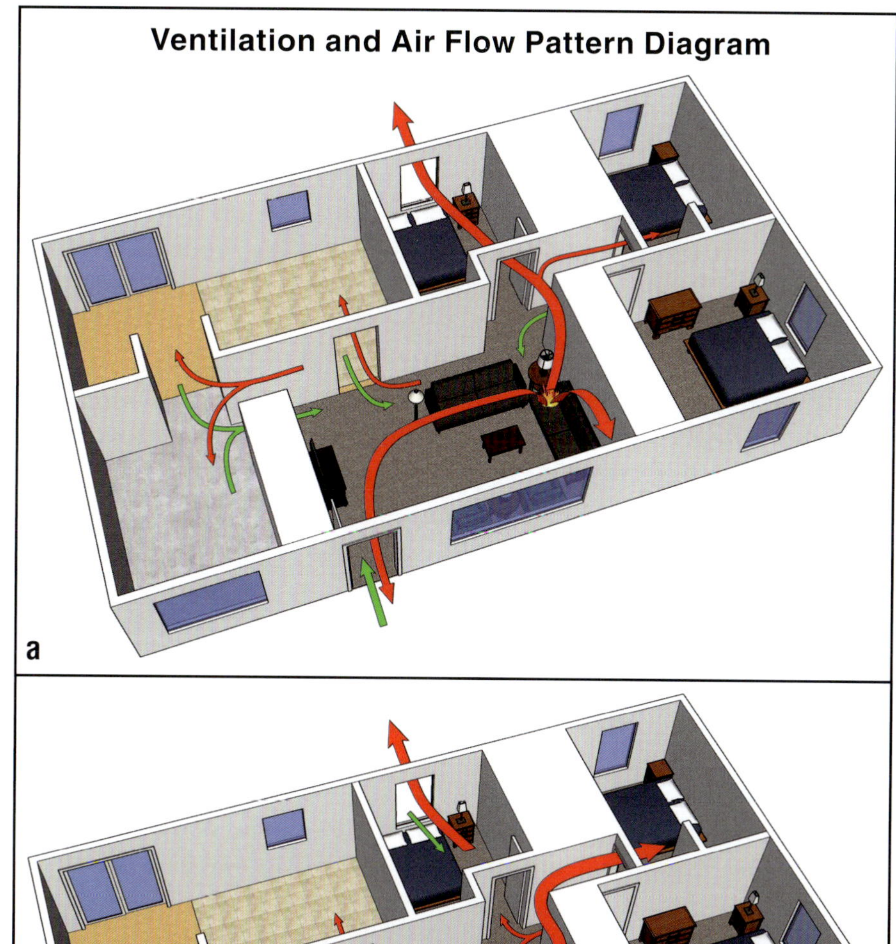

Figure 14.6 a-b Diagrams used in a report can show context, change over time, and effects of changes in the scene. *Courtesy of Dan Madrzykowski, NIST.*

- Heat and flame vector analysis
- Explosion effect analysis (including the location of items from explosion debris)

Scene Evaluation Notes

NFPA 1033 (2022): 4.3.3, 4.4.2

While investigators evaluate a scene, they should remember that their work may be reviewed and potentially challenged in a court of law. Any fire scene will present a substantial amount of evidence. An investigator's scene evaluation notes should be detailed enough to help the investigator recall the incident scene and/or produce a report at a later date. Because these notes are for the investigator's personal use, they may include shorthand.

When selecting information from notes to include in the written report, the investigator should be careful to include only relevant details. For the purposes of the report, the investigator must be able to recognize different types of evidence and determine whether a particular piece of evidence is critical to the investigation.

Any conclusions stated in the report must be correlated to observations and actions taken during scene examination. For example, scene examination field notes may include several steps taken to examine the front door and its mechanisms. In the report itself, the investigator may simply indicate the position of the door and whether it appeared forced or otherwise damaged.

Interviews / Witness Information / Electronic Data

Relevant information from interviews obtained during the investigation should be included in the report. Some interviews will already be in a written statement or transcript. Audio and video interviews may also be included and referenced in the report.

In the report, the investigator normally summarizes information obtained from interviewing witnesses. If an investigator wants to use direct quotes from interviewees in the report, the quotes should be attributed to the correct individual and restated exactly.

Information from electronic data that may have been collected as part of the investigation should also be included in the report. Surveillance camera footage and security alarm data are examples of electronic data that may be relevant to the investigation of the origin or cause of the fire.

Fire Dynamics and Patterns

The report should contain information on the analysis of the fire's effects and patterns. This information may include arc mapping that is relevant to the origin of the fire and has been collected during the scene examination.

The investigative report should also include an analysis of the fire dynamics that were present during the fire and that assisted the investigator with determining the origin of the fire. This analysis should explain the fire's origin, growth, and spread. Fire patterns and dynamics are discussed extensively in Chapters 3, Fire Dynamics, and 11, Explosion Dynamics and Investigation.

Legal Proceedings

NFPA 1033 (2022): 4.1.5, 4.3.3, 4.7.3

All investigations should be conducted under the premise that the investigator will eventually provide sworn testimony in criminal proceedings. Investigations of noncriminal actions, such as code violations, product defects, and/or follow-up investigations to criminal proceedings, may result in **civil proceedings**. The following sections highlight the litigation process.

Civil Proceedings

A civil proceeding may result if one or more parties alleges a **civil wrong** or *tort*, a breach of contract, or product liability against another person or entity. For example, in the case of a defective product that resulted in injury, the owner may decide to initiate a civil proceeding.

Civil Proceeding Example

John Doe purchased a new electric heater from a local retail mall. After years of use, the heater suddenly malfunctioned, ignited, and damaged the home. Mr. Doe was also burned as he attempted to extinguish the fire. John Doe later filed a civil lawsuit against ABC Corporation, the manufacturer of the heater, and XYZ Corporation, the retail store that sold him the appliance. The civil case alleged that the ABC Corporation manufactured an unsafe product and that XYZ Corporation sold that unsafe product.

In civil proceedings, the **plaintiff** is the person or party filing the lawsuit, and the **defendant** is the person or party being sued. The plaintiff is normally required to establish through a **preponderance of evidence** that the defendant did, in fact, commit the alleged wrongdoing, which resulted in damage to the plaintiff.

Discovery is the pretrial disclosure of pertinent facts or documents by one or both parties. Plaintiffs may use discovery documents to prove an allegation. Defendants may use these documents to disprove the allegation and show they are not responsible. Time frames for the discovery process vary among jurisdictions, so investigators should consult local attorneys for specific information.

Rules governing the disclosure documents also vary among jurisdictions. Although some state jurisdictions do not require either party to provide reports from their expert witness(es), U.S. Federal Rule 26 requires a written report from each testifying expert involved in the action. In Canada, civil procedure rules vary by province. Investigators in each province should consult with a representative of the courts to obtain a copy of applicable civil procedure rules.

Rule 26, Duty to Disclose; General Provisions Governing Discovery

Rule 26 (2020) presents the legal requirements for introducing expert testimony in U.S. federal court. The rule outlines the requirements attorneys must satisfy in order to have an expert, such as a fire investigator, take the stand. Generally speaking, this rule applies most frequently to private investigators asked to give their expert opinion during civil proceedings.

One section of the document has particular bearing for fire investigators. According to Rule 26(a)(2), Disclosure of Expert Testimony, a written report is required to accompany the disclosure of expert testimony:

"Unless otherwise stipulated or ordered by the court, this disclosure must be accompanied by a written report — prepared and signed by the witness — if the witness is one retained or specially employed to provide expert testimony in the case or one whose duties as the party's employee regularly involve giving expert testimony. The report must contain:

(i)   a complete statement of all opinions the witness will express and the basis and reasons for them;

(ii) the facts or data considered by the witness in forming them;

(iii) any exhibits that will be used to summarize or support them;

(iv) the witness's qualifications, including a list of all publications authored in the previous 10 years;

(v)   a list of all other cases in which, during the previous 4 years, the witness testified as an expert at trial or by deposition; and

(vi) a statement of the compensation to be paid for the study and testimony in the case."

Depositions

A **deposition** is an out-of-court testimony made under oath and recorded for later use in court. Most jurisdictions provide rules indicating that each party to the suit may depose the opposing party's witnesses.

Deposition questions are typically based on the facts surrounding the allegation. However, if the witness is an expert, deposition questions may focus on opinions about the allegation.

Each party uses the deposition testimony to prepare for the trial. The deposition testimony itself may be used during the trial if the witness is not available, or if trial testimony differs from the deposition testimony. The process of showing the judge or jury that a witness has changed his or her opinion or prior testimony is known as **impeachment**.

NOTE: Canada does not use the term deposition to describe a portion of the discovery process. The deposition process falls under the more general category of "discoveries."

Trials

During the trial phase of the civil proceeding, the judge makes determinations based on law as to the evidence that will be presented to the jury. Based on their individual interpretation of the law, judges may exclude all or part of a person's testimony and/or items of physical evidence.

In a **bench trial**, both parties in the case must mutually agree to have a judge serve as the sole **trier of fact** (**Figure 14.7**). If the parties do not agree to a bench trial, the decision will go to a **jury trial**. The number of jurors impaneled and the number required for a **verdict** vary by jurisdiction. Most jurisdictions do not require a unanimous decision from the jury.

Figure 14.7 A judge can serve as the sole trier of fact, given the appropriate conditions.

As the plaintiff has the burden of proof, the plaintiff's evidence must be presented first. Next, the defense may provide its evidence. The matter is presented to the trier of fact (judge or jury) for a determination or verdict. As previously stated, the plaintiff must establish its allegation by a preponderance of evidence. The final outcome of the case may be a judgment for either side.

Criminal Proceedings and Criminal Classifications

Criminal proceedings vary according to jurisdiction. Investigators should consult the prosecuting attorney's office for complete information regarding the local process. Criminal proceedings are usually initiated as a result of:

- Probable cause arrest
- Grand jury indictment
- Service of an arrest warrant

These actions are the result of an investigation establishing that a crime has occurred and that sufficient probable cause is present to believe that the arrested or indicted party committed the crime. Probable cause is a logical belief that sufficient information or facts indicate that the party is responsible. This level of proof is not to be confused with the level necessary for a finding of guilt by the trier of fact. Whether the crime is a **misdemeanor** or a **felony** determines the potential punishment a party could receive if found guilty and how the case will proceed through the justice system.

NOTE: Canada and some states in the U.S. do not have grand juries. Instead, the police must present their case to the office of the prosecutor who will decide if there is enough evidence to proceed.

In most jurisdictions, investigators should be familiar with types of criminal violations including:

- **Summary offenses**
- Misdemeanors (designation in U.S. only)
- Hybrid offenses that may be prosecuted as either summary offense or indictable offense (designation in Canada only)
- Felonies (designated an indictable offense in Canada)

The fire investigators should clarify whether they have legal authority to write a citation. If not, the investigator may need to call a police officer to the scene to perform the task.

NOTE: In Canada, the counterpart to police officers for most functions are peace officers.

Summary Offense

A summary offense is the lowest form of offense in most legal systems. Summary offenses are minor infractions of laws or local ordinances. Persons who commit a summary offense typically receive a written citation of the violation at the moment the violation occurs. They have the choice of paying the fine that accompanies the citation or challenging the citation before a judge. The judge may uphold the citation or dismiss it.

The most common forms of summary offenses include speeding or parking violations and similar offenses. In some jurisdictions, unauthorized controlled burns (trash or leaves) are considered summary offenses, and the fire investigator may be involved in writing the citations.

Misdemeanor (USA Only)

A misdemeanor is generally viewed as a lesser crime. Misdemeanors are punishable by a fine or a term of less than one year in jail or prison. An arrested person at this level of charge is given an initial court appearance and an opportunity to plead guilty or not guilty. An arrested person who pleads guilty may be sentenced to a fine or incarceration. An arrested person who pleads not guilty will be given a date for a trial by the judge or jury.

Felony (Canada – Indictable Offense)

A felony, or indictable offense, is a serious crime. Felonies are punishable by a fine, incarceration, and/or death, depending on the severity of the offense and the jurisdiction. A person charged with a felony may be entitled to a **preliminary hearing** at which time the prosecution must establish that a crime has been committed and that there is probable cause to believe the defendant committed the crime. Individuals charged with a felony have a right to a jury trial unless they waive that right.

Testimony at Court Proceedings

NFPA 1033 (2022): 4.6.3, 4.7.3

Testimony is an affirmation or declaration made under oath, typically in a court of law. Sworn testimony is verbal information attested to be true under penalty of **perjury**. Witness information is normally divided into categories based on the type of information they present in court.

Fact Witness

Fact witnesses answer questions posed by an attorney regarding what they saw, heard, touched, smelled, or tasted. These witnesses are not allowed to provide their own interpretation of those facts. Fact witnesses are not allowed to provide opinion testimony. For example, a **fact testimony** (also known as *lay testimony*) may be a witness's description of looking through a front window and seeing a fire on a couch in a corner of the living room.

Expert Testimony

NFPA 1033 (2022): 4.6.3

An expert witness is generally defined as a person with sufficient skill, knowledge, or experience in a given field to be trained in drawing inferences or reaching conclusions or opinions that an average person would not be competent to reach. The expert's opinion aids the trier of fact in understanding the facts at issue and assists them in the search for the truth, also known as a verdict. In contrast to the fact or lay witness, an expert witness is allowed at the discretion of the court to provide opinion testimony. For example, an expert testimony may be an investigator's assertion that the point of origin was the living room couch and the ignition source was electrical as a result of the couch leg resting on the lamp extension cord **(Figure 14.8)**.

Figure 14.8 An expert witness can offer an opinion as credible testimony.

The investigator's evidence must be relevant and reliable, thereby allowing it to be admissible. The trial judge has sole discretion as to what evidence is relevant and/or reliable. What is admissible may be open to interpretation between courts or jurisdictions.

United States Supreme Court decisions including the Daubert Decision (1993) and the Carmichael Decision (1999) have particular relevance to the fire investigator. Although these Supreme Court decisions are federal cases, they are widely accepted as a standard within many state courts.

A trial court judge that recognizes the Daubert and Carmichael Decisions has the responsibility as a gatekeeper to allow or disallow expert testimony utilizing accepted criteria. That criteria may include the reliability of the methodology used by the investigator and the general acceptance within the fire investigation community of that methodology.

Review Questions

1. How should an investigator prepare for a media interview?
2. How should an investigator prepare for a press conference?
3. What are the parts of the media release?
4. What data analysis and interventions are informed by investigative files and reports?
5. What are the components of an investigative file?
6. What are written reports required for?
7. How should the written report style be conceived?
8. How do the photographs complement the written report?
9. How should the scene evaluation notes be used to support the investigative report?
10. How is witness interview information used in the investigative report?
11. In a civil proceeding, who is the plaintiff?
12. In a civil proceeding, who is the defendant?
13. Who is normally required to establish, through a preponderance of evidence, what the defendant did?
14. During a deposition, what are the questions typically based on?
15. During trials, how do judges make determinations?
16. How is a criminal proceeding initiated?
17. What are the types of criminal violations?
18. What is a summary offense?
19. What is a misdemeanor?
20. How is a felony punished?
21. What is a testimony?
22. What is a fact witness?
23. What is an expert testimony?

Chapter References

Daubert v Merrell Dow Pharmaceuticals, Inc. 1993. 509 U.S. 5709.

Kumho Tire Co. vs Carmichael, 526 U.S. 137 (1999).

"Rule 26. Duty to Disclose; General Provisions Governing Discovery." Revised 2020. Federal Rules of Civil Procedure, Title V Disclosures and Discovery. Accessed online.

Chapter Key Terms

Bench Trial — Court proceeding in which the judge alone acts as the trier of fact.

Civil Proceeding — Legal action to enforce or declare a right, or to recover money or property.

Civil Wrong — Wrongdoing for which an action for damages may be brought. *Also known as* Tort.

Defendant — Party accused of alleged wrongdoing in a civil proceeding, or the party accused of a crime in a criminal proceeding.

Deposition — Process during which the witness answers questions under oath posed by the attorneys for each party; sworn testimony taken out of court. Not specifically delineated as a separate portion of discovery in Canada.

Expert Witness — Person with sufficient skill, knowledge, or experience in a given field so as to be capable of drawing inferences or reaching conclusions or opinions that an average person would not be competent to reach.

Fact Testimony — Presentation of personal observations by a witness. *Also known as* Lay Testimony.

Felony — Serious crime punishable by a fine, incarceration, and/or death, depending on the severity of the crime and the jurisdiction. *Also known as* Indictable Offense.

Impeachment — To challenge the credibility or validity of a witness or evidence.

Jury Trial — Court proceeding in which a jury acts as the trier of fact.

Misdemeanor — Lesser crime usually punishable by a fine or a term of less than one year in jail or prison.

Opinion Testimony — When a credentialed expert takes the stand to present not just factual information but interpretation of those facts based upon his or her area of expertise.

Perjury — Lying under oath in a court or legal proceeding.

Plaintiff — Party filing suit in a civil proceeding.

Preliminary Hearing — Court proceeding in which the prosecution must establish that a crime has been committed and there is probable cause to believe the defendant committed the crime.

Preponderance of Evidence — In civil proceedings, collection of evidence that proves an alleged civil wrong is probable.

Public Information Officer (PIO) — (1) Member of the command staff responsible for interfacing with the media, public, or other agencies requiring information direct from the incident scene. *Also known as* Information Officer (IO). (2) Person or individual who has demonstrated the ability to conduct media interviews, prepare news releases, and provide advisories.

Summary Offense — Lowest form of offense in most legal systems; minor infractions of laws or local ordinances. Offender is typically served a citation where the violation occurred.

Trier of Fact — Party that determines the facts in a legal case. In a jury trial, the jury is the trier of fact; in a bench (non-jury) trial, the judge is the trier of fact.

Verdict — Decision made by a jury on matters submitted to them by a judge. *Also known as* Discovery of Truth.

NOTE: This skill sheet provides basic criteria for report writing. The format of the report may differ based on local SOPs. Policies and procedures for presenting investigative findings vary greatly according to the AHJ and the factors involved in the incident itself.

Step 1: Evaluate investigative notes, photographs, sketches, interviews, and any other incident documentation.

Step 2: Write the report introduction.

Step 3: Describe the methodology used to reach conclusions.

Step 4: Summarize investigative findings.

Step 5: Select the appropriate photographs to include with the report and complete a photo log.

Step 6: Include any applicable scene sketches or diagrams.

Step 7: Include any applicable witness interview information.

Step 8: State investigator's conclusions.

Step 9: Review the report for clarity, grammar, and spelling errors.

Step 10: Sign the report.

Step 11: Verbally and visually present the report and any related investigative information in a way that is appropriate for the audience.

NOTE: This may include testimony during legal proceedings.

Photo courtesy of Donny Howard, Agent, Oklahoma State Fire Marshal's Office.

Contents

NFPA 1033 Competencies	Chapter References	Page References
4.1.1	1, 10	14-16, 287-288
4.1.2	1, 7	9-11, 14-16, 205-208
4.1.3	2	22-36, 51
4.1.4	1	12-14
4.1.5	1, 14	14-17, 426-429
4.1.6	1, 2	9-11, 21-22
4.1.7	1, 2, 3, 6, 8, 9, 11, 12, 13	9-11, 36-48, 55-92, 167-200, 229-231, 257-278, 281, 315-338, 347-350, 396-398
4.2.1	2, 7, 9, 10, 11	22-48, 208-213, 219-223, 225, 257-261, 264-267, 285-286, 332-333
4.2.2	2, 3, 4, 5, 7, 9, 10, 12	22-24, 36-44, 55-92, 99-135, 148-151, 154-160, 208-213, 219-223, 257-278, 285-286, 289-291, 311-312, 347-350, 361-376
4.2.3	3, 4, 5, 7, 9, 10, 12, 13	55-92, 99-135, 148-151, 154-160, 208-223, 257-278, 291-293, 311-312, 347-350, 361-376, 386-396, 413
4.2.4	3, 4, 10, 11, 12, 13	55-92, 128-135, 307-308, 311, 361-376, 380, 396-398, 413
4.2.5	3, 4, 10, 12, 13	55-92, 99-135, 307-308, 311-312, 351-376, 380, 400-406, 413
4.2.6	3, 8, 9, 10, 13	55-62, 229-231, 236-249, 257-278, 281, 287-288, 302-307, 384-398, 413
4.2.7	10, 12, 13	307-308, 361-377, 380, 383-384
4.2.8	2, 4, 5, 6, 10, 12, 13	45-48, 113-128, 141-163, 167-200, 289-291, 311-312, 361-369, 369-376, 380, 386-396, 413
4.2.9	11, 12, 13	315-338, 376-377, 380, 413
4.3.1	8, 13, 14	229-236, 252, 253, 413, 420-426
4.3.2	8, 13, 14	236-249, 252, 253, 413, 420-426
4.3.3	8, 14	231-232, 234-249, 420-429
4.4.1	7, 9, 10	219-223, 257-278, 287-288, 294-302, 312
4.4.2	5, 7, 9, 10, 11, 13, 14	154-163, 208-213, 257-278, 281, 285-286, 291-293, 302-307, 332-333, 413, 420-426
4.4.3	9, 13	261-264, 267-278, 281, 400-406, 413
4.4.4	9, 10, 13	257-264, 267-278, 281, 294-302, 312, 413
4.4.5	9	261-264, 281
4.5.1	5, 7, 13	161-163, 213-218, 413
4.5.2	7, 8, 13	213-218, 249-250, 413
4.5.3	7, 13	213-218, 413
4.6.1	5, 13	154-163, 400-406, 413
4.6.2	13, 14	413, 420-426
4.6.3	9, 10, 13, 14	281, 287-288, 400-406, 413, 429-430
4.6.4	13	407-411, 413
4.6.5	7, 10, 12, 13	205-208, 213-218, 307-308, 345-350, 376-377, 380, 383, 398-406, 413
4.7.1	14	420-426
4.7.2	14	417-420
4.7.3	6, 14	199-200, 426-430

FESHE Course Outcomes, Fire Investigation I (C0283)	Chapter References
1. Demonstrate the importance of documentation, evidence collection, and scene security process needed for successful resolution.	7, 8, 9, 10
2. Understand and demonstrate the process of conducting fire origin and cause investigation.	1, 13
3. Identify the responsibilities of a firefighter when responding to the scene of a fire, including scene security and evidence preservation.	7, 9
4. Describe the implications of constitutional amendments as they apply to fire investigations to include case law decisions that have affected fire investigations.	14
5. Define the common terms used in fire investigations.	1
6. Explain how the basic elements fire dynamics, construction, and fire protection systems as to how they affect origin and cause determination.	3, 4, 5, 10
7. Discuss the basic principles and identify cause and origin of fires.	11, 12, 13
8. Recognize potential health and safety hazards.	2

FESHE Course Outcomes, Fire Investigation II (C0284)	Chapter References
1. Recognize the need for the use of the scientific method for investigations.	1
2. Explain the rule of law as it pertains to arrest, search, and seizure.	14
3. Describe the nature and behavior of fire as it relates to fire dynamics.	3
4. Analyze and determine the causes of fires and contributing factors.	1, 6, 7
5. Evaluate the use of incendiary devices, explosives, and bombs.	10, 11, 12
6. List the procedures for fire scene documentation, including sources and technology available for fire investigations.	7, 8
7. Explain the role of the fire investigator in courtroom demeanor and testifying.	14

FESHE Course Outcomes, Fire Investigation and Analysis (C0285)	Chapter References
1. Demonstrate a technical understanding of the characteristics and impacts of fire loss and the crime of arson, which is necessary to conduct competent fire investigation and analysis.	1
2. Document the fire scene in accordance with best practice and legal requirements.	6, 7, 8, 10, 12, 13, 14
3. Analyze the fire scenario utilizing the scientific method, fire science, and relevant technology.	3, 7, 10, 11, 12, 13, 14
4. Analyze the legal foundation for conducting a systematic incendiary fire investigation and case preparation.	1, 7, 8, 10, 11, 12, 13, 14
5. Design and integrate a variety of arson-related intervention and mitigation strategies.	1

Appendix B
Common Construction Terminology

The following terms are taken from **Building Construction Related to the Fire Service, Fourth Edition**. These terms supplement the Key Terms that appear in this edition of **Fire Investigator**. This collection should help investigators navigate terminology they may hear in the course of an investigation regarding particular aspects of a given structure.

A

Admixture — Ingredients or chemicals added to concrete mix to produce concrete with specific characteristics.

Aesthetics — Branch of philosophy dealing with the nature of beauty, art, and taste.

Aftershock — A smaller earthquake that occurs in the same area and similar timeframe.

Aggregate — (1) Gravel, stone, sand, or other inert materials used in concrete. These materials may be fine or coarse. (2) Term used in fire prevention and building codes to describe the sum total of individual parts or components of an assembly or feature, such as in units of exit.

Air-Inflated Structure — Membrane structure that uses air pressure to develop its initial shape, but may not use air pressure throughout the entire high-profile occupancy; enclosed columns or tubes may be inflated to hold the shape of the structure. This type of structure is often intended to be temporary or movable.

Air-Supported Structure — Membrane structure that is fully or partially held up by interior air pressure.

Alloy — Substance or mixture composed of two or more metals (or a metal and nonmetallic elements) fused together and dissolved into each other to enhance the properties or usefulness of the base metal.

American Society for Testing and Materials (ASTM) — Voluntary standards-setting organization that sets guidelines on characteristics and performance of materials, products, systems and services; for example, the quality of concrete or the flammability of interior finishes.

Americans with Disabilities Act (ADA) of 1990 – Public Law 101-336 — Federal statute intended to remove barriers, physical and otherwise, that limit access for individuals with disabilities.

Arch — Curved structural member in which the interior stresses are primarily compressive. Arches develop inclined reactions at their supports.

Area of Refuge — (1) Space where a victim may find safety from hazards at an incident. *Also known as* Sheltered Areas. (2) Space protected from fire in the normal means of egress either by an approved sprinkler system, separation from other spaces within the same building by smokeproof walls, or location in an adjacent building. (3) Two-hour-rated building compartment containing one elevator to the ground floor and at least one enclosed exit stairway. (4) Area where persons who are unable to use stairs can temporarily wait for instructions or assistance during an emergency building evacuation.

ASTM E-84 — Standard test used to measure the surface-burning characteristics of various materials. *Also known as* Steiner Tunnel Test *or* Tunnel Test.

Atrium — Open area in the center of a building, extending through two or more stories, similar to a courtyard but usually covered by a skylight, to allow natural light and ventilation to interior rooms.

Authority Having Jurisdiction (AHJ) — An organization, office, or individual responsible for enforcing the requirements of a code or standard, or approving equipment, materials, an installation, or a procedure.

Axial Load — Load applied to the center of the cross section of a member and perpendicular to that cross section. It can be either tensile or compressive and creates uniform stresses across the cross section of the material.

B

Bar Joist — Open web truss constructed entirely of steel, with steel bars used as the web members.

Base Isolation — A system of structural elements that create a joint between a building and its base to minimize seismic force effects on the main structure. The type of system may be customized to the type of seismic forces expected in an area.

Beam — Structural member subjected to loads, usually vertical loads, perpendicular to its length.

Bearing Wall Structure — Common type of structure that uses the walls of a building to support spanning elements such as beams, trusses, and precast concrete slabs.

Bending Moment — A reaction within a structural component that opposes a vertical load. When the bending moment is exceeded, the component will fail. Bending stress can be calculated from the bending moment.

Bending Stress — Compressive and tensile stresses in a beam. When the stresses are not held in equilibrium, the beam will bend and ultimately fail. Bending stresses are calculated from the bending moment.

Blind Hoistway — Used for express elevators that serve only upper floors of tall buildings. There are no entrances to the shaft on floors between the main entrance and the lowest floor served.

Board of Appeals — Group of people, usually five to seven, with experience in fire prevention, building construction, and/or code enforcement, who are legally constituted to arbitrate differences of opinion between fire inspectors and building officials, property owners, occupants or builders.

Bowstring Truss — Lightweight truss design noted by the bow shape, or curve, of the top chord.

British Thermal Unit (Btu) — Amount of heat energy required to raise the temperature of 1 lb of water 1°F. 1 Btu = 1.055 kilo joules (kJ).

Building Code — Body of local law, adopted by states/provinces, counties, cities, or other governmental bodies to regulate the construction, renovation, and maintenance of buildings.

Building Permit — Authorization issued from the appropriate authority having jurisdiction (AHJ) before any new construction, addition, renovation, alteration, or demolition of buildings or structures occurs.

Butt Joint — Connection between two parts made by simply securing end surfaces together without additional shaping at the ends; a simple but weak joint.

Buttress — Structure projecting from a wall designed to receive lateral pressure action at a particular point.

C

Cable — A flexible structural member used to support roofs, brace tents, and restrain pneumatic structures.

Cable Membrane Structure — Freestanding structure that uses suspension cables for support. Also known as Cable Covered Structure.

Caisson — Protective sleeve used to keep water out of an excavation for a pier.

Calcined — Process that heats a substance to a high temperature but below the melting or fusing point, causing loss of moisture, reduction or oxidation, and decomposition of carbonates and other compounds.

Cantilever — (1) Projecting beam or slab supported at one end. (2) Type of collapse void in which one end of a floor or roof section that has collapsed remains suspended and unsupported.

Capital — Broad top surface of a column or pilaster, designed to spread the load held by a column.

Cast-in-Place Concrete — Common type of concrete construction. Refers to concrete that is poured into forms as a liquid and assumes the shape of the form in the position and location it will be used.

Cement — Any adhesive material or variety of materials which can be made into a paste with adhesive and cohesive properties to bond inert aggregate materials into a solid mass by chemical hardening. For example, portland cement is combined with sand and/or other aggregates and water to produce mortar or concrete.

Chord — Top or bottom longitudinal member of a truss; main members of trusses, as distinguished from diagonals.

Cladding — Exterior finish or skin.

Cold Rolled Steel (CRS) — Commercial and drawing steels; shaped by being passed between a series of rollers to reduce the thickness incrementally.

Collar Tie — Horizontal roof framing member in the top third of the framing system; braces the roof framing against the uplift of wind.

Column — Vertical supporting member.

Column Footing — Square pad of concrete that supports a column.

Common Truss — Truss structure with the chords and diagonal members arranged in parallel planes. Also known as Monoplane Truss.

Compartment — Any enclosed space without internal fire barriers.

Compensated System — Stairwell pressurization system that can modulate the pressure in the stairwell in relation to the interior of the building, or vent excess pressure.

Composite Panels — Produced with parallel external face veneers bonded to a core of reconstituted fibers. Also known as Sandwich Panel.

Concentrated Load — Load that is applied at one point or over a small area.

Concrete — Strong, hard building material produced from a mixture of portland cement and an aggregate filler/binder to which water is added to form a slurry that sets into a rigid building material.

Concrete Block — Rectangular block of a standard size used in building construction; the most common type is the hollow concrete block. Also known as Concrete Masonry Unit (CMU).

Concrete Block Brick Faced (CBBF) — Wall construction system that includes one wythe of concrete blocks with a brick wythe attached to the outside.

Conflagration — Large, uncontrollable fire covering a considerable area and crossing natural fire barriers such as streets; usually involving buildings in more than one block and causing a large fire loss. Forest fires can also be considered conflagrations.

Controlled Collapse Demolition — Deliberate demolition process that uses large machinery and equipment to reduce an entire structure to ground level.

Convenience Stair — Stair that usually connects two floors in a multistory building.

Conventionally Framed Roofs — Roofing system constructed on site; often uses dimensional lumber and nails/screws but can also use preengineered components.

Cooling Tower — Rooftop or independent unit that ejects waste heat into the atmosphere to lower the temperature in a system. Commonly used in HVAC systems.

Corbel — Bracket or ledge made of stone, wood, brick, or other building material projecting from the face of a wall or column used to support a beam, cornice, or arch.

Corbelling — Use of a corbel to provide additional support for an arch.

Cornice — Concealed space near the eave of a building; usually overhanging the area adjacent to exterior walls.

Corrugated — Formed into ridges or grooves; serrated.

Course — Horizontal layer of individual masonry units.

Criterion-Referenced Testing (CRT) — Measurement of one component's tested performance against a set standard or criteria, not against similar components or assemblies. Similar to Criterion-Referenced Assessment.

Cross-Section — Theoretical slice of a 3-dimensional structural component to enable area and stress calculations.

Curtain Wall — Nonload-bearing exterior wall attached to the outside of a building with a rigid steel frame. Usually the front exterior wall of a building intended to provide a certain appearance.

D

Damping Mechanism — Structural element designed to control vibration.

Design Principles — Guidelines applied to basic units of a project that cause the items to work together as a unified, completely finished item that serves a purpose within established parameters. Units can include the materials, concepts, and setting.

Design-Build — The use of a single organization to both design and build a facility to minimize risks for the project owner. May also refer to a firm specializing in design-build.

Dewatering — Process of removing water from a vessel or building.

Dielectric — Material that is a poor conductor of electricity; usually applied to tools that are used to handle energized electrical wires or equipment.

Dimensional Lumber — Lumber with standard, nominal measurements for use in building construction. Dimensional lumber is also available in rough, green components with actual dimensions that match the nominal dimensions.

Door Closer — Mechanical device that closes a door. *Also known as Self-Closing Door.*

Door Hold-Open Device — Mechanical device that holds a door open and releases it upon a signal. Usually a magnetic device used to hold fire or smoke door assemblies in the open position.

Draft Curtain — Noncombustible barriers or dividers hung from the ceiling in large open areas that are designed to minimize the mushrooming effect of heat and smoke and impede the flow of heat. *Also known as Curtain Boards and Draft Stops.*

Drop Panel — Type of concrete floor construction in which the portion of the floor above each column is dropped below the bottom level of the rest of the slab, increasing the floor thickness at the column.

Ductility — A measure of a metal's ability to be drawn, hammered thin, or rolled into shapes without breaking. The high ductility of steel makes it very versatile for use in constructing buildings.

Dynamic Load — Loads that involve motion, including impact from wind, falling objects, and vibration. *Also known as Shock Loading.*

E

Earthquake — A sudden release of energy in the Earth's crust that creates seismic forces that shake and sometimes disrupt the ground. Earthquakes are associated with volcanic activity, landslides, and tsunamis.

Eccentric Load — Load perpendicular to the cross-section of the structural member, but which does not pass through the center of the cross-section. An eccentric load creates stresses that vary across the cross-section, and may be both tensile and compressive.

Elastomer — Generic term for rubber-like materials including natural rubber, butyl rubber, neoprene, and silicone rubber used in facepiece seals, low-pressure hoses, and similar SCBA components.

Elevator — Mechanical system that travels vertically and is used to transport people and items in a multistory building.

Elevator Pit — Depression at the base of an elevator hoistway that contains equipment and maintenance access.

Engineered Wood — A material manufactured by bonding pieces of wood with glue or resin to form finished shapes.

Equilibrium — When the support provided by a structural system is equal to the applied loads.

Escalator — Belt-driven moving stairs that move in one direction at a fixed rate of speed.

Exhaust System — Ventilation system designed to remove stale air, smoke, vapors, or other airborne contaminants from an area.

Expanded Polystyrene (EPS) — Closed-cell foam used for a growing number of purposes including building insulation. Properties include rigidity, low weight, and formability.

Expansion Joint — Flexible joint in concrete used to prevent cracking or breaking because of expansion and contraction due to temperature changes.

Exposure — Structure surfaces or separate parts of the fireground to which a fire or products of combustion could spread.

Exterior Insulation and Finish Systems (EIFS) — Nonload-bearing, exterior wall cladding or covering system composed of an adhesively or mechanically attached foam insulation board, reinforcing mesh, a base coat, and an outer finish coat. *Also known as* Synthetic Stucco.

Exterior Stairs — Stairs separated from the interior of a building by walls.

F

Façade — Fascia added to some buildings with flat roofs to create the appearance of a mansard roof. *Also known as* False Roof *or* Fascia.

Factor of Safety — Ratio of the failure point of a material to the maximum design stress; indicates the strength of a structure beyond the expected or actual loads.

Failure Point — Point at which material ceases to perform satisfactorily; depending on the application, this can involve breaking, permanent deformation, excessive deflection, or vibration.

Fascia — Broad flat surface over a storefront or below a cornice.

Fast-Track Construction — Strategy to reduce the overall time for completion of a project by merging the design and construction phases. Often used in conjunction with design-build.

Fault — Area of discontinuity in the Earth's crust associated with movement by tectonic plates.

Federal Emergency Management Agency (FEMA) — Agency within the U.S. Department of Homeland Security (DHS) that is responsible for emergency preparedness, mitigation, and response activities for events including natural, technological, and attack-related emergencies.

Finger Joint — Connection between two parts made by cutting complementary mating parts, and then securing the joint with glue.

Fire Area — One of a set of sections in a building separated from each other by fire-resistant partitions.

Fire Department Connection (FDC) — Point at which the fire department can connect into a sprinkler or standpipe system to boost the water pressure and flow in the system. This connection consists of a clapper valve with two or more 2½-inch (65 mm) intakes or one large-diameter (4-inch [100 mm] or larger) intake. *Also known as* Fire Department Sprinkler Connection.

Fire Door — Specially constructed, tested, and approved fire-rated assembly designed and installed to prevent fire spread by automatically sealing an opening in a fire wall to block the spread of fire.

Fire Escape — Means of escaping from a building in case of fire; usually an interior or exterior stairway or slide, independently supported and made of fire-resistive material.

Fire Flow — The amount of water required to extinguish a fire in a timely manner.

Fire Load — Maximum amount of heat that can be released if all fuel in a given area is consumed; expressed in pounds per square foot and obtained by dividing the amount of fuel present by the floor area. Used as a measure of the potential heat release of a fire within a compartment. *Similar to* Fuel Load *and* Heat of Combustion.

Fire Partition — Fire barrier that extends from one floor to the bottom of the floor above or to the underside of a fire-rated ceiling assembly; provides a lower level of protection than a fire wall. An example is a one-hour rated corridor wall.

Fire Resistance Rating — Rating assigned to a material or assembly after standardized testing by an independent testing organization; identifies the amount of time a material or assembly will resist a typical fire, as measured on a standard time-temperature curve.

Fire Retardant — Any substance, except plain water, that when applied to another material or substance will reduce the flammability of fuels or slow their rate of combustion by chemical or physical action.

Fire Spread — The movement of fire from one material (source) to another (exposure). May occur within a compartment or across a break.

Fire Wall — Fire-rated wall with a specified degree of fire resistance, built of fire-resistive materials and usually extending from the foundation up to and through the roof of a building, that is designed to limit the spread of a fire within a structure or between adjacent structures.

Firefighter's Smoke Control Station (FSCS) — Interface between the smoke management system and the fire response forces.

Flame Spread — The movement of fire from one material (source) to another (exposure). May occur within a compartment or across a break.

Flange — Single or paired external ridges or rims on a beam that do most of the work of supporting a load.

Flat Plate — Plain floor slab about 8 inches (200 mm) thick that rests on columns spaced up to 22 feet (6.5 m) apart and depends on diagonal and orthogonal patterns of reinforcing bars for structural support because the slab lacks beams; simplest and most economical floor system.

Flat-Slab Concrete Frame — Construction technique using concrete slabs supported by concrete columns.

Floating Foundation — Foundation for which the volume of earth excavated will approximately equal the weight of the building supported. Thus, the total weight supported by the soil beneath the foundation remains about the same, and settlement is minimized because of the weight of the building.

Footing — Part of the building that rests on the bearing soil and is wider than the foundation wall.

Force — (1) Simple measure of weight, usually expressed in pounds (kilograms). (2) In physics: Any interaction that may change the motion of an object.

Forced-Air System — A building heating and cooling system that uses air as the heat transfer medium.

Foundation Wall — Vertical element of a foundation; rests on the foundation footers. May be full-story height as in a basement, or partial height. Materials often include poured concrete, or mortar elements such as block, brick, or stone.

Frame — Internal system of structural supports within a building.

Frame Membrane Structure — Structure supported primarily by a frame or skeleton rather than by load-bearing walls. *Also known as* Frame Covered Structure.

Freestanding Walls — Self-supporting fire walls independent of the structure's frame. Must resist a lateral load of 5 pounds per square foot (.25 kPa per square meter).

Frost Line — Common depth at which ground water in soil will freeze. Influential variables include climate, soil properties, and nearby heat sources.

G

Gentrification — Process of restoring rundown or deteriorated properties by more affluent people, often displacing poorer residents.

Glazing — Glass or thermoplastic panel in a window that allows light to pass through.

Glue-Laminated Beam — (1) Wooden structural member composed of many relatively short pieces of lumber glued and laminated together under pressure to form a long, extremely strong beam. (2) Term used to describe wood members produced by joining small, flat strips of wood together with glue. *Also known as* Glued-Laminated Beam *or* Glulam Beam.

Grain — Direction of growth of a tree. Loads aligned perpendicular to the grain are more sturdily supported; lumber will split more easily when cut parallel to the grain.

Gravity (G) — Force acting to draw an object toward the earth's center; force is equal to the object's weight.

Green Design — Incorporation of such environmental principles as energy efficiency and environmentally friendly building materials into design and construction.

Green Roof — Roof of a building that is partially or completely covered with vegetation and a growing medium, planted over waterproof roofing elements. Term can also indicate the presence of green design technology including photovoltaic systems and reflective surfaces.

Grillage Footing — Footing consisting of layers of beams placed at right angles to each other and usually encased in concrete.

Grout — A mixture of cement, aggregate, and water that hardens over time; used to embed reinforcement materials in masonry walls. Similar to mortar.

H

Hardware — General term for small pieces of equipment made with metal, such as hand tools.

Heat of Hydration — During the hardening of concrete, heat is given off by the chemical process of hydration.

Heat Transfer — Flow of heat from a hot substance to a cold substance; may be accomplished by convection, conduction, or radiation.

Heaving — Upward deformation of a building's structural elements.

High-Rise Building — Building that requires fire fighting on levels above the reach of the department's equipment. The *Uniform Building Code (UBC)* defines a high-rise building as one greater than 75 feet (23 m) in height, but other fire and building codes may define the term differently. *Also known as* High-Rise.

Hoistway — The vertical shaft in which the elevator car travels; includes the elevator pit.

Horizontal Motion — Side-to-side, swaying motion.

Hot Work — Any operation that requires the use of tools or machines that may produce a source of ignition.

Hydronic System — A building heating and cooling system that uses water as the heat-transfer medium.

Hygroscopic — Ability of a substance to absorb moisture from the air.

I

Ignition Source — Mechanism or initial energy source employed to initiate combustion, such as a spark that provides a means for the initiation of self-sustained combustion.

I-Joist — Engineered wood joists with an "I" shaped cross section. Commonly used in modern roof and floor construction.

Institutional Sprinklers — Low profile sprinkler system and pendant used with concealed piping in correctional facilities and institutions where tampering of the system must be discouraged or prevented.

Insulated Concrete Form (ICF) Construction — Construction technique that uses hollow foam blocks with predetermined sizes and shapes. The blocks lock together and are filled with concrete to form structural supports.

International Building Code® (IBC®) — Code that is dedicated to providing safety regulations for life safety, structural, and fire protection issues that occur throughout the life of a building.

International Code Council (ICC) — Organization that develops the *International Building Code® (IBC®)* and the *International Fire Code® (IFC®)*, for city and state adoption. Was formed by the merger of the Building Officials and Code Administrators (BOCA) International, Inc., the International Conference of Building Officials (ICBO), and the Southern Building Code Congress International (SBCCI).

Intumescent Coating — Coating or paintlike product that expands when exposed to the heat of a fire; creates an insulating barrier that protects the material that is underneath.

Inverted Truss — Truss support system that is constructed with a deep triangular portion projecting down instead of up, and the portions of a standard truss are under compression instead of tension.

J

Joist — Horizontal structural members used to support a ceiling or floor. Drywall materials are nailed or screwed to the ceiling joists, and the subfloor is nailed or screwed to the floor joists.

L

Lamella Arch — Special type of arch constructed of short pieces of wood called lamellas.

Laminated Wood — Material made of wood strips and resin, shaped, and bonded with heat and/or pressure.

Landing — Horizontal platform where a flight of stairs begins or ends.

Lateral Displacement — Sideways deformation of a building's structural elements.

Lateral Load — Load that exerts a horizontal force against a structure. Calculated as a live load; includes seismic activity and soil pressure against vertical restraints such as retaining walls and foundations.

Ledger Board — Horizontal framework member, especially one attached to a beam side that supports the joists. *Also known as* Ribbon Board.

Lintel — Support for masonry over an opening; usually made of steel angles or other rolled shapes, singularly or in combination.

Listed — Refers to a device that has been tested by the Underwriters' Laboratories Inc. Factory Mutual System and certified as having met minimum criteria.

Live Load — (1) Items within a building that are movable but are not included as a permanent part of the structure; merchandise, stock, furnishings, occupants, firefighters, and the water used for fire suppression are examples of live loads. (2) Force placed upon a structure by the addition of people, objects, or weather.

Load — (1) The sum of the wattages of the various devices being served by a circuit. (2) Any effect that a structure must be designed to resist, such as gravity, wind, earthquakes, or soil pressure.

Load-Bearing Wall — Wall that supports itself, the weight of the roof, and/or other internal structural framing components, such as the floor beams and trusses above it; used for structural support. *Also known as* Bearing Wall.

Louvers — A series of horizontal slats that are angled to permit easy ventilation in one direction of flow and restricted ventilation in the opposite direction. Louvers are commonly used in applications where the restrictive side blocks sunshine, rain, or products of combustion.

Lumber — Lengths of wood cut and prepared for use in construction.

M

Manufactured Components — Structural elements constructed in a factory and shipped to the construction site.

Masonry — Bricks, blocks, stones, and unreinforced and reinforced concrete products.

Mastic — Heat resistant construction adhesive that bonds with most materials; can be used as a fire-retardant coating.

Mat Slab Foundation — Thick slab beneath the entire area of a building; thicker and more reinforced than a simple slab-on-grade foundation.

Means of Egress — Safe, continuous path of travel from any point in a structure to a public way. Composed of three parts: exit access, exit, and exit discharge.

Membrane Ceiling — Usually refers to a suspended, insulating ceiling tile system.

Membrane Structure — (1) Structure with an enclosing surface of a thin stretched flexible material. Examples include a simple tent or an air-supported structure. (2) Weather-resistant, flexible or semiflexible covering consisting of layers of materials over a supporting framework.

Metal-Clad Door — Door with a metal exterior; may be flush type or panel type. *Also known as* Kalamein Door.

Mortar — Cement-like liquid material that hardens and bonds individual masonry units into a solid mass.

Multiple-Injection System — Stairwell pressurization system that uses an air supply shaft that discharges supply air at a uniform rate along several points within the stairwell.

Mushrooming — Tendency of heat, smoke, and other products of combustion to rise until they encounter a horizontal obstruction; at this point they will spread laterally until they encounter vertical obstructions and begin to bank downward.

N

Nailability — Property of a material that allows it to accept a fastener, such as a nail. Nailable materials include wood, gypsum, and some thin metals.

Negative Pressure — Artificially lowering the pressure inside a structure so the fresh air from outside moves in more quickly.

NFPA 265 — Large scale test used to evaluate the performance of textile wall coverings under fire conditions. Older test, succeeded by NFPA 286. *Similar to* ASTM E-84.

NFPA 286 — Large scale test used to evaluate the performance of textile wall coverings under fire conditions. Designed to accommodate materials that may not remain in place during ASTM E-84 testing. Also includes the capacity of attaching materials to the ceiling. Newer test, preceded by NFPA 265. *Similar to* ASTM E-84.

Nominal Dimension of Lumber — Named dimensions of softwood members, which may not match the actual dimensions of processed lumber, within defined parameters. Historically, the two sets of dimensions were identical.

Nonload-Bearing Wall — Wall, usually interior, that supports only its own weight. These walls can be breached or removed without compromising the structural integrity of the building.

Nonveneered Panel — Lightweight wood construction panel manufactured from wood chips, strands, wafers, or sawdust and a bonding agent such as glue or resin. Used as sheathing, reinforcement of structural elements, and sub-flooring. Includes OSB, particleboard, waferboard.

O

Occupancy — (1) General fire and emergency services term for a building, structure, or residency. (2) Building code classification based on the use to which owners or tenants put buildings or portions of buildings. Regulated by the various building and fire codes. *Also known as* Occupancy Classification.

Oriented Strand Board (OSB) — Wooden structural panel formed by gluing and compressing wood strands together under pressure. This material has replaced plywood and planking in the majority of construction applications. Roof decks, walls, and subfloors are all commonly made of OSB.

Overhead Door — Door that opens and closes above a large opening, such as in a warehouse or garage, and is usually of the rolling, hinged-panel, or slab type.

Overpressure — Air pressure above normal or atmospheric pressure.

P

Parapet Wall — Vertical extension of an exterior wall, and sometimes an interior fire wall, above the roofline of a building.

Particleboard — Wooden structural panel formed from wood particles and synthetic resins. *Also known as* Flakeboard, Chipboard, *or* Shavings board.

Passive Smoke Control — Smoke control strategies that incorporate fixed components that provide protection against the spread of smoke and fire. Passive smoke control components include fire doors, fire walls, fire stopping of barrier penetrations, and stair and elevator vestibules.

Phase I Operation — Emergency operating mode for elevators. Phase I operation recalls the car to a certain floor and opens the doors.

Phase II Operation — Emergency elevator operating mode that allows emergency use of the elevator with certain safeguards and special functions.

Piecemeal Demolition — Demolition process that uses hand tools and machines to gradually decrease the height of the structure.

Pier — Deep foundation type that uses beams mounted on concrete wedges/blocks to support loads. *Similar to* Caissons *and* Belled Piers.

Pilaster — Rectangular masonry pillar that extends from the face of a wall to provide additional support for the wall. Decorative pilasters may not provide any support.

Piles — Wooden or steel beams used to support loads; piles are driven into the ground and develop their load-carrying ability either through friction with the surrounding soil or by being driven into contact with rock or a load-bearing soil layer.

Pipe Chase — Concealed vertical channel in which pipes and other utility conduits are housed. Pipe chases that are not properly protected can be major contributors to the vertical spread of smoke and fire in a building. *Also known as* Chase.

Plywood — Wood sheet product made from several thin veneer layers that are sliced from logs and glued together.

Polychlorinated Biphenal (PCB) — Toxic compound found in some older oil-filled electric transformers.

Polyurethane — A polymer formed by reacting an isocyanate with a polyol; used in many applications including floating insulating foams and floating ropes.

Portland Cement — Most commonly used cement, consisting chiefly of calcium and aluminum silicate. It is mixed with water to form a paste that hardens, and is therefore known as a hydraulic cement.

Post and Beam Construction — Construction style using vertical elements to support horizontal elements. Associated with heavy beams and columns; historically constructed of wood.

Posttensioned Reinforcement (Concrete) — Concrete reinforcement method. Reinforcing steel strands placed in protective sleeves in the concrete are tensioned after the concrete has hardened.

Precast Concrete — Concrete reinforcement method. Reinforcing steel strands placed in protective sleeves in the concrete are tensioned after the concrete has hardened.

Preincident Planning — Act of preparing to manage an incident at a particular location or a particular type of incident before an incident occurs. *Also known as* Prefire Inspection, Prefire Planning, Preincident Inspection, Preincident Survey, or Preplanning.

Preincident Survey — Assessment of a facility or location made before an emergency occurs, in order to prepare for an appropriate emergency response. *Also known as* Preplan.

Pressure-Reducing Valve — Valve installed at standpipe connection that is designed to reduce the amount of water pressure at that discharge to a specific pressure, usually 100 psi (700 kPa).

Prestressing — Compressive force induced in the concrete before the load is applied by placing the reinforcing bars in concrete beams in tension before the concrete is poured, so that the member will develop greater strength after the concrete has set. This "pre" stress is applied by tightening or "pre" loading the reinforcing steel; the preloading of the steel creates compressive stresses in the concrete that counteract the tensile stresses, which result when loads are applied.

Pretensioned Reinforcement (Concrete) — Concrete reinforcement method. Reinforcing steel strands are stretched, producing a tensile force in the steel. Concrete is then placed around the steel strands and allowed to harden.

Purlin — Horizontal member between trusses that support the roof.

R

Rafter — Inclined beam that supports a roof, runs parallel to the slope of the roof, and to which the roof decking is attached.

Rafter Tie — Horizontal roof framing member at the bottom of the roof framing system; helps keep walls from spreading due to the weight of the roof.

Rated Assembly — Assemblies of building components such as doors, walls, roofs, and other structural features that may be, because of the occupancy, required by code to have a minimum fire-resistance rating from an independent testing agency. *Also known as* Labeled Assembly.

Rated Fire Door Assembly — Door, frame, and hardware assembly that has a fire-resistive rating from an independent testing agency.

Recirculation — Movement of smoke being blown out of a ventilation opening only to be drawn back inside by the negative pressure created by the ejector because the open area around the ejector has not been sealed.

Reinforced Concrete — Concrete that is internally fortified with steel reinforcement bars or mesh placed within the concrete before it hardens. Reinforcement allows the concrete to resist tensile forces.

Reinforcing Bars (Rebar) — Short for reinforcing bar. These steel bars are placed in concrete forms before the cement is poured. When the concrete sets (hardens), the rebar within it adds considerable strength and reinforcement.

Remodel — Restructuring of a building's spaces and occupancy features.

Renovate — Restoring or updating a building's features including finishing materials, furnishing, and overall appearance.

Resonance — Movements of relatively large amplitude resulting from a small force applied at the natural frequency of a structure.

Return-Air Plenum — Unoccupied space within a building through which air flows back to the heating, ventilating, and air-conditioning (HVAC) system; normally immediately above a ceiling and below an insulated roof or the floor above.

Ridge Beam — Highest horizontal member in a pitched roof to which the upper ends of the rafters attach. *Also known as* Ridge Board *or* Ridgepole.

Rigid Frame — Load-bearing system constructed with a skeletal frame and reinforcement between a column and beam.

Rise — Vertical distance between the treads of a stairway, or the height of the entire stairway.

Riser — (1) Vertical part of a stair step. (2) Vertical water pipe used to carry water for fire protection systems above ground, such as a standpipe riser or sprinkler riser.

Roll Roofing — Roof covering made of flexible material that may be applied to the roof deck as a continuous sheet. Commonly used on shallow pitch roofs.

Rolling (Steel) — Process of forming metal stock into shapes, including sheets, by passing thick bars of metal between a pair of rollers.

Run — The horizontal measurement of a stair tread or the distance of the entire stair length.

R-Value — A measure of the ability of a material to insulate. Used in structural engineering and construction. Insulators with higher R-values are more effective.

S

Scarf Joint — Connection between two parts made by the cutting of overlapping mating parts and securing them by glue or fasteners so that the joint is not enlarged and the patterns are complementary.

Seismic Effect — Movement of a shock wave through the ground or structure after a large detonation; may cause additional damage to surrounding structures.

Seismic Force — Force produced by earthquakes; the most complex forces that can be exerted on a building.

Seismic Load — Application of forces caused by earthquakes.

Self-Closing Door — Door equipped with a door closer.

Setback — Distance from the street line to the front of a building.

Settlement — Downward deformation of a building's structural elements. *Also known as* Settling.

Shear Stress — Stress resulting when two forces act on a body in opposite directions in parallel adjacent planes.

Shear Wall — Wall panels that are braced against lateral loads. May be load-bearing or nonload-bearing.

Shell Structure — Rigid, three-dimensional structure having an outer "skin" thickness that is small compared to other dimensions.

Shelter-in-Place — Having occupants remain in a structure or vehicle in order to provide protection from a rapidly approaching hazard, such as a fire or hazardous gas cloud. *Opposite of* evacuation. *Also known as* Protection-in-Place, Sheltering, *and* Taking Refuge.

Shoring — General term used for lengths of timber, screw jacks, hydraulic and pneumatic jacks, and other devices that can be used as temporary support for formwork or structural components or used to hold sheeting against trench walls. Individual supports are called shores, cross braces, and struts. Commonly used in conjunction with cribbing.

Shunt Trip — A circuit breaker used as a safety device in an elevator system. When electrical current surges, the device disconnects the power source.

Single-Injection System — Stairwell pressurization system that uses one point of supply air; pressurization can be lost if the system becomes unsealed through the use of doors.

Slab and Beam Frame — Construction technique using concrete slabs supported by concrete beams.

Slenderness Ratio — Comparison of the height or length of a structural component and the width/thickness of the component. Used to determine the load that can be supported by the component; lower ratios indicate components are more stable.

Sliding Door — Door that opens and closes by sliding across its opening, usually on rollers.

Slump Test — Method of evaluating the moisture content of wet concrete by measuring the amount that a small, cone-shaped sample of the concrete settles or "slumps" after it is removed from a standard-sized test mold.

Smoke Developed Rating — The measure of the relative visual obscurity created during the testing process by a known material.

Smokeproof Stair Enclosures — Stairways that are designed to limit the penetration of smoke, heat, and toxic gases from a fire on a floor of a building into the stairway, and that serve as part of a means of egress.

Soil Property — Physical qualities of the materials at the surface of the earth. Affects a building's foundation and size. Influential variables include texture, structure, density, porosity, and consistency.

Space Frame — Aluminum skeleton that is similar to aircraft frames, upon which the aluminum, plastic, or composite skin of the vehicle's body is attached; the internal structure of these space frames provides the structural support for the vehicle, while the skin provides aerodynamics, styling, and protection from the elements.

Spec Building — Building built without a tenant or occupant. *Spec* is short for *speculation.*

Spray-Applied Fire Resistive Material (SRFM) — Coating used to increase the fire resistance rating of structural components. Materials commonly include mineral fiber or aggregates such as vermiculite and perlite.

Standpipe System — Wet or dry system of pipes in a large single-story or multistory building, with fire hose outlets installed in different areas or on different levels of a building to be used by firefighters and/or building occupants. This system is used to provide for the quick deployment of hoselines during fire fighting operations.

Static Load — Load that is steady, motionless, constant, or applied gradually.

Stationary Storage Battery System — A system including a battery, a charger, and electrical equipment for a particular application. This type of system can include a lead-acid battery or a safer type of battery.

Steel — An alloy of iron and carbon; proportions and additional elements affect the characteristics of the finished material. Used widely in the construction of buildings and other infrastructure.

Steiner Tunnel — Test apparatus used in the determination of flame spread ratings; consists of a horizontal test furnace 25 feet (7.6 m) long, 17½ inches (445 mm) wide, and 12 inches (305 mm) high that is used to observe flame travel. A 5,000 Btu (5 270 kJ) flame is produced in the tunnel, and the extent of flame travel across the surface of the test material is observed through ports in the side of the furnace.

Stratum — Sheet-like layer of rock or earth; numerous other layers, each with different characteristics, are typically found above and below. *Plural:* Strata.

Stress — Factors that work against the strength of any piece of apparatus, equipment, or structural support. Measurement of intensity is calculated as force divided by area.

Structural Insulated Panel (SIP) — A composite panel used in structural applications; made of plastic foam between two outer wood panels, often oriented strand board (OSB).

Structural Stiffness — The use or addition of structural supports to improve the ability of a structure to withstand forces imposed by loads. Often indicates supplemental reinforcement to accommodate specific types of loads, such as earthquake forces. *Also known as* Stiffening.

Stud — Vertical structural member within a wall in frame buildings; most are made of wood, but some are made of light-gauge metal.

Superplasticizer — Admixture used with concrete or mortar mix to make it workable, pliable, and soft while using relatively little water.

Surface Systems — System of construction in which the building consists primarily of an enclosing surface, and in which the stresses resulting from the applied loads occur within the surface-bearing wall structures.

Surface-Burning Characteristic — Speed at which flame will spread over the surface of a material.

Swinging Door — Door that opens and closes by swinging from one side of its opening, usually on hinges. *Also known as* Hinged Door.

T

Tensile Stress — Stress in a structural member that tends to stretch the member or pull it apart; often used to denote the greatest amount of tensile force a component can withstand without failure.

Thermal Radiation — Transmission or transfer of heat energy, from one body to another body at a lower temperature, through intervening space by electromagnetic waves similar to radio waves or X-rays.

Tied Wall — Fire wall connected to a line of columns or steel structural supports with the same degree of fire resistance. Must resist lateral collapse on either side of the structure.

Tilt-Up Construction — Type of construction in which concrete wall sections (slabs) are cast on the concrete floor of the building, then tilted up into the vertical position. *Also known as* Tilt-Slab Construction.

Tin-Clad Door — Similar to a metal-clad door, except covered with a lighter-gauge metal.

Torsional Load — Load offset from the center of the cross section of the member and at an angle to or in the same plane as the cross section; produces a twisting effect that creates shear stresses in a material.

Transverse Load — Structural load that exerts a force perpendicular to structural members.

Tread — Horizontal face of a step.

Truss — Structural member used to support a roof or floor with triangles or combinations of triangles to provide maximum load-bearing capacity with a minimum amount of material. Connections are likely to fail in intense heat.

Two-Way Slab Construction — Concrete construction framework type that uses reinforcing steel placed on the bottom of the framework that provides reinforcement in two directions. *Also known as* Waffle Construction.

U

Underpinning — The use of permanent supports to strengthening an existing foundation.

Unprotected Steel — Steel structural members that are not protected against exposure to heat.

Utility Chase — Vertical pathway (shaft) in a building that contains utility services such as laundry or refuse chutes, and grease ducts.

V

Veneered Walls — Walls with a surface layer of attractive material laid over a base of a common material.

Voice over Internet Protocol (VoIP) — Communication services that utilize an Internet connection to transmit telephone signals.

W

Wall Footing — Type of shallow foundation that includes a wide, thick area to distribute the weight of a wall on the bearing soil. *Also known as* Strip Footing.

Water Table — The highest level of ground water saturation of subsurface materials. Influential variables include the season, soil properties, and topography.

Web — (1) Wide vertical part of a beam between thick horizontal flanges at the top and bottom of the beam. (2) Secondary member of a truss contained between the chords. *Also known as* Diagonals.

Wildland/Urban Interface — Line, area, or zone where an undeveloped wildland area meets a human development area. *Also known as* Urban/Wildland Interface.

Wind — Horizontal movement of air relative to the surface of the earth.

Wired Glass — Flat sheet glass containing an embedded wire mesh that increases its resistance to breakage and penetration; installed in exterior doors, windows, and skylights to increase interior illumination without compromising fire resistance and security. When the glass is exposed to the heat of a fire, it will typically break or crack due to thermal stresses; the wires give the glass dimensional stability, permitting it to act as a barrier to the fire even when the glass has failed. Can be either transparent or translucent.

Wythe — Single vertical row of a series of rows of masonry units in a wall; usually brick or concrete block.

Appendix C
Federal Constitutional Search and Seizure Issues in Fire Scene Investigations

This document is by David M. Bessho, Georgia State University, College of Law, Atlanta, Georgia.

Prepared for Michael A. McKenzie; Cozen and O'Connor; Suite 200; One Peachtree Street, NE; Atlanta, Georgia 30308.

Copyright © 1995 Cozen and O'Connor

Introduction

This memorandum summarizes the state of the law with respect to federal constitutional search and seizure issues in fire scene investigations. It does not address search and seizure issues with respect to State law, which might afford more protection to property owners and place greater restraint on fire investigators than does the federal constitution. See *Mills v. Rogers*, 457 U.S. 291, 303 (1982). However, because the federal constitution provides the minimum level of search and seizure protection below which the States may not stray, this summarization provides the minimum level of protection generally applicable in the United States. *Id.*

The Law

The Supreme Court outlined the Fourth Amendment search and seizure protections applicable to fire scene investigations in the cases of *Michigan v. Tyler*, 436 U.S. 499 (1978), and *Michigan v. Clifford*, 464 U.S. 287 (1984). In these cases, the Court held that the conduct of fire scene investigations fall within the strictures of the Fourth Amendment's protection against unreasonable searches and seizures. *Tyler*, 436 U.S. 499, 506 (1978). Accordingly, the general rule is that fire investigators may not make a nonconsensual entry onto private property without a duly authorized search warrant. *Tyler*, 436 U.S. at 506. However, the Court granted fire investigators some latitude to conduct warrantless searches of fire scenes by allowing fire officials to remain on private property for a reasonable time after the fire has been extinguished, and by allowing the use of administrative search warrants for routine fire investigations.

The Court has recognized that "[a] burning building clearly presents an exigency of sufficient proportions to render a warrantless entry 'reasonable.'" *Id.* at 509. Because the firefighting function encompasses "not only … extinguishing fires, but with finding their causes," fire officials can remain on private property for a "reasonable time" after the fire has been extinguished to determine the origin of the blaze. *Id.* at 510. However, what constitutes a "reasonable" time is a very fact specific determination that may vary with the circumstances of each case. *Id.* at 510 n. 6.

In order to obtain a search warrant for the purpose of conducting a routine investigation into the origin of an unexplained fire, fire officials need only establish the level of probable cause required for administrative searches. *Michigan v. Clifford*, 464 U.S. 287, 294 (1984). This level of probable cause is established upon a showing that "reasonable legislative or administrative standards for conducting an … inspection are satisfied with respect to a particular dwelling." *Camara v. Municipal Court*, 387 U.S. 523, 538. As applied in the fire investigation context, this standard requires fire investigators to "show only that a fire of undetermined origin has occurred on the premises, that the scope of the proposed search is reasonable and will not intrude unnecessarily on the fire victim's privacy, and that the search will be executed at a reasonable and convenient time." *Michigan v. Clifford*, 464 U.S. 287, 294 (1984). However, to search for evidence of arson, officials must establish the traditional level of probable cause required of searches for evidence of crime. *Tyler*, 436 U.S. at 512. Criminal search warrants require a showing that the facts and circumstances of which investigators have "reasonably trustworthy information" are sufficient to cause a person of "reasonable caution" to believe that an offense has been committed. *Draper v. United States*, 358 U.S. 307, 313 (1959).

Supreme Court Cases

Michigan v. Tyler, 436 U.S. 499 (1978)

In *Tyler*, a fire department responding to a midnight fire in a furniture store was able to control the blaze by approximately 2:00 a.m. As the firefighters were "watering down smoldering embers," the Fire Chief arrived and entered the smoking building to examine two containers of flammable liquid that had been found in the store. Believing the fire "could possibly have been arson," the Fire Chief called a detective who arrived at approximately 3:30 a.m. *Tyler* at 502. The detective entered the building to begin his investigation, but was forced to abandon the effort because of steam and smoke in the building. The firefighters completely extinguished the fire and departed by 4:00 a.m. At approximately 8:00 a.m., the Chief returned with the Assistant Chief for a brief examination of the scene. The Assistant Chief returned again at 9:00 a.m. with the detective. They found and seized evidence of arson, including what appeared to be a fuse trail burned into the carpet. Several weeks later, other investigators returned to the scene and found further evidence of arson. Although the building owners did not object to theses entries and seizures at the time, they objected when the evidence was introduced against them at trial. The Michigan Supreme Court held that all the entries made by fire investigators after the flames had been extinguished were illegal warrantless searches, ruled inadmissible any evidence found as a result of these entries, and reversed the convictions of the defendants.

The U.S. Supreme Court held that a property owner's reasonable expectation of privacy is not destroyed simply because his property has been damaged by fire. *Tyler*, 436 U.S. at 505. Having established that an expectation of privacy remained, the property owner was protected by the Fourth Amendment's prohibition on unreasonable searches and seizures. *Id.* at 506. However, the Court also recognized that there are circumstances in which official action is compelled, but there is no time to obtain a search warrant. Such compelling circumstances can arise in either a criminal investigation or administrative inspection context. *Id.* at 509.

The Court held that a "burning building clearly presents an exigency of sufficient proportions to render a warrantless entry [to put out the blaze] 'reasonable.' . . . [O]nce in the building for this purpose, firefighters may seize evidence of arson that is in plain view." *Id.* at 509. In addition, because "[fire officials are charged not only with extinguishing fires, but with finding their causes ... [they] need no warrant to remain in a building for a reasonable time to investigate the cause of a blaze after it has been extinguished." *Id.* at 510. What constitutes "a reasonable time" is a very fact specific determination. The Court commented in a footnote that a number of factors go into the "reasonableness" determination, including the type of structure, the size of the fire, as well as the individual's reasonable expectation of privacy. *Id.* at 510 n. 6. In this case, the Court decided that the warrantless re-entries of the property the following morning were justified as "no more than an actual continuation of the first [entry], and the lack of a warrant did not invalidate the resulting seizure of evidence." *Id.* at 511. However, the entries onto the property weeks later were found to be "clearly detached from the initial exigency." *Id.* at 511. Because investigators obtained neither administrative nor criminal search warrants for these later entries, the evidence obtained was inadmissible at trial. *Id.* at 511.

Michigan v. Clifford, 464 U.S. 287 (1984)

After an early morning fire at the Cliffords' residence was extinguished at 7:04 a.m., all police and firefighters left the scene. Arson investigators, having been notified that the fire was suspicious, arrived at the scene five hours later. They found a work crew busy, on the vacationing Cliffords' instructions, securing the house and pumping water from the basement. While waiting for the water to be pumped, the investigators found a fuel can in the driveway which had been removed by firefighters and seized it as evidence. Although the investigators knew that the Cliffords' had given instructions to secure the house, they entered the residence without consent or a warrant and conducted a thorough search. In the basement they found two more fuel cans and a crock pot attached to a timer, which they also seized as evidence. They then continued their search into the living areas of the residence and found further evidence of arson. Before trial, the Cliffords' moved to exclude all evidence that was seized during this search of their residence.

The Supreme Court rejected the State's assertion that all postfire administrative searches should be exempt from warrant requirements, and affirmed the principle that, absent exceptional circumstances, all nonconsensual searches require warrants. *Clifford*, 464 U.S. at 291-92. The Court also rejected the State's suggestion that the Tyler principle allowed the warrantless postfire search as a continuation of the entry made by the firefighters. The Court distinguished the two cases not on legal principle, but on the particular facts of each case.

In *Tyler*, investigators were forced to call off the initial attempt at investigation because of smoke and darkness, but resumed the search as soon as practicable. *Tyler*, at 296. The Court found that the Clifford's efforts to secure their residence during the time between the departure of the firefighters and the arrival of the arson investigators separated the entry of the firefighters and the entry of the investigators into two different events and precluded considering the investigators' entry a continuation of the firefighters valid entry.

The Court also found that the Cliffords' privacy interest in their residence, particularly because they had taken steps to secure it, was greater than the owner's interest in the furniture store in *Tyler*. The Court held that "[a]t least where a homeowner has made a reasonable effort to secure his fire damaged home … we hold that a subsequent postfire search must be conducted pursuant to a warrant, consent, or the identification of some new exigency. So long as the primary purpose is to ascertain the cause of the fire, an administrative warrant will suffice." *Clifford*, at 297. Because no warrant was obtained before the investigators entered the Clifford home, all evidence found during their entry was inadmissible as evidence. *Id.* at 287.

While not stating it explicitly, the Court implied that even though the investigators were suspicious of arson, an administrative warrant would have sufficed for the initial basement search. The Court noted that even if the investigator's initial search of the basement had been pursuant to a duly authorized administrative search warrant, continuation of the search into the living quarters of the residence was not authorized. *Id.* at 287. However, because the purpose of an administrative search is to determine the origin of the fire and preclude its rekindling, "not [to] give fire officials license to roam freely throughout the fire victim's private residence," once investigators had determined the cause of the fire was the crock pot and timer they would have exhausted their search authority. *Id.* at 297-98.

Nevertheless, the one fuel-can seized from the driveway by investigators was admissible as evidence. *Id.* at 299. Because this can was seen in plain view when the investigators arrived at the scene and had been in plain view when the firefighters had entered the house to extinguish the blaze, it was admissible whether it had been seized by the firefighters or the investigators. *Id.*

This Appendix is a lengthy, but not necessarily comprehensive, list of questions an investigator may wish to reference when interviewing individuals during an investigation. These questions may be asked in any order; investigators may need to ask some questions before others.

Topics to Cover When Interviewing Police Officers

Alarm Information
- Describe the alarm.
- Can you identify the reporting party or parties?
- What was the time of the alarm?
- Do you have any other comments regarding the alarm?

Arrival at the Scene
- What was your time of arrival?
- What were your observations during travel to the fire scene?
- Provide a general description of the property upon your arrival.
- Describe the fire and/or smoke.
- Did you notice anything unusual about the structure?
- Were persons/vehicles arriving or leaving the fire scene?
- Did you take any statements or hear any comments by persons at the fire scene?
- Did you recognize anyone who is common to fire or other emergency incidents?
- Were there any distractions upon arrival or during scene security?

Activities at the Scene
- Did you arrive before or after the fire department?
- Were there any rescues or attempted rescues during the incident?
- Were the doors and windows opened or closed, locked or unlocked?
- Did an officer conduct any activities? If so, explain.
- Were there area(s) where the fire appeared to be more intense?
- Describe the area(s) where the fire had burned through the exterior walls or roof.
- Did the police or fire department secure the scene? If so, explain.
- Did you conduct canvas neighborhood interviews?
- What is your opinion as to the area of origin?
- Did you observe any unusual odors?
- What were your observations of the flame spread and intensity?
- Did you notice any vehicles that passed the scene more than one time?
- If so, did you observe any unusual actions by person(s) in the vehicle?
- Were any arrests made at or near the scene?
- Did you receive any domestic disturbance calls at or near this fire scene?

- Do you have any comments about items not mentioned?
- Did you see the fire? If so, where did you first see the flames?
- Did you take any photographs or video recordings at the scene (Dashboard cams, camera photography, etc.)?
- Did you observe anyone else taking photographs or video of the scene?

Other Observations

- Did firefighters or police officers make any comments?
- Is there a history of police calls to this location?
- Did the police have concerns regarding this neighborhood?

Topics to Cover When Interviewing Firefighters

Alarm Information

- What method (9-1-1 or walk-in) was used to report this alarm?
- Can you identify the reporting party or parties?
- What was the time of the alarm?
- Did you notice any normal or unusual situations regarding the alarm received?

Arrival at the Scene

- What was your time of arrival?
- Who performed the initial size-up of the scene?
- What were your observations during travel to the fire scene?
- Describe the fire and/or smoke.
- Did you notice the fire venting? If so, where was it venting?
- Were there any hydrant problems?
- Describe any problems with the fire department connections or private fire system, if applicable.
- Was there fire alarm/smoke detector activation?
- Did you notice anything unusual about the structure?
- Upon arrival, were persons/vehicles at the scene? If so, were they arriving or leaving?
- Did you take any statements or hear any comments by persons at the fire scene?

Suppression Activities and Observations

- Were the doors and windows opened or closed, locked or unlocked?
- Did you force any doors or windows? Which ones?
- Who were the first-in firefighters? What were their observations?
- Explain the activities during the fire incident.
- Was your department using an offensive or defensive attack?
- Describe areas where attack lines were used.
- What size attack lines were used?
- Did flashover occur?
- Was there any evidence of an explosion?
- Were there any rescues or attempted rescues during the incident?

- Describe the location and types of ventilation used.
- Were usual or unusual suppression activities used?
- Which area(s) are the most difficult to extinguish?
- Were any rekindles experienced?
- Did you observe any low-level flames?
- Do you have an opinion as to the area of origin?
- Did you observe any unusual odors?
- Did you observe any unusual flame spread or intensity?
- Were there any unusual observations of the fire/smoke?
- Describe the conditions of the utilities at the fire scene.
- Do you have any comments about items not mentioned?
- Where and how did you make entry into the building?
- Did you or other responding personnel have any contact with surrounding exposures?

Other Observations

- Did the witnesses and/or firefighters make any comments?
- Did the owner(s) or occupant(s) make any comments?
- Did the owner(s) or occupant(s) sign a consent form?
- Did you notice persons who have been seen at other fire occurrences?
- Did you find or seize any physical evidence?
- Did you take any photographs or video recordings at the scene (dashboard cams, camera photography, etc.)?
- Did you observe anyone else taking photographs or video of the scene?

Interview with Owners, Witnesses and/or Suspect(s)

Before the start of the interview, the investigator should record (in written format, audiotape, or videotape) the following information:

- Date/time of interview
- Location of interview
- Case number
- Tape/side number if a tape was used
- File name of a digital recording
- Date of fire
- Location of fire
- Acknowledgment of recording device
- Status of the interviewee: owner, renter, neighbor, etc.
- Persons present at interview

Ask Interviewee to State:

- Full name, and spell each
- Date of birth
- Social security number. (U.S. only: Optional - cannot be required)

- Driver's License Number
- Current address and length of occupancy
- Phone number — home/work/cellular/other

The investigator reads Miranda or The Charter Warning if applicable.

Owner/Occupant Interview

Ask the Following Questions of Owners/Occupants as they Apply:

- Can you tell me what happened?
- Do you own this house/building?
- Who is your landlord?
- Do you have homeowner's insurance?
- Do you have renter's insurance?
- Who is your insurance company? Agent?
- Where were you at the time of the fire?
- Have you ever started a fire? Have you ever started an unsafe fire?
- Have you ever had any fire losses? If yes, when and where?
- Was anything moved before this fire?
- How do you think the fire started?
- How did you find out about the fire?
- Is there a history of fire occurrence at this location?
- Were you or any member of the household present when the fire occurred?
- If yes, where was the fire first observed?
- What time did the last person leave the house?
- Who was the last one to leave the house?
- Were the doors and windows locked and secured?
- Who had keys to the property?
- What were the weather conditions at the time of the fire?
- Were there any problems with the gas or electric utilities or appliances? If yes, describe.
- Have you purchased or received any new appliances within 90 days before this fire loss?
- If yes, please list appliance(s) and place of purchase(s).
- Have you had any utilities or appliances repaired within 90 days before this fire loss? If yes, describe repairs and list the repair person or company.
- What was the primary heating system (central heat, gas, electric, or other)?
- Were any auxiliary-type heating devices used (wood stoves, kerosene heaters, electric space heaters, and/or gas heaters)? If yes, please list type and location.
- Do any occupants of the house smoke? If yes, where were they smoking before the fire?
- Were there any flammable products, such as gasoline, kerosene, camping fuel, etc., stored inside the structure? If yes, what and where was it located?
- Were there any weapons, especially firearms, in the structure? If yes, what type and where were they located?
- Who discovered and reported the fire, if you know?

* What time was the fire reported, if you know?

* Have you ever had any theft or vandalism losses? If yes, when and where?

* Have you determined if anything is missing from the structure? If yes, what is missing and from what area?

* What was burning when the fire was discovered, if you know?

* Do you have any opinions as to what may have caused the fire? If yes, explain.

* Were the utilities connected and turned on at the time of the loss? If no, what was off? Do you know the reason for the disconnection?

Owner/Occupant Interview – Financial Aspects

Ask the Following Questions of Owners/Occupants as they Apply:

* What type of business was involved in the fire?

* How long have you been in business?

* Do you own this house/building?

* Who did you buy the business from and what were the terms of the sale? (Assume that the property is owned, but if it is leased, get data on terms of lease.)

* What was the form of business organization — sole proprietor, partnership, or corporation?

* Are any other locations used in conjunction with your business for storage operations?

* Do you have any other business(es)? What type?

* What was your initial investment in the business and have you made any additional deposits to support the business operation?

* What type of products/services do you sell?

* Has the product mix of your sales changed recently (for example, from the top of the line products to the cheaper discount goods)?

* Who sold the structure to you?

* What agencies were trying to sell the structure for you?

* Have you made any recent repairs or renovations to the property?

* Did an outside service contractor make these repairs/renovations? (If so, obtain details of contract.)

* Are you planning to rebuild?

* Was the business for sale and, if so, for how much?

* If you were to sell the business, what would be the estimated selling price?

* Do you have an accountant? (If so, obtain details.)

* What services does the accountant provide?

* Have you filed federal and state or provincial income tax returns for the past five years?

* Who prepared these returns?

* Do you or your accountant have copies?

* Did your accountant prepare financial statements? If so, for what years and are they available?

* What was the financial condition of the business at the time of the fire?

* Was the financial condition of the business improving or declining at the time of the fire?

* What were the average monthly gross receipts? Expenses? Net income?

* Were there seasonal fluctuations in your business?

* Were you paying all your bills in a timely manner?

- Did any suppliers have you on a cash-on-delivery system?
- Did you maintain an adequate supply of inventory items to support your sales?
- Did you have any old and/or obsolete inventory?
- What products were involved/damaged in the fire and what was the approximate inventory value?
- Please provide us with a complete list of all your main suppliers and contractors.
- Have any judgments, liens, or lawsuits been filed against you or your business?
- Have you ever filed or are you in the process of filing for bankruptcy?
- Did you draw any type of a salary from the business?
- Do you have any other sources of income? (If yes, explain.)
- Is your spouse employed? (If yes, obtain particulars as to salary, employer, etc.)
- Do you have a business checking account? (If yes, obtain account number, bank, etc.)
- Do you have any other accounts?
- Are all business deposits and disbursements made to these accounts?
- Do you ever commingle any of your personal funds with the business accounts?
- Have you ever loaned any of your personal funds to your business and if so, how much, when, etc.?
- Do you have any outstanding loans, mortgages, etc.? (Obtain complete loan data as to amounts, terms, banks, etc.)
- Has any lending institution ever turned down any of your loan requests?
- Are any of your loans in default?
- Do you have any automobile or student loans not previously mentioned?
- Do you have any credit cards? (If so, obtain type of card, bank, account number, and outstanding balance.)
- Have you been paying all your obligations in a timely manner?
- Have suppliers been paid in a timely manner? If not, why and what is the outstanding balance?
- How does the total amount that you currently owe to suppliers compare to this same time last year? Two years ago?
- Do any of your customers owe you a significant amount of money? If so, who are they, and why have you not been paid?
- Do you have any employees? Who are they?
- Do all of your current and former employees have good standing with the business?

Witness/Suspect Interview

Ask the Following Questions of Witnesses or Suspects as they Apply:

- Are you under the influence of any medication at this time?
- Are you under the influence of alcohol or other substance(s) that may affect your answers?
- Do you have tattoos or other obvious markings? If so, identify them.
- Do you have an arrest history?
- Do you use any aliases or nicknames?
- Have you ever been investigated?
- How long have you lived in this city?
- Where were you at the time of the fire?

- Have you ever started a fire? Do you know how fires are started?
- Was anything moved before the fire?
- How do you think the fire started?
- How did you find out about the fire?
- Is there a history of fire occurrence at this location?
- What were your actions upon finding out about the fire?
- Who do you think had the best opportunity to set this fire?
- Did you like this house or business?
- Did you dislike something about this house or business?
- Who helped you move the contents out of the structure?
- Were the electricity/gas/water turned on at the time of the fire?
- Were any appliances on at the time of the fire?
- Have you been inside the house after the fire?
- Was anything missing from the structure?
- Identify and give locations for any personal pictures inside the structure.
- Identify and give locations for guns inside the structure.
- Were there any problems with utilities, appliances, or other items?
- Had the utilities, appliances, or other ignition source items been repaired?
- Do you have conflict with neighbors or family?
- Where was the last place you purchased your gasoline?
- Did you talk to anyone before talking with me?
- Have there been any repairs/renovations to the house (business)?
- Provide financial account location and numbers. Include mortgage history and information on property or properties.
- What is the monthly income from employment?
- What is the total family income?
- Have you ever been late with payment of expenses?
- What was your financial condition prior to the fire?
- What do you consider your financial condition to be after the fire?
- Have you contacted your insurance company?
- What do you think should happen to the person who set this fire (if applicable)?
- Who was the last person inside the house or business?
- Were normal procedures used for closing or securing the house (business)?
- Were doors/windows locked at time of fire?
- Who did you ask to set this fire for you?
- Is there any reason why someone would tell me that they saw you at the house (business) just before the fire?
- Tell me the reason why your fingerprints were found at the scene.
- Have you answered all my questions truthfully?
- Will you agree to a polygraph test?
- How do you think you will (would) do on a polygraph test?

- Who do you think set this fire?
- Did you set this fire?
- Have you made a list of items lost in the fire for the insurance company?
- Where do you usually buy gasoline (diesel) for your vehicle(s)?
- Will you vouch for anyone you feel did not set this fire?
- What was the condition of the house before the fire?
- Where did you first see the fire?
- What were the occupant activities before the fire?

Questions about Fire Suppression Activities

- Where did you first see fire?
- How did the fire department extinguish the fire?
- What fire department activities did you observe?

Research the Following Information as it Applies:

- History of previous losses from insurance companies
- Insurance coverage/building content and additional living expenses
- Replacement cost for loss
- Any changes in the policy
- Agent and insurance identity
- History of entire family regarding fire occurrences
- History of residency
- Educational history
- History of service in armed forces
- Political ideologies
- Marriage and family history
- Employment history
- Lightning reports or weather records
- Real estate agencies
- Court records

Other Investigative Topics

- Allow interviewee to clarify or comment on items not questioned on.
- Follow up with questions relating to answers that appear deceptive.
- Change chronological order of activities of suspects.

Tools of the Profession

NFPA 1033 (2022): 4.1.7

Fire investigators use a wide variety of tools and equipment in the performance of their duties. The equipment ranges from very basic hand tools used to excavate the scene and collect samples to canine ignitable liquid detection teams and at times very sophisticated computer-modeling programs. In addition, the investigative methodology itself can be used as a tool. While an investigator may not personally use all of the equipment that is available, there are some basic items that any investigator will need. An investigator must also understand what is available to assist with an investigation and when and how to access special services.

Standard Tools

NFPA 1033 lists the following items as the standard equipment and tool list for a fire investigator:

- Broom
- Evidence collection equipment and supplies (see Chapter 10, Scene Documentation)
- Flashlight
- Hand tools
- High-resolution camera, flash, or media
- Safety clothing and equipment (see Chapter 2, Safety)
- Shovel
- Tape measure or other measuring equipment

NOTE: This list of equipment is the minimum required. Additional equipment is listed in NFPA 921.

All tools used by a fire investigator should be kept in good condition and cleaned after each use to prevent contamination. Proper cleaning methods and materials are discussed in this manual. An investigator should also have a supply of barrier tape and other materials for marking a fire scene and protecting potential evidence.

Fire Investigator's Hand Tools

As fire investigators conduct scene examinations and collect evidence, they often require a set of hand tools to do the job properly. These tools should be used specifically for investigations, and to avoid contamination, they should be kept separate from those tools used by fire-suppression crews for forcible entry and overhaul. The following are examples of tools that may be included in an investigator's tool kit:

- Brushes (paint or whisk broom)
- Char depth-gauge tool (such as a tire depth gauges, calipers, or specifically modified metal rulers)
- Claw hammer
- Cold chisel
- Colored electrical tape or wire markers
- Hacksaw (extra blades)
- Hatchet
- Keyhole saw (extra blades)
- Marking pens
- Mason trowel

- Multimeter (volt/ohm)
- Pencil magnet
- Pencil scribe
- Pliers
- Pry bar
- Screwdrivers (multiple sizes and types)
- Six-inch (150 mm) ruler
- Tape measure
- Toolbox large enough to store and protect the tools
- Utility knife (extra blades)
- Wire cutters
- Wood chisel

Specialized Tools

NFPA 1033 also includes a provision for special tools, which it defines as "Tools of a specialized or unique nature that may not be required for every fire investigation ...." Examples of this type of equipment include heavy excavation equipment, ladders, rope, or portable lighting equipment, or 3D scanner. Depending on the scene, an investigator may require special equipment to safely and efficiently conduct an investigation.

NOTE: Canine ignitable liquid detection teams and electronic hydrocarbon detectors would also fall into the category of special equipment. Both may be valuable investigative resources. These tools may be useful for locating potential sites to collect samples for laboratory analyses.

A fire investigator should know the policies of the organization regarding the use of specialized equipment and where it can be obtained when needed. In many cases, other agencies such as the state/provincial fire marshal's office or federal law enforcement agencies can assist with special equipment for major incidents. In the case of major losses involving commercial, industrial, or similar properties, the insurance carrier or carriers may be willing to provide financial and other forms of support to accomplish the investigation.

A

Index by Meredith Murray